D1261570

Ultraviolet Germicidal Irradiation Handbook

Wladyslaw Kowalski

Ultraviolet Germicidal Irradiation Handbook

UVGI for Air and Surface Disinfection

 Springer

Wladyslaw Kowalski
Immune Building Systems, Inc.
575 Madison Ave.
New York, NY 10022
drkowalski@ibsix.com

ISBN 978-3-642-01998-2   e-ISBN 978-3-642-01999-9
DOI 10.1007/978-3-642-01999-9
Springer Heidelberg Dordrecht London New York

Library of Congress Control Number: 2009932132

Cover design: WMXDesign GmbH, Heidelberg

Printed on acid-free paper

Springer is part of Springer Science+Business Media (www.springer.com)

I dedicate this book to Dr. Thomas Whittam and to his family. Tom passed away all too suddenly during the writing of this book. He was my mentor and tutor in the field of microbiology and genetics, and taught me laboratory techniques as well as inspiring me with the desire to achieve excellence. It had always seemed to me that scientists, unlike so many musicians, artists, writers, and actors, were always allowed to complete their greatest works, and so it is especially sad to see anyone leave this world when they are in the midst of doing great research and useful science, and when they had so much more to give.

Preface

This book is a comprehensive source for technical information regarding ultraviolet germicidal irradiation (UVGI) and its application to air and surface disinfection for the control of pathogens and allergens. The primary focus is on airborne microbes and surface contamination applications. Water-based applications are not addressed here except incidentally as they relate to air and surface microbes, since many adequate texts on water-based UV disinfection are available. All aspects of UVGI systems, including design methods, modeling, safety, installation, guidelines, and disinfection theory are addressed in sufficient detail that no additional sources need be consulted. It is hoped by the author that providing this information in one single volume will simplify the design and installation of UVGI systems, help guarantee effective performance of new systems, and facilitate their use on a wide scale for the purpose of improving human health and controlling epidemic disease. This book is organized to provide systematic coverage of all essential issues and will serve equally well as both a textbook and a handbook for general reference.

Any readers who find technical errors or omissions in this book may to send them to me at drkowalski@aerobiologicalengineering.com. Errata will be posted at http://www.aerobiologicalengineering.com/UVGI/errata.htm. All other correspondence should be sent to drkowalski@ibsix.com.

Acknowledgements

I gratefully acknowledge all those who assisted me in the research, preparation, and review of the chapters in this book, including William Bahnfleth, Steve Martin, Chuck Dunn, Jim Freihaut, Dave Witham, Ed Nardell, Richard Vincent, Mark Hernandez, Ana Nedeljkovic-Davidovic, Clive Beggs, Renzhou Chen, Normand Brais, Katja Auer, Warren Lynn, Mike Sasges, Bill Carey, Tatiana Koutchma, Forrest Fencl, Russ Briggs, Josephine Lau, Carlos Gomes, Fahmi Yigit, Herbert Silderhuis, Joe Ritorto, Merrill Ritter, Brad Hollander, Scott Prahl, Karl Linden, William Balch, Atanu Sengupta, M.D. Lechner, Ketan Sharma, Donald Milton, and especially Mary Clancy all the members of the IUVA Air Treatment Group who supported the UV Guidelines project, and also Linda Gowman, Jim Bolton and everyone in the International Ultraviolet Association who sponsored the UV Air Treatment Group. A special thanks goes to Ali Demirci and Raymond Schaefer for their contributions to the chapter on Pulsed UV Systems. I especially thank my parents, Stanley J. Kowalski and Maryla Kowalski, and my sister Victoria Chorpenning for their unwavering support and encouragement during these past few years as I recovered my health and returned to research.

Contents

Chapter 1

Introduction

1.1 Ultraviolet Germicidal Irradiation (UVGI)

Ultraviolet germicidal irradiation (UVGI) is defined as the use of ultraviolet (UV) wavelengths of light in the germicidal range (200–320 nm) for the disinfection of air and surfaces. The term 'UVGI' was originally coined by the International Commission on Illumination (CIE) and adopted later by the Centers for Disease Control (CDC), and this term distinguishes disinfection applications from the non-germicidal UVA wavelengths of black lights and suntan lamps (320–400 nm). UVGI is also used to distinguish air and surface disinfection applications from those in water (CIE 2003). Throughout this book the terms 'UVGI' and 'UV' will be used interchangeably, with the understanding that in every context, unless otherwise noted, both terms refer to the germicidal wavelengths of UVC (200–280 nm) and UVB (280–320 nm). UV radiation below 320 nm is actinic, which means it causes photochemical reactions. UVA radiation (320–400 nm) is not considered germicidal and is not specifically addressed in this book (except in relation to pulsed light). Table 1.1 summarizes the definitions of the primary bands of UV radiation. Previously, UVA was considered to extend down to 315 nm, but the range 315–320 nm was found to have some minor germicidal effects. The UVA band, and therefore the UVB band also, have been redefined by various organizations such that all actinic UV radiation is now contained strictly within the UVB and UVC range. This division allows UVA to be completely non-germicidal and conveniently places all the germicidal UV into the UVB and UVC bands.

The definitions of the UV bands UVA, UVB, and UVC given in Table 1.1 have not yet been fully incorporated into every relevant guideline or adopted by every agency, but are likely to be adopted eventually and universally. VUV (Vacuum Ultraviolet) does not transmit through air since it is rapidly absorbed and therefore the VUV band is of no interest in air and surface disinfection, and is not addressed further in this book.

This introductory chapter presents some other basic definitions and subjects which provide a background for the subsequent chapters, where these concepts

W. Kowalski, *Ultraviolet Germicidal Irradiation Handbook*,
DOI 10.1007/978-3-642-01999-9_1, © Springer-Verlag Berlin Heidelberg 2009

Table 1.1 Primary bands of ultraviolet radiation

Band	Wavelength, nm	Type and classification		
UVA	320–400	Non-germicidal (Near-UV, Blacklight)		
UVB	280–320	Erythemal	Germicidal	Actinic
UVC	200–280	Ozone-producing		
VUV	100–200	Vacuum ultraviolet		

will be addressed in greater detail. The chapters in this book are arranged to first address the background information, including theory and mathematical modeling, then equipment and design methods, and finally testing and applications. In the Appendices are provided tabulations of data and information that are useful and that will be referenced throughout this book.

1.2 Brief History of Ultraviolet Disinfection

The story of ultraviolet light begins with Isaac Newton and his contemporaries. In 1672, Isaac Newton published a series of experiments with prisms that resolved sunlight into its constituents colors, red through violet. The effects of sunlight on man, microorganisms, and chemicals became a matter of great interest and experimentation in the 1800s. In 1814, Fraunhofer mapped over 500 bands of sunlight, some of which were in the ultraviolet region. In 1842, Becquerel and Draper each independently showed that wavelengths between 340 and 400 nm induced photochemical changes on daguerreotype plates (Hockberger 2002).

UV Lamp development predates sunlight studies on bacteria. In 1835, Wheatstone invented the first mercury (Hg) vapor arc lamp, but it was unstable and short-lived. Fizeau and Foucault (1843) reported problems with their eyes after experimenting with a carbon arc lamp, and speculated that it was due to 'chemical rays.' In 1850, Stokes used aluminum electrodes to produce a 'closed' arc lamp in a quartz tube that emitted UV rays to 185 nm (Hockberger 2002).

The earliest scientific observations of the germicidal effects of ultraviolet radiation began with Downes and Blunt (1877) who reported that bacteria were inactivated by sunlight, and found that the violet-blue spectrum was the most effective. In 1885, Arloing and Duclaux demonstrated that sunlight had a killing effect on *Bacillus anthacis* and *Tyrothrix scaber*, respectively. Widmark (1889, 1889a) published studies confirming that UV rays from arc lamps were responsible for skin burns, using a prism to separate the UV spectrum and water to block the infrared rays. It was demonstrated in 1892 that ultraviolet light was responsible for this action with tests on *Bacillus anthracis* (Ward 1892). Also in 1892, Geisler used a prism and a heliostat to show that sunlight and electric arc lamps were lethal to *Bacillus typhosus*. Finsen (1900) performed the first rigorous analysis of the effects of UV light. The UV spectrum around 250 was identified as biocidal by Barnard and Morgan (1903), and the range was narrowed by Newcomer (1917), and isolated to

253.7 nm by Ehrismann and Noethling (1932). Table 1.2 summarizes most of the critical developments in the history of UV research and applications.

The first use of UV to disinfect drinking water is said to have been in 1906 according to von Recklinghausen (1914). In 1909/1910 the first water disinfection

Table 1.2   Chronology of critical events in UV history

Year	Event	Reference
1814	Fraunhofer maps spectral bands of sunlight	Hockberger (2002)
1835	Wheatstone invents first mercury vapor arc lamp	Hockberger (2002)
1850	Stokes invents quartz arc lamp that produces to 185 nm	Hockberger (2002)
1842	Becquerel and Draper find 340–400 nm light photoreactive	Hockberger (2002)
1877	Bactericidal effects of sunlight first demonstrated	Downes and Blunt (1877)
1889	UV light demonstrated to be erythemal	Widmark (1889)
1892	UV component of sunlight identified as biocidal	Ward (1892)
1892	Geissler demonstrates lethality of arc lamps to *B. typhosus*	Hockberger (2002)
1903	UV spectrum from 226 to 328 nm found to be germicidal	Barnard and Morgan (1903)
1904	First quartz lamp for UV developed	Lorch (1987)
1906	UV first used to disinfect drinking water	von Recklinghausen (1914)
1909	First European applications for UV water disinfection	AWWA (1971)
1912	Henri found shorter UV wavelengths don't penetrate	Henri (1912)
1916	First USA applications of UV for water disinfection	AWWA (1971)
1921	UV photoreactivity with TiO_2 first demonstrated	Renz (1921)
1925	UV photodegradation of materials first demonstrated	Luckiesh and Taylor (1925)
1927	First erythemal action spectrum published	Hausser and Vahle (1927)
1927	Bactericidal action of UV first quantified scientifically	Bedford 1927, Gates (1929)
1928	Virucidal action of UV first quantified scientifically	Rivers and Gates (1928)
1929	Fungicidal action of UV first quantified scientifically	Fulton and Coblentz (1929)
1932	UV germicidal peak at 253.7 nm isolated	Ehrismann and Noethling (1932)
1932	Erythemal spectrum of UV first quantified	Coblentz et al. (1932)
1936	First overhead UV system in hospitals	Wells and Wells (1936), Hart (1936)
1936	UV photoreactivation phenomena first identified	Prat (1936)

Table 1.2 (continued)

Year	Event	Reference
1937	First upper air application in schools	Wells (1938)
1938	First fluorescent gas discharge UV lamp	Whitby and Scheible (2004)
1940	UV first applied to air conditioning systems	Rentschler and Nagy (1940)
1942	First UV air disinfection sizing guidelines	Luckiesh and Holladay (1942)
1942	Upper and lower UV applied to Army/Navy barracks	Wells et al. (1942)
1950	First catalog sizing methods	Buttolph and Haynes (1950)
1954	First air conditioner application	Harstad et al. (1954)
1954	Faulty British study concludes UV is ineffective	MRC (1954)
1957	Riley proves effectiveness of UV for TB control	Riley et al. (1957)
1974	First microbial growth control systems	Grun and Pitz (1974)
1985	Cooling coil UV systems in use in European breweries	Philips (1985)
1994	CDC acknowledges UV effectiveness for TB control	CDC (2005)
1996	First cooling coil irradiation system in US	Scheir (2000)
1997	First UV light emitting diodes (LEDs) at 265 nm	Guha and Bojarczuk (1998)
1999	WHO recommends UVGI for TB control	WHO (1999)
2000	US army recommends UVGI for disease isolation	USACE (2000)
2003	CDC formally sanctions UVGI use in hospitals	CDC (2003)
2003	FEMA sanctions UVGI as a biodefense option for buildings	FEMA (2003)
2003	First in-duct UVGI system demonstrated to reduce illness symptoms and airborne contamination	Menzies et al. (2003)
2003	ASHRAE forms UV air and surface treatment committee	Martin et al. (2008)
2005	Federal government specifies UV for cooling coil disinfection	GSA (2003)
2007	Overhead UV system proven to reduce SSIs in ORs	Ritter et al. (2007)

system was operated at Marseilles, France. The first evidence that UV light produced photochemical effects on microorganisms was presented by Henri (1914). In 1916 the first US system for water disinfection was tested at Henderson, KY (AWWA 1971). In 1921, Renz demonstrated that UV could cause photoreactions with titanium oxide (TiO2). Hausser and Vahle (1927) produced the first detailed action spectrum for erythema. Bedford (1927) and Gates (1929) were among the first to establish UV dosages necessary for bacterial disinfection. Fungal disinfection dosages were first published by Fulton and Coblentz (1929). The first studies on UV irradiation of viruses appear to have been those published by

Rivers and Gates (1928) and Sturm et al. (1932). Coblentz et al. (1932) refined the erythemal action spectrum.

The 1930s saw the first applications of UV systems in hospitals to control infections (Wells and Wells 1936, Hart and Sanger 1939, Robertson et al. 1939, Kraissl et al. 1940, Overholt and Betts 1940). The first Upper Room UV systems appear to be those installed by Wells (1938). In the 1940s the first detailed design and analysis of UV air disinfection were published along with basic guidelines for applying UV in ventilation systems (Rentschler and Nagy 1940, Sharp 1940, Wells 1940, Buchbinder and Phelps 1941, DelMundo and McKhann 1941, Luckiesh and Holladay 1942, Sommer and Stokes 1942, Henle et al. 1942, Hollaender 1943). The first attempts to use UV systems to control respiratory infections in schools and barracks occurred shortly thereafter (Wells et al. 1942, Wells 1943, Schneiter et al. 1944, Wheeler et al. 1945, Perkins et al. 1947, Higgons and Hyde 1947). Several early attempts were made to develop rigorous sizing methods and engineering guidelines for UV applications (Luckiesh and Holladay 1942a, Luckiesh 1945, 1946).

By the 1950s it had been well-established that UV irradiation was effective at disinfecting both air and surfaces, and new engineering applications were being developed. General Electric catalogs detailed a wide variety of UV applications including various methods of installing UV lamps inside ducts and air conditioners (Buttolph and Haynes 1950, GE 1950). Harstad et al. (1954) demonstrated that installation of UV lamps in air conditioners would reduce airborne contamination, and that microorganisms were impinging upon and collecting on internal AHU surfaces. Bacterial growth on cooling coils had been recognized as a potential health problem as early as 1958 (Walter 1969). The first evidence that air cooling equipment could actually cause respiratory infections was presented by Anderson (1959) when an air cooling apparatus was found to be contaminated with microbial growth. This very same concern had been raised in hospital environments since about 1944 but the possibility of growth of bacteria on air-conditioning cooling coils wasn't conclusively demonstrated until 1964 (Cole et al. 1964). The growth of microbes on other equipment like filters and dust inside air-conditioning ducts was first demonstrated by Whyte (1968). The dissemination of microbes by ventilation systems and their potential to cause respiratory infections became widely recognized in the late 1960s in both the medical and engineering professions (Banaszak et al. 1970, Schicht 1972, Zeterberg 1973). It was understood at this time that microbial growth could occur anywhere that air came into contact with moisture (Gunderman 1980, Ager and Tickner 1983, Spendlove and Fannin 1983). The first UVGI system designed specifically for disinfecting the surfaces of air handling equipment, including humidifier water and filters, was detailed by Grun and Pitz (1974). Luciano (1977) detailed many applications of UVGI, including hospital applications in which the UV lamps are specifically placed upstream of the cooling coils and downstream of the filters. In 1985 Phillips published a design guide in which the first definitive description of applications of UV lamps for the control of microbial growth were presented (Philips 1985). The Philips design guide mentions European installations that were already in operation prior to 1985. In January of 1996 the first UVGI system in the US designed for controlling microbial growth on cooling coils was installed by Public Service of Omaha (PSO) in Tulsa (Scheir 2000). In the same year, the Central and Southwest Corporation followed the PSO example and began realizing considerable energy savings (ELP 2000).

Although UVGI systems had been in use in hospitals since 1936 but it wasn't until 2003, some sixty years later, that the CDC formally acknowledged that UV systems were effective and could be used in hospitals with one caveat – UV Upper Room and in-duct systems could only be used to supplement other air cleaning systems (CDC 2003). In 1957, Riley and associates successfully completed a demonstration of how UV air disinfection could control the spread of tuberculosis (TB) in hospital wards (Riley et al. 1957). It wasn't until 1994, over forty years later, that the Centers for Disease Control (CDC) acknowledged that UV could be effective for controlling TB, in response to the growing worldwide TB epidemic which had resisted control by traditional methods (CDC 2005). In 2003, The influential American Society of Heating Refrigerating and Air Conditioning Engineers (ASHRAE) formed a task group to focus on UV air and surface treatment (TG2.UVAS) which became the standing Technical Committee TC 2.9 in 2007 (Martin et al. 2008).

1.3 Units and Terminology

A variety of units have been used in UV disinfection for the irradiance and the UV dose. The irradiance, sometimes called intensity, has the preferred units of W/m^2 in air and surface disinfection. The UV dose (aka fluence rate) has the preferred units of J/m^2 in air and surface disinfection. Conversion factors for the various units that have been used in the literature are provided in Table 1.3. The use of Table 1.3 is straightforward as shown in the following examples. Note that a joule (J) is equivalent to a watt-second (W-s), and that a W/m^2 is equal to a $\mu W/mm^2$.

Example 1: Convert the irradiance 144 mW/cm^2 (milliwatts per square centimeter) to units of W/m^2 (watts per square meter)

Answer: Read downwards from the first column, mw/cm^2 to the gray box and then over to the second column, W/m^2, where the conversion factor is seen to be 10. Multiply $144 \times 10 = 1440$ W/m^2.

Example 2: Convert the UV dose 33 $\mu J/cm^2$ (microJoules per square centimeter) to units of J/m^2 (Joules per square meter).

Answer: Read up from the fourth column, $\mu J/cm^2$, to the gray box, and then over to the second column, J/m^2, where the conversion factor is seen to be 0.01. Multiply $33 \times 0.01 = 0.33$ J/m^2.

The term 'UVC' is often used today to encompass all applications of germicidal UV but the correct definition of this term is, of course, the band of UV wavelengths between 200 and 280 nm, and this strict definition is the one used throughout this book. For example, a UV lamp could refer to any lamp that produces any UV wavelengths, including black light and suntan lamps (although in this book only

Table 1.3 Units of Irradiance and UV Dose

Conversion factors for irradiance (Read down from units to gray block and then horizontally for unit equivalence.)					
mW/cm^2	W/m^2 $\mu W/mm^2$ J/m^2-s	erg/mm^2-s	$\mu W/cm^2$	erg/cm^2-s	W/ft^2
1	10	100	1000	10000	1.07639
0.1	1	10	100	1000	0.107639
0.01	0.1	1	10	100	0.010764
0.001	0.01	0.1	1	10	0.001076
0.0001	0.001	0.01	0.1	1	0.000108
0.929	9.29	92.90	929.0	9290	1
mJ/cm^2 mW-s/cm^2	μW-s/mm^2 W-s/m^2 J/m^2	erg/mm^2	$\mu J/cm^2$ μW-s/cm^2	erg/cm^2	W-s/ft^2
Conversion factors for UV Dose (Exposure time = 1 second) (Read up from units to gray block and then horizontally for unit equivalence.)					

germicidal UV lamps are addressed). The term 'UVGI lamp' specifically refers to lamps that produce UV wavelengths in the actinic 'broad-band' range between 200 and 320 nm (excluding black lights and sunlamps). The term 'UVC lamp' specifically refers to lamps that produce UVC wavelengths in the 'narrow-band' range 200–280 nm. In current and common usage, the term 'UVC' often implies that the UVC band is the only contributor to the germicidal effect, but this implication is often incorrect and such usage is avoided in this text – wherever the term 'UVC' is used herein it refers specifically to the UVC band of radiation. Other terms such as UVR (ultraviolet radiation) and GUV (germicidal ultraviolet) have also been used in a germicidal context in the past.

The term 'germicidal' implies that these UV systems destroy, kill, or inactivate microorganisms such as viruses, bacteria, and fungi. Technically, viruses are molecules, and so it customary to refer to viruses as being inactivated rather than killed. In all cases, germicidal action means disinfection, and disinfection implies a reduction in the microbial population, whether in air, water, or on surfaces. Microbial populations are measured in terms of cfu, or colony-forming units (i.e. grown on petri dishes). In the case of viruses, the appropriate measure of viral populations is pfu, or plaque-forming units. However, wherever the term 'cfu' is used in this book, it will be considered to apply to both viruses and bacteria (as a matter of convenience), with the reader's understanding that the correct terminology for viruses is pfu, whether it is used or not. The density of microbes in air is always given in units of cfu/m^3 (although some older texts use cfu/ft^3). The density of microbes on surfaces is given in cfu/cm^2 (in older texts it is cfu/in^2). The disinfection of air will therefore be measured in terms of a reduction in the airborne density in cfu/m^3, and the disinfection of surfaces is measured in terms of the reduction of cfu/cm^2.

Sterilization is a related term that implies the complete elimination of a microbial population. It is difficult, however, to actually demonstrate the complete elimination of a microbial population since any microbiological test will have some limit of

accuracy. As a practical convenience in air disinfection applications, and for analytical purposes, 'mathematical sterilization' can be assumed to represent a disinfection rate of six logs or better, when there are no survivors. This definition may
not apply to water UV disinfection, but in air disinfection applications it would be
most unusual to have airborne densities on the order of a million cfu/m^3, and typical
airborne densities are in the range of a few thousand cfu/m^3. Obviously, with densities of a few thousand cfu/m^3, a three or four log reduction (1000–10,000 cfu/m^3)
would result in sterilization, and a six log reduction is an adequate definition of
sterilization for air disinfection applications. Surface disinfection is another matter
altogether and the densities of surface contamination are neither well understood
nor well defined, and so the 'mathematical definition' of sterilization for surface
disinfection is left undefined in this text.

1.4 Air vs. Water Disinfection

The design of UV systems for water disinfection differs from that of air and surface
disinfection applications and therefore the cumulative knowledge accrued in the
water industry is of limited direct use for air and surface disinfection applications.
UV rays are attenuated in water and this process has no parallel in air disinfection,
even with saturated air. The attenuation of UV irradiance in water occurs within
about 15 cm and this necessitates both higher UV power levels and closely packed
arrays of UV lamps. The estimates of UV doses required for water disinfection are
on the order of ten times higher than those needed in air disinfection applications,
and this difference distorts any attempt to use water UV system sizing methods to
design air disinfection systems. Furthermore, the array of particular microorganisms
of concern in the water industry differs considerably from those found in air and
therefore water-based UV rate constants are of use only where the microbial agent
is both airborne and waterborne (i.e. *Legionella*), or is also surface-borne, and for
theoretical analysis. Some overlap in waterside and airside UV applications also
exists in the area of foodborne pathogens, where certain foodborne pathogens may
become airborne, and where they may exist as surface contamination amenable to
UV disinfection.

Although the UV exposure dose in air is a simple function of airflow and exposure time, and the UV irradiance field in air is not too difficult to define, the susceptibility of airborne microbes is a complex function of relative humidity and
species-dependent response. It has often been thought that the UV susceptibility of microbes in air at 100% relative humidity (RH) should correspond to their
susceptibility in water, but this proves to be overly simplistic and it can only be
said that UV susceptibility at high RH approaches that in water. As a result of these
various differences between water-based UV disinfection and UVGI air and surface disinfection, research into the former provides limited benefits to research into
air-based disinfection, and the subject of water disinfection is not addressed in this
book except insofar as it has some specific impact on air and surface disinfection and in the matter of their common theoretical aspects. The UV rate constant

database in the Appendices, addresses all known UV studies on microbial disinfection, including those for waterborne and foodborne pathogens and allergens. There are a wide variety of detailed texts on the subject of water disinfection with UV (many times more than for airside UV disinfection) and readers who wish to familiarize themselves with waterside UV technology and methods should consult these texts directly (see for example, Bolton and Cotton 2008). However, the information provided herein on UVGI theory and UV inactivation rate constants may be of no little interest to those involved in water disinfection.

1.5 Surface Disinfection

Surface disinfection refers to either the disinfection of building and ventilation system internal surfaces, or the disinfection of equipment and material surfaces, such as dental and medical equipment. Like water systems, surface UV disinfection systems have a long history of success. The design and operation of surface UV systems, however, have much more in common with air disinfection than do water systems. Contaminated surfaces are often a source of airborne microbes, and airborne microbes often produce surface contamination. The interaction of air and surface contamination processes makes the issue of air vs. surface disinfection almost inseparable for some applications, such as in the health care industry and the food industry. The disinfection of cooling coil surfaces, for example, removes mold spores from the coils and prevents subsequent aerosolization, thereby helping keep the air clean. UV systems for coil disinfection applications also disinfect the air directly and so such systems often perform simultaneous air and surface disinfection functions.

One of the main differences between air and surface disinfection with UV is that the relevant UV rate constants differ under these two types of exposure – airborne rate constants tend to be higher in air, under normal humidity. That is, microbes are more vulnerable in air, whereas microbes on surfaces appear to have a certain degree of inherent protection. Although the matter remains to be resolved by future research, the available database for UV rate constants for microbes on surfaces is useful as a conservative estimate of airborne rate constants, as are water-based rate constants, whenever airborne rate constant studies do not exist.

Since airborne microbes are often surface-borne, and vice-versa, and for the various reasons mentioned above, the overlap between these topics, air and surface UV disinfection, is extensive. In fact, the subject of surface disinfection has considerably more in common with air disinfection than water disinfection and often the two technologies are inter-related. It is appropriate, therefore to treat air and surface UV disinfection together, as in this text, and it should be understood that most air disinfection systems will simultaneously perform some surface disinfection function, intentionally or otherwise. Similarly, many UV surface disinfection systems (i.e. Lower Room systems, Overhead Surgical systems) will also perform some air disinfection functions, by design or otherwise.

1.6 Air Disinfection

Airborne pathogens and allergens present a much greater threat to human health than water-based microbes, in terms of total incidence and net costs of health care, but there are far fewer air disinfection systems in place than water disinfection systems, and much less information is available for airborne UV disinfection than for water applications. The success of UVGI for air disinfection application has also been subject to much interpretation and even outright dismissal, in spite of repeated demonstrations of its effectiveness. After decades of research, the field of airborne UV disinfection remains fraught with unknowns and misconceptions, and applications are far less numerous than they perhaps should be. The subsequent chapters attempt to consolidate the entire knowledge base relevant to UVGI and to demonstrate how the careful application of proper design principles and new approaches can produce results as predictable and reliable as those of water disinfection systems. New computational methods, combined with a wealth of recent design and installation experience and ongoing research, now allows systems to be installed with fairly high levels of confidence in terms of their performance.

Methods for demonstrating in-place performance, along with various guidelines and standards for such installations, have brought the field of UVGI from an uncertain art to a nearly complete science. The key missing component at this time is conclusive evidence that UVGI air disinfection reduces the incidence of airborne disease, a matter that will require years of data collection, once there are sufficient and adequate installations available for monitoring. Some current studies are exploring this avenue of research and preliminary results suggest that UVGI is, as theory and analysis predicts, effective in reducing both the symptoms and incidence of various airborne diseases. As applications increase, this database should eventually provide reliable evidence that may lead to full economic justification of UVGI installations in health care facilities, schools, and other types of buildings. The widespread use of UV for air disinfection in buildings is likely an eventuality that will pay economic and health dividends to future generations.

1.7 Air Disinfection Field Studies

The use of UV to disinfect air goes back some eighty years, and yet applications are still far from being common in modern buildings. Although nine out of ten UVGI field studies had positive results, the few that did not meet grand expectations were cited most often as proof of failure. In 1936 Hart used an array of UV lamps to sterilize supply air in a surgical operating room (Hart 1937). In 1937 the first installation of UV lamps in a school ventilation system dramatically reduced the incidence of measles, and subsequent applications enjoyed similar successes (Riley 1972). In the late 1940s, Wells and his associates installed UV systems across entire communities and demonstrated reductions in community disease transmission rates (Wells and Holla 1950). In the late 1950s experiments using guinea pigs demonstrated the elimination of tuberculosis (TB) bacilli from hospital ward exhaust air

(Riley and O'Grady 1961). A plethora of designs that were more imitative then engineered followed these early applications. The result was a mixture of successes and failures. In one poorly planned study by the MRC, investigators failed to achieve statistically significant reductions of disease and concluded erroneously that UV failed to reduce disease incidence (MRC 1954). The MRC study put a severe damper on further development of UV technology in the 1950 s and 1960 s, and as a result of this study being widely quoted (and in spite of it being widely criticized by experts), brought a near halt to further implementations and slowed research (Riley 1980, 1980a). This setback of the UV industry cannot be measured in terms of lost development, lost field data, lost health care costs, or lost lives, but there are those who today continue to insist that UVGI is an unproven 'snake oil' technology. We must convince them otherwise.

As a result of an apparently imperfect record of success, UVGI was widely ignored in various guidelines and standards, particularly in health care settings, where it should have been directly addressed. Without the sanction of the highest health authorities, UVGI languished and was largely ignored in health care settings where it would have done the most good. This checkered history is reflected in various guidelines that decline to mention the use of UVGI. Only a select few of the most well-informed hospitals and researchers took the time to investigate and adopt UV technologies, often with great success (Goldner and Allen 1973, Goldner et al. 1980, Nardell 1988). Recently, new research has reaffirmed what the original studies had always said, that UV technology can have a major impact on the reduction of various types of nosocomial and community-acquired infections (Menzies et al. 2003, Ritter et al. 2007, Escombe et al. 2009).

1.8 Pathogens and Allergens

Pathogens are any microbes that cause infections in humans and animals, and these include viruses, bacteria, and fungi. Some larger microbes, like protozoa, may also cause infections but these parasites are generally too large to be airborne. Therefore, this book primarily addresses only viruses, bacteria, and fungi, including bacterial spores and fungal spores, as air and surface contaminants. Insects, like dust mites, are not eradicable by UV and are not addressed in this text. All of the viruses listed in Appendix B are pathogens or bacteriophages, as are most of the bacteria in Appendix A and some of the fungi in Appendix C.

Allergens are microbes, biological products, and compounds that induce allergic reactions in atopic, or susceptible, individuals. Compounds and biological products (i.e. VOCs and pet dander) are generally not very susceptible to UV destruction (although they are easily removed by filters) and so they are not specifically addressed in this book. The allergens addressed in this book are strictly fungi and bacteria – there are no viral allergens. Some pathogens may also be allergens. Almost all of the fungi listed in Appendix C are allergens, and many of these are also pathogens. Almost all of the bacteria listed in Appendix A are pathogens, and some of these are also allergens.

Some bacteria, and virtually all fungi, can form spores. In the normal or growth state (called 'vegetative') bacteria exist as cells, and fungi exist as cells or yeast. Spore-forming (sporulating) bacteria and fungi will form spores under the right conditions (usually adverse conditions). Spores are dormant forms, usually more compact than the cell forms, round or ovoid shaped, and can resist heat, dehydration, and cold much better than the cell (or vegetative) form. They also tend to be resistant to UV exposure. Since they are typically smaller than the original cells, spores tend to become airborne easily and can be transported outdoors. Spores require only warmth, moisture, and shade to germinate, or return to a vegetative state, and require only nutrients to grow and multiply.

Some bacteria and fungi produce toxins, including endotoxins and exotoxins. UV has a limited effect on toxins but prolonged exposure can reduce toxin concentrations (Anderson et al. 2003, Asthana and Tuveson 1992, Shantha and Sreenivasa 1977). High levels of growth are required, usually under adverse conditions, for microbes to produce toxins, and since UV destroys toxin-producing bacteria and fungi, it is unlikely that sufficient levels of toxins will remain after UV disinfection to pose toxic hazards.

1.9 Current Research

In spite of the extensive research results available on UVGI air and surface disinfection, much work remains to be done. Recent studies have demonstrated the effectiveness of UVGI and have hinted at the possible reduction of airborne disease in commercial office buildings. The ability of UVGI to save costs in cooling coil maintenance has been fairly well established. There is also a need for more research to determine UV rate constants for a wider array of pathogens. Towards this end has been provided a guideline for laboratory testing (included in this book) that should facilitate the production of reproducible test results, something that has not always been the case in the past.

In addition to providing the most up-to-date information from current literature of UVGI, this text also provides the fruits of the author's research into UV susceptibility, including a model for relative humidity effects in air, and a genomic model for predicting the UV susceptibility for viruses (addressed in Chaps. 2 and 4). Also presented here for the first time is the author's research on pulsed UV light modeling.

Perhaps the most important applications of UVGI today is in the health care industry, which is in dire need of solutions to the problem of hospital-acquired (nosocomial) infections. Such infections have now spread outwards to become community-acquired infections and it is likely this pattern will keep repeating with new and emerging pathogens and drug-resistant strains until a more effective solution is implemented on a wide scale. UVGI can play a major role in limiting the spread of nosocomial infections but what is needed most is not new technology (adequate technology exists in the present) but encouragement and support in terms of guidelines and recommendations on UV technology from those authorities who have until recently been somewhat reticent and noncommittal on the matter.

1.10 UVGI and the Future of Disease Control

UVGI has a definite future in the control of contagious diseases and if applied on a widespread basis, it may be the key to controlling epidemics and pandemics. No other current technology has the capability, the adaptability, and the favorable economics to make it viable for an extremely wide variety of disease control applications. It is already used extensively and effectively in water applications and in surface disinfection applications. In combination with air filtration, it is the most effective and economic technology for disinfecting air. From health care applications to schools and residential environments, UVGI holds the promise of one day contributing in a major fashion to the eradication of many contagious diseases. The advent of multidrug-resistant microbes like MRSA and XTB, and emerging pathogens like SARS and Avian Influenza is likely to stimulate the increased use of UVGI systems in an ever wider number of applications. The contribution UV technology can make to the control of epidemics is amenable to analysis by the statistical models of epidemiology, which have demonstrated the potential for widespread use of UVGI systems to theoretically halt contagious airborne disease epidemics (Kowalski 2006). Perhaps continued research and development will ultimately lead to UVGI becoming a standard component of ventilation systems in all indoor environments and this age of airborne epidemics will come to an end.

References

Ager BP, Tickner JA. 1983. The control of microbiological hazards associated with air-conditioning and ventilation systems. Ann Occup Hyg 27(4):341–358.

Anderson K. 1959. *Pseudomonas pyocyanea* disseminated form an air cooling apparatus. Med J Aus 1:529.

Anderson WB, Huck PM, Dixon DG, Mayfield CI. 2003. Endotoxin inactivation in water by using medium-pressure UV lamps. Appl Environ Microbiol 69(5):3002–3004.

Asthana A, Tuveson RW. 1992. Effects of UV and phototoxins on selected fungal pathogens of citrus. Int J Plant Sci 153(3):442–452.

AWWA. 1971. Water Quality and Treatment. The American Water Works Association I, editor. New York: McGraw-Hill.

Banaszak EF, Thiede WH, Fink JN. 1970. Hypersensitivity pneumonitis due to contamination of an air conditioner. New England J Med 283(6):271–276.

Barnard J, Morgan H. 1903. The physical factors in phototherapy. Brit Med J 2:1269–1271.

Bedford THB. 1927. The nature of the action of ultra-violet light on micro-organisms. Brit J Exp Path 8:437–441.

Bolton JR, Cotton CA. 2008. Ultraviolet disinfection handbook. Denver, CO: American Water Works Association.

Buchbinder L, Phelps EB. 1941. Studies on microorganisms in simulated room environments. II. The survival rates of streptococci in the dark. J Bact 42:345–351.

Buttolph LJ, Haynes H. 1950. Ultraviolet Air Sanitation. Cleveland, OH: General Electric. Report nr LD-11.

CDC. 2003. Guidelines for environmental infection control in health-care facilities. MMWR 52(RR-10).

CDC. 2005. Guidelines for preventing the transmission of *Mycobacterium tuberculosis* in health-care facilities. In: CDC, editor. Federal Register. Washington: US Govt. Printing Office.

CIE. 2003. Ultraviolet Air Disinfection. Vienna, Austria: International Commission on Illumination. Report nr CIE 155:2003.

Coblentz WW, Stair R, Hogue JM. 1932. Spectral erythemic reaction of the untanned human skin to ultraviolet radiation. J Res Nat Bur Stand 8:541–547.

Cole WR, Bernard HR, Dunn B. 1964. Growth of bacteria on direct expansion air-conditioning coils. Surgery 55(3):436–439.

DelMundo F, McKhann CF. 1941. Effect of ultra-violet irradiation of air on incidence of infections in an infant's hospital. Am J Dis Child 61:213–225.

Downes A, Blunt TP. 1877. Research on the effect of light upon bacteria and other organisms. Proc Roy Soc London 26:488–500.

Ehrismann O, Noethling W. 1932. Uber die bactericide wirkung monochromatischen lichtes. Ztschr Hyg Infektionskr 113:597–628.

ELP. 2000. How UV-C lamps saved one company $58,000. Electric Light & Power 78(2).

Escombe A, Moore D, Gilman R, Navincopa M, Ticona E, Mitchell B, Noakes C, Martinez C, Sheen P, Ramirez R and others. 2009. Upper-room ultraviolet light and negative air ionization to prevent tuberculosis transmission. PLoS Med 6(3):312–322.

FEMA. 2003. Reference Manual to Mitigate Potential Terrorist Attacks Against Buildings: Federal Emergency Management Agency. Report nr FEMA 426.

Finsen N. 1900. Die Resultate der Behandlung des lupus vulgaris durch knzentrierte chemische Lichtstrahlen. Allg Wien Med Z A5:60–65.

Fizeau H, Foucault L. 1843. Observations concernant l'action des rayons rouges sur les plaques degueeriennes. C R Habd Seances Acad Sci 23:679–682.

Fulton HR, Coblentz WW. 1929. The fungicidal action of ultraviolet radiation. J Agric Res 38:159.

Gates FL. 1929. A study of the bactericidal action of ultraviolet light. J Gen Physiol 13:231–260.

GE. 1950. Germicidal Lamps and Applications. USA: General Electric. Report nr SMA TAB: VIII-B.

Goldner JL, Allen BL. 1973. Ultraviolet light in orthopedic operating rooms at Duke University. Clin Ortho 96:195–205.

Goldner JL, Moggio M, Beissinger SF, McCollum DE. 1980. Ultraviolet light for the control of airborne bacteria in the operating room. In: Kundsin RB, editor. Airborne Contagion, Annals of the New York Academy of Sciences. New York: NYAS, pp. 271–284.

Grun L, Pitz N. 1974. U.V. radiators in humidifying units and air channels of air conditioning systems in hospitals. Zbl Bakt Hyg B159:50–60.

GSA. 2003. The Facilities Standards for the Public Buildings Service. Washington: Public Buildings Service of the General Services Administration.

Guha S, Bojarczuk N. 1998. Ultraviolet and violet GaN light emitting diodes on silicon. Appl Phys Lett 72:415.

Gunderman KO. 1980. Spread of microorganisms by air-conditioning systems – especially in hospitals. In: Kundsin RB, editor. Airborne Contagion. Boston: New York Academy of Sciences, pp.209–217.

Harstad JB, Decker HM, Wedum AG. 1954. Use of ultraviolet irradiation in a room air conditioner for removal of bacteria. Am Ind Hyg Assoc J 2:148–151.

Hart J. 1936. Sterilization of air in operating room by special bactericidal radiant energy. J Thoracic Surg 6:45–81.

Hart D. 1937. Operating room infections: preliminary report. Arch Surg 34:874–896.

Hart D, Sanger PW. 1939. Effect on wound healing of bactericidal ultraviolet radiation from a special unit: experimental study. Arch Surg 38(5):797–815.

Hausser K, Vahle W. 1927. Sonnenbrand und sonnenbraunung. In: Reiter T, Gabor D, editors. Zellteilung und Strahlung. Berlin: Springer, pp. 101–120.

Henle W, Sommer HE, Stokes J. 1942. Studies on air-borne infection in a hospital ward: II. Effects of ultra-violet irradiation and propylene glycol vaporization upon the prevention of experimental air-borne infection of mice by droplet nuclei. J Pediat 21:577–590.

Henri V. 1912. Comparaison de l'action des rayons ultrav-violets sur les organismes avec les reactions photochemique simples et complexes. Compt Rend Soc Biol 73:323–325.

Henri V. 1914. Etude de l'action metabolique des rayons ultraviolets. CR Acad Sci (Paris) 159:340–343.

Higgons RA, Hyde GM. 1947. Effect of ultra-violet air sterilization upon incidence of respiratory infections in a children's institution. New York State J Med 47(7).

Hockberger P. 2002. A history of ultraviolet photobiology for humans, animals, and microorganisms. Photochem Photobiol 76(6):561–579.

Hollaender A. 1943. Effect of long ultraviolet and short visible radiation (3500 to 4900) on *Escherichia coli*. J Bact 46:531–541.

Kowalski WJ. 2006. Aerobiological Engineering Handbook: A Guide to Airborne Disease Control Technologies. New York: McGraw-Hill.

Kraissl CJ, Cimiotti JG, Meleney FL. 1940. Considerations in the use of ultra-violet radiation in operating rooms. Ann Surg 111:161–185.

Lorch W. 1987. Handbook of Water Purification. Chichester: Ellis Horwood Ltd.

Luciano JR. 1977. Air Contamination Control in Hospitals. New York: Plenum Press.

Luckiesh M, Taylor AH. 1925. Fading of colored materials in daylight and artificial light. Trans Illum Eng Soc 20:1078.

Luckiesh M, Holladay LL. 1942. Tests and data on disinfection of air with germicidal lamps. Gen Electric Rev 45(4):223–231.

Luckiesh M, Holladay LL. 1942a. Designing installations of germicidal lamps for occupied rooms. General Electric Review 45(6):343–349.

Luckiesh M. 1945. Disinfection with germicidal lamps: Air – II. Electrical World Oct.13:109–111.

Luckiesh M. 1946. Applications of Germicidal, Erythemal and Infrared Energy. New York: D. Van Nostrand Co.

Martin S, Dunn C, Friehaut J, Bahnfleth W, Lau J, Nedeljkovic A. 2008. Ultraviolet germicidal irradiation: current best practices. ASHRAE J August:28–36.

Menzies D, Popa J, Hanley JA, Rand T, Milton DK. 2003. Effect of ultraviolet germicidal lights installed in office ventilation systems on workers' health and wellbeing: double-blind multiple crossover trials. The Lancet 362(November 29):1785–1791.

MRC. 1954. Air Disinfection with Ultra-violet Irradiation; Its Effect on Illness among School-Children by the Air Hygiene Committee. London: Medical Research Council, Her Majesty's Stationary Office. Report nr 283.

Nardell EA. 1988. Chapter 12: Ultraviolet air disinfection to control tuberculosis. In: Kundsin RB, editor. Architectural Design and Indoor Microbial Pollution. New York: Oxford University Press, pp. 296–308.

Newcomer H. 1917. The abiotic action of ultraviolet light. J Exp Med 26:841.

Overholt RH, Betts RH. 1940. A comparative report on infection of thoracoplasty wounds. J Thoracic Surg 9:520–529.

Perkins JE, Bahlke AM, Silverman HF. 1947. Effect of ultra-violet irradiation of classrooms on the spread of measles in large rural central schools. Am J Pub Health 37:529–537.

Philips. 1985. UVGI Catalog and Design Guide. Netherlands: Catalog No. U.D.C. 628.9.

Prat S. 1936. Strahlung und antagonistische wirkungen. Protoplasma 26:113–149.

Rentschler HC, Nagy R. 1940. Advantages of bactericidal ultraviolet radiation in air conditioning systems. HPAC 12:127–130.

Renz C. 1921. Photo-reactions of the oxides of titanium, cerium, and earth-acids. Helv Chim Acta 4:961.

Riley R, Wells W, Mills C, Nyka W, McLean R. 1957. Air hygiene in tuberculosis: Quantitative studies of infectivity and control in a pilot ward. Am Rev Tuberc Pulmon 75:420–431.

Riley RL, O'Grady F. 1961. Airborne Infection. New York: The Macmillan Company.

Riley RL. 1972. The ecology of indoor atmospheres: Airborne infection in hospitals. J Chron Dis 25:421–423.

Riley RL. 1980. Part I. History and epidemiology: Historical background. In: Kundsin RB, editor. Airborne Contagion, Annals of the New York Academy of Sciences. New York: NYAS, pp. 1–9.

Riley RL. 1980a. Prevention and control of airborne infection in the community. In: Kundsin RB, editor. Airborne Contagion, Annals of the New York Academy of Sciences. New York: NYAS, pp. 331–339.

Ritter M, Olberding E, Malinzak R. 2007. Ultraviolet lighting during orthopaedic surgery and the rate of infection. J Bone Joint Surg 89:1935–1940.

Rivers T, Gates F. 1928. Ultra-violet light and vaccine virus. II. The effect of monochromatic ultra-violet light upon vaccine virus. J Exp Med 47:45–49.

Robertson EC, Doyle ME, Tisdall FF, Koller LR, Ward FS. 1939. Air contamination and air sterilization. Am J Dis Child 58:1023–1037.

Scheir R. 2000. Electric utility solves IAQ problem with UVC electrical energy. HPAC Eng 69(5):28–29, 85, 87.

Schicht HH. 1972. The diffusion of micro-organisms by air conditioning installations. The Steam and Heating Engineer October:6–13.

Schneiter R, Hollaender A, Caminita BH, Kolb RW, Fraser HF, duBuy HG, Neal PA, Rosenblum HG. 1944. Effectiveness of ultra-violet irradiation of upper air for the control of bacterial air contamination in sleeping quarters. Am J Hyg 40:136.

Shantha T, Sreenivasa M. 1977. Photo-destruction of aflatoxin in groundnut oil. Indian J Technol 15:453.

Sharp G. 1940. The effects of ultraviolet light on bacteria suspended in air. J Bact 38:535–547.

Sommer HE, Stokes J. 1942. Studies on air-borne infection in a hospital ward. J Pediat 21:569–576.

Spendlove JC, Fannin KF. 1983. Source, significance, and control of indoor microbial aerosols: Human health aspects. Public Health Rep 98(3):229–244.

USACE. 2000. Guidelines on the Design and Operation of HVAC Systems in Disease Isolation Areas. U.S. Army Center for Health Promotion and Preventive Medicine. Report nr TG252.

von Recklinghausen M. 1914. The Ultra-Violet rays and their application for the sterilization of water. J Franklin Institute 1057–1062:681–704.

Walter CW. 1969. Ventilation and air conditioning as bacteriologic engineering. Anesthesiology 31:186–192.

Ward M. 1892. Experiments of the action of light on *Bacillus anthracis*. Proc Roy Soc Lond 52:393–403.

Wells W, Wells M. 1936. Air-borne infection. JAMA 107:1698–1703.

Wells WF. 1938. Air-borne infections. Mod Hosp 51:66–69.

Wells WF. 1940. Bactericidal irradiation of air, physical factors. J Franklin Inst 229:347–372.

Wells WF, Wells MW, Wilder TS. 1942. The environmental control of epidemic contagion; I – An epidemiologic study of radiant disinfection of air in day schools. Am J Hyg 35:97–121.

Wells WF. 1943. Air disinfection in day schools. Am J Pub Health 33:1436–1443.

Wells WF, Holla WA. 1950. Ventilation in flow of measles and chickenpox through community. Progress report, Jan. 1, 1946 to June 15, 1949, airborne infection study. J Am Med Assoc 142:1337–1344.

Wheeler SM, Ingraham HS, Hollaender A, Lill ND, Gershon-Cohen J, Brown EW. 1945. Ultra-violet light control of airborne infections in a naval training center. Am J Pub Health 35:457–468.

Whitby G, Scheible O. 2004. The history of UV and wastewater. IUVA News 6(3):15–26.

WHO. 1999. Guidelines for the Prevention of Tuberculosis in Health Care Facilities in Resource Limited Settings. Geneva: World Health Organization. Report nr WHO/CDS/TB/99.269.

Whyte W. 1968. Bacteriological aspects of air-conditioning plants. J Hyg 66:567–584.

Widmark E. 1889. Uber den einfluss des lichtes auf die haut. Hygiea 3:1–23.

Widmark J. 1889a. De l'influence de la lumiere sur la peau. Biol Foren Forhandl Verhandlungen Biolog Vereins 1:9–13, 131–134.

Zeterberg JM. 1973. A review of respiratory virology and the spread of virulent and possibly antigenic viruses via air conditioning systems. Ann Allergy 31:228–299.

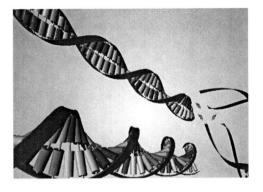

Chapter 2
UVGI Disinfection Theory

2.1 Introduction

Ultraviolet Germicidal Irradiation (UVGI) is electromagnetic radiation that can destroy the ability of microorganisms to reproduce by causing photochemical changes in nucleic acids. Wavelengths in the UVC range are especially damaging to cells because they are absorbed by nucleic acids. The germicidal effectiveness of UVC peaks at about 260–265 nm. This peak corresponds to the peak of UV absorption by bacterial DNA. The germicidal effectiveness of UVC radiation can vary between species and the broader range wavelengths that include UVB also make a small contribution to inactivation (Webb and Tuveson 1982). Although the methods and details of disinfection with ultraviolet light are fairly well understood, to the point that effective disinfection systems can be designed and installed with predictable effects, the exact nature of the effect of ultraviolet light on microorganisms at the molecular level is still a matter of intensive research. This chapter examines the fundamentals of the complex interaction between UV irradiation and cell DNA at the molecular level and provides detailed background information to aid in the understanding of the various biophysical processes that are involved in microbial inactivation.

2.2 UV Inactivation

The spectrum of ultraviolet light extends from wavelengths of about 100–400 nm. The subdivisions of most interest include UVC (200–280 nm), and UVB (280–320 nm). Although all UV wavelengths cause some photochemical effects, wavelengths in the UVC range are particularly damaging to cells because they are absorbed by proteins, RNA, and DNA (Bolton and Cotton 2008, Rauth 1965). The germicidal effectiveness of UVC is illustrated in Fig. 2.1, where it can be observed that germicidal efficiency reaches a peak at about 260–265 nm. This corresponds to the peak of UV absorption by bacterial DNA (Harm 1980). The germicidal effectiveness of UVC and UVB wavelengths can vary between species. Low pressure

W. Kowalski, *Ultraviolet Germicidal Irradiation Handbook*,
DOI 10.1007/978-3-642-01999-9_2, © Springer-Verlag Berlin Heidelberg 2009

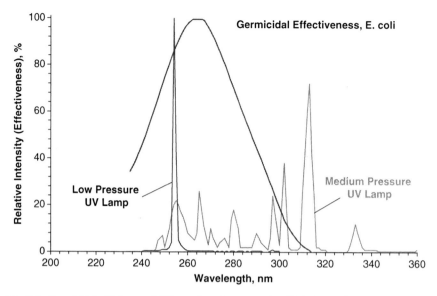

Fig. 2.1 Germicidal efficiency of UV wavelengths, comparing High (or medium) and Low pressure UV lamps with germicidal effectiveness for *E. coli*. Based on data from Luckiesh (1946) and IESNA (2000)

mercury vapor lamps radiate about 95% of their energy at a wavelength of 253.7 nm, which is coincidentally so close to the DNA absorption peak (260–265 nm) that it has a high germicidal effectiveness (IESNA 2000).

If we assume the LP and MP lamps in Fig. 2.1 produce the same total UV wattage, and multiply spectrum by the germicidal efficiency at each wavelength, we find the LP lamp has a net germicidal efficiency of 84% vs. 79% for the MP lamp. The optimum wavelength for inactivating *E. coli*, about 265 nm, is about 15% more effective than the UVC peak of 254 nm. The optimum wavelength for inactivating *Bacillus subtilis* is 270 nm, and this is about 40% more effective than 254 nm (Waites et al. 1988). The optimum wavelength for destroying *Cryptosporidium parvum* oocysts is 271 nm and this is about 15% more effective than 254 nm (Linden 2001). Although UVC is responsible for the bulk of the germicidal effects of broad-spectrum UV, the effects of UVB wavelengths cannot be discounted altogether. In a study by Elasri and Miller (1999) it was found that UVB had about 15% of the effect of UVC on *Pseudomonas aeruginosa*.

THE STRUCTURE OF DNA

Deoxyribonucleic acid (DNA) is a large, high molecular weight macromolecule composed of subunits called nucleotides. Each nucleotide subunit has three parts: deoxyribose, phosphate, and one of four nitrogenous bases (nucleic acid bases). The four bases are thymine (T), adenine (A), cytosine (C), and guanine (G). These four bases form base pairs of either thymine bonded to adenine or cytosine bonded to guanine. Since thymine always pairs with adenine, there will be equal amounts of thymine and adenine. Likewise, cytosine will always exist in amounts equal to guanine. The specific sequences formed by these base pairs make up the genetic code that forms the chemical basis for heredity (Atlas 1995). Nucleotides are the basic repeating unit of DNA and they are composed of nitrogenous bases called purines and pyrimidines. These bases are linked to pentoses to make nucleosides. The nucleosides are linked by phosphate groups to make the DNA chain.

DNA forms a double helix, as shown in the figure above, in which two complementary strands of nucleotides coil around each other. The two outside helices of DNA form a backbone that is held together by strong covalent bonds, locking in the stability of the hereditary macromolecule. Each helix terminates in a free hydroxyl group at one end, and a free phosphate group at the other, conferring directionality. The two halves of the DNA molecule run in opposite directions and coil around each other. Supercoiling may also occur as long chains of DNA fold and pack into the available space (i.e. in a cell or viral capsid). Several million nucleotides may be held together in sequence and they establish the genetic code for each species.

The two complementary chains of the DNA double helix are held together by hydrogen bonding between the chains. Two of the nitrogenous bases(C and T) are single-ring structures called pyrimidines and the other two (A and G) are double-ring structures called purines. The internal hydrogen bonds between the base pairs, which hold the entire structure together, have only about 5% of the strength of the covalent bonds in the outer helix. Thymine forms two hydrogen bonds with adenine, while cytosine forms three hydrogen bonds with guanine. The thymine/adenine bond, therefore, represents the weakest link in the structure.

Hydrogen bonds between complementary bases are not the primary stabilizing force of DNA since the energy of a hydrogen bond (2–4 kcal/bond) is insufficient to account for the observed stability of DNA. Ionic bonds between the negative phosphate groups and positive cations reduce the electrostatic repulsion between the negative charges of the sugar-phosphate backbone. The stability of DNA is also accounted for by the hydrophobic forces associated with stacking of the bases, which is due to mutual interactions of the bases and geometrical considerations (Guschlbauer 1976). In polynucleotide chains, this interaction results in a compact stack of bases that is restricted by the sugar-phosphate backbone and results in a narrow range of possible overlap angles between the bases (36° in DNA). The stacked bases form a hydrophobic core which favors hydrogen bonding between the complementary strands (see Fig. 2.2). The stacking and twisting of base pairs creates channels in which water may bond or be excluded depending of DNA conformation (Neidle 1999).

UVA was also found to have a lethal effect on *P. aeruginosa* although considerably more lamp power was needed (Fernandez and Pizarro 1996). The UVA inactivation effect, however, is relatively insignificant and may involve non-actinic effects (no photochemical changes). Throughout the following discussion only the actinic bands of UVC and UVB are considered to be operative.

Two general types of nucleic acids exist, ribonucleic acid (RNA) and deoxyribonucleic acid (DNA). Viruses contain DNA or RNA, but not both. During UV irradiation and inactivation, the most sensitive target of microorganisms is the DNA of bacteria, the DNA of DNA viruses, the RNA of RNA viruses, and the DNA of

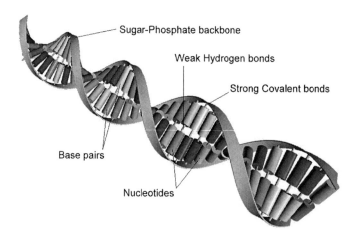

Fig. 2.2 The two helical backbones of DNA are connected by hydrogen bonds between the nucleotides

fungi. RNA has D-ribose as its main constituent and adenine, cytosine, guanine, and uracil as bases. DNA has 2-deoxy-D-ribose as its main constituent and adenine, cytosine, guanine, and thymine as bases. Hydrogen bonds link the bases. UV radiation can cause a crosslink between two thymine bases that is more stable than a hydrogen bond (Casarett 1968). Bacteria and fungi have DNA, while viruses may have either DNA or RNA. DNA and RNA are responsible for microbial replication and protein synthesis and damage to these nucleic acids results in inactivation or the failure to reproduce.

UV wavelengths inactivate microorganisms by causing cross-links between constituent nucleic acids. The absorption of UV can result in the formation of intrastrand cyclobutyl-pyrimidine dimers in DNA, which can lead to mutations or cell death (Harm 1980, Koller 1952, Kuluncsics et al. 1999). Pyrimidines are molecular components in the biosynthesis process and include thymine and cytosine (see Fig.2.2). Thymine and cytosine are two of the base pair components of DNA, the others being adenine and guanine. The primary dimers formed in DNA by UV exposure are known as thymine dimers (see the page on the Structure of DNA). The lethal effect of UV radiation is primarily due to the structural defects caused when thymine dimers form but secondary damage is also produced by cytosine dimers (David 1973, Masschelein 2002). Various other types of photoproducts are also formed that can contribute to cell death. Photohydration reactions can occur under UV irradiation in which the pyrimidines cytosine and uracil bond with elements of water molecules. The same reaction does not occur with thymine. The photohydration yield is independent of wavelength.

Double stranded RNA has a higher resistance to UV irradiation than single stranded RNA, and this may be due to various factors, including more structural stability (Becker and Wang 1989) and the redundancy of information in the complementary strands (Bishop et al. 1967). Ultraviolet light also causes photochemical reactions in proteins in the cell other than DNA and UV absorption in proteins peaks at about 280 nm. There is also some absorption in the peptide bonds within proteins at wavelengths below 240 nm.

Figure 2.3 illustrates how UV absorption can lead to cross-linking between adjacent thymine molecules and the formation of thymine dimers. When thymine bases happen to sit next to each other, the pair is called a doublet. The dimerization of thymine doublets by UV can lead to inactivation of the DNA, or RNA, with the result that cell may be unable to reproduce effectively.

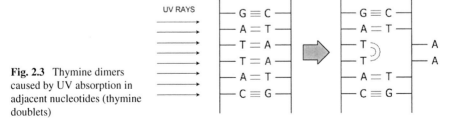

Fig. 2.3 Thymine dimers caused by UV absorption in adjacent nucleotides (thymine doublets)

The exact mechanism by which UV causes thymine dimers is not completely understood. Bhattacharjee and Sharan (2005) demonstrated that exposing *E. coli* DNA to UVC irradiation induced sparsely placed, dose-dependent, single strand breaks, and proposed that the conformational relaxation generates negative super-coiling strain on the DNA backbone.

It has been repeatedly demonstrated that thymine dimers produced by UV exposure result in the inactivation of bacteria and DNA viruses. A dose of 4.5 J/m^2 is reported to cause 50,000 pyrimidine dimers per cell (Rothman and Setlow 1979). It has been reported that 100 J/m^2 induces approximately seven pyrimidine dimers per viral genome in SV40, which is sufficient to strongly inhibit viral DNA synthesis (Sarasin and Hanawalt 1980). Thymine dimers form within 1 picosecond of UV excitation provided the bases are properly oriented at the instant of light absorption (Schreier et al. 2007). Only a few percent of the thymine doublets are likely to be favorably positioned for reaction and dimerization at the time of UV excitation. Figure 2.4 illustrates the dimerization process for a thymine doublet with the appropriate orientation.

The two most common conformations of DNA are called A-DNA and B-DNA. Molecular orientations can vary due to A and B conformation and vibrational or other movement in the DNA molecule. The average twist angle between successive base pairs differs between the A conformation and the B conformation is only a few degrees. The smaller amount of conformational variation in A-DNA vs. B-DNA explains the greater resistance of A-DNA to cyclobutane pyrimidine dimer

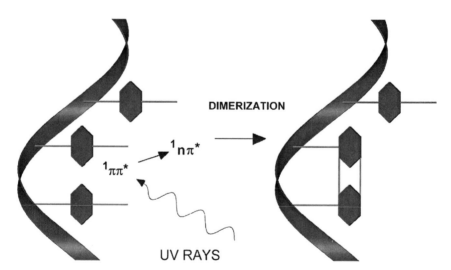

Fig. 2.4 The photodynamics of dimerization. A single strand of the DNA sugar-phosphate backbone is shown with thymine nucleotides. UV excitation populates the singlet state $\pi\pi^*$, which decays into the singlet $n\pi^*$ state (*left*). All energy is converted internally to form a thymine double hydrogen bond (*right*)

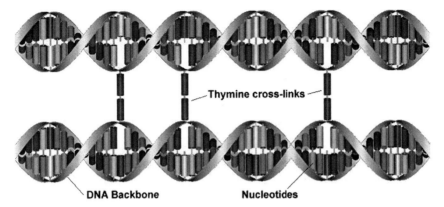

Fig. 2.5 Cross-linking between thymine nucleotides (or uracil nucleotides in the case of RNA viruses) can occur between adjacent strands of DNA (or RNA). It can also occur between the DNA (RNA) and the proteins of the capsid, for viruses

formation (Schreier et al. 2007). That is to say, the more dense packing of bases and lower flexibility in the B-DNA form ensures a higher probability that thymine doublets will be available for dimerization.

The exact sequences of thymines, cytosines, adenines, and guanines in DNA can directly impact the probability of dimerization. Adjacent pyrimidines (thymine and cytosine) are considerably more photoreactive than adjacent purines (adenine and guanine). Becker and Wang (1989) found that 80% of pyrimidines and 45% or purines form UV photoproducts in double-stranded DNA.

In addition to cross-links between adjacent thymines, UV may also induce cross-links between non-adjacent thymines, as illustrated in Fig.2.5. Cross-linking can also occur between the nucleotides and the proteins in the capsid of viruses, damaging the capsid of DNA viruses.

Cross-linking can also occur with cytosine and guanine, but the energy required is higher due to their having three hydrogen bonds instead of two for thymine/adenine bonds, and so thymine dimers predominate. Besides cross-links with adjacent thymine nucleotides and with thymine in adjacent strands of DNA, thymine may also form links with proteins, including proteins in the capsid (in the case of viruses) as shown in Fig. 2.6. Other biological molecules with unsaturated bonds like coenzymes, hormones, and electron carriers may be susceptible to UV damage.

In RNA, whether in prokaryotic cells, eukaryotic cells, or viruses, uracil takes the place of thymine. Inactivation of RNA viruses involves cross-linking between the uracil nucleotides and the creation of uracil dimers (Miller and Plageman 1974). Uracil dimers may also damage the capsid of RNA viruses. Some limited quantitative data is available on the specific nature of DNA damage produced by UV absorption. Miller and Plageman (1974) demonstrated that ultraviolet exposure of Mengovirus caused rapid formation of uracil dimers and that this appeared to be the

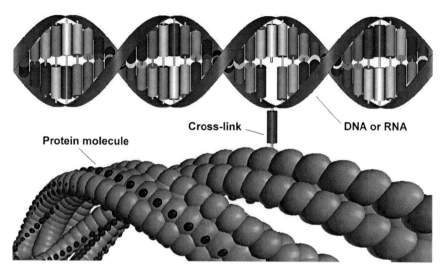

Fig. 2.6 Thymine dimerization can also occur between DNA/RNA and adjacent protein molecules, such as in cell cytoplasm or the capsid of a virus

primary cause of virus inactivation (i.e. loss of infectivity). A maximum of about 9% of the total uracil bases of the viral DNA formed dimers within 10 min of UV irradiation. Results also indicated that viral RNA became covalently linked to viral protein as a result of irradiation. A slower process of capsid destruction also occurred in which capsid proteins were modified and photoproducts were formed. UV irradiation of the virus also caused covalent linkage of viral RNA to viral polypeptides, apparently due to close proximity between the RNA and proteins in the capsid. The amount of protein covalently linked to the RNA represented not more than 1.5% of the total protein capsid. Smirnov et al. (1992) studied Venezuelan equine encephalitis (VEE) under UV irradiation and found evidence suggesting that the formation of uracil dimers led to extensive contacts of the RNA with protein in the nucleocapsid.

Viruses containing many thymine dimers may still be capable of plaque formation (Rainbow and Mak 1973). An *E. coli* chromosome exposed to UVB produced pyrimidine photoproducts in the following proportions: 59% thymine dimers, 34% thymine-cytosine dimers, and 7% cytosine-cytosine dimers (Palmeira et al. 2006).

Figure 2.7 shows the rate at which uracil dimers form under irradiation, shown in terms of the uracil bonds remaining intact in RNA. This plot is shown alongside the decay rate for Mengovirus. It can be observed that the virus is rapidly inactivated while the formation of uracil dimers proceeds relatively slowly. The scale of the chart is limited, but the virus goes through six logs of reduction before 9% of the uracil is cross-linked. Clearly, it takes but little cross-linking to inactivate a virus. The ratio of the microbial inactivation rate to the dimer production rate should be a constant for any given species. Theoretically, each species should have a characteristic inactivation rate that is a function of the dimerization probability.

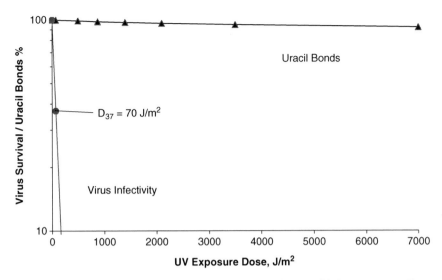

Fig. 2.7 Survival of Mengovirus under UV irradiation, plotted along with the percentage of intact uracil bonds in viral RNA. Based on data from Miller and Plageman (1974)

2.3 UV Absorption Spectra

An absorption spectrum is a quantitative description of the absorptive capacity of a molecule over some specified range of electromagnetic frequencies. The absorption of ultraviolet light by a molecule will result in altered electronic configuration, conversion into radiant energy, rotation, and vibration. When these energy levels are at a minimum the molecule is in a ground state. The energy imparted to a molecule by UV absorption produces an excited state. The capacity of a molecule to absorb UV energy over a band of wavelengths is described in terms of an absorption spectrum. The intensity of absorption is generally expressed in absorbance or optical density. The intensity of an absorption band is directly related to the probability that the particular transition will take place when a photon of the right energy comes along. Figure 2.8 represents the absorption spectra for the four DNA bases, which have peaks in the UVC band, and also below 220 nm, which is in the VUV range. Thymine and cytosine both have strong peaks near 265 nm.

Pyrimidines (thymine, cytosine, and uracil) absorb about ten times more UV than purines. The quantum yield at 254 nm is $\phi \sim 10^{-3}$ for pyrimidines and for purines $\phi \sim 10^{-4}$. The capacity of a molecule to absorb light of a particular wavelength depends on both the electronic configuration of the molecule and on its available higher energy states (Smith and Hanawalt 1969). An absorption spectrum may be regarded as a summation of a series of individual absorption bands, each corresponding to a transition between two particular electronic configurations (Hollaender 1955). This transition typically occurs when an orbital electron is raised from the normal ground state to an excited state. These transitions occur only in

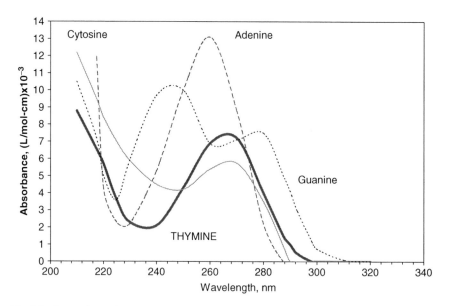

Fig. 2.8 Comparison of response spectra for the four main nucleotides

discrete jumps, and therefore only a specific quantum of energy can be absorbed. The electronic configuration of an excited molecules can be a very transitory event, and only those excitations of sufficient duration will have a high probability of absorption.

An absorption band may be described by the width of the band (or the range of wavelengths) and the degree of absorption (absorptivity). The width of the band is defined as the spectral separation between the points of half-maximum (50%) absorption. The width of a band is inversely dependent upon the duration of the excited electronic state. Integration of the absorption band over the width determines the probability that the particular transition will occur when a photon of the right energy comes along.

The absorption spectrum is typically measured by beaming light through a transparent solution containing microbes or molecules and comparing it against the pure solution. The transmittance (or transmissivity), T, of a solution is defined as:

$$T = \frac{I}{I_0} \tag{2.1}$$

where

I = irradiance of light exiting the solution, W/m^2
I_0 = irradiance of light entering the solution, W/m^2

Beer's Law establishes a relationship between the transmittance in Eq. (2.1) and the absorbance, A, as follows:

$$T = \frac{I}{I_0} = 10^{-A} \qquad (2.2)$$

The absorbance (or absorptivity), also called the optical density, OD, is defined as:

$$A = \varepsilon l c \qquad (2.3)$$

where

  ε = molar absorptivity, liters/mole-cm
  l = thickness of the solution, cm
  c = concentration of solute, moles/liter

An alternate form of Beer's Law is:

$$\frac{I}{I_0} = e^{-nsl} \qquad (2.4)$$

where

  n = number of molecules per unit volume
  s = absorption cross-section, m^2 or μm^2

The absorption cross-section represents the product of the average cross-sectional area of the molecule and the probability that a photon will be absorbed. The absorption cross-section is related to the molar absorptivity by the following relation:

$$s = 3.8x10^{-21}\varepsilon \qquad (2.5)$$

Strong absorption bands in the ultraviolet region correlate with molecular structures containing conjugated double bonds. Ring structures, such as the pyrimidines and purines, exhibit particularly strong absorption and can define the overall absorption spectrum of DNA. Von Sonntag (1986) reports that DNA has a peak of UV absorption not only at 265 nm but also at 200 nm. Most of the absorption at 200 nm occurs in the DNA backbone molecules of ribose and phosphate. At 265 nm, most of the absorption occurs at the nucleotide bases, thymine and adenine, and cytosine and guanine, but dimers of thymine are by far the most common UV photoproducts. In RNA-based microbes, uracil is also involved in UV absorption in place of thymine but not necessarily to the same degree. Figure 2.9 compares the absorption spectrum of uracil with that of thymine. It can be observed that not only are the absorption spectra very similar for these nucleotides, but that the mercury emission line at 254 nm is more nearly aligned with the peak absorption of uracil.

Carbohydrates make up about 41% by weight of nucleic acids, but they show essentially no UV absorption above about 230 nm and would not be expected to

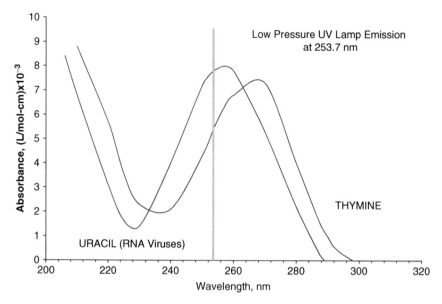

Fig. 2.9 Comparison of thymine UV absorption spectra with uracil, the nucleotide that takes the place of thymine in RNA viruses

participate in photochemical reactions at around 254 nm. However, certain photochemical processes that produce uracil radicals can result in chemical alterations to the carbohydrates of nucleic acid (Smith and Hanawalt 1969).

The ultraviolet absorption spectrum of a polymer is not necessarily the linear sum of its constituents. This nonadditivity is referred to as *hyperchromicity*. If the absorbance of a given oligonucleotide is higher than its constituents molecules, it is hyperchromic. Hyperchromicity is largely explained by the coulombic interaction of the ordered bases in the polymer. Hyperchromicity is a kind of resonant electronic effect in which the partial alignment of transition moments by base stacking results in coupled oscillation. A relatively small number of bases in a DNA strand are required for such coupling and about 8–10 base pairs can exhibit roughly 80% of the hyperchromism of an infinite helix. Becker and Wang (1989) present data that indicates that hyperchromicity may add about twice the number of photoproducts when strings of eight or more thymines occur sequentially.

2.4 UV Photoprotection

Microbes have various mechanisms by which they can protect themselves from UV exposure, including nucleocapsids and cytoplasm which may contain UV absorbing proteins (i.e. dark proteins). The absorption of UV in any surrounding complex of proteins will reduce the density of photons reaching the nucleic acid and thereby provide photoprotection. Comparisons of virus inactivation with inactivation of purified DNA show the absorption spectrums are not identical, the implication being that UV is absorbed in the envelope, the nucleocapsid, or other protein-laden constituents of the viroid, although in some cases the nucleic acid is more resistant in isolation (Zavadova et al. 1968, Furuse and Watanabe 1971, Bishop et al. 1967). In bacteria, the cytoplasm may offer photoprotection due to its UV absorptivity. Unrau et al. (1973) have suggested that there is a *in vivo* shielding effect in *Bacillus subtilis* since dimer formation is doubled when its DNA is irradiated separately, although they do not attribute this effect to the cytoplasm. Fungal spores are among the most resistant microbes and they often have melanin-containing dark pigmented conidia. The photoprotective component melanin increases the survival and longevity of fungal spores (Bell and Wheeler 1986). *Aspergillus niger* conidia are more resistant to UV due to the high UV absorbance of their melanin pigments (Anderson et al. 2000). Various studies on fungi have suggested that lighter-pigmented thin-walled conidia are more susceptible to UV than thicker-walled dark-pigmented conidia (Boyd-Wilson et al. 1998, Durrell and Shields 1960, Valero et al. 2007). UV scattering (addressed later) can also contribute to photoprotection.

Figure 2.10 illustrates photoprotection mechanisms in viruses – UV scattering by the envelope, UV absorption by the envelope, and UV absorption by the nucleocapsid the latter being mostly negligible. UV scattering, occurs when the particle is in the Mie scattering size range, and the effective scattering cross-section my be much larger than the actual physical cross-section of the particle (see Sect. 2.13).

Chromophores are chemical groups in molecules that are capable of absorbing photons. Polyatomic molecules have fairly broad absorption bands. In proteins, the molecular groupings which give rise to absorption are principally amino acids,

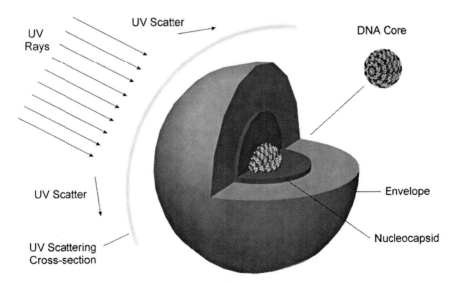

Fig. 2.10 Schematic illustration of the levels of photoprotection of an enveloped virus, including the UV scattering cross-section, the envelope, and the nucleocapsid

which have absorption peaks at about 280 nm (Webb 1965). For the nucleic acids, the chromophores responsible for the absorption peak around 265 nm are the bases, the purines and pyrimidines, the dimers of which are considered primary products of the biocidal action of UV. The chromophores that are likely to confer protection to these bases are those in the cytoplasm or cell wall of bacteria, in the capsid or protein coat, if any, of the virus, and the spore coat and cortex of a spore.

Some amino acids are optically transparent above UV wavelengths of about 240 nm, while others, the chromophores, have high molar absorptivities at or near 253.7 nm. The molar absorptivity of a compound is the probability that the wavelength will be absorbed. Figure 2.11 shows the molar absorptivity of amino acids at 253.7. These might be called 'dark proteins' because of their relatively high absorbance for UV wavelengths. Microbes that contain higher proportions of chromophores will likely absorb UV photons that would otherwise be absorbed by DNA and cause dimerization. However, degradation of the proteins in a cell may also contribute to cell death.

Enzymes are proteins that function as efficient biological catalysts that increase the rate of a reaction. Biological systems depend on enzymes to lower the activation energy of a chemical reaction and thereby facilitate processes of growth, and repair (Atlas 1995). Enzymes consist of various proportions of the amino acids including those in Fig. 2.11 and their quantum yield will vary accordingly. The quantum yield indicates the probability that absorbed UV light will induce a chemical change. Table 2.1 lists several enzymes, their chromophore constituents, and their measured quantum yields, based on data from Webb (1965).

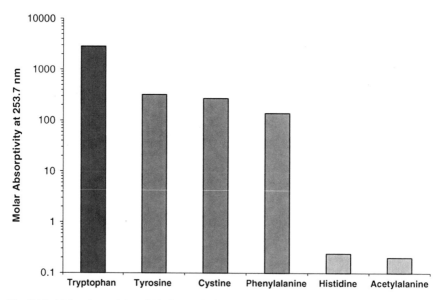

Fig. 2.11 Molar absorptivity of 'dark proteins' or amino acids with a relatively high probability of absorbing light at 253.7 nm. Based on data from Webb (1965)

Table 2.1 Quantum yields for enzyme inactivation by UV at 253.7 nm

Protein	Chromophores	Quantum Yield
Ribonuclease	$Cys_4 \cdot His_4 \cdot Phe_3 \cdot Tyr_{16}$	0.027
Lyzsozyme	$Cys_5 \cdot His_1 \cdot Phe_3 \cdot Tyr_8 \cdot Tyr_2$	0.024
Trypsin	$Cys_6 \cdot His_1 \cdot Phe_3 \cdot Tyr_4 \cdot Tyr_4$	0.015
Insulin	$Cys_{18} \cdot His_{12} \cdot Phe_{18} \cdot Tyr_{24}$	0.015
Subtilisin A	$His_5 \cdot Phe_4 \cdot Try_1 \cdot Tyr_{13}$	0.007
Japaneses Nagarse	$His_6 \cdot Phe_3 \cdot Try_4 \cdot Tyr_{10}$	0.007
Subtilisin B	$His_6 \cdot Phe_3 \cdot Try_4 \cdot Tyr_{10}$	0.006
Chymotrypsin	$Cys_5 \cdot His_2 \cdot Phe_6 \cdot Tyr_7 \cdot Tyr_4$	0.005
Pepsin	$Cys_3 \cdot His_2 \cdot Phe_9 \cdot Tyr_4 \cdot Tyr_{16}$	0.002
Carboxypeptidase	$Cys_2 \cdot His_8 \cdot Phe_{15} \cdot Tyr_6 \cdot Tyr_{20}$	0.001

Since enzymes are catalysts, they are not consumed during normal biological processes and are relatively few in number, and may therefore contribute little protective effect. However, their destruction will inhibit repair processes during or after UV exposure and may limit the effective UV rate constant. Inactivation of enzymes can be higher at wavelengths other than 253.7 and broadband UV irradiation is reported to be more effective at eliminating repair enzymes than narrow band UVC (Zimmer and Slawson 2002, Hu et al. 2005). Enzymes are associated with bacterial cells and not with viruses, although some viruses (i.e. bacteriophages) may employ enzymes for self-repair from the cells they parasitize.

Powell (1959) used optical density measurements to estimate the reduction of UV absorption at 265 nm in Herpes Simplex virus due to shielding by the host cell. The cells had a radius of 6 μm and this thickness was estimated via the Beer-Lambert law to result in a transmission, including corrections for scattering, of 40%, which is an attenuation of 60%. Such levels of photoprotection may be possible for other bacterial cells in this size range. Viruses, however, have such relatively thin coats that it seems unlikely that any chromophores present would provide any significant photoprotection.

2.5 Covalent Bonding and Photon Interaction

Chemical bonding between atoms occurs when a single electron is shared between more than one atomic nucleus. The wave function between the two atomic orbitals is called a molecular orbital. Two kinds of bonding molecular orbitals may be involved in complex molecules, σ (sigma) orbitals, and π (Pi) orbitals. The σ orbitals are localized around two nuclei and the π orbitals are nonlocalized and may involve two or more nuclei. The larger the nonlocalized orbital the more spread out is the electron probability distribution, and the longer the wavelength for electrons in that orbital. The longer wavelength means lower energy and more stability. Conjugated ring structures like the pyrimidines and the purines have large nonlocalized π orbitals and stable structures (Smith and Hanawalt 1969). A single bond is usually a σ bond while double bonds may involve both σ and π orbitals. There can be no free rotation about a double bond.

An incoming UV photon will promote an electron to an orbital of higher energy, which may be a new set of antibonding orbitals called σ* or π* orbitals, which may result in weakening of the bonds. A ππ* transition involves the excitation of a π electron into a π* state. The ππ* transitions are responsible for the most intense absorption bands in molecular spectra.

An ultraviolet photon at 253.7 nm has an energy of 4.9 eV and if this were totally converted to vibrational energy it would be sufficient to break chemical bonds, but the energy becomes distributed over many possible vibrational modes. Upon absorption of a UV photon, which may take 10^{-15} s, molecules may briefly exist in an excited state before the energy is dissipated, either by re-emission or by vibration and rotational modes. Differential quantized modes of vibration can be represented as levels of potential energy, the first of which is the singlet state. The triplet state may also be stimulated and it is one in which the system has two electrons with unpaired spins. The triplet state may persist for 10^{-3} s. The triplet state does not, as a rule, degrade directly back to the ground state, but it allows more time for photochemistry to occur and the probability of a chemical reaction is briefly increased.

2.6 UV Photoproducts

Thymine dimers are formed when two thymine molecules are cross-linked between their respective 5 and 6 carbon atoms, forming a cyclobutane ring. There are six possible isomers of the thymine dimer. Dimers can both be formed by UV exposure and separated or monomerized by UV. At longer UV wavelengths (about 280 nm) the formation of the dimer is favored while at shorter wavelengths (about 240 nm) monomerization occurs due to differences in the absorption spectra of thymine and its dimer and in the quantum yields for the formation and splitting of the dimer. The maximum yield of cyclobutane dimers is dependent on equilibrium between the formation and splitting of dimers. The reversal of dimerization by wavelengths of UV or visible light is known as photoreactivation, as is the repair of dimers by enzymes.

Thymine dimers can be formed by wavelengths of light that are not directly absorbed if the thymine molecule is in close proximity to other molecules that absorb these wavelengths. This process is called molecular photosensitization and it requires that the triplet state of the absorbing species (the photosensitizer) be slightly higher in energy than the triplet state of the thymine. Upon absorption, the triplet energy of the photosensitizer is transferred to the thymine molecule where it may induce dimerization. There are at least five other dimers of the natural pyrimidines, including cytosine dimers, cytosine-thymine dimers, uracil dimers, uracil-thymine dimers, and uracil-cytosine dimers. Cytosine dimers are formed at lower rates than thymine dimers and are readily converted to uracil dimers.

Cytosine hydrate, a water addition photoproduct, can be formed in RNA and single-stranded DNA but is not commonly found in irradiated double-stranded DNA (Smith and Hanawalt 1969). Uracil hydrates can be formed from the excited singlet state. Uracil can be derived from the monomerization of cytosine dimers. The formation of hydrates is greatly favored in single-stranded DNA.

Many other pyrimidine photoproducts, besides hydrates and cyclobutane dimers, can be produced and may be at least partly responsible for damage to nucleic acids or to a cell. Chief among these is the spore photoproduct, also called the azetane thymine dimer. The spore photoproduct, so named because it was first noted in spores, can be formed from as much as 30% of the thymine. The spore photoproduct is a type of thymine dimer that cannot be photoreversed (although it can be repaired) and the yield of this product can approach the maximum determined from the number of thymines that are nearest neighbors in DNA. In the normal B conformation of DNA the planes of the bases are parallel to each other and perpendicular to the helical backbone, and the cyclobutane dimers are favored. In the dehydrated A conformation, which is the form in which DNA is held by spores, the planes of the bases are parallel but they are inclined at an angle of 70° to the axis of the helix, a conformation which favors the spore photoproduct.

DNA cross-linking can occur under UV irradiation and this apparently involves cyclobutane dimers. Cross-linking can be highly fatal to DNA but such lesions do not appear to play a major role in UV inactivation since the cyclobutane dimers and other photoproducts are largely responsible for the inactivation effect. Per Edenberg (1983), the hypothesis that DNA replication forks are halted upon encountering thymine dimers in the template strand is consistent with data on inhibition of Simian virus replication by ultraviolet light. Per Stacks et al. (1983), the percentage of repaired and completed molecules containing dimers increases with time after irradiation ceases, and they postulate that the cellular replication machinery can accommodate limited amounts of UV-induced damage and that the progressive decrease in simian virus 40 DNA synthesis after UV irradiation is due to the accumulation in the replication pool of blocked molecules containing levels of damage greater than that which can be tolerated.

DNA may also cross-link to proteins in the cell wall, nucleocapsid, or cytoplasm, forming potentially fatal lesions. Amino acids that may contribute to photoreactivity in DNA and that may impact cross-linking include cysteine, cystine, tyrosine, serine, methionine, lysine, arginine, histidine, tryptophan, and phenylalanine. Under dry conditions (A DNA) the yield of thymine dimers is greatly decreased but there is an increase in the amount of DNA cross-linked to protein (Smith and Hanawalt 1969). Per Becker and Wang (1989), the ability of UV to damage a given base in DNA by inducing dimers or photoproducts is determined by two factors, the DNA sequence and the flexibility of DNA. Upon absorption of a UV photon, only those bases that are in a geometry capable of easily forming a photodimer can photoreact.

2.7 DNA Conformation

DNA molecules can exist in two conformations, A or B (Eyster and Prohofsky 1977). The UV susceptibility differs between the conformations. In the A conformation the bases are tilted with respect to the helix axis. In the B conformation the bases are roughly perpendicular to the double helix axis. The interaction of electro-static and van der Waals forces at the molecular level are influenced by the presence of water. The B conformation is fully hydrated (i.e. in solution or even in air at 100% relative humidity) and the A conformation could therefore be considered to be the dehydrated state. The dry A conformation shrinks in length in comparison with the wet DNA, and transitions through a phase when the population is mixed with cells in both A and B conformations. In general, microbes in high relative humidity or in water (B-DNA), have a higher resistance to UV (Peccia et al. 2001). Microbes transition from A to B when humidity or moisture increases and it is possible that the more compact A conformation (see Fig. 2.12) lends itself to more cross-links, but it is also possible that the presence of moisture or bound water provides extra protection or improved self-repair mechanisms, or that a combination of these factors is responsible for the difference in UV susceptibility between the A and B conformation.

DNA undergoes conformational transition from the B form to a disordered form as the relative humidity is lowered from about 75% to 55%. At the lower relative humidity, the dry condition, the bases are no longer stacked one above the other but are slightly angled with reference to the helix and DNA films equilibrated between 75 and 100% RH show no conformational changes and are assumed to be entirely in the B-form (Rahn and Hosszu 1969). The yield of thymine dimers remains constant

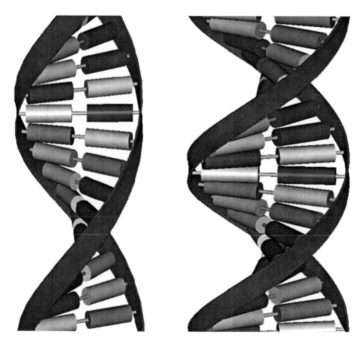

Fig. 2.12 DNA can exist in two states, the hydrated B conformation (*left*), and the dehydrated A conformation (*right*) with tighter packing of the nucleotides

in this range and is the same as that found in solution. Although most air-based UV rate constants in the range of 75–100% RH tend to converge towards water-based UV rate constants, they do not appear to become equivalent. One possible reason for this nonequality is that the refractive index of UV in air is different from that in water, causing differences in the photoprotective effect due to UV scattering.

At lower relative humidities, DNA transitions to the A form with more order and less probability of contact between thymine bases during irradiation, and there is a reduction in the rate of thymine dimer formation. The bases have different affinities for water and can trap available water molecules. The purines have two principal hydration sites in each of the major and minor DNA grooves, while the pyrimidines have only one hydration site in each groove (Neidle 1999). The individual hydration sites for bases in the A and B conformations are much the same, the major difference being that in B-DNA water is found in both grooves equally while in A-DNA more water is found in the major groove than in the minor groove.

The B form of DNA contains more bound water molecules, including those that attach to the internal grooves of DNA. The A form of DNA leads to the exposure of more hydrophobic portions of the sugar units of the backbone compared to the B form (Neidle 1999). In A-DNA, water molecules are displaced from the shallow groove, creating a local environment of low water content that favors and stabilizes the A form. The mere presence of water is insufficient to induce a conversion from A-DNA to B-DNA, instead, the water molecules must be able to contact the DNA directly over its entire length. In spores, the DNA is typically maintained in a tightly packed hydrophobic environment which prevents the DNA from going into the B-form even under high humidity or in solution which partly explains their higher UV resistance. The interaction of water molecules with DNA indicates that water forms an integral part of DNA structure and stability, and can impact UV inactivation rates.

2.8 Photon Density and Single-Hit Concepts

It can be informative to consider UV energy incident upon a microbe in terms of the number of photons, or the photon density per unit surface area. Each photon carries an amount of energy called a quantum, ϵ, determined from quantum mechanics as (Modest 1993):

$$\varepsilon = h v \tag{2.6}$$

where

h = Planck's constant, 6.626×10^{-34} Js
v = frequency, cycles per sec or Hz

The energy of a mole of photons is called an Einstein. It is defined as:

$$Einstein = Nh v \tag{2.7}$$

where N = Avogadro's number, 6.022×10^{23}

The frequency of UVC light at a wavelength of 253.7 nm is 1.18×10^{15} Hz, and the energy of UVC is computed to be 7.819×10^{-19} J/photon. Inverting this value gives us 1.279×10^{18} *photons/Joule*. A UV dose of 10 J/m² produces 1.279×10^{19} photons per m². A virus of 0.1 micron diameter has a cross-sectional area of 3.14×10^{-14} m and will be subject to the passage of about 401,000 photons when exposed to 10 J/m², which is sufficient to highly inactivate most viruses.

Despite of the vast number of photons passing through a virus, only an extremely small number are absorbed. Klein et al. (1994) report that Vaccinia virus experienced some 15 dimers per genome after a dose of 8 J/m². Miller and Plageman (1974) found that 1.7 uracil dimers were formed per PFU of inactivated Mengovirus. Based on data from Rainbow and Mak (1973), 100 J/m² produced about 102 dimers in Adenovirus Type 1, and a lethal hit (D_{37}) involved 30 thymine dimers and one single strand break. Per Ryan and Rainbow (1977), 0.3 dimers and 3.5 uridine hydrates were formed per three lethal hits in herpes simplex virus. Cornelis et al. (1981) reports that UV dosing of Parvovirus H-1 produced 10 dimers per genome, and that 80 dimers were formed in Simian virus SV40. Peak and Peak (1978) report that a frequency of 0.3 single-strand breaks occurs per lethal hit in phage T7. Sarasin and Hanawalt (1978) report that a 100 J/m² dose results in 7 pyrimidine dimers per the SV40 genome. Studies with phage T7 DNA suggest a rate of damage of 0.21 sites per 10,000 base pairs per 10 J/m² (Hanawalt et al. 1978). Clearly, very few photons out of the total interact photochemically with the nucleotides, implying that virions are virtually transparent to UV.

The First Law of Photochemistry (Grotthus-Draper Law) states that light must be absorbed by a molecule before any photochemical reactions can occur. The Second Law of Photochemistry (Stark-Einstein Law) states that absorbed light may not necessarily result in a photochemical reaction but if it does, then only one photon is required for each molecule affected (Smith and Hanawalt 1969). Since not every quantum of incident energy is absorbed by a molecule, there is an absorption efficiency that describes photochemical absorptivity. This efficiency is called the quantum yield, ϕ, and it is defined as:

$$\Phi = \frac{N_c}{N_p} \tag{2.8}$$

where

N_c = Number of molecules reacting chemically
N_p = Number of photons absorbed

The number of photons absorbed is sometimes specified in Einsteins, or moles of photons as defined in Eq. (2.7). Quantum yields may be extremely low for macromolecules of low absorptivity. Since most of these photochemical excitations of molecules do not lead to chemical reactions, energy is dissipated by various means. Light may be re-emitted at a different wavelength, and energy absorption may result in molecular vibrations that translate into heat.

The inactivation of nucleic acids involves quantum yields on the order of 10^{-3} to 10^{-4} and at the UV doses typically applied, it is clear that a relatively small number of photons photoreact, and that they are absorbed at discrete genomic sites – normally those bases that produce dimers. Based on UV inactivation studies of *E. coli*, only about 0.025% of the DNA molecule is photochemically altered at the point that 99% of *E. coli* are killed (Smith and Hanawalt 1969). A UV dose of 180 J/m^2 dimerizes only 0.1% of the total thymine. This dose represents $(180 \text{ J/m}^2)(1.279 \times 10^{18}$ photons/J) $= 2.3 \times 10^{20}$ photons/m^2. With a diameter of about 0.5 microns, and a DNA size of 5490 kb, this would imply that 1.8×10^8 photons impinged upon the bacterial cell to produce about 1372 photochemical reactions. Even if we ignore the cell space that is not occupied by the bacterial DNA, it is clear that it takes a relatively large number of photons to induce a relatively limited number of photoreactions sufficient to inactivate a cell. Figure 2.13 shows a comparison of the typical number of photons necessary for certain processes to provide some further perspective.

Rauth (1965) measured the inactivation and absorption cross-sections of several viruses and computed the quantum yields to be in the range of $5–65 \times 10^{-4}$. The quantum yield is computed from the inactivation cross-section divided by the absorption cross-section as follows:

$$\Phi = \frac{\sigma}{S} \tag{2.9}$$

where

σ = inactivation cross-section, m^2/photon
S = absorption cross-section, m^2/photon

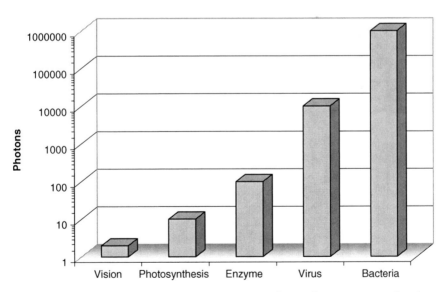

Fig. 2.13 Typical numbers of absorbed photons necessary for certain processes or to inactivate enzymes, bacteria, and viruses. Based on Setlow and Pollard, 1962

The absorption cross-section can be determined by various means but the measurement typically involves putting specific densities of cells (or virions) in solution and measuring the difference in irradiance that passes through the solution containing cells versus a solution containing no cells. The inactivation cross-section is equivalent to the UV rate constant, which is normally given in units of m^2/J.

2.9 Photochemistry of RNA Viruses

RNA viruses are characteristically in the A conformation and this partly dictates the type of photoproducts produced under UV irradiation. The main photoproducts produced are pyrimidine hydrates and cyclobutadipyrimidines, while other photoproducts, like altered purines and pyrimidine dimers, occur at much lower rates, if at all (Fraenkel-Conrat and Wagner 1981). Remsen et al. (1970) found that inactivation of R17 phage at 280 nm was a log-linear function of the number of uridine hydrates formed and that no cyclobutapyrimidines were formed. RNA animal viruses and phages demonstrate little or no photoreactivation, and the photoreactive effects that have been observed are attributed to host cells (Fraenkel-Conrat and Wagner 1981). RNA viruses have no repair enzymes and any photoreactive effects may be due strictly to thermal, visible light, or near-UV light effects.

Much information on the photochemistry of RNA viruses has come from studies of Tobacco Mosaic Virus (TMV), and although plant viruses are not the subject of this text some useful information can be garnered from UV studies regarding the protective effect of protein coats. The quantum yield for inactivation of the whole TMV virus is barely 1% that of RNA at longer wavelengths of the absorption spectrum (Fraenkel-Conrat and Wagner 1981). The relative insensitivity to UV of the whole virus is a property of the coat protein, which modifies the UV photoproducts formed in RNA. Protein inhibits the formation of pyrimidine photoproducts (cyclobutadipyrimidines) and inhibits the formation of other photoproducts by reducing the quantum yield for photoreactions. This could occur through shielding of the RNA, through quenching of the excited states of RNA, and by surrounding the bases with a hydrophobic environment and limiting the mobility of the individual bases. Although the protein coat reduces the overall rate of photoreactions, it allows the formation of noncyclobutane-type dipyrimidines and of uridine hydrates. In irradiated TMV, the number of uridine hydrates formed was about two per lethal hit (D_{37} or 37% survival), while only about one dimeric photoproduct was formed.

RNA bacteriophages, which are viruses that infect bacteria, typically consist of a capsid composed principally of one protein, with small numbers of one or two other proteins. The action spectra of several RNA phages has been studied by Rauth (1965), who showed that the quantum yield for virus inactivation was approximately the same as the quantum yield for RNA inactivation, which suggests that not only is the RNA the primary target but that the protein coats of the phages studied contribute little protective effect. Under UV irradiation, a lethal hit for mengovirus (at 70 J/m^2) produced 1.7 uracil dimers but no apparent structural damage to the RNA (Miller and Plageman 1974).

Most RNA virus coats consist of one major protein and sometimes one or two other proteins. There are five known viral proteins, M, G, L, N, and NS. The protein of the common strain of TMV consists of three tryptophan and four tyrosine residues, both of which have high molar absorptivities (see Fig. 2.11). The coat protein itself may suffer UV photodamage and may become cross-linked to RNA, but the extent to which this contributes to overall inactivation may be of limited significance (Fraenkel-Conrat and Wagner 1981).

Double stranded RNA viruses tend to be much more resistant to UV exposure than single stranded RNA viruses, by almost an order of magnitude. Zavadova (1971) showed that the D_{90} for double stranded encephalomyocarditis virus RNA was about six times that for the single stranded version.

2.10 Photochemistry of DNA Viruses

UV irradiation of DNA produces photoproducts called pyrimidine dimers as well as non-dimer photoproducts. Dimers produced in DNA can consist of thymine:thymine, thymine:cytosine, and cytosine:cytosine. Thymine dimers, the most common photoproducts, were the focus of much early investigation in UV irradiation experiments. The number of thymine dimers produced per lethal hit in the DNA of phage ϕX174 is about 0.3 (David 1964). For coliphage lambda, about two dimers were produced per lethal hit (Radman et al. 1970). For phage T4, 10.2 dimers were formed per lethal hit (Meistrich 1972). For a given dose, more dimers are produced when AT-rich DNA is irradiated than GC-rich DNA, but this difference is not more than twofold. The relative efficiency of dimer formation in DNA is in the order TT > CT > CC (Setlow and Carrier 1966). However, this can vary with the thymine content since, for viruses with high GC content, the number of CC dimers produced under UV exposure can exceed the number of TT dimers (Matallana-Surget et al. 2008). Photoproducts other than dimers are also produced, including pyrimidine adducts, which occur at about a tenth of the frequency of dimers, and others which occur at even lower frequencies (Wang 1976). Cytidine-derived photoproducts include cytidine hydrate (or the deamination product, uridine hydrate) and cytosine dimer (deamination product, uracil dimer).

2.11 Photochemistry of Bacteria

The kinetics of bacterial inactivation by ultraviolet light are much the same as in DNA viruses since they contain DNA, except that many bacteria have more photoprotection and often have the ability to photorecover or photoreactivate. About 65 dimers are produced per every 10^7 nucleotides in the DNA of E. coli, for every J/m^2 of UV irradiation (Fraenkel-Conrat and Wagner 1981). The number of dimers formed varies from one species to another but the ultraviolet sensitivities of bacteria with varying GC content are not directly proportional to the TT frequency, indicating that thymine dimers are not the sole cause of lethality. David (1973) inferred that for a constant G+C content (or T+A content), the sensitivity to UV radiation is a reciprocal function of the molecular weight of the genome, suggesting that the smaller the DNA molecule, the higher the probability that a hit would be lethal.

Bacteria invariably contain enzymes and other repair mechanisms that may allow for photoreactivation and photorecovery from UV exposure (Atlas 1995). The quantum yield for inactivation of an enzyme is approximately proportional to its cystine content and is roughly inversely proportional to its molecular weight (Smith and Hanawalt 1969). The latter is explained by the fact that the cystine content of proteins is inversely proportional to their molecular weight. The action spectrum for an enzyme can be resolved into the contributions from its constituent chromophores.

2.12 Photoreactivation

Photoreactivation is a natural process in which bacterial cells can partially recover from ultraviolet damage when visible and UV wavelengths of light reverse DNA damage by monomerizing cyclobutane pyrimidine dimers. It was first identified in *E. coli* by Prat (1936) and later demonstrated by Kelner (1949), and has since been noted to occur in many other bacteria. Photoreactivation is an effect that primarily operates on bacteria and spores. Viruses and certain bacteria seem to have very limited capability to self-repair or photoreactivate, including *Haemophilus influenzae, Diplococcus pneumoniae, Bacillus subtilis*, and *Deinococcus radiodurans* (Masschelein 2002). David et al. (1971) reports photoreactivation rates of 40–56% in mycobacteria. Little evidence exists for the photoreactivation of animal viruses since they lack enzymes although they may be photoreactivated by host-cell repair mechanisms (Samad et al. 1987). Photoreactivation has never been observed in animal virus RNA (Bishop et al. 1967). The photoreactivation effect may be dependent on RH, with the effect possibly absent when RH is less than approximately 65%. Evidence suggests that the conformation change in DNA that occurs at higher RH may allow microbes to experience photoreactivation (Rahn and Hosszu 1969, Munakata and Rupert 1974).

Many bacterial cells possess repair enzymes that can repair gaps and defects in the DNA. Thymine dimers formed by UV irradiation of DNA are hydrolyzed by specific DNases and are replaced with correct sequences by repair enzymes (Guschlbauer 1976). The maximum yield of thymine dimers in irradiated bacterial cells depends on the wavelength as well as the conditions (i.e. RH%). After a sufficient UV dose has been imparted to the bacteria a steady state is reached in which the relative numbers of dimers do not change (Smith and Hanawalt 1969). Dimer formation is a reversible process and thymine dimers may revert to free thymines via the absorption of UV and visible light. Photoreactivation cannot completely reverse damage to DNA since UV may cause other types of photoproducts but it can effectively limit UV damage.

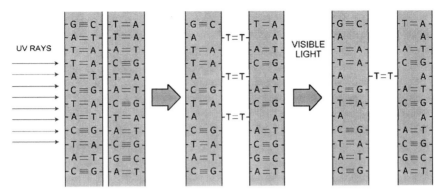

Fig. 2.14 After inactivation by UV irradiance, exposure to visible light for 2–3 h may produce photoreactivation, in which many broken thymine links (thymine dimers) are repaired by enzymes

Thymine dimers absorb light in the visible range (blue light) and this leads to self-repair of the nucleotide bonds, as illustrated in Fig. 2.14 . Reactivation can also occur under conditions of no visible light, or what is called dark repair. The ability to self-repair can depend on the biological organization of the microorganism, as well as the amount of UV damage inflicted on the cell.

Photoreactivation can be catalyzed by enzymes, which are commonly present in bacterial cells. The process occurs in two stages; the first involves the production of an enzyme-substrate complex at the DNA lesion site in the absence of light, and the second is a photolytic reaction in which light energy is absorbed and the lesion is repaired (Fletcher et al. 2003). Enzymatic photoreactivation is facilitated by visible light and results in the splitting of pyrimidine dimers, called monomerization. Thymine dimers are more efficiently eliminated than other types. Only polynucleotide strands containing adjacent pyrimidines are photoreactivable and a minimum length of about nine bases appears to be necessary for the enzyme to attach and excise dimers.

In photoreactivation, repair is due to an enzyme called photolyase. Photolyase reverses UV-induced damage in DNA. In dark repair the damage is reversed by the action of a number of different enzymes. All of these enzymes must initially be activated by an energy source, which may be visible light (300–500 nm) or nutrients that exist within the cell. Masschelein (2002) suggests that the enzymatic repair mechanism requires at least two enzyme systems: an exonuclease systems (i.e. to disrupt the thymine-thymine linkage), and a polymerase system to reinsert the thymine bases on the adenosine sites of the complementary strain of DNA.

In DNA repair mechanisms, the damaged strand is excised by the enzyme and then the complementary strand of DNA is used as a template for inserting the correct nucleotides. Failure to repair UV damaged DNA can result in errors during the replication process during which base substitutions can occur and result in the development of mutants. The most favored sites for base substitutions leading to mutants involve transitions from GC to AT at sites with adjacent pyrimidines (Miller 1985).

Enzymes can be damaged by broadband UV wavelengths other than 253.7 nm, and it has been reported that the use of medium pressure UV lamps inhibits photoreactivation due to the fact that broadband wavelengths inflict damage on photorepair enzymes (Kalisvaart 2004, Quek and Hu 2008).

No types of DNA damage other than that produced by UV can be photoreactivated. In cell systems with efficient dark repair mechanisms, like *D. radiodurans*, little or no photoreactivation occurs. UV damage produced at 253.7 nm can be attenuated by exposure to wavelengths between 330 and 480 nm (Hollaender 1955). The enzymatic monomerization of pyrimidine dimers operates when pyrimidine dimers are the primary type of photodamage, and when photodamage is due to other photoproducts, like the spore photoproduct, photoreactivation appears to be absent.

2.13 UV Scattering

Another kind of photoprotection, other than shielding or photoreactivation, occurs when light is scattered from microbes. Scattering of UV light from microbes is a phenomenon that is routinely observed during the measurement of optical density of microbes in solution to obtain absorption spectra, and during which corrections for scattering must often be made (Holler et al. 1998). Luria et al. (1951) used corrections of 10–20% for scattering at 260 nm while Zelle and Hollaender (1954) found the absorbance corrections for phages T2 and T7 were somewhat greater than 20%. In studies on phage T2 (0.065–0.095 μm), Dulbecco (1950) found that for wavelengths longer than 320 nm the absorption closely followed Rayleigh's law of scattering, and that the photosensitive pigments were part of the phage (but not the DNA) and tended to darken after exposure. Rauth (1965) found that for small viruses like MS2 and ϕX174, the corrections for scattering are almost negligible and only become appreciable above about 280 nm, where they can approach 20–25% depending on virus size. Powell (1959) found that UV scattering effects accounted for no more than 25% attenuation in water. According to Jagger (1967) the UV transmission through an *E. coli* cell is only 70% at 254 nm, leaving a maximum of 30% to be scattered or absorbed.

Scattering may cause appreciable loss of light when the exposed microbes have dimensions comparable with UV wavelengths (Hollaender 1955). The scattering effect is reduced as the index of refraction of the microbe approaches the index of refraction of the medium (i.e. air or water). The scattering effect increases, however, when the size parameter (a function of the diameter) approaches the wavelength of the ultraviolet light (van de Hulst 1957).

Mie scattering is the dominant form of light scattering in the micron-size range of viruses and small airborne bacteria (Bohren and Huffman 1983). Scattering can have significant impact on the amount of UV that actually reaches the nucleocapsid or DNA of a microbe in air and the effect appears to become significant at diameters of about 0.03 microns and greater. Scattering is a protective effect and not dependent on the protein content of nucleocapsids or cell walls, since most microbes appear to have similar indices of refraction (i.e. about 1.05–1.08).

Absorption of photons takes place as ultraviolet radiation penetrates a particle. Light that is not absorbed may be scattered from a particle in the virus and bacteria size range (0.02–20 μm) by three different mechanisms: (1) reflection of photons from the particle, (2) refraction of photons that pass unabsorbed through the particle, and (3) diffraction of photons that pass through or near a particle. Diffraction may alter the path of photons even though they are not in the direct path of the particle. This latter phenomena can result in a particle scattering more light than it would actually intercept due to its physical size alone (Modest 1993). The interaction between ultraviolet wavelengths and the particle is a function of the relative size of the particle compared with the wavelength, as defined by the size parameter:

$$x = \frac{2\pi a}{\lambda} \qquad (2.10)$$

where

a = the effective radius of the particle
λ = wavelength

If the size parameter, x << 1, then Rayleigh scattering dominates and for simple spherical particles of diameters less than λ/10 the scattering will approximately vary with the inverse of the wavelength raised to the fourth power ($1/\lambda^4$). If x>>1, the principles of normal geometric optics may be applied. If $x \approx 1$, Mie scattering dominates, and this is the case for small viruses and bacteria. For Mie scattering in air, the size parameter can be written as follows (Chen et al. 2003):

$$x = \frac{\pi d n_m}{\lambda} \qquad (2.11)$$

where

n_m = refractive index of the medium (air)
d = particle diameter (typically nanometers)

For nonspherical microbes where the length is significantly greater than the diameter (i.e. aspect ratio > 5), the size parameter for rods may be used. In such cases the length is merely substituted for the diameter (from Stacey 1956). The scattering of light is due to differences in the refractive indices between the medium and the particle (Modest 1993, Garcia-Lopez et al. 2006). The scattering properties of a spherical particle in any medium are defined by the complex index of refraction:

$$m = n - i\kappa \qquad (2.12)$$

where

n = real refractive index
κ = imaginary refractive index (absorptive index or absorption coefficient)

Since the refractive index of ultraviolet light approaches 1 in air (or about 1.00029 for visible light), Eqs. (2.10) and (2.11) are virtually identical. If scattering is not affected by the presence of other surrounding particles, and this is generally the case for airborne microbes since concentrations will never be so high as to even be visible, the process is known as *independent scattering*. The process of independent Mie scattering is also governed by the *relative refractive index*, defined with the same symbol (m) as follows:

$$m = \frac{n_s}{n_m} \tag{2.13}$$

where n_s = refractive index of the particle (microbe)

Water has a refractive index of approximately $n_m = 1.4$ in the ultraviolet range and about 1.33 in the visible range. The refractive index of microbes in visible light has been studied by several researchers. Balch et al. (2000) found the median refractive index of four viruses to be 1.06, with a range of 1.03–1.26. Stramski and Keifer (1991) assumed viruses to have a refractive index of 1.05. Biological cells were assumed by Mullaney and Dean (1970) to have *relative refractive indices* of about 1.05 in visible light. Klenin (1965) found *S. aureus* to have a refractive index in the range 1.05–1.12. Petukhov (1964) gives the refractive index of certain bacteria in the limits of 1.37–1.4. There are no studies that address the real refractive index of bacteria or viruses at UV wavelengths except Hoyle and Wickramasinghe (1983) who suggest $n_s = 1.43$ as a reasonable choice for coliform bacteria. Garcia-Lopez et al. (2006) state that for soft-bodied biological particles n is between 1.04 and 1.45. For the imaginary refractive index (the absorptive index or absorption coefficient) in the UV range no information is currently available. Per Garcia-Lopez et al. (2006), hemoglobin has a κ of 0.01–0.15, while polystyrene has a κ of 0.01–0.82.

The mathematical solution of Mie scattering is so complicated as to generally require the use of advanced computational methods. For details of these solutions see van de Hulst (1957), Bohren and Huffman (1983), and Modest (1993). A variety of software packages are freely available for solving the scattering problem for small particles in air, such as DDSCAT, and tables have also been published for use (Draine and Flatau 2004).

Per Eq. (2.13) the refractive index for microbes in air would be about $(1.05)(1.33) = 1.4$. Figure 2.15 shows two examples of scattering effects in microbes of 0.2 μm (small virus) and 1 μm (large virus or small bacteria) when light is incident from the left passing to the right. The scattering was evaluated for a 253 nm wavelength, a real refractive index of 1.4, an imaginary refractive index of −1.4, a medium refractive index of 1.0003 (air), and wide dispersion of particles (negligible concentration). Computations were performed using the Mie Scattering Calculator (Prahl 2009).

The amount of scattering and absorption by a particle is defined by the scattering cross-section, C_{sca}, and the absorption cross-section, C_{abs}. The scattering cross section is defined as the area which when multiplied by the incident irradiance gives the total power scattered by the particle. The absorption cross section is the area which when multiplied by the incident irradiance gives the total power absorbed.

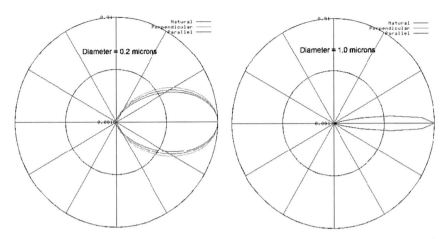

Fig. 2.15 Angular UV light scattering functions for spherical microbes in air, with diameters as indicated. Plots reprinted courtesy of Scott Prahl, Oregon Medical Laser Center

The total amount of absorption and scattering is the extinction cross-section, C_{ext}, defined as the area which when multiplied by the incident irradiance gives the total power removed from the incident wave by scattering and absorption.

$$C_{ext} = C_{abs} + C_{sca} \tag{2.14}$$

The fraction of UV that is scattered from the total incident irradiation, S_{uv}, can be computed as follows (Kowalski et al. 2009):

$$S_{uv} = \frac{C_{sca}}{C_{abs} + C_{sca}} = \frac{C_{sca}}{C_{ext}} \tag{2.15}$$

Equation (2.15) effectively defines the correction factor (as a complement) for UV incident on a particle that scatters UV. Efficiency factors used in Mie scattering are the cross-sections divided by the area, and include the absorption efficiency factor, Q_{abs}, the scattering efficiency factor, Q_{sca}, and the extinction efficiency factor, Q_{ext}, defined as follows:

$$Q_{abs} = \frac{C_{abs}}{\pi a^2} \tag{2.16}$$

$$Q_{sca} = \frac{C_{sca}}{\pi a^2} \tag{2.17}$$

$$Q_{ext} = \frac{C_{ext}}{\pi a^2} \tag{2.18}$$

The extinction efficiency factor is equal to the sum of the other two factors:

$$Q_{ext} = Q_{abs} + Q_{sca} \tag{2.19}$$

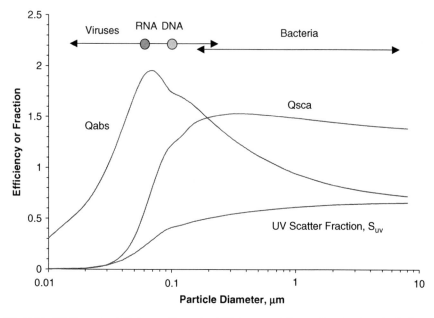

Fig. 2.16 UV Scattering Efficiency, Absorption Efficiency, and Scatter Fraction for spherical particles in air. Real refractive index = 1.4, imaginary absorptive index = −1.4, n_m = 1.0003. Dots show approximate logmean diameters of RNA and DNA viruses

The degree of scattering increases from small viruses up to small bacteria. Figure 2.16 shows an example of the scattering efficiency, absorption efficiency, and scatter fraction of spherical viruses and bacteria in the $0.02 - 8\ \mu m$ size range in air, based on computations performed using Mie scattering software (Prahl 2009). The refractive index (real component) used for this example is n=1.4, while the absorptive index (the imaginary component) is assumed the same as water, κ=1.4. Note that the scattering efficiency increases dramatically above about $0.06\ \mu m$. The range of sizes for viruses and bacteria are also shown, and it can be seen that DNA viruses and bacteria would be most impacted by scattering effects. This graph is revisited for water in Chap. 4, where it will be seen that although the medium changes the efficiency factors, it has little effect on the scattering fraction.

The refractive index and the absorptive index of individual species of microbes may be somewhat different from the values assumed above, but testing of alternate values indicates that the general pattern of behavior observed in Fig. 2.16 remains essentially unchanged for all possible microbial values previously cited. The UV scatter fraction confers some significant UV photoprotection, especially to larger viruses and to bacteria. Such photoprotection translates into an effective UV dose lower than that to which the microbe is exposed. The subject of UV scattering as a mechanism of photoprotection, and how it relates to predicting UV susceptibility will be revisited in Chap. 4.

References

Anderson J, Rowan N, Macgregor S, Fouracre R, Farish O. 2000. Inactivation of food borne enteropathogenic bacteria and spoilage fungi using pulsed-light. IEEE Trans Plasma Sci 28:83–88.

Atlas RM. 1995. Microorganisms in Our World. St. Louis: Mosby.

Balch WM, Vaughn J, Novotny J, Drapeau D, Vaillancourt R, Lapierre J, Ashe A. 2000. Light scattering by viral suspensions. Limnol Oceanogr 45(2):492–498.

Becker MM, Wang Z. 1989. Origin of ultraviolet damage in DNA. J Mol Biol 210:429–438.

Bell A, Wheeler M. 1986. Biosynthesis and function of melanin. Ann Rev Phytopathology 24:411–451.

Bhattacharjee C, Sharan RN. 2005. UV-C radiation induced conformational relaxation of pMTa4 DNA in *Escherichia coli* may be the cause of single strand breaks. Int J Radiat Biol 81(12):919–927.

Bishop JM, Quintrell N, Koch G. 1967. Poliovirus double-stranded RNA: Inactivation by ultraviolet light. J Mol Biol 24:125–128.

Bohren C, Huffman D. 1983. Absorption and Scattering of Light by Small Particles. New York: Wiley & Sons.

Bolton JR, Cotton CA. 2008. Ultraviolet Disinfection Handbook.American Water Works Association. Denver, CO:

Boyd-Wilson K, Perry J, Walter M. 1998. Persistence and survival of saprophytic fungi antagonistic to *Botrytis cinerea* on kiwifruit leaves. Proc of the 51st Conf of the New Zealand Plant Prot Soc, Inc.:96–101.

Casarett AP. 1968. Radiation Biology. Englewood: Prentice-Hall.

Chen K, Kromin A, Ulmer M, Wessels B, Backman V. 2003. Nanoparticle sizing with a resolution beyond the diffraction limit using UV light scattering spectroscopy. Optics Commun 228:1–7.

Cornelis JJ, Su ZZ, Ward DC, Rommelaere J. 1981. Indirect induction of mutagenesis of intact parvovirus H-1 in mammalian cells treated with UV light or with UV-irradiated H-1 or simian virus 40. Proc Natl Acad Sci 78(7):4480–4484.

David, CN. 1964. UV inactivation and thymine dimerization in bacteriophage phiX. Z. Verberbungsl. 95:318–325.

David HL, Jones WD, Newman CM. 1971 Ultraviolet light inactivation and photoreactivation in the mycobacteria. Infect and Immun 4:318–319.

David HL. 1973. Response of mycobacteria to ultraviolet radiation. Am Rev Resp Dis 108:1175–1184.

Draine B, Flatau P. 2004. User guide for the discrete dipole approximation code DDSCAT 6.1.

Dulbecco R. 1950. Experiments on photoreaction of bacteriophages inactivated with ultraviolet radiation. J Bact 59:329–347.

Durrell L, Shields L. 1960. Fungi isolated in culture from soil of the Nevada test site. Mycologia 52:636–641.

Edenberg HJ. 1983. Inhibition of simian virus 40 DNA replication by ultraviolet light. Virology 128(2):298–309.

Elasri MO, Miller RV. 1999. Response of a biofilm bacterial community to UV Radiation. Appl Environ Microbiol 65(5):2025–2031.

Eyster JM, Prohofsky EW. 1977. On the B to A Conformation Change of the Double Helix. Biopolymers 16:965–982.

Fernandez RO, Pizarro RA. 1996. Lethal effect induced in *Pseudomonas aeruginosa* exposed to ultraviolet-A radiation. Photochem Photobiol 64(2):334–339.

Fletcher LA, Noakes CJ, Beggs CB, Sleigh PA, Kerr KG. The ultraviolet susceptibility of aerosolised microorganisms and the role of photoreactivation; 2003; Vienna. IUVA.

Fraenkel-Conrat H, Wagner RR. 1981. Comprehensive Virology. New York: Plenum Press.

Furuse K, Watanabe I. 1971. Effects of ultraviolet light (UV) irradiation on RNA phage in H2O and in D2O. Virol 46:171–172.

Garcia-Lopez A, Snider A, Garcia-Rubio L. 2006. Rayleigh-Debye-Gans as a model for continuous monitoring of biological particles: Part I, assessment of theoretical limits and approximations. Optics Express 14(19):17.

Guschlbauer W. 1976. Nucleic Acid Structure. New York: Springer-Verlag.

Hanawalt PC, Friedberg EC, Fox CF, editors. 1978. DNA Repair Mechanisms. New York: Academic Press.

Harm W. 1980. Biological Effects of Ultraviolet Radiation. New York: Cambridge University Press.

Hollaender A. 1955. Radiation Biology, Volume II: Ultraviolet and Related Radiations. New York: McGraw-Hill.

Holler S, Pan Y, Bottiger JR, Hill SC, Hillis DB, Chang RK. Two-dimensional angular scattering measurements of single airborne micro-particles; 1998; Boston, MA. SPIE. pp 64–72.

Hoyle F, Wickramasinghe C. 1983. The ultraviolet absorbance spectrum of coliform bacteria and its relationship to astronomy. Astrophysics and Space Science 95:227–231.

Hu JY, Chu SN, Quek PH, Feng YY, Tang XL. 2005. Repair and regrowth of *Escherichia coli* after low- and medium-pressure ultraviolet disinfection. Wat Sci Technol 5(5):101–108.

IESNA. 2000. Lighting Handbook: Reference & Application IESNA HB-9-2000. New York: Illumination Engineering Society of North America.

Jagger J. 1967. Ultraviolet Photobiology. Englewood Cliffs: Prentice-Hall, Inc.

Kalisvaart BF. 2004. Re-use of wastewater: Preventing the recovery of pathogens by using medium-pressure UV lamp technology. Wat Sci Technol 33(10):261–269.

Kelner A. 1949. Effect of visible light on the recovery of *Streptomyces griseus* conidia from ultraviolet irradiation injury. Proc Nat Acad Sci 35(2):73–79.

Klein B, Filon AR, vanZeeland AA, vanderEb AJ. 1994. Survival of UV-irradiated vaccinia virus in normal and xeroderma pigmentosum fibroblasts; evidence for repair of UV-damaged viral DNA. Mutat Res 307(1):25–32.

Klenin V. 1965. The problem concerning the scattering of light by suspensions of bacteria. Biofizika 10(2):387–388.

Koller LR. 1952. Ultraviolet Radiation. New York: John Wiley & Sons.

Kowalski W, Bahnfleth W, Hernandez M. 2009. A Genomic Model for the Prediction of Ultraviolet Inactivation Rate Constants for RNA and DNA Viruses; 2009 May 4–5; Boston, MA. International Ultraviolet Association.

Kuluncsics Z, Perdiz D, Brulay E, Muel B, Sage E. 1999. Wavelength dependence of ultraviolet-induced DNA damage distribution: Involvement of direct or indirect mechanisms and possible artefacts. J Photochem Photobiol 49(1):71–80.

Linden KG. 2001. Comparative effects of UV wavelengths for the inactivation of Cryptosporidium parvum oocysts in water. Wat Sci Technol 34(12):171–174.

Luckiesh M. 1946. Applications of Germicidal, Erythemal and Infrared Energy. New York: D. Van Nostrand Co.

Luria SE, Williams RC, Backus RC. 1951. Electron micrographic counts bacteriophage particles. J Bact 61:179–188.

Masschelein WJ. 2002. Ultraviolet Light in Water and Wastewater Sanitation. Boca Raton: Lewis Publishers.

Matallana-Surget S, Meador J, Joux F, Douki T. 2008. Effect of the GC content of DNA on the distribution of UVB-induced bipyrimidine photoproducts. Photochem Photobiol Sci 7:794–801.

Meistrich ML. 1972. Contribution of thymine dimers to the ultraviolet light inactivation of mutants of bacteriophage T4. J Mol Biol 66:97.

Miller RL, Plageman PGW. 1974. Effect of Ultraviolet Light on Mengovirus: Formation of Uracil Dimers, Instability and Degradation of Capsid, and Covalent Linkage of Protein to Viral RNA. J Virol 13(3):729–739.

Miller J. 1985. Mutagenic specificity of ultraviolet light. J Mol Biol 182:45–65.

Modest MF. 1993. Radiative Heat Transfer. New York: McGraw-Hill.

Mullaney P, Dean P. 1970. The small angle light scattering fo biological cells. Biophys J 10: 764–772.

Munakata N, Rupert CS. 1974. Dark Repair of DNA Containing "Spore Photoproduct" in *Bacillus subtilis*. Molec Gen Genet 130:239–250.

Neidle S. 1999. Oxford Handbook of Nucleic Acid Structure. New York: Oxford University Press.

Palmeira L, Guegen L, Lobry J. 2006. UV-targeted dinucleotides are not depleted in light-exposed prokaryotic genomes. Mol Biol Evol 23:2214–2219.

Peak MJ, Peak JG. 1978. Action spectra for the ultraviolet and visible light inactivation of phage T7: Effect of host-cell reactivation. Radiat Res 76:325–330.

Peccia J, Werth HM, Miller S, Hernandez M. 2001. Effects of relative humidity on the ultraviolet induced inactivation of airborne bacteria. Aerosol Sci Technol 35:728–740.

Petukhov V. 1964. The feasibility of using the Mie theory for the scattering of light from suspensions of spherical bacteria. Biofizika 10(6):993–999.

Powell WF. 1959. Radiosensitivity as an index of herpes simplex virus development. Virology 9(1–19).

Prahl S. 2009. Mie Scattering Calculator. Portland, OR: Oregon Medical Laser Center.

Prat S. 1936. Strahlung und antagonistische wirkungen. Protoplasma 26:113–149.

Quek PH, Hu J. 2008. Indicators for photoreactivation and dark repair studies following ultraviolet disinfection. J Ind Microbiol Biotechnol 35(6): 533–541.

Radman M, Cordone L, Krsmanovic-Simic D, Errera M. 1970. Complementary action of recombination and excision in the repair of ultraviolet irradiation damage to DNA. J Mol Biol 49:203.

Rahn RO, Hosszu JL. 1969. Influence of relative humidity on the photochemistry of DNA films. Biochim Biophys Acta 190:126–131.

Rainbow AJ, Mak S. 1973. DNA damage and biological function of human adenovirus after U.V. irradiation. Int J Radiat Biol 24(1):59–72.

Rauth AM. 1965. The physical state of viral nucleic acid and the sensitivity of viruses to ultraviolet light. Biophys J 5:257–273.

Remsen JF, Miller N, Cerutti PA. 1970. Photohydration of uridine in the RNA of coliphage R17. II. The relationship between ultraviolet inactivation and uridine photohydration. Proc Natl Acad Sci USA 65:460–466.

Rothman R, Setlow R. 1979. An action spectrum for cell killing and pyrimidine dimer formation in Chinese hamster V-79 cells. Photochem Photobiol 29:57–61.

Ryan DKG, Rainbow AJ. 1977. Comparative studies of host-cell reactivation, cellular capacity and enhanced reactivation of herpes simplex virus in normal, xeroderma pigmentosum and Cockayne syndrome fibroblasts. Photochem Photobiol 26:263–268.

Samad SA, Bhattacharya SC, Chatterjee SN. 1987. Ultraviolet inactivation and photoreactivation of the cholera phage 'Kappa'. Radiat Environ Biophys 26:295–300.

Sarasin AR, Hanawalt PC. 1978. Carcinogens enhance survival of UV-irradiated simian virus 40 in treated monkey kidney cells: Induction of a recovery pathway? Proc Natl Acad Sci 75(1): 356–350.

Sarasin AR, Hanawalt PC. 1980. Replication of ultraviolet-irradiated Simian Virus 40 in monkey kidney cells. J Mol Biol 138:299–319.

Schreier WJ, Schrader TE, Koller FO, Gilch P, Crespo-Hernandez CE, Swaminathan VN, Carell T, Zinth W, Kohler B. 2007. Thymine dimerization in DNA is an ultrafast photoreaction. Science 315:625–629.

Setlow RB, Pollard EC. 1962. Molecular Biophysics. Reading, MA: Addison-Wesley.

Setlow RB, Carrier WL. 1966. Pyrimidine dimers in ultraviolet-irradiated DNA's. J Mol Biol 17:237–254.

Smirnov Y, Kapitulez S, Kaverin N. 1992. Effects of UV-irradiation upon Venezuelan equine encephalomyelitis virus. Virus Res 22(2):151–158.

Smith KC, Hanawalt PC. 1969. Molecular Photobiology: Inactivation and Recovery. New York: Academic Press.

Stacey K. 1956. Light-Scattering in Physical Chemistry. London: Butterworths Scientific Publications.

Stacks PC, White JH, Dixon K. 1983. Accomodation of Pyrimidine Dimers During Replication of UV-Damaged Simian Virus 40 DNA. Molec Cell Biol 3(8):1403–1411.

Stramski D, Keifer D. 1991. Light scattering by microorganisms in the open ocean. Prog Oceanogr 28:343–383.

Unrau P, Wheatcroft R, Cox B, Olive T. 1973. The formation of pyrimidine dimers in the DNA of fungi and bacteria. Biochim Biophys Acta 312:626–632.

Valero A, Begum M, Leong S, Hocking A, Ramos A, Sanchis V, Marin S. 2007. Fungi isolated from grapes and raisins as affected by germicidal UVC light. Lett Appl Microbiol 45:238–243.

Van de Hulst H. 1957. Light Scattering by Small Particles. New York: Chapman & Hall, Ltd.

Von Sonntag C. 1986. Disinfection by free radicals and UV-radiation. Water Supply 4:11–18.

Waites WM, Harding SE, Fowler DR, Jones SH, Shaw D, Martin M. 1988. The destruction of spores of *Bacillus subtilis* by the combined effects of hydrogen peroxide and ultraviolet light. Lett Appl Microbiol 7:139–140.

Wang SY. 1976. Photochemistry and Photobiology of Nucleic Acids. New York: Academic Press.

Webb SJ. 1965. Bound Water in Biological Integrity. Springfield, IL: Charles C. Thomas.

Webb RB, Tuveson TW. 1982. Differential sensitivity to inactivation of NUR and NUR* strains of *Escherichia coli* at six selected wavelengths in the UVA, UVB, and UVC ranges. Photochem Photobiol 36:525–530.

Zavadova Z, Gresland L, Rosenbergova M. 1968. Inactivation of single- and double-stranded ribonucleic acid of encephalomyocarditis virus by ultraviolet light. Acta Virol 12:515–522.

Zavadova Z. 1971. Host-cell repair of vaccinia virus and of double stranded RNA of encephalomyocarditis virus. Nature (London) New Biol 233:123.

Zelle MR, Hollaender A. 1954. Monochromatic ultraviolet action spectra and quantum yields for inactivation of T1 and T2 *Escherichia coli* bacteriophages. J Bact 68:210–215.

Zimmer JL, Slawson RM. 2002. Potential repair of *Escherichia coli* following exposure to UV radiation from both medium- and low-pressure UV sources used in drinking water treatment. Appl Environ Microbiol 68(7):3293–3299.

Chapter 3

Mathematical Modeling of UV Disinfection

3.1 Introduction

Mathematical modeling of the UV disinfection process provides a basis for sizing ultraviolet disinfection equipment and for interpreting test results. It also allows for adaptation of UV systems to specific disinfection processes and for the disinfection of specific microorganisms. Disinfection is generally modeled in terms of the survival, or its converse the inactivation rate, and may be rendered as a fraction or a percentage. Disinfection is invariably a logarithmic process, as is microbial growth. The disinfection rate of the microorganisms varies widely and is subject to many complexities, including shoulder effects, second stage decay, relative humidity effects, and photoreactivation. Not all these processes need be considered for every design application, but a familiarity with these effects is essential for understanding UV disinfection. Furthermore, not all of these processes can be completely and accurately modeled at present, but enough is known to adequately design reliable UV disinfection systems. All of these processes can be modeled with basic exponential equations and by using computational methods if necessary. For more detailed background information on theoretical decay models for general disinfection purposes see Kowalski (2006), Chick et al. (1963), or Hiatt (1964).

This chapter addresses the fundamental equations and modeling methods of UVGI disinfection processes, and these form a basis for the subsequent chapters. The modeling of UV equipment and UV lamps is addressed separately and in detail in Chap. 7, while the modeling of complete UV systems such as Upper Room systems and UV air disinfection systems is treated in later chapters.

3.2 UV Disinfection Modeling

The ultraviolet disinfection process may involve simple exponential decay, or a more complex function composed of two or more decay processes, a shoulder or delayed response, and photoreactivation. The entire process may also be subject to relative

W. Kowalski, *Ultraviolet Germicidal Irradiation Handbook*,
DOI 10.1007/978-3-642-01999-9_3, © Springer-Verlag Berlin Heidelberg 2009

humidity effects. In addition, the exposure dose itself may be subject to variations from an uneven irradiance field (in air or on surfaces), and in the case of air disinfection there may be airflow irregularities. Each of these components of the disinfection process is describable with basic mathematical models as detailed in the following sections.

3.3 UV Exposure Dose (Fluence)

Microbes exposed to UV irradiation are subject to an exposure dose (fluence) that is a function of the irradiance multiplied by the exposure time, as follows:

$$D = E_t \cdot I_R \tag{3.1}$$

where

$D =$ UV exposure dose (fluence), J/m^2
$E_t =$ exposure time, sec
$I_R =$ Irradiance, W/m^2

The parameter I_R can be used to refer to either irradiance, which is the radiative flux through a flat surface, or the fluence rate, which is the radiative flux through an external surface (i.e. a spherical microbe). In the latter case it is often called spherical irradiance. The same units apply in both cases and the choice of which term to use depends on the context. The fact that both types of irradiance have the units of W/m^2 is an artifact of the method of measurement – in reality both the irradiation field and the microbial mass absorbing the dose exist in a volume and the irradiance should more properly have units of W/m^3, and the proper units for UV rate constants would then be m^3/J.

When the UV dose results in a 90% disinfection rate (10% survival), it is known as a D$_{90}$. The D$_{90}$ value is commonly used as an indicator of system size and can be used to assess the survival rate of individual microbes. Also common is the D$_{99}$, or the dose that results in 99% inactivation.

In the case of surface exposure, the irradiance field may be relatively constant and Eq. (3.1) can be applied directly to get the UV dose. This is generally the case for equipment exposed inside a UV chamber, where the irradiance can be established with a high degree of certainty through measurement. If the irradiance forms a contour across a surface, such as a wall, floor, or cooling coil, then the surface can be subdivided and the total UV dose can be summed per unit area to get an area-weighted average dose. Since the irradiance field of a UV lamp on a surface can be subject to exponential variations, in which the near-field has a wide range of irradiance while the far field becomes relatively constant, it is often necessary to either use a very fine grid (i.e. 50 × 50 or 100 × 100) or to arrange the grid logarithmically so as to concentrate more gridlines in near-field areas of great variation.

For air disinfection, the same problem is encountered – large variation of irradiance in the near field of the UV lamp and gradual drop-off in the far field. In this case the same approach is used except that the irradiance field must be quantified in three dimensions instead of two. A computational model for a three-dimensional grid, as well as for two-dimensional grids, is presented in Chap. 7.

3.4 Single Stage Decay

The primary model used to evaluate the survival of microorganisms subject to UV exposure is the classical exponential decay model. This is a first-order decay rate model and is generally adequate for most UVGI design purposes provided the UV dose is within first order parameters. This is because disinfection rates of 90–99% can generally be achieved in the first stage of decay, and this is adequate for most design purposes. With few exceptions, a D_{90} value defines the first stage of decay for bacteria and viruses. The D_{90} value typically remains accurate up to a D_{99} or even higher, but extrapolation beyond this point is not always valid.

The single stage decay equation for microbes exposed to UV irradiation is:

$$S = e^{-kD} \tag{3.2}$$

where

   $S = $ Survival, fractional
   $k = $ UV rate constant, m^2/J

This decay equation applies as long as shoulder effects can be ignored and the inactivation rate does not extend into the second stage. When exponential decay involves on a single stage, it is referred to as log-linear, since it is linear on a logarithmic scale. Figure 3.1 illustrates the simple exponential decay of a bacteria exposed to UV irradiation. This data displays relatively log-linear decay and the UV rate constant computed from the exponential curve fit (the line in the figure) is seen to be 0.0701 m^2/J.

The UV rate constant computed from the data in Fig. 3.1 will provide reasonable predictions within the upper limit of the test data, 40 J/m^2, which represents a survival of about 7% (or a 93% inactivation rate). The D_{90} is seen to be about 33 J/m^2. Extrapolating performance beyond the upper limit is not recommended as there may be a second stage. If a second stage exists then any extrapolation will produce non-conservative predictions.

The UV rate constant is always assumed to be single stage unless otherwise noted. High values of rate constants imply fast decay and rapid disinfection. Low rate constants imply UV resistance. Many bacteria and viruses have high rate constants, while fungal spores have low rate constants. Appendices A, B, and C may be consulted for representative values of k, which all represent single stage rate constants subject to the indicated upper limit (UL).

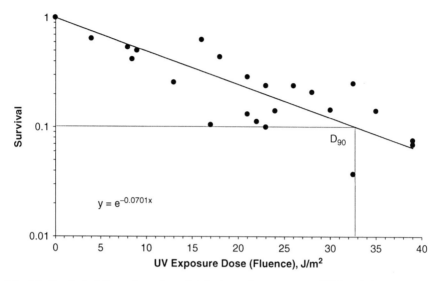

Fig. 3.1 Survival of *Corynebacterium diphtheriae* under exposure to UV irradiation. Based on data from Sharp (1939). *Line* is curve fit of the indicated exponential decay equation

3.5 Two Stage Decay

It is commonly observed in most methods of disinfection that a tiny fraction of the microbial population exhibits a higher level of resistance, and the same is true in UV disinfection (Chick et al. 1963). When the exposure dose is sufficient to cause several logs of reduction (i.e. 99% disinfection or higher) in the microbial population, the surviving population is often an order of magnitude more resistant to UV. That is, the UV rate constant for the resistant population may be ten times lower than for the first stage. This effect will, of course, only be apparent if the disinfection rate is very high, sometime as much as six logs of disinfection. In effect, most microbial populations behave as if two separate populations were present – one relatively susceptible and one relatively resistant. The first stage of decay (fast decay) will then be defined by the susceptible portion of the population and the second stage of decay (slow decay) will be defined by the resistant population. Since the resistant fraction is often on the order of about 1% or less, the second stage only becomes manifest at about the D_{99} value or higher. An alternate model for two stage curves (or tailing effects) has been proposed in Hiatt (1964).

We define the resistant fraction as 'f' and the fast decay fraction is the complement, (1–f). We define the first stage (fast decay) rate constant as k_1 and the second stage (slow decay) rate constant as k_2. Note that $k_1 > k_2$ and that k_1 is not necessarily the same as k in Eq. (3.2) due to the additive effect of the second stage. The survival of the two populations is simply the sum of each decay rate computed per each contribution, as follows.

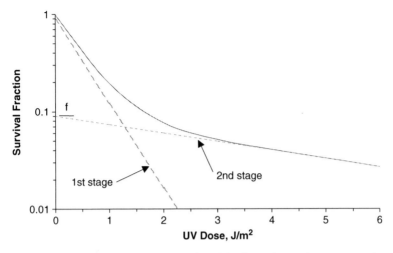

Fig. 3.2   A two stage decay curve is the summation of the first and second stages proportioned by the resistant fraction f

$$S = (1 - f)e^{-k_1 D} + fe^{-k_2 D} \qquad (3.3)$$

where

f = UV resistant fraction (slow decay)
k_1 = first stage rate constant, m^2/J
k_2 = second stage rate constant, m^2/J

Figure 3.2 illustrates Eq. (3.3) for a figurative two stage decay curve, showing the first and second stages separately. Note that the point at which the second stage intercepts the y-axis is equivalent to the resistant fraction f. When actual data is plotted and separated, the second stage can be extrapolated to the y-axis to determine the resistant fraction f. The slope of the two stages can be estimated by dividing the decay curve into two separate sections and computing each per Eq. (3.2). The complete two stage curve can then be developed by incorporating the k values into Eq. (3.3) and adjusting k_1 and k_2 to fit the data by trial and error (and to minimize the R^2 value).

Table 3.1 summarizes the first and second stage rate constants and resistant fractions for a number of microbes. The original single stage rate constant (from Appendices A, B, and C) is provided for comparison. The first stage rate constant k_1 will differ from the original single stage rate constant k because the resistant fraction of the second stage, when added to the first stage per Eq. (3.3), will change the effective rate constant of the first stage to a degree that depends on the resistant fraction. Note that the value $(1-f)$ almost always exceeds 90% (the D_{90}) and often exceeds 99% (the D_{99}).

Table 3.1 Two stage inactivation rate constants

Microorganism	Type	Media	k original μW/cm²	Two stage curve k₁ μW/cm²	(1-f) Susc.	k₂ μW/cm²	f Resist.	Reference
Adenovirus Type 2	Virus	W	0.00470	0.00778	0.99986	0.00500	0.00014	Rainbow and Mak (1973)
Bacillus anthracis spores	Bacteria	S	0.0031	0.0042	0.9984	0.00060	0.0016	Knudson (1986)
Bacillus subtilis spores	Bacteria	W	0.0288	0.035	0.99997	0.00350	0.00003	DeGuchi et al. (2005)
Coxsackievirus A-9	Virus	W	0.015900	0.01600	0.9807	0.01250	0.0193	Hill et al. (1970)
Coxsackievirus B-1	Virus	W	0.020200	0.02480	0.7378	0.00881	0.2622	Hill et al. (1970)
Escherichia coli	Bacteria	S	0.092700	0.80980	0.9174	0.39470	0.0826	Sharp (1939)
Escherichia coli	Bacteria	W	0.18	0.22	0.999	0.04500	0.001	Harris et al. (1987)
Fusarium oxysporum	Fungi	W	0.0142	0.0155	0.999	0.00370	0.001	Asthana and Tuveson (1992)
Herpes simplex virus Type 1	Virus	W	0.1105	0.08	0.996	0.01300	0.004	Albrecht (1974)
Herpes simplex virus Type 1	Virus	W	0.0326	0.058	0.96	0.00650	0.04	Bockstahler and Lytle (1976)
Human Cytomegalovirus	Virus	W	0.01740	0.062	0.95	0.01600	0.05	Albrecht (1974)
Legionella pneumophila	Bacteria	S	0.44613	0.45	0.999	0.14000	0.001	Knudson (1985)
Mycobateriophage D29	phage	W	0.143	0.18	0.99	0.01900	0.01	David (1973)
Mycobacterium smegmatis	Bacteria	W	0.034	0.038	0.999	0.00800	0.001	Boshoff et al. (2003)
Mycobacterium tuberculosis	Bacteria	W	0.031	0.04	0.997	0.01150	0.003	Boshoff et al. (2003)
Penicillium italicum	Fungi	W	0.0114	0.017	0.996	0.00500	0.004	Asthana and Tuveson (1992)
Poliovirus type 1	Virus	W	0.0345	0.0365	0.997	0.01250	0.003	Chang et al. (1985)
Reovirus	Virus	W	0.00853	0.014	0.92	0.00430	0.08	McClain and Spendlove (1966)
Rhizopus nigricans spores	Fungi	S	0.0133	0.0285	0.92	0.00203	0.08	Kowalski (2001)
Staphylococcus aureus	Bacteria	W	0.08531	0.15	0.9998	0.00700	0.0002	Chang et al. (1985)
Streptococcus pyogenes	Bacteria	S	0.061600	0.28700	0.8516	0.01670	0.1484	Lidwell and Lowbury (1950)

3.6 Shoulder Curves

The exponential decay of a microbial population in response to biocidal factors like UV is often subject to a slight delay called a shoulder because of the shape (Cerf 1977, Munakata et al. 1991, Pruitt and Kamau 1993). Shoulder curves start out with a horizontal slope before developing into full exponential decay. The lag in response to the stimulus implies that either a threshold dose is necessary before measurable effects occur or that that repair mechanisms actively deal with low-level damage at low doses (Casarett 1968). Once the threshold is passed the exponential decay curve becomes fully developed (as a single stage or two stage curve). The effect is species dependent and also appears to be a function of the intensity of irradiation. In many cases it can be neglected, especially for susceptible microbes or for high doses. However, for low irradiance levels and for spores and certain resistant bacteria the shoulder can be significant and prolonged. There are at least two separate mathematical models that can deal with shoulder effects – the classic model, and the multihit target model (Kowalski et al. 2000). Various similar target models have been proposed, including recovery models, split-dose recovery models, and empirical models (Russell 1982, Harm 1980, Casarett 1968), but the multihit model, is the most convenient approach for most applications and it is the one addressed here. The multihit target model (Severin et al. 1983) can be written as follows:

$$S(t) = 1 - \left(1 - e^{-kD}\right)^{n} \tag{3.4}$$

where $n = $ multitarget exponent

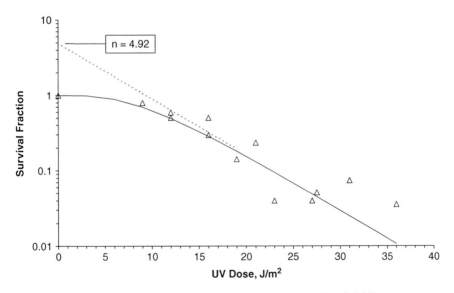

Fig. 3.3 Survival curve for *Staphylococcus aureus* illustrating a shoulder. *Solid line* represents a curve fit to Eq. (3.3). *Dotted line* represents an extrapolation to the intercept, $n = 4.92$. Based on data from Sharp (1939)

The parameter n can be found from extrapolating the first stage data to the y-intercept. It presumably represents the number of discrete critical sites that must be hit to inactivate the microorganism. In theory n is an integer, but in practice this is not always the case. The value of n is unique to each species and values may range from near-unity to over 1000. Figure 3.3 shows data for UV-irradiated *Staphylococcus* in which a shoulder is evident. The rate constant, $k = 0.1702$ m^2/J and the exponent value is $n = 4.92$. The extrapolated first stage is shown with a dotted line, where it can be seen that the y-axis is intercepted at the exponent of $n = 4.92$.

3.7 Two Stage Shoulder Curves

Occasionally, a data set will exhibit both a shoulder and two stages of decay. The mathematical model for this combined curve is simply an incorporation of Eq. (3.4) in Eq. (3.3) as follows:

$$S(t) = (1 - f)[1 - \left(1 - e^{-k_1 D}\right)^{n_1}] + f[1 - \left(1 - e^{-k_2 D}\right)^{n_2}] \qquad (3.5)$$

where

$n_1 =$ exponent of fast decay population
$n_2 =$ exponent of resistant fraction

It can be assumed that $n_1 = n_2$, and this would make sense if we assume that the number of 'targets' is a constant for any species, but the contribution due to the second stage (the resistant fraction) is so typically insignificant that it can virtually always be ignored. That is, it can be assumed that $n_2 = 1$ without significant loss of accuracy, and Eq. (3.5) can be written as:

$$S(t) = (1 - f)[1 - \left(1 - e^{-k_1 D}\right)^{n_1}] + fe^{-k_2 D} \qquad (3.6)$$

The complete model given by Eq. (3.6), which includes the shoulder and the second stage, requires two rate constants, one multihit exponent, and a population resistant fraction. Limited data is available in the literature to define such values for any but a handful of microbes, but Appendices A, B, and C note studies in which two stages and shoulders occur, and these may be consulted for data. Table 3.2 lists a few examples of two stage curves with shoulders and the associated parameters. In general, the multihit exponent for the second stage rate constant has little or no effect and the indicated values for n_2 in may be assumed to be unity without loss of predictive accuracy.

The first and second stage rate constants computed from data can differ somewhat from the single stage rate constants due to the fact that the second stage and the shoulder, if any, will affect the first stage rate constant. This effect can be seen by the two stage curve data in Fig. 3.4. In Table 3.2, the rate constant for *Mycobacterium tuberculosis* based on fitting all the data to a single stage curve without a shoulder

Table 3.2 Two stage and multihit parameters

Microorganism	k_1 m²/J	k_2 m²/J	Res. Pop. (f)	Multihit Exponent n_1	Multihit Exponent n_2	References
Adenovirus Type 2	0.0048	0.7784	0.0001	1.29	1.29	Rainbow and Mak (1973)
Bacillus anthracis spores	0.0042	0.0006	0.0016	2.6	2.6	Knudson (1986)
Mycobacterium smegmatis	0.0380	0.0080	0.00062	6	1	Boshoff et al. (2003)
Mycobacterium tuberculosis	0.0400	0.0115	0.012	30	1	Boshoff et al. (2003)
Staphylococcus aureus	0.0500	0.0108	0.0860	4.9	4.9	Sharp (1939)

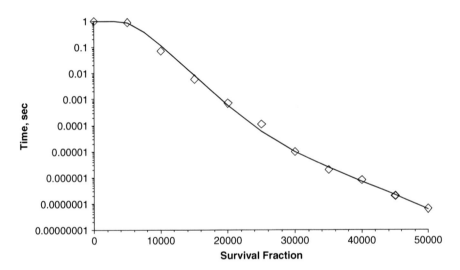

Fig. 3.4 UV survival curve for *Mycobacterium tuberculosis*. *Line* represents a curve fit to a two stage multihit model. Based on data from Boshoff et al. (2003)

proved to be 0.035 m²/J. When a two stage curve with a shoulder is fitted to the same data, the first stage rate constant proves to be approximately 0.055 m²/J.

3.8 Relative Humidity Effects

In air and surface disinfection the relative humidity may impact the UV disinfection rate and therefore will affect the UV rate constant. The effect appears to be different in viruses than in bacteria and therefore these topics are treated separately in the following sections. Insufficient data is available to evaluate fungal and bacterial spores, but a summary is also provided for this topic.

3.8.1 RH Effects on Viruses

The effect of relative humidity on virus UV susceptibility is much less pronounced than it is in bacteria. Table 4.7 summarizes all the available data on RH effects for viruses and provides a ratio for dry to wet conditions. Except for Vaccinia, the ratio between Dry (Lo RH) and Wet (HI RH) conditions is not as great as it is in bacteria. In fact, these differences may be barely beyond the testing accuracy in most cases and therefore the conclusion that virus sensitivity decreases as RH increases must be considered a tenuous one until such time as more definitive data is accumulated on a larger number of viruses. Although DNA viruses go though a conformational change from the A form to the B form as RH increases, RNA viruses may not. RNA viruses are typically in the A conformation and if the conformation change is entirely responsible for RH effect, it could be expected that RNA viruses would not respond to RH changes. There is, however, a possible secondary effect from RH – increased bonding of water to the virus may increases the effective diameter and causes more UV scattering. Table 3.3 shows two complete RNA virus test results but they are hardly conclusive.

Figure 3.5 illustrates the results of Table 3.3 where it can be observed that, except for Vaccinia, the UV susceptibilities show barely any significant response to increased RH. The decrease in UV rate constants from high RH to that in water are clearly more significant.

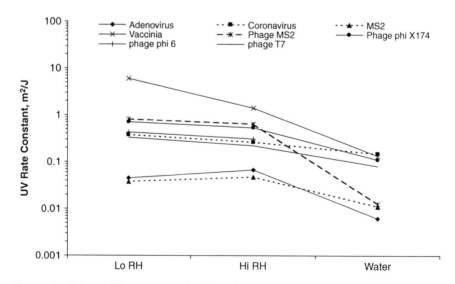

Fig. 3.5 Variation of UV rate constant for viruses in Low to High RH, and compared to Water as an endpoint. Based on data from Table 3.2

Table 3.3 Virus UV rate constants versus relative humidity

| Microbe | Type | UV k, m²/J | | | Ratio Dry/Wet | Hi RH effect on UV k | Reference |
		<68%RH	>75%RH	Water			
Adenovirus	dsDNA	0.039	0.068	0.006	0.6	Increase	Walker and Ko (2007)
Vaccinia	dsDNA	6.01	1.42	0.134	4.2	Decrease	McDevitt et al. (2007)
phage T7	dsDNA	0.33	0.22	0.082	1.5	Decrease	Tseng and Li (2005)
phage phi 6	dsRNA	0.43	0.31	unk	1.4	Decrease	Tseng and Li (2005)
phage phi X174	ssDNA	0.71	0.53	0.111	1.3	Decrease	Tseng and Li (2005)
Coronavirus	ssRNA	0.377	–	0.149	–	–	Walker and Ko (2007)
MS2	ssRNA	0.038	0.048	0.013	0.8	Increase	Walker and Ko (2007)
MS2	ssRNA	0.81	0.64	0.013	1.3	Decrease	Tseng and Li (2005)

3.8.2 RH Effects on Bacteria

Various sources state that increased Relative Humidity (RH) decreases the decay rate under UVGI exposure (Riley and Kaufman 1972). Lidwell and Lowbury (1950) showed the rate constant for *Serratia marcescens* decreasing with increasing RH. Rentschler and Nagy (1942) showed the rate constant for *Streptococcus pyogenes* increasing with higher RH. Fletcher et al. (2003), Fletcher (2004) showed the rate constant for *Burkholderia cepacia* decreases at higher RH. One study on three bacteria species indicates that the decay rate decreases with higher RH (Peccia et al. 2001a). Figure 3.6 shows the effect of relative humidity on the rate constant of *Serratia marcescens*.

Lai et al. (2004) obtained a similar response to high RH for *S. marcescens* as Peccia et al. (2001a) and others but demonstrated that the suspending solution used for aerosolization solution had a significant effect on the rate constant. They recommended that a synthetic saliva would more accurately account for real-world conditions.

In indoor environments, ASHRAE (1999) defines comfort zones as having an RH below 60% and this would be the design operating RH range of any UVGI system that recirculated room air or disinfected return air. In an air handling unit however, the RH could vary greatly depending on where the UVGI system was located (i.e. upstream or downstream of the cooling coils). Obviously, the nature of the RH effect on UV susceptibility may dictate the preferred location of any installed UVGI system, and may even allow a means of boosting UVGI efficiency through RH control, but further quantitative results on RH effects are needed.

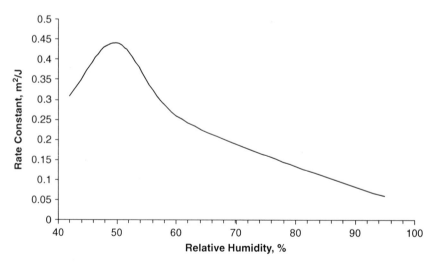

Fig. 3.6 Effect of relative humidity on the rate constant of *Serratia marcescens*. Based on data from Peccia et al. (2001a)

Table 3.4 summarizes the 13 complete studies on bacteria, of which 11 showed a statistically significant decrease in UV rate constants with increasing RH. The effect is broken down into categories of high (Hi) RH and low (Lo) RH to simplify interpretation. The two studies that imply an increase in UV rate constant showed an essentially flat response, within experimental error, and so are listed as having no effect.

Table 3.4 also compares the results of these studies on airborne rate constants with averages for water-based rate constants. Figure 3.7 illustrates the results of Table 3.4 in the form of a plot as the relative humidity goes from low to high, and then to the water-based rate constant as an endpoint. It has been suggested in the literature that airborne rate constants at 100% RH converge towards water rate constants. Although it would be convenient if the UV rate constant at 100% RH equaled the UV rate constant in water there is no conclusive evidence that this occurs. In fact, photoprotective effects due to UV scattering in air (refractive index = 1.0003) must be different from those in water (refractive index = 1.33) and therefore even at 100% RH the UV rate constant in air is unlikely to be identical to that in water. It can be concluded that, in general, UV rate constants in air at high RH converge towards those in water, but are *not* necessarily identical at 100% RH.

Figure 3.8 presents a detailed plot summarizing the results for the various studies on *Serratia marcescens*. It can be seen that there are wide variations in the measured UV rate constants in these studies and that although most of them indicate an increased susceptibility with increasing RH, one of them indicates the opposite. In

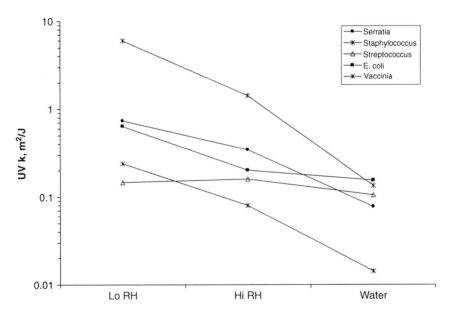

Fig. 3.7 Plot of bacteria UV rate constant variation between low and high RH, and in water as an endpoint. Based on data from Table 3.3

Table 3.4 Bacterial UV rate constants versus RH effects

Microbe	UV k, m²/J			Ratio Dry/Wet	Hi RH effect on UV k	Ref
	<68% RH	>75%RH	Water			
E. coli	0.218583	0.180527	0.154	1.2	Decrease	Rentschler and Nagy (1942)
E. coli	0.965188	0.204733	0.154	4.7	Decrease	Koller (1939)
E. coli	0.72303	0.217862	0.154	3.3	Decrease	Webb and Tai (1970)
M. parafortuitum	0.18	0.05	unk	3.6	Decrease	Peccia and Hernandez (2001)
M. parafortuitum	0.061333	0.0415	unk	1.5	Decrease	Peccia et al. (2001a)
Francisella tularensis	0.008953	0.00795	unk	1.1	None	Beebe (1959)
Serratia marcescens 1	0.595	0.864	0.077	0.7	Increase	Riley and Kaufman (1972)*
Serratia marcescens 2	0.45	0.07	0.077	6.4	Decrease	Peccia and Hernandez (2001)
Serratia marcescens 3	2.2	0.92	0.077	2.4	Decrease	Lai et al. (2004)
Serratia marcescens 4	0.9389	0.0949	0.077	9.9	Decrease	Fletcher et al. (2003)
Serratia marcescens 5	0.3	0.1	0.077	3.0	Decrease	Peccia et al. (2001a)
Serratia marcescens 6	0.57	–	0.077	–	None	Ko et al. (2000)
Serratia marcescens 7	0.113	0.01835	0.077	6.2	Decrease	Webb and Tai (1970)
Streptococcus pyogenes	0.146667	0.16	0.105	0.9	None	Lidwell and Lowbury (1950)
Staphylococcus epidermis	0.024	0.008	0.014	3.0	Decrease	VanOsdell and Foarde (2002)
Average for bacteria	**0.4996**	**0.2098**	–	**3.4**	Decrease	*(revised by author)*

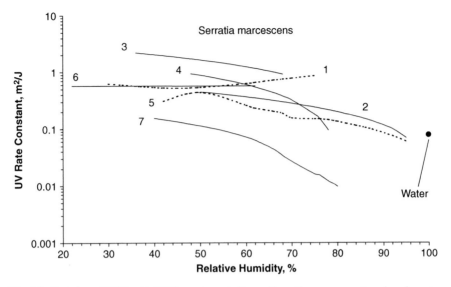

Fig. 3.8 Variation of UV k with RH for various studies on *Serratia marcescens* (numbered as per Table 3.4). Average rate constant in water is shown as a single point at 100% RH

this case, when the UV rate constant for water is placed at the 100% RH point, it is suggestive of the possibility that the UV rate constant for *Serratia* converges to that of water and will equal it at 100% RH, although this has yet to be conclusively demonstrated.

The data sets presented previously indicate that the response to RH may be one that increases, decreases, remains unchanged, or even presents a maxima. These variations in responses have posed difficulties when it comes to modeling, but the key to accurately modeling these RH effects lies in defining a model composed of components, each of which may rise or decrease in response to RH. Since the RH effect is generally understood to be mainly a function of both thymine dimers and spore photoproducts, it can be modeled as two components that will sum to produce an overall response curve. In fact, there are probably additional components, but as a first order model, if we assume that the RH effect is due only to these two factors, the combined equation for the rate constant can be written as follows:

$$k = k_t + k_s \tag{3.7}$$

where

k_t = contribution to k from thymine dimers, m^2/J
k_s = contribution to k from spore photoproducts, m^2/J

Each of these contributions can be modeled as simple sigmoid curves, as a function of humidity, H. The contribution to the rate constant from thymine is:

$$k_t = k_{min} + (k_{max} - k_{min})^* B_t^{0.01\left(\frac{H-M}{M}\right)} \tag{3.8}$$

where

k_{min}, k_{max} = minimum k value, maximum k value
M = Mean humidity value
B_t = species constant for thymine dimers

The contribution to the rate constant from spore photoproduct is:

$$k_s = k_{max} - (k_{max} - k_{min})^* B_s^{0.01\left(\frac{H-M}{M}\right)} \tag{3.9}$$

where B_s = species constant for spore photoproduct

Figure 3.9 shows an example of the RH effects on *Mycobacterium parafortuitum*, with two sigmoid curves fitted such that the summation of the contribution to the UV rate constant value matches the measured values for UV k.

The previous model is simplistic but does provide a means to explain much of the existing data on RH effects. Whether or not this theoretical model for relative humidity effects on RH is valid or not remains to be verified by future

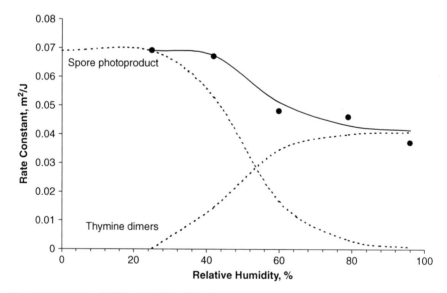

Fig. 3.9 Variation of UV k with RH for *Mycobacterium parafortuitum*, based on data from Peccia et al. (2001a). Theoretical contributions from spore photoproduct and thymine dimers are shown in *dotted lines*. *Solid line* is a summation of the contributions

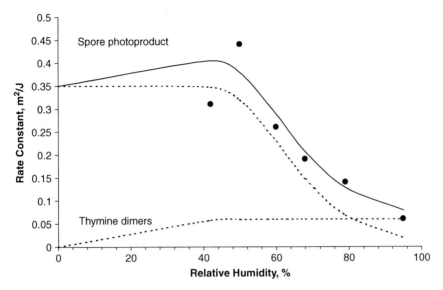

Fig. 3.10 Variation of UV k with RH for *Serratia marcescens*, based on data from Peccia et al. (2001a). Theoretical contributions from spore photoproduct and thymine dimers are shown in *dotted lines*. *Solid line* is a summation of the contributions

research. It should be noted, however, that this model has only addressed two components, thymine dimers and the spore photoproduct, whereas there may actually be some additional components (i.e. cytosine dimers) that also contribute to relative humidity effects, and a more complete multifunctional model can be developed when sufficient data becomes available.

3.8.3 RH Effects on Bacterial and Fungal Spores

There is little data available that would allow generalization of the effects of relative humidity in fungal spores. Zahl et al. (1939) reports no relative humidity effects on *Aspergillus niger*, and it is likely this is a general characteristic of all fungal spores since in fungal spores the DNA is maintained in the A conformation and it could be expected that, if DNA conformation plays a major role in RH response, that fungi would have little or no response to changing RH levels. VanOsdell and Foarde (2002), however, report that the UV susceptibility of *Aspergillus niger* spores decreases by a factor of two from low humidity to high humidity.

Two studies on *Bacillus subtilis* spores have indicated they have no response to increases in RH (Peccia et al. 2001a, VanOsdell and Foarde 2002). It is likely that bacterial spores are maintained in a dry (A conformation) state in the same way as fungal spores and therefore it could be expected that RH effects would not alter the UV susceptibility. But, the lack of any further data on the effects of RH on

UV susceptibility of spore prevents any definitive conclusions form being drawn at this time.

3.9 Modeling Photoreactivation

Another factor that impacts disinfection modeling is the phenomenon of photoreactivation. Photoreactivation occurs when microorganisms are exposed to visible light during or after UV irradiation (Setlow and Carrier 1966, Fletcher et al. 2003). This process can result in self-repair of damaged microbes and can cause a fraction of the population to recover from UV inactivation. Photoreactivation has been studied at length in water-based UV experiments but the data for photoreactivation of airborne microbes is limited at present (Linden and Darby 1994, Masschelein 2002). One study indicates the decay rate of *Mycobacterium parafortuitum* under UV exposure in liquid suspension is effectively decreased by simultaneous exposure to visible light (Peccia and Hernandez 2001). The same study suggests that airborne microbial populations can recover significantly if allowed sufficient time. Knudson (1986) investigated the photoreactivation of *Bacillus anthracis* and found that although vegetative cells had a 1–2 log higher survival rate after exposure to photoreactivating light, the spore forms had no detectable photoreactivation.

Some general considerations for the design of UVGI systems can be given based on what is known about the photoreactivation effect. Ideally, the RH would be kept below approximately 65% in any UVGI system, and the UVGI enclosure should admit little or no internal visible light. Studies on photoreactivation generally involve narrow band UVC lamps, whereas the evidence suggests that broad band UV lamps inhibit photoreactivation (Masschelein 2002). Damage to the enzymes needed for photoreactivation may occur more easily under broadband UVC/UVB radiation, since the peak absorption efficiency for enzymes is at higher wavelengths than that for DNA. Harm (1980) showed that absorption by proteins in the UVA and UVB region equaled the absorption by DNA at 265 nm. Hu et al. (2005) have shown that the DNA of *Escherichia coli* was repaired following irradiation with a low pressure, narrow-band UV lamp while there was no photorepair following irradiation with a medium pressure, broad-band UV lamp. Zimmer and Slawson (2002) found that high levels of photorepair, peaking after 2–3 hours, occurred in *E. coli* after exposure from low pressure lamps, virtually no photorepair occurred after exposure from medium pressure lamps.

Theoretically, photorepair can be preempted by using high enough levels of UV exposure (even with low pressure UV lamps) to cause such extensive damage that no photorepair mechanisms will operate. Coliform bacteria experience little or no repair after exposure to higher UV doses (Linden and Darby 1994).

Photoreactivation effects are generally measured by comparing the survival curves of microbes irradiated under dark and light conditions. The increased survival rates of photoreactivated microbe can be compared directly in terms of their respective UV rate constants. This approach assumes that the inactivation curve consists of two parts – one that can be photoreactivated and one that cannot. Zelle et al. (1958) described the photoreactivation of *E. coli* in terms of the ratio of photoreactivation in the dark to that in light. Hollaender (1955) compared the absolute slope (ignoring the shoulder) of the survival curves of photoreactivated phages with a ratio called the photoreactivable sector, a, defined as follows:

$$a = \frac{k_p}{k_d} \tag{3.10}$$

where

k_p = absolute slope of survival curve after photoreactivation, m^2/J
k_d = absolute slope of survival curve in darkness, m^2/J

The photoreactivable sectors for T1 and T2 phages were relatively high (measured as 0.68 and 0.56 respectively), while lower values were obtained for T4 and T7 phages (0.20 and 0.35 respectively). Sanz et al. (2007) investigated the photoreactivation of coliforms in water and has developed a model in which photoreactivation is quantified in terms of specific parameters that include a reactivation rate constant. Quek and Hu (2008) found that photorepair in *E. coli* occurred at a rate of 10–85% and defined the percent photorepair (%P) as follows:

$$\%P = \frac{N_t - N_0}{N_{init} - N_0} \tag{3.11}$$

where

N_t = concentration at time of exposure, cfu/ml
N_0 = concentration following UV inactivation, cfu/ml
N_{init} = initial concentration before disinfection, cfu/ml

Nebot et al. (2007) modeled photoreactivation in terms of first and second order reactivation constants. They found that photoreactivation was less than 0.1% in the dark, and modeled this type of photoreactivation with a zero-order rate constant. Photorepair is generally found to be higher at lower UV doses (Hu et al. 2005). The degree of photoreactivation depends on the available light energy after UV exposure as well as the relative humidity. Tosa and Hirata (1999) found that photoreactivation was achieved in *E. coli* at 2 W/m^2 and varied from 20 to 78%. Peccia and Hernandez (2001) found that photoreactivation of *Mycobacterium parafortuitum* was only observed when RH above 80% and that no photoreactivation was detectable below 65% RH. Fletcher et al. (2003) found that photoreactivation of *Serratia marcescens* occurred over a wide range of UV doses and increased at lower RH. Photoreactivation occurred at a rate of 56% in *M. tuberculosis* and at 40% in *M. marinum*, at UV doses that produced 2–4 logs of inactivation, by David et al. (1971). Harris et al. (1987) demonstrated photoreactivation in *E. coli* and *Streptococcus faecalis* and stated that the UV doses for 99.9% inactivation were approximately double when photoreactivation occurred.

Jagger et al. (1970) investigated the action spectrum for photoreactivation in *S. griseus* and found that the effect was wavelength-dependent and temperature dependent with peaks at 313 nm and 436 nm. A second microbe, *S. coelicolor*, was found to have a photoreactivation peak at 313 nm but no temperature dependence. Knudson (1985) found that photoreactivation in various species of *Legionella* increased survival about 2–4 times when inactivation levels were 90–99.9%. Oguma et al. (2001) exposed *Cryptosporidium parvum* to fluorescent light after UV inactivation and determined that although pyrimidine dimers were continuously repaired under the near-UV light, the infectivity of the pathogen did not recover.

In conclusion, if photoreactivation occurs, it is essentially incorporated into the measured rate constant of any microbe, depending on test conditions, and therefore ignoring the effect may be a necessity most of the time. For modeling purposes, the photoreactivation effect can be viewed as either an increase in the effective rate constant or as a separate growth rate curve added to the decay curve.

3.10 Air Temperature Effects

The air temperature has a negligible impact on microbial survival during UVGI irradiation provided that neither heat damage nor freezing occurs (Rentschler et al. 1941). It is possible that high indoor temperatures combined with high humidity may alter the effects of UV to some unknown degree. Most building ventilation systems maintain air temperatures in a narrow range between approximately 13–27°C (55–80°F) and moderate relative humidity (<70%), and therefore the effect of temperature on microbes can be ignored. However, air temperature may impact UV lamp output by overcooling the lamp, especially when air velocity is beyond design limits. See Chap. 5 for information on modeling lamp cooling effects.

References

Albrecht T. 1974. Multiplicity reactivation of human cytomegalovirus inactivated by ultra-violet light. Biochim Biophys Acta 905:227–230.
ASHRAE. 1999. Handbook of Applications. Atlanta: ASHRAE.
Asthana A, Tuveson RW. 1992. Effects of UV and phototoxins on selected fungal pathogens of citrus. Int J Plant Sci 153(3):442–452.
Beebe JM. 1959. Stability of disseminated aerosols of *Pastuerella tularensis* subjected to simulated solar radiations at various humidities. J Bacteriol 78:18–24.
Bockstahler LE, Lytle CD, Stafford JE, Haynes KF. 1976. Ultraviolet enhanced reactivation of a human virus: Effect of delayed infection. Mutat Res 35:189–198.
Boshoff HIM, Reed MB, Barry CE, Mizrahi V. 2003. DnaE2 polymerase contributes to in vivo survival and the emergence of drug resistance in *Mycobacterium tuberculosis*. Cell 113:183–193.
Casarett AP. 1968. Radiation Biology. Englewood: Prentice-Hall.
Cerf O. 1977. A review: Tailing of survival curves of bacterial spores. J Appl Bact 42:1–19.
Chang JCH, Ossoff SF, Lobe DC, Dorfman MH, Dumais CM, Qualls RG, Johnson JD. 1985. UV inactivation of pathogenic and indicator microorganisms. Appl & Environ Microbiol 49(6):1361–1365.

Chick EW, A.B. Hudnell J, Sharp DG. 1963. Ultraviolet sensitivity of fungi associated with mycotic keratitis and other mycoses. Sabouviad 2(4):195–200.

David HL, Jones WD, Newman CM. 1971. Ultraviolet light inactivation and photoreactivation in the mycobacteria. Infect and Immun 4:318–319.

David HL. 1973. Response of mycobacteria to ultraviolet radiation. Am Rev Resp Dis 108: 1175–1184.

Deguchi K, Yamaguchi S, Ishida H. 2005. UV-disinfection reactor validation by computational fluid dynamics and relation to biodosimetry and actinometry. J Wat Technol Environ Technol 3(1):77–84.

Fletcher LA, Noakes CJ, Beggs CB, Sleigh PA, Kerr KG. 2003. The ultraviolet susceptibility of aerosolised microorganisms and the role of photoreactivation; Vienna. IUVA.

Fletcher L. 2004. The influence of relative humidity on the UV susceptibility of airborne gram negative bacteria. IUVA News 6(1):12–19.

Harm W. 1980. Biological effects of ultraviolet radiation. New York: Cambridge University Press.

Harris GD, Adams VD, Sorenson DL, Curtis MS. 1987. Ultraviolet inactivation of selected bacteria and viruses with photoreactivation of the bacteria. Water Res 21(6):687–692.

Hiatt C. 1964. Kinetics of the inactivation of viruses. Bact Rev 28(2):150–163.

Hill WF, Hamblet FE, Benton WH, Akin EW. 1970. Ultraviolet devitalization of eight selected enteric viruses in estuarine water. Appl Microb 19(5):805–812.

Hollaender A. 1955. Radiation Biology, Volume II: Ultraviolet and Related Radiations. New York: McGraw-Hill.

Hu JY, Chu SN, Quek PH, Feng YY, Tang XL. 2005. Repair and regrowth of *Escherichia coli* after low- and medium-pressure ultraviolet disinfection. Wat Sci Technol 5(5):101–108.

Jagger J, Takebe H, Snow JM. 1970. Photoreactivation of killing in *Streptomyces*: Action spectra and kinetic studies. Photochem Photobiol 12:185–196.

Knudson GB. 1985. Photoreactivation of UV-irradiated *Legionella pneumophila* and other *Legionella* species. Appl & Environ Microbiol 49(4):975–980.

Knudson GB. 1986. Photoreactivation of ultraviolet-irradiated, plasmid-bearing, and plasmid-free strains of *Bacillus anthracis*. Appl & Environ Microbiol 52(3):444–449.

Ko G, First MW, Burge HA. 2000. Influence of relative humidity on particle size and UV sensitivity of *Serratia marcescens* and *Mycobacterium bovis* BCG aerosols. Tuber Lung Dis 80(4/5): 217–228.

Koller LR. 1939. Bactericidal effects of ultraviolet radiation produced by low pressure mercury vapor lamps. J Appl Phys 10:624.

Kowalski WJ, Bahnfleth WP, Witham DL, Severin BF, Whittam TS. 2000. Mathematical modeling of UVGI for air disinfection. Quant Microbiol 2(3):249–270.

Kowalski WJ. 2001. Design and Optimization of UVGI Air Disinfection Systems [PhD]. State College: The Pennsylvania State University.

Kowalski WJ. 2006. Aerobiological Engineering Handbook: A Guide to Airborne Disease Control Technologies. New York: McGraw-Hill.

Lai KM, Burge, H, First, MW. 2004. Size and UV germicidal irradiation susceptibility of *Serratia marcescens* when aerosolized from different suspending media. Appl Environ Microbiol 70(4):2021–2027.

Lidwell OM, Lowbury EJ. 1950. The survival of bacteria in dust. Ann Rev Microbiol 14:38–43.

Linden KG, Darby JL. 1994. Ultraviolet disinfection of wastewater: effect of dose on subsequent reactivation. Water Res 28:805–817.

Masschelein WJ. 2002. Ultraviolet Light in Water and Wastewater Sanitation. Boca Raton: Lewis Publishers.

McClain ME, Spendlove RS. 1966. Multiplicity reactivation of Reovirus particles after exposure to ultraviolet light. J Bact 92(5):1422–1429.

McDevitt JJ, Lai KM, Rudnick SN, Houseman EA, First MW, Milton DK. 2007. Characterization of UVC light sensitivity of vaccinia virus. Appl Environ Microbiol 73(18):5760–5766.

Munakata N, Saito M, Hieda K. 1991. Inactivation action spectra of *Bacillus subtilis* spores in extended ultraviolet wavelengths (50–300 nm) obtained with synchrotron radiation. Photochem & Photobiol 54(5):761–768.

Nebot SE, Davila IS, Andrade JAA, Alonso JMQ. 2007. Modelling of reactivation after UV disinfection: effect of UV-C dose on subsequent photoreactivation and dark repair. Water Res 41(14):3141–3151.

Oguma K, Katayama H, Mitani H, Morita S, Hirata T, Ohgaki S. 2001. Determination of pyrimidine dimers in *Escherichia coli* and *Cryptosporidium parvum* during UV light inactivation, photoreactivation, and dark repair. Appl Environ Microbiol 67(10):4630–4637.

Peccia J, Hernandez M. 2001. Photoreactivation in airborne *Mycobacterium parafortuitum*. Appl Environ Microbiol 67:4225–4232.

Peccia J, Werth HM, Miller S, Hernandez M. 2001a. Effects of relative humidity on the ultraviolet induced inactivation of airborne bacteria. Aerosol Sci & Technol 35:728–740.

Pruitt KM, Kamau DN. 1993. Mathematical models of bacterial growth, inhibition and death under combined stress conditions. J Ind Microb 12:221–231.

Quek PH, Hu J. 2008. Indicators for photoreactivation and dark repair studies following ultraviolet disinfection. J Ind Microbiol Biotechnol 35(6):533–541.

Rainbow AJ, Mak S. 1973. DNA damage and biological function of human adenovirus after UV irradiation. Int J Radiat Biol 24(1):59–72.

Rentschler HC, Nagy R, Mouromseff G. 1941. Bactericidal effect of ultraviolet radiation. J Bacteriol 42:745–774.

Rentschler HC, Nagy R. 1942. Bactericidal action of ultraviolet radiation on air-borne microorganisms. J Bacteriol 44:85–94.

Riley RL, Kaufman JE. 1972. Effect of relative humidity on the inactivation of airborne *Serratia marcescens* by ultraviolet radiation. Appl Microbiol 23(6):1113–1120.

Russell AD. 1982. The Destruction of Bacterial Spores. New York: Academic Press.

Sanz EN, Davila IS, Balao JAA, Alonso JMQ. 2007. Modelling of reactivation after UV disinfection: Effect of UV-C dose on subsequent photoreactivation and dark repair. Wat Res 41:3141–3151.

Setlow JK. 1966. Photoreactivation. Radiat Res Suppl 6:141–155.

Setlow RB, Carrier WL. 1966. Pyrimidine dimers in ultraviolet-irradiated DNA's. J Mol Biol 17:237–254.

Severin BF, Suidan MT, Englebrecht RS. 1983. Kinetic modeling of UV disinfection of water. Water Res 17(11):1669–1678.

Sharp G. 1939. The lethal action of short ultraviolet rays on several common pathogenic bacteria. J Bact 37:447–459.

Tosa K, Hirata T. 1999. Photoreactivation of enterohemorrhagic Escherichia coli following UV disinfection. Wat Res 33(2):361–366.

Tseng C-C, Li C-S. 2005. Inactivation of virus-containing aerosols by ultraviolet germicidal irradiation. Aerosol Sci Technol 39(1136–1142).

VanOsdell D, Foarde K. 2002. Defining the Effectiveness of UV Lamps Installed in Circulating Air Ductwork. Arlington, VA: Air-Conditioning and Refrigeration Technology Institute. Report nr ARTI-21CR/610-40030-01.

Walker CM, Ko G. 2007. Effect of ultraviolet germicidal irradiation on viral aerosols. Environ Sci Technol 41(15):5460–5465.

Webb SJ, Tai CC. 1970. Differential, lethal and mutagenic action of 254 nm and 320–400 nm radiation on semi-dried bacteria. Photochem Photobiol 12:119–143.

Zahl PA, Koller LR, Haskins CP. 1939. The effects of ultraviolet radiation on spores of the fungus *Aspergillus niger*. J Gen Physiol 16:221–235.

Zelle MR, Ogg JE, Hollaender A. 1958. Photoreactivation of induced mutation and inactivation of *Escherichia coli* exposed to various wave lengths of monochromatic ultraviolet radiation. J Bact 75(2):190–198.

Zimmer JL, Slawson RM. 2002. Potential repair of *Escherichia coli* following exposure to UV radiation from both medium- and low-pressure UV sources used in drinking water treatment. Appl Environ Microbiol 68(7):3293–3299.

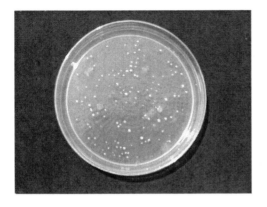

Chapter 4

UV Rate Constants

4.1 Introduction

Microbial susceptibility to ultraviolet light varies widely between species of microbes. Bacteria, viruses, and fungal spores respond to UV exposure at rates defined in terms of UV rate constants. Other parameters used to define UV susceptibility include the Z value or Z_{eff} (same as UV rate constant), the inactivation cross-section, the D_{90} (UV dose to inactivate 90%), and variations of the D_{90} (i.e. D_{99}, $D_{99.9}$ etc.). The classic lethal hit dose, D_{37} or lethe, does not follow the same convention, and refers to a 37% survival dose (e^{-1}), or 63.2% inactivation (Fraenkel-Conrat and Wagner 1981, Hollaender 1955, Casarett 1968, Wells 1940). The UV rate constant refers to either broadband UVB/UVC spectrum (200–320 nm) or to narrow-band UVC (253.7 nm). It will be shown in this chapter that the differences between the UV rate constants defined in these separate contexts are both qualitative and relatively minor in most cases.

The UV rate constant can vary with the spectrum of UV (see Chap. 2), the relative humidity, the DNA conformation (A or B), and can be impacted by photoreactivation and even the irradiance level. Furthermore, microbes tend to have survival curves that reflect two rate constants, one for the majority subject to fast decay, and one for the minority resistant fraction. These topics are addressed here along with an extensive database of UV rate constants for microbes in air and on surfaces. The effects of media are compared, as are the relative susceptibilities of various species. A complete summary of UV rate constants and D_{90} values, a rate constant database, is provided in Appendices A, B, and C, for bacteria, viruses, and fungi respectively. The cited sources for the UV rate constants are provided at the end of this chapter. This chapter also provides a predictive model for UV rate constants.

W. Kowalski, *Ultraviolet Germicidal Irradiation Handbook*,
DOI 10.1007/978-3-642-01999-9_4, © Springer-Verlag Berlin Heidelberg 2009

4.2 Rate Constants and Survival Curves

Microbial populations decay exponentially under UV exposure, as described in the previous chapter (see Eq. 3.2). The slope of the logarithmic decay curve is defined by the rate constant, which is designated as k. The UV rate constant k has units of m^2/J, and is also known as the UV susceptibility. Since there may be two stages of decay in a curve, and there are limits on the range in which each rate constant applies, it can be somewhat more convenient and definitive to use the dose for 90% inactivation, D_{90}, as the primary indicator of UV susceptibility. The D_{90} represents 10% survival and can be computed from the rate constant as follows:

$$D_{90} = \frac{-\ln(1 - 0.9)}{k} = \frac{2.3026}{k} \qquad (4.1)$$

The values for D_{99}, $D_{99.9}$ or any other inactivation rate dose can be computed in a similar manner. The D_{90} value will generally place inactivation sufficiently beyond the shoulder and solidly in the first stage of decay. Note that the D_{90} does not describe the actual shape of the decay curve, which may be shouldered, but merely defines the dose to achieve 90% inactivation, and is therefore less ambiguous when decay curves include shoulders or second stages. For this reason, and for clarity, many of the charts and tables in this chapter will refer to D_{90} values rather than the corresponding rate constants. In Appendices A, B, and C, the UV rate constant is provided along with both D_{90} values and the experimental upper limits (UL) within which the k value applies. The UV rate constants in Appendix A, B, and C are subject to considerable variation due to various factors such as the type of test equipment, the type of UV lamps employed, the dose range used in the test, culture media and methods, specific strains used, and ambient conditions during the test. Test results performed using collimated beam apparatus are considered to be the most accurate but even these appear to be subject to wide variations (Hijnen et al. 2006).

When a microbe is exposed for extended periods of time or to high UV doses, the survival curve often displays evidence of a second stage (sometimes called tailing). In addition, survival curves may display a shoulder, or a period of delayed response, especially at low levels of UV dose. The mathematical equations for these curves have been detailed previously in Chap. 3. Throughout much of the discussion that follows, the UV rate constant is always assumed to be the first stage rate constant, unless otherwise noted.

4.3 UV Rate Constant Database

A summary of over 600 UV rate constant studies for bacteria, viruses, and fungi are provided in Appendices A, B, and C respectively. Appendix C also includes other microbes such as protozoa. Data are from studies in air (A), water (W), and on surfaces (S). The Type of bacteria is identified as either a spore (Spore), or vegetative (Veg), while the viruses are identified as either double stranded DNA (dsDNA), single stranded DNA (ssDNA), double-stranded RNA (dsRNA), or single stranded RNA (ssRNA). Fungi are identified as either spores, vegetative (Veg), or yeast (VegY). D_{90} values are identified along with the upper limit Dose (UL) for which the rate constants apply. Exceptions include two-stage curves, which are identified in the column labeled "St" for the stages. Curves that display a shoulder are identified in the column labeled "Sh" for shoulder. For airborne studies the relative humidity conditions are identified if known. Some studies are differentiated by an "LP" (low pressure UV lamp) or an "MP" (medium pressure UV lamp) when the results differ between the lamp types. Several additional studies could not be included due to UV doses specified as an upper limit, or as zero survival, or, in the case of *E. coli*, due to excessive redundancy of test results – see Chevrefils et al. (2006) for a summary of these additional studies. Also see Hijnen et al. (2006) for a summary of collimated beam tests.

Table 4.1 summarizes overall average UV rate constants and D_{90} values for bacteria, viruses, and fungi in air, water, and on surfaces, from Appendices A, B, and C. The term "Lo RH" refers to relative humidity below about 68% while "Hi RH" implies an RH of about 68% or higher. Based on these overall averages, it would appear that viruses are over twice as resistant to UV as bacteria, while fungal spores are about three times more resistant than bacteria. The D_{90} values in Table 4.1 are computed from the average rate constants (not averaged from the database).

Table 4.1 Overall average rate constants for microbial groups

Microbe	Type	Water		Surface		Air – Lo RH		Air – Hi RH	
		UV k m^2/J	D_{90} J/m^2	UV k m^2/J	D_{90} J/m^2	UV k m^2/J	D_{90} J/m^2	UV k m^2/J	D_{90} J/m^2
Bacteria	Veg	0.08463	27	0.14045	16	0.38887	6	0.07384	31
Viruses	All	0.05798	40	0.03156	73	0.39985	6	0.29050	8
Bacterial spores	Spores	0.01439	160	0.01823	126	0.02566	90	0.02600	89
Fungal cells and yeast	Veg	0.01008	229	0.00700	329	0.09986	23	–	–
Fungal spores	Spores	0.00916	251	0.00789	292	0.00730	315	0.00472	488

4.3.1 Bacteria Rate Constants

Bacteria present a diverse range of UV susceptibilities, and bacterial spores are as resistant as fungal spores. Table 4.1 lists the air, water, and surface rate constants, for all those for which the media was identified, summarized from

Appendix A in terms of averages when more than one data set was available, or representative values otherwise. Bacteria span a wide range of UV sensitivities, and although viruses are on the average more resistant, there is considerable overlap, with some bacteria being more resistant to UV than most viruses, and vice-versa. One bacteria in Appendix A, *Deinococcus radiodurans*, has such unusual resistance to UV that the extended shoulder makes it infeasible to model the UV susceptibility as a single stage rate constant – see Setlow and Duggan (1964) for the experimental results. The number of blanks in Table 4.2 make it clear how much research remains to be done in air and surface disinfection. The bulk of the data available pertains to waterborne microbes.

The media in which bacteria are exposed have a major impact on their UV susceptibility. UV rate constants are about ten times higher (meaning higher susceptibility or lower resistance) for bacteria in air at low RH than in water. In terms of the D_{90}, the dose to inactivate 90%, bacteria in water is almost ten times higher than in air at low RH, and the D_{90} in water is about twice as high as in air at high RH. Figure 4.1 illustrates these ratios. It has been suggested that the UV rate constant in water is a limit to which the rate constant in air will converge at high humidity, and this would certainly seem to be the case in general. However, UV rate constants at 100% relative humidity do not, in general, match UV rate constants in water. Comparisons of the surface D_{90} values are not included in Fig. 4.1 due to the fact that there is limited correspondence between the species studied in water and those studied on surfaces. The surface rate constants, or D_{90} values, appear to be very species-dependent. Figure 4.2 illustrates specific examples of bacteria and their associated D_{90} values in various media, where it can be observed that the UV susceptibility on surfaces can be much higher or lower than in water, depending on species.

4.3.2 Virus Rate Constants

Viruses show UV susceptibilities that range from very susceptible to very resistant, and on the average viruses are more resistant to UV inactivation than bacteria. Viruses may have DNA or RNA and they may be single or double stranded. Two main types of viruses are addressed here, animal viruses and bacteriophages. A bacteriophage, or phage, is a virus that infects bacterial cells. Although animal viruses are or primary concern for air and surface disinfection, phages are extremely similar and provide important information on UV susceptibility. Table 4.3 summarizes the UV rate constant data for animal viruses from Appendix B, averaged for each species and separated by media type. Table 4.4 provides the same summary for phages. In air at high RH, viruses require less than half the UV dose for inactivation as do bacteria. Viruses do not appear to respond significantly to increased RH in air and most viruses have approximately the same susceptibility at high and low RH. On surfaces, viruses appear to be about five times

Table 4.2 Average UV rate constants for bacteria

Bacteria	Type	Water D_{90} J/m²	Water UVGI k m²/J	Surface D_{90} J/m²	Surface UVGI k m²/J	Air - Lo RH D_{90} J/m²	Air - Lo RH UVGI k m²/J	Air Hi RH D_{90} J/m²	Air Hi RH UVGI k m²/J
B. atrophaeus spores	Sp	1323	0.00174			144	0.01600		
Bacillus anthracis spores	Sp	411	0.00560	85	0.02702				
Bacillus cereus spores	Sp			204	0.01126				
Bacillus megatherium	Sp			273	0.00843				
Bacillus pumilis spores	Sp	50	0.04600						
Bacillus subtilis spores	Sp	131	0.01763	88	0.02620	95	0.02413	89	0.02600
Bacillus thuringiensis	Sp	2303	0.00100						
Acinetobacter baumannii	Veg	12	0.19200	18	0.12800				
Aeromonas	Veg	11	0.20310						
Aeromonas hydrophila	Veg	16	0.14100						
Bacillus megatherium	Veg			113	0.02038				
Bacillus subtilis	Veg	25	0.09210			14	0.16858		
Burkholderia cenocepacia	Veg	58	0.03956						
Burkholderia cepacia	Veg					11	0.21150	22	0.10520
Campylobacter jejuni	Veg	16	0.14436						
Citrobacter diversus	Veg	32	0.07140						
Citrobacter freundii	Veg	44	0.05246						
Clostridium perfringens	Veg	38	0.06000						
Corynebacterium diphtheriae	Veg			33	0.07010				
Coxiella burnetii	Veg	15	0.15350						
Deinococcus radiodurans	Veg	365	0.00630						
Enterobacter cloacae	Veg	64	0.03598						
Escherichia coli	Veg	26	0.08767	22	0.10575	5	0.50628	11	0.21400
Francisella tularensis	Veg					256	0.00900	288	0.00800
Haemophilus influenzae	Veg	13	0.17700	38	0.05990				
Halobacterium	Veg			25	0.09210				
Helicobacter pylori	Veg	33	0.06900						
Klebsiella pneumoniae	Veg	52	0.04435						
Klebsiella terrigena	Veg	33	0.07000						
Legionella dumoffi	Veg			24	0.09594				
Legionella bozemanii	Veg	13	0.17400	15	0.15351				
Legionella gormanii	Veg			26	0.08856				
Legionella jordanis	Veg			11	0.20933				
Legionella longbeach	Veg			11	0.20933				
Legionella micdadei	Veg			15	0.15351				
Legionella oakridgensis	Veg			22	0.10466				
Legionella pneumophila	Veg	14	0.16178	5	0.44613				
Legionella wadsworthii	Veg			4	0.57565				
Listeria monocytogenes	Veg	181	0.01270	19	0.12255				
Micrococcus candidus	Veg			60	0.03806				
Micrococcus piltonensis	Veg			81	0.02843				
Micrococcus sphaeroides	Veg			100	0.02303				
Moraxella	Veg	10466	0.00022						
Mycobacterium avium	Veg	52	0.04387						
Mycobacterium bovis BCG	Veg			22	0.10550	13	0.18100	33	0.07000
Mycobacterium flaviscens	Veg	120	0.01919						
Mycobacterium fortuitum	Veg	80	0.02895						
Mycobacterium kansasii	Veg	80	0.02880						
Mycobacterium marinum	Veg	138	0.01670						
Mycobacterium parafortuitum	Veg					15	0.15000	46	0.05000
Mycobacterium phlei	Veg	76	0.03030			25	0.09217		
Mycobacterium smegmatis	Veg	120	0.01917			12	0.19000		
Mycobacterium terrae	Veg	50	0.04610						
Mycobacterium tuberculosis	Veg	48	0.04773	11	0.21320	5	0.47210		
Mycoplasma arthritidis	Veg			7	0.31240				
Mycoplasma fermentans	Veg			9	0.25220				
Mycoplasma hominis	Veg			7	0.32710				
Mycoplasma Orale type 1	Veg			11	0.21800				

Table 4.2 (continued)

Bacteria	Type	Water		Surface		Air - Lo RH		Air Hi RH	
		D_{90} J/m²	UVGI k m²/J	D_{90} J/m²	UVGI k m²/J	D_{90} J/m²	UVGI k m²/J	D_{90} J/m²	UVGI k m²/J
Mycoplasma Orale type 2	Veg			6	0.38760				
Mycoplasma pneumoniae	Veg			8	0.27910				
Mycoplasma salivarium	Veg			11	0.21140				
Myxobolus cerebralis	Veg	10011	0.00023						
Neisseria catarrhalis	Veg			44	0.05233				
Nocardia asteroides	Veg			280	0.00822				
Phytomonas tumefaciens	Veg			44	0.05233				
Proteus mirabilis	Veg	8	0.28900						
Proteus vulgaris	Veg			30	0.07675				
Pseudomonas aeruginosa	Veg	26	0.08706	18	0.12802	4	0.57210		
Pseudomonas diminuta	Veg	96	0.02391						
Pseudomonas fluorescens	Veg			35	0.06579	5	0.47730		
Pseudomonas maltophilia	Veg	70	0.03294						
Pseudomonas putrefaciens	Veg	86	0.02662						
Rickettsia prowazekii	Veg	13	0.17600						
Salmonella anatum	Veg	60	0.03840						
Salmonella derby	Veg	36	0.06360						
Salmonella enteritidis	Veg	33	0.07010	10	0.22100				
Salmonella infantis	Veg	20	0.11510						
Salmonella spp.	Veg	11	0.21380						
Salmonella typhi	Veg	16	0.14672	21	0.10760				
Salmonella typhimurium	Veg	34	0.06805						
Sarcina lutea	Veg			197	0.01169				
Serratia indica	Veg	209	0.01100						
Serratia marcescens	Veg	36	0.06342	14	0.16226	3	0.70657	37	0.06167
Shigella dysenteriae	Veg	18	0.13080						
Shigella paradysenteriae	Veg			17	0.13706				
Shigella sonnei	Veg	18	0.12500						
Spirillum rubrum	Veg			44	0.05233				
Staphylococcus albus	Veg			24	0.09746	32	0.07175		
Staphylococcus aureus	Veg	40	0.05688	32	0.07132	4	0.59570		
Staphylococcus epidermis	Veg	161	0.01433			18	0.12670	288	0.00800
Streptococcus agalactiae	Veg					5	0.43420		
Streptococcus faecalis	Veg	46	0.04960						
Streptococcus faecium	Veg	45	0.05100						
Streptococcus haemolyticus	Veg			22	0.10660				
Streptococcus lactis	Veg			62	0.03744				
Streptococcus pneumoniae	Veg			468	0.00492				
Streptococcus pyogenes	Veg			37	0.06161	1	1.56100		
Streptococcus viridans	Veg			20	0.11513				
Streptomyces coelicolor	Veg	60	0.03840						
Streptomyces griseus	Veg	82	0.02810						
Vibrio cholerae	Veg	17	0.13400						
Vibrio ordalii	Veg	18	0.12560						
Vibrio parahaemolyticus	Veg	8	0.30700						
Yersinia enterocolitica	Veg	15	0.15279						

more resistant to UV inactivation than bacteria, although this could be due to the limited virus surface data sets.

The data provided in Table 4.3 on airborne virus inactivation may be limited but it does provide some insight into the response of viruses irradiated in air. Like bacteria, viruses are more susceptible in air than in water. Figure 4.3 illustrates the overall averages of UV susceptibility for all viruses in the database. It would seem that relative humidity has a limited effect on phage inactivation but of course there are very few airborne inactivation data sets. Double stranded RNA viruses typically remain in the A conformation and they might have less response to RH than DNA

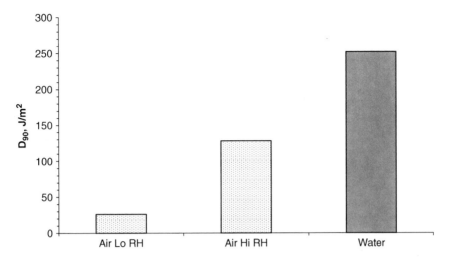

Fig. 4.1 Comparison of overall averages of bacteria D_{90} values in various media

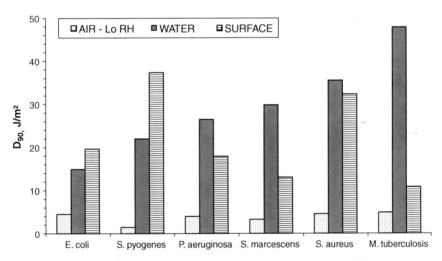

Fig. 4.2 Comparison of various vegetative bacteria D_{90} values in air at low RH, water, and on surfaces

viruses, but the data is limited (Fasman 1996). The differences between the surface and water averages in Fig. 4.3 are also probably due to the limited number of surface data sets.

There are no significant differences between the UV inactivation rate constants of phages and animal viruses regardless of media and therefore it is acceptable to use phages as substitutes for animal viruses in inactivation experiments.

Table 4.3   Average UV rate constants for animal viruses and phages

Virus	Type	Water		Surface		Air - Lo RH		Air Hi RH	
		D_{90}	UVGI k	D_{90}	UVGI k	D_{90}	UVGI k	D_{90}	UVGI k
		J/m^2	m^2/J	J/m^2	m^2/J	J/m^2	m^2/J	J/m^2	m^2/J
Adenovirus	dsDNA	903	0.00255			49	0.04700	34	0.06800
Adenovirus type 1	dsDNA	322	0.00714						
Adenovirus type 15	dsDNA	396	0.00581						
Adenovirus type 2	dsDNA	324	0.00711	400	0.00576				
Adenovirus type 4	dsDNA	921	0.00250						
Adenovirus type 40	dsDNA	546	0.00422	300	0.00768				
Adenovirus type 41	dsDNA	515	0.00447	236	0.00976				
Adenovirus type 5	dsDNA	522	0.00441						
Adenovirus type 6	dsDNA	395	0.00583						
Avian Influenza virus	ssRNA	25	0.09140						
Avian Leukosis virus (RSA)	ssRNA	631	0.00365						
Avian Sarcoma virus	ssDNA	220	0.01047						
B. subtilis phage 029	dsDNA	70	0.03289						
B. subtilis phage SP02c12	dsDNA	100	0.02303						
B. subtilis phage SPP1	dsDNA	195	0.01181						
Bacteriophage B40-8	dsDNA	137	0.01679						
Bacteriophage F-specific	dsRNA	292	0.00789						
Bacteriophage MS2	ssRNA	182	0.01268			5	0.42400	7	0.34400
Bacteriophage Qβ	ssRNA	235	0.00980						
Berne virus	ssRNA	13	0.18420						
BLV	ssRNA	394	0.00584						
Borna virus	ssRNA	79	0.02920						
Bovine Calicivirus	ssDNA	95	0.02420						
Bovine Parvovirus	ssDNA	35	0.06580						
Canine Calicivirus	ssRNA	67	0.03450						
Canine hepatic Adenovirus	dsDNA	265	0.00869						
Cholera phage Kappa	dsDNA	634	0.00363						
Coliphage f2	ssRNA	310	0.00743						
Coliphage fd	ssDNA	23	0.09940						
Coliphage φ X-174	ssDNA	25	0.09292			3	0.71000	4	0.53000
Coliphage lambda	dsDNA	78	0.02953						
Coliphage PRD1	dsDNA	20	0.11500	87	0.02650				
Coliphage T1	dsDNA	14	0.16257						
Coliphage T2	dsDNA	9	0.25243						
Coliphage T3	dsDNA	10	0.23100						
Coliphage T4	dsDNA	13	0.17575						
Coliphage T7	dsDNA	28	0.08152			7	0.33000	10	0.22000
Coronavirus	ssRNA	21	0.11059			6	0.37700		
Coxsackievirus	ssRNA	81	0.02834			21	0.11100		
Echovirus	ssRNA	83	0.02786						
Encephalomyocarditis virus	ssRNA	55	0.04220						
Epstein-Barr virus (EBV)	ssDNA	162	0.01420						
Equine Herpes virus	dsDNA	25	0.09210						
Feline Calicivirus (FeCV)	ssRNA	64	0.03610						
Friend Murine Leukemia v.	ssRNA	320	0.00720						
Frog virus 3	dsDNA	25	0.09210						
Hepatitis A virus	dsDNA	66	0.03513						
Herpes simplex virus Type 1	dsDNA	36	0.06325						
Herpes Simplex virus Type 2	dsDNA	35	0.06569						
HIV-1	ssRNA	280	0.00822						
HP1c1 phage	dsDNA	40	0.05760						
HTLV-1	ssRNA	20	0.11510						
Human Cytomegalovirus	dsDNA			93	0.02478				
Influenza A virus	ssRNA	23	0.10103			19	0.11900		
Kemerovo (R-10 strain)	dsRNA	230	0.01000						
Kilham Rat Virus (parvovirus)	ssDNA	30	0.07650						
Lipovnik (Lip-91 strain)	dsRNA	299	0.00770						
Measles virus	ssRNA	22	0.10510						
Mengovirus	dsRNA	162	0.01420						

Table 4.3   (continued)

Virus	Type	Water		Surface		Air - Lo RH		Air Hi RH	
		D_{90} J/m^2	UVGI k m^2/J	D_{90} J/m^2	UVGI k m^2/J	D_{90} J/m^2	UVGI k m^2/J	D_{90} J/m^2	UVGI k m^2/J
Minute Virus of Mice (MVM)	ssDNA	21	0.10850						
Moloney Murine Leukemia v.	ssRNA	201	0.01148						
Murine Cytomegalovirus	dsDNA	46	0.05000						
Murine Norovirus (MNV)	ssRNA	76	0.03040						
Murine sarcoma virus	ssRNA	207	0.01113						
Mycobacteriophage D29	dsDNA	44	0.05290						
Mycobacteriophage D32	dsDNA	354	0.00650						
Mycobacteriophage D4	dsDNA	245	0.00940						
Mycoplasmavirus MVL	dsDNA	105	0.02200						
Newcastle Disease Virus	ssRNA	14	0.16355	16	0.14400				
Parvovirus H-1	ssDNA	25	0.09200						
phage B40-8 (B. fragilis)	dsDNA	75	0.03070						
phage GA	ssRNA	200	0.01150						
phage phi 6	dsRNA	5	0.43000						
phage phi 6	dsRNA	7	0.31000						
Poliovirus	dsRNA	85	0.02694	42	0.05425				
Poliovirus type 2	dsRNA	121	0.01910						
Poliovirus type 3	dsRNA	103	0.02240						
Polyomavirus	dsDNA	564	0.00408						
Porcine Parvovirus (PPV)	ssDNA	23	0.10230						
Pseudorabies (PRV)	dsDNA	34	0.06760						
Rabies virus (env)	ssRNA	10	0.21930						
Rauscher Murine Leukemia v.	ssRNA	236	0.00975	959	0.00240				
Reovirus	dsRNA	148	0.01556						
Reovirus 3	dsRNA	334	0.00690						
Rotavirus	dsRNA	200	0.01150						
Rotavirus SA11	dsRNA	89	0.02580						
Rous Sarcoma virus (RSV)	ssRNA	360	0.00640	200	0.01150				
S. aureus phage	dsDNA	65	0.03542	79	0.02900				
Semliki forest virus	ssRNA	25	0.09210						
Simian virus 40	dsDNA	83	0.02768						
Sindbis virus	ssRNA	66	0.03501			22	0.10400		
Vaccinia virus	dsDNA	18	0.12454			2	1.34650		
VEE	ssRNA	55	0.04190						
Vesicular Stomatitis virus	ssRNA	12	0.19440						
WEE	ssRNA	54	0.04300						

4.3.3 Fungi Rate Constants

Fungi can exist in at least two major forms – spores and vegetative cells or yeast. Fungal spores are resistant to UV and have low UV rate constants while the vegetative forms are about as susceptible to UV as bacterial cells. Table 4.4 lists the various studies on fungi in water, air, and on surfaces. There are no quantified high relative humidity studies on fungi, and only low RH results are shown. Little or no cyclobutane pyrimidine dimers are formed in UV-irradiated spores but azetane, the spore photoproduct, is the dominant photoproduct.

Figure 4.4 illustrates the variation of UV susceptibility in terms of the D_{90} values in various media. Fungal spores are about nine times more resistant than vegetative fungi and yeast in air, and almost three times more resistant than vegetative fungi and yeast in water. Since there are a limited number of data sets for surface inactivation, the generalized average for surfaces in Fig. 4.8 may not be representative.

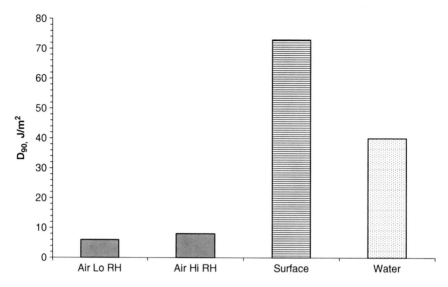

Fig. 4.3 Comparison of overall averages for virus D_{90} values in various media

4.4 Surface UV Rate Constants

The usefulness of surface rate constants is evident for equipment sterilization and for disinfection of floors, walls, ductwork, and cooling coils. The inactivation rates of microbes on surfaces may vary depending on the type of surface. Some metals, for example, have biocidal effects. Many published studies on UV inactivation of microbes involved irradiation on petri dishes or other solid materials, and therefore these results can be considered applicable to surface disinfection. In fact, since many surface rate constant tests were performed in open air, they tend to resemble airborne rate constants more than those in water. Even so, the presence of microbes adjacent to any surfaces appears to confer some additional protection above that of airborne microbes. The question of whether microbes on surfaces are more UV resistant than those in water is difficult to answer with the current database since there is limited correspondence between those species tested on surfaces and those tested in water. Additionally, the protective effect of surfaces is partly dependent on the type of surface. This effect was illustrated previously in Fig. 4.3, and further comparisons are shown in Fig. 4.5 where it can be observed that the UV susceptibility on surfaces can be greater or lesser than in water.

Although the UV susceptibility of microbes on surfaces may be higher or lower than it is in water, for any given species, the differences do not seem so great that it would invalidate the use of water-based rate constants to predict surface disinfection rates. That is, water-based UV rate constants are a reasonable substitute for surface rate constants when the latter are unavailable.

Table 4.3 previously summarized the average or representative UV rate constants on surfaces for a number of bacteria, including many airborne bacteria. Refer to

Table 4.4   Average UV rate constants for fungi

Fungi	Type	Water		Surface		Air - Lo RH		Air Hi RH	
		D_{90} J/m^2	UVGI k m^2/J	D_{90} J/m^2	UVGI k m^2/J	D_{90} J/m^2	UVGI k m^2/J	D_{90} J/m^2	UVGI k m^2/J
Aspergillus amstelodami	Sp	700	0.00329	258	0.00892	669	0.00344		
Aspergillus flavus	Sp	853	0.00270	349	0.00660				
Aspergillus fumigatus	Sp			866	0.00266				
Aspergillus fumigatus	Veg			560	0.00411				
Aspergillus niger	Veg	1245	0.00185						
Aspergillus niger	Sp	1251	0.00184			3984	0.00058		
Aspergillus versicolor	Sp					158	0.01453	384	0.00600
Blastomyces dermatitidis	VegY	140	0.01645						
Botrytis cinerea	Sp	250	0.00920						
Candida albicans	VegY	285	0.00808	374	0.00615				
Candida parapsilosis	VegY	98	0.02360						
Cladosporium herbarum	Sp	50	0.04605	189	0.01220	622	0.00370		
Cladosporium trichoides	Veg			748	0.00308				
C. sphaerospermum	Sp					1096	0.00210		
Cladosporium wernecki	Sp			997	0.00231				
Cryptococcus neoformans	Sp			185	0.01246				
Curvularia lunata	Veg			560	0.00411				
Eurotium rubrum	Sp	434	0.00531						
Fusarium oxysporum	Sp	162	0.01420						
Fusarium solani	Sp	313	0.00735						
Fusarium spp.	Sp	560	0.00411						
Fusarium spp.	Veg	1120	0.00206						
Histoplasma capsulatum	Veg	140	0.01645						
Monilinia fructigena	Sp	167	0.01380						
Mucor mucedo	Sp	600	0.00384	180	0.01280	577	0.00399		
Mucor spp.	Sp	187	0.01234						
Penicillium chrysogenum	Sp	400	0.00576	148	0.01560	750	0.00307		
Penicillium corylophilium	Sp	381	0.00604						
Penicillium digitatum	Sp	321	0.00718						
Penicillium italicum	Sp	202	0.01140						
Penicillium spp.	Sp			2240	0.00103				
Penicillium spp.	Veg			280	0.00822				
Rhizopus nigricans	Sp	3000	0.00077	173	0.01330	267	0.00861		
Rhizopus oryzae	Sp	4480	0.00051						
Rhodotorula spp.	VegY	1120	0.00206						
Scopulariopsis brevicaulis	Sp	125	0.01840	226	0.01020			669	0.00344
Sporotrichum schenkii	VegY			280	0.00822				
Stachybotrys chartarum	Sp			5575	0.00041				
Torula bergeri	Veg			4480	0.00051				
Torula sphaerica	VegY					23	0.09986		
Torula sphaerica	VegY			78	0.02940				
Trichophyton rubrum	Veg			560	0.00411				
Trichophyton rubrum	Sp			560	0.00411				
Ustilago zeae	VegY			1120	0.00206				

Appendix A for the reference sources from which the values or averages were computed. The surfaces in which these microbes were tested are usually Petri dishes, but in some cases other types of surfaces, such as metal, skin, or various media, were used and readers should consult the reference sources for additional information on surface type. Certain surfaces, such as copper or silver, are naturally biocidal and their effects may be additive with UV exposure (Thurman and Gerba 1989). Some surfaces have irregularities and roughness at the microscopic level and these may offer limited protection against UV exposure. The surfaces of food may offer varying degrees of UV protection to microbes (Yaun and Summer 2002). Some surfaces, like organic paints, may actually encourage surface colonization by fungi (Ahearn et al. 1991).

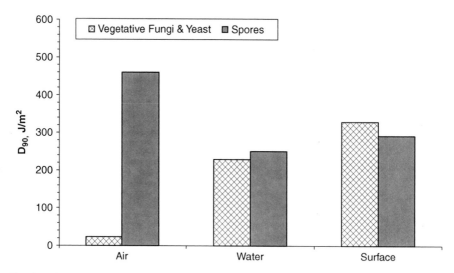

Fig. 4.4 Comparison of D_{90} values for fungal spores and vegetative fungi (including yeast) in various media

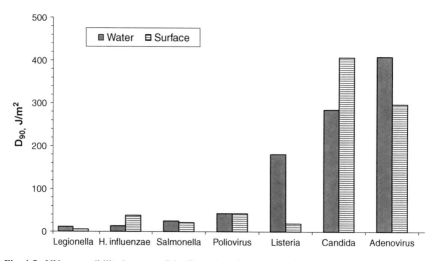

Fig. 4.5   UV susceptibility in terms of the D_{90} values for various microbes in water and on surfaces

4.5 Water Rate Constants

The susceptibility of microbes in water and in other solutions has been studied extensively and the UV rate constants for many microorganisms have been established with fair reproducibility. Table 4.2 previously summarizes the average or representative UV rate constants in water for a number of bacteria, including many airborne bacteria. Figure 4.6 illustrates the differences between the D_{90} values

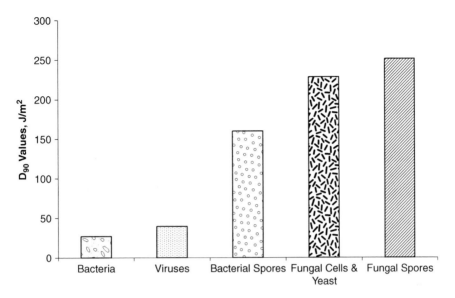

Fig. 4.6 Comparison of the D_{90} values for various microbe types in water

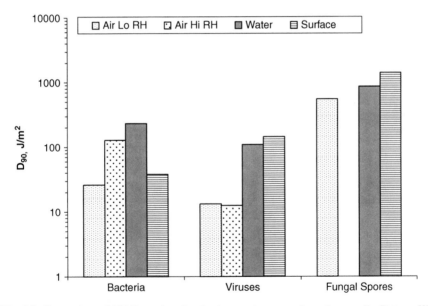

Fig. 4.7 Comparison of UV D_{90} values for the three major groups in various media. Data on Hi RH in air for fungal spores is lacking

(computed from average rate constants) for the microbial groups. The value for the bacterial spores may be anomalous due to the paucity of data for bacterial spores in water.

Figure 4.7 shows the variation of rate constants in different media. It can be seen that the D_{90} is less in air than in water, and that surface D_{90} values are similar to water values, except that the scarcity of surface rate constant data renders the comparisons somewhat uncertain.

4.6 Airborne Rate Constants

The sensitivity of airborne microbes to UV is much greater than that of microbes in suspensions or in films on the surface of agar plates (Webb 1965). For the representative examples shown in Fig. 4.8, bacteria are about five times more resistant in water than in air at low humidity while viruses are about three times more resistant. The exact ratios can vary considerably between microbial species.

The reasons that microbes are more susceptible in air than in water may involve several factors. Water absorbs UV but water-based rate constant studies generally account for the absorptivity of water. There is considerably more turbulence and diffusion in air than in water due to the much lower friction and this would cause tumbling and more mixing, which would ensure that airborne populations are more evenly exposed. The very process of aerosolization may reduce microbial survival potential through physical damage. Oxygenation in air may contribute to increased vulnerability, and so will dehydration at low humidity. Humidity itself plays a part in determining the DNA conformation and therefore affects the survival of bacteria. Relative humidity is addressed in more detail in the following section.

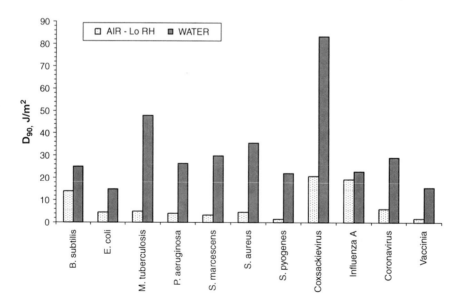

Fig. 4.8 Comparison of air and water UV susceptibility for representative bacteria and viruses

Regardless of the exact mechanism by which microbes are more vulnerable to UV inactivation in air, the studies summarized in the Appendices show that in virtually all cases, and with only one or two anomalous exceptions, that UV rate constants are lower in water than in air, even at 100% RH, often by factors of two or more. Therefore, it is conservative to use water-based UV rate constants as substitutes for airborne rate constants if the latter are unavailable. Similarly, surface rate constants may also be used conservatively in place or airborne data when such is unavailable.

4.7   Narrow Band UVC vs. Broadband UVB/UVC

The inactivation of microbes by UV radiation is a wavelength dependent phenomenon with a peak occurring at about 265 nm. Low pressure (LP) mercury lamps produce a narrow band of UVC at about 254 nm, while medium pressure (MP) mercury lamps produce a broader but flatter spectrum that extends from about 200 nm to 400 nm. Narrow band LP lamps are often referred to as monochromatic while broadband MP lamps are called polychromatic. Wavelengths above and below the peak of 265 nm are capable of inducing inactivation to a degree that depends on the absorption spectrum. The UV spectrum of lamps can influence photoreactivation in bacteria, but not viruses, which have no significant capacity for independent photoreactivation (Bishop et al. 1967).

UV rate constants are generally treated without regard to the spectrum of the source, and this approach usually results in no significant differences in inactivation rates, but there are some qualitative differences between the effects of narrow band UVC and broadband UVB/UVC exposure, especially for bacteria. The broad spectrum of medium pressure UV lamps is capable of damaging enzymes in bacteria that facilitate photoreactivation, with the result that the apparent or effective UV rate constant is higher. Zimmer and Slawson (2002) and Oguma et al. (2001) studied the effects of low pressure and medium pressure UV lamps on *E. coli* photoreactivation. Both studies demonstrated photorepair in *E. coli* after irradiation with low pressure lamps but showed virtually no photorepair after exposure to the broad band medium pressure lamps. Bohrerova and Linden (2006) studied *Mycobacterium terrae* under both LP and MP lamps and found no significant differences in inactivation or in their photoreactivation. Shin (2008) studied *Mycobacterium avium* under MP and LP lamps and found very little difference in UV susceptibility. Linden et al. (2007) reports that medium pressure UV lamps were four times more effective than low pressure lamps in inactivating adenoviruses. The latter effect was explained in part by the increased action spectrum below 230 nm.

It has been demonstrated in at least one study that UVB (280–320 nm) can inactivate viruses at doses that are similar to UVC inactivation doses. Data from Duizer et al. (2004) indicates a UVB D_{90} for canine calicivirus and feline calicivirus at about 113 J/m^2 for both. The UVC D_{90} for these viruses averages to 67 J/m^2 for the former and 185 J/m^2 for the latter.

Inspection of the sources for the UV rate constants tabulated in Appendices A through C indicates that the overwhelming number of tests used low pressure lamps with narrow band output around 253.7 nm. In these cases the D_{90} values and UV rate constants are therefore relative to LP lamps, but it would also appear that using these values for MP lamps may be conservative in some cases, and not significantly different in others. Until more studies are available detailing the differences between LP and MP lamps on a larger array of specific microorganisms, it is difficult to say whether any one type of lamp should be preferred over another or to what degree it may affect results.

4.8 Relative Humidity Effects

The effect of relative humidity (RH) on the UV susceptibility of microorganisms has been studied for decades but no general explanation has emerged that applies to all microbes, and no predictive methods have previously been developed. Most bacteria tested tend to experience decreased UV susceptibility at high RH, but some show the opposite effect, or no significant effect at all. Viruses show mixed results, with some experiencing a small increase in UV susceptibility with increasing RH. Webb (1965) found that airborne Pigeon pox virus was extremely hardy and resisted inactivation from variations in RH, while Rous sarcoma virus was largely inactivated at 30% humidity. Both viruses survived well at 80% RH. Spores show little, if any, response to changes in RH, although data is still quite limited.

Aerosolized microorganisms invariably lose water until they reach equilibrium with the ambient water content of the air, as measured by the relative humidity. Desiccation from low RH can cause the death of bacteria due to the removal of water, which is essential for normal bacterial cellular functions. Death from dehydration is due to a complex interplay of factors and dried bacteria die from simultaneous stresses that mainly relate to oxidation processes (Mitscherlich and Marth 1984). Due to the effects of relative humidity on conformational stability in DNA, microbes may go through a phase transition in which survival is lower in midrange humidities (Cox 1987). The A DNA conformation is preferentially formed under low humidity and this change results in water molecules being desorbed from DNA due to the hydrophobic nature of the tighter internal channels where water molecules normally attach in the B DNA conformation (Neidle 1999). All of these processes are, to some degree, species-dependent.

Microbes can increase in size and mass under higher humidity due to internal absorption (or desorption) and the binding of water molecules to the cell surface (Reponen et al. 1996). Peccia et al. (2001a) found no increase in aerodynamic diameter of *S. marcescens* with increasing RH, only a mass increase, but Lai et al. (2004) and Ko et al. (2000) found the mean diameter increased with RH for the same microbe. Johnson et al. (1999) evaluated the size distributions of aerosolized microbes and found that the effect on *Bacillus subtilis* was much greater than for *Pseudomonas fluorescens* under identical conditions, suggesting a species-related difference in hygroscopic growth. Kundsin (1966) shows data on the count median diameter of aerosolized *Mycoplasma pharyngis*, which suggests an increase in size between 25% and 60% RH. Harper (1961) studied the effects of RH on aerosolized viruses, and found that Vaccinia, Influenza, and VEE experienced small increases in die-off at high humidity, while the die-off for poliomyelitis peaked at about 50% RH and decreased above and below this point. Based on this limited data, the effect of RH on the natural decay of viruses in air would appear to be a species-dependent phenomenon.

The medium in which a microbe is aerosolized, such as water, buffered solution, saliva, etc.) can impact its response to changes in RH, and some substances, such as inositol, can protect against both RH effects and combined RH-UV effects (Webb 1965). Natural decay constants for microbes in air tend to be low at normal humidities and can generally be neglected when evaluating UV effects. Die-away rate constants in air for viruses are on the order of 0.0047–2.52% per h (Harper 1961). Per Peccia et al. (2001a), natural decay rates were on the order of 2.8 % per h for *M. parafortuitum* and *Serratia marcescens,* and about half that value for *Bacillus subtilis* spores.

Some suggestions have been made as to why RH impact UV susceptibility. The earliest proposition was that moisture in the air absorbed UV as it does in water, but this view has been put to rest by laboratory studies as well as by analysis – there is simply not enough water in air at 100% RH to absorb UV, even over several meters distance. A more plausible hypothesis is that since DNA undergoes a transition from conformation A to conformation B as the RH is increased above about 70%, it may be the conformation change that is responsible for the decrease in the UV rate constant (decrease in susceptibility). However, the DNA conformation change alone cannot account for the effect over the full range of relative humidity.

According to Rahn and Hosszu (1969), DNA films irradiated with UV at RH above 65% show the same photochemical behavior as in solution, while below 65% there is a major decrease in thymine dimers and a concomitant increase in spore photoproduct. These changes are consistent with the conformational change as DNA transitions from the B form to the A form as the RH is reduced. These transitions are sigmoid in appearance and can be modeled separately. Figure 4.9 shows the data on spore photoproduct and thymine dimer yield for DNA films, compared with curve-fit sigmoid models. The mathematical models used to fit the data in Fig. 4.9 are based on Eqs. (3.8) and (3.9) from Chap. 3.

Although there are additional photoproducts that may occur in solution and in the dry state, such as protein crosslinks, that are not included in these models, these will either be minor for most microbes, or they can be modeled with additional terms. High concentrations of spore photoproduct can inhibit dimer formation, and therefore there may be additional species-dependent parameter in Eq. (4.1) that will decrease the yield at low RH. The ratio of thymine dimers to spore photoproducts varies with species and with DNA conformation. Nicholson et al. (1991) found the ratio of spore photoproducts to thymine dimers was 1.12 at low humidity versus 0.27 at high humidity. Rahn and Hosszu (1969) found the ratio to be 1.52 at 0% RH and 0.09 at 100% RH. It should be reiterated that relative humidity may be a function of more than just thymine dimers and spore photoproducts, and that additional types

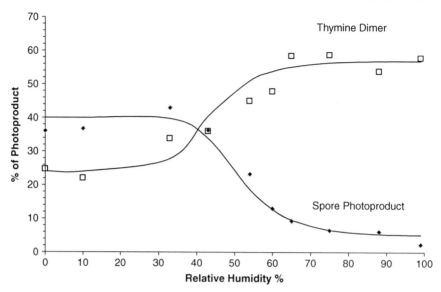

Fig. 4.9 Plots of thymine dimer and spore photoproduct yield in DNA films as a function of RH, based on data from Rahn and Hosszu (1969). *Lines* show sigmoid models that have been curve-fit to the data

of dimers (i.e. cytosine) may have to be considered to create a complete model of RH effects.

An alternate or additional explanation for the decrease in UV rate constants with RH observed for some microbes is that the absorption of water and the layers of bound water that form at high RH produces a protective effect due to the increased scattering of UV light waves. Higher RH may also increase clumping, which may also impact light scattering as well as provide photoprotection to internal cells. For a particle that is already near the size range for Mie scattering, any increase in the size of an airborne microbe, whether due to swelling from water absorption or from clumping could cause a major change in the amount of absorbed UV radiation.

4.9 Rate Constant Determinants

Various intrinsic factors will determine the sensitivity of a given microbe to UV exposure under any set of constant ambient conditions of temperature and humidity (i.e. the normal indoor design range). For non-spore bacteria and viruses these include, but may not be limited to, the following species-dependent properties:

- Physical size
- Molecular weight of DNA or RNA
- DNA Conformation (A or B)
- G+C% and T+A%

- Percentage of Potential Pyrimidine or Purine Dimers
- Presence of chromophores or UV absorbers
- Propensity for clumping or agglutination
- Presence of repair enzymes or dark/light repair mechanisms
- Hydrophilic surface properties
- Relative Index of Refraction in air

The physical size of a microbe bears an uncertain relationship with the UV rate constant, as shown in Fig. 4.10. There would appear to be no definitive relationship for viruses, but for bacteria a loose relationship exists between size and UV rate constant – as bacteria size increases, the UV rate constant seemingly decreases. There is no apparent relationship for fungi. It might be expected that physical size would confer photoprotection through thickness alone, but it seems clear that the composition of the microbial exterior, whether it is a chromophore or not, is a more important determinant. UV scattering effects, which are related to size, are also a factor for large viruses and small bacteria in the Mie scattering range.

Molecular weight has sometimes been cited as a factor in UV susceptibility (David 1973). However, it has been demonstrated that shearing DNA molecules to half size and then irradiating them does not alter the number of pyrimidine dimers or viral DNA inactivation, and this could be considered proof that UV-induced damage to DNA is independent of molecular weight (Scholes et al. 1967). Figure 4.11 illustrates this effect on a macroscopic scale – there is no definitive relationship

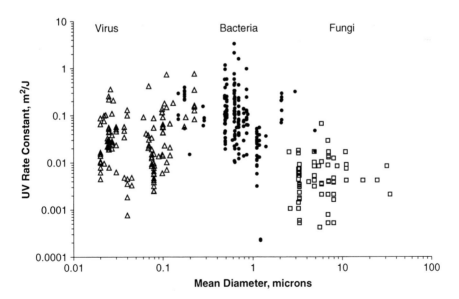

Fig. 4.10 Plot of UV rate constants vs. mean diameter, in which no apparent general relationship between physical size and UV susceptibility can be discerned

between molecular weight (or genome size) and the UV rate constants for the entire array of viruses, bacteria, and fungi that have been studied so far.

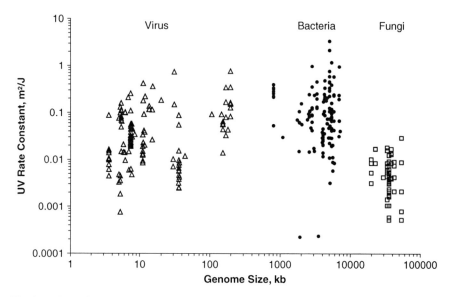

Fig. 4.11 Plot of UV rate constants vs. genome size, in which no general relationship between genome size and UV susceptibility can be discerned

Figure 4.12 shows all viruses from Appendix B in a scatter plot of D_{90} values versus molecular weight in kilobases (where the latter is known). The two major types, single-stranded RNA and double-stranded DNA, are shown together with the minor types, single-stranded DNA and double-stranded RNA. Even if the plot is broken down by type, there is no discernable pattern or relationship with molecular weight within any of the four groups except for dsDNA, for which there is a slight decrease in average D_{90} with molecular weight. The lack of any correlation between molecular weight and UV susceptibility is in accord with various studies like that of Scholes et al. (1967), who demonstrated that there was no relationship between molecular weight and UV-induced damage to DNA. Gerba et al. (2002) suggests that double stranded DNA viruses are likely to be the most resistant to UV. Capsid structure, as well as nucleic acid size, render double-stranded DNA less susceptible to UV inactivation (Thurston-Enriquez et al. 2003). Based on the extensive UV rate constant data in Appendix B, this does appear to be the case, with dsDNA and dsRNA viruses having almost half the UV rate constant of ssRNA and ssDNA viruses. Van der Eb and Cohen (1967) demonstrated that the double stranded version of Polyoma virus DNA was four times more resistant to UV inactivation.

The next criteria worth examining is the genomic G+C or T+A content. The genomic GC content of RNA viruses varies from about 30 to 60%, while that of DNA viruses varies from about 30 to 75%, based on Appendix B. The DNA of bacteria varies from about 25 to 75% (Mooers and Holmes 2000). Since an increased

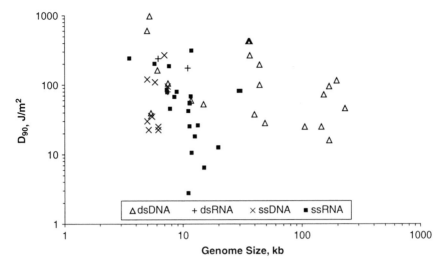

Fig. 4.12 Scatter plot of all virus D_{90} values vs. genome size. Based on Appendix B data for all viruses with known molecular weight

thymine content will likely result in a proportional increase in photodimers of the TT and CT variety, it could be expected that there must be some statistical relationship between G+C% (or conversely with T+A%) and UV susceptibility. Figure 4.13 shows the results of this comparison for single-stranded RNA viruses and double-stranded DNA viruses irradiated in water. These comparison represent averages for virus species – 29 RNA viruses and 24 DNA viruses. The RNA viruses show a correlation with an R^2 of about 24%, but the DNA viruses show no correlation at all. Analysis of 58 bacteria (not shown) from Appendix A produce an R^2 of about 9%. One study on UVB irradiation of bacteria reports a strong correlation between the formation of cytosine-containing photoproducts with increasing GC

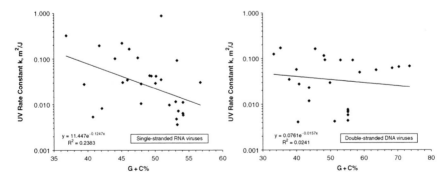

Fig. 4.13 Plots of G+C% vs. UV susceptibility for ssRNA viruses (*Left*) and dsDNA viruses (*Right*). Averages for virus species from Appendix B

content (Matallana-Surget et al. 2008). However, GC content analysis alone would appear to have limited usefulness as a predictor of UV susceptibility.

Although thymine dimers are considered largely responsible for inactivation, the presence of thymine doublets (TT), in which two thymines are adjacent, would seem to be a likely determinant, but it neglects to consider the other doublets (CT, CC) and the triplets. The actual relationship between genomic content and UV susceptibility is somewhat more complicated, and involves the other doublets, triplets, as well as the quadruplets onwards, as will be shown in Sect. 4.11.

4.10 UV Scattering in Water

The relative index of refraction of microbial-sized particles causes incident UV light to be scattered, as was discussed in detail in Chap. 2 (see Sect. 2.13). The reduction of incident UV irradiance for particles whose size parameter is near that of the UVC wavelength (253.7 nm) can be determined from the particle size and Mie scattering models. Since the predictive model (addressed in the following section) is based on data for UV irradiation in water, it is necessary to develop the correction factors for scattering in water before applying it to the predictive model. Table 4.5 summarizes the data used in Fig. 4.14 and provides values for the scattering fraction S_{uv}. The complement of the scattering fraction is a correction factor

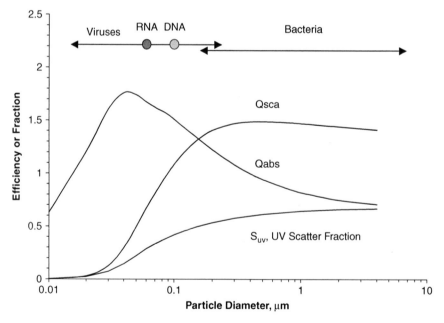

Fig. 4.14 Plot of scattering efficiency (Q_{sca}), absorption efficiency (Q_{abs}), and S_{uv}, the fraction of incident UV scattered in water. *Dark circles* represent logmean diameters for RNA and DNA viruses

Table 4.5 Mie scattering parameters for UV in water

Diameter μm	Size para x	Q_{sca}	Q_{ext}	Q_{abs}	C_{sca} μm²	C_{ext} μm²	C_{abs} μm²	S_{uv} (UV scatter)
0.01	0.17336	0.0020	0.6328	0.630759	1.56E-07	0.000050	0.00005	0.00
0.02	0.34673	0.02950	1.236	1.206502	9.27E-06	0.000388	0.0004	0.02
0.03	0.52009	0.12498	1.7312	1.60622	0.000088	0.00122	0.0011	0.07
0.04	0.69345	0.29531	2.0564	1.76109	0.000371	0.00258	0.0022	0.14
0.05	0.86682	0.4937	2.237	1.7433	0.000969	0.00439	0.0034	0.22
0.06	1.0402	0.6704	2.3456	1.6753	0.001895	0.00663	0.0047	0.29
0.07	1.2135	0.8089	2.431	1.6221	0.0031	0.0094	0.0062	0.33
0.08	1.3869	0.9173	2.5039	1.5866	0.0046	0.0126	0.0080	0.37
0.09	1.5603	1.0074	2.5585	1.5511	0.0064	0.0163	0.0099	0.39
0.1	1.7336	1.0841	2.5938	1.5097	0.0085	0.0204	0.0119	0.42
0.12	2.0804	1.2005	2.6316	1.4311	0.0136	0.0298	0.0162	0.46
0.15	2.6004	1.3103	2.6499	1.3396	0.0232	0.0468	0.0237	0.49
0.2	3.4673	1.4048	2.6283	1.2235	0.0441	0.0826	0.0384	0.53
0.3	5.2009	1.4695	2.5501	1.0806	0.1039	0.1803	0.0764	0.58
0.4	6.9345	1.4855	2.4824	0.9969	0.1867	0.3120	0.1253	0.60
0.5	8.6682	1.4881	2.4298	0.9417	0.2922	0.4771	0.1849	0.61
0.8	13.869	1.4782	2.3289	0.8507	0.7430	1.1706	0.4276	0.63
0.9	15.603	1.4738	2.3064	0.8326	0.9376	1.4673	0.5297	0.64
1	17.336	1.4697	2.2873	0.8176	1.1543	1.7964	0.6421	0.64
2	34.673	1.4392	2.1854	0.7462	4.5214	6.8656	2.3442	0.66
4	69.345	1.411	2.1179	0.7069	17.731	26.614	8.8830	0.67

which is applied to the incident irradiance (actually to the UV dose) to determine the effective UV dose to which they are subject (Kowalski et al. 2009a). These correction factors are used in the predictive model detailed in the following section. It should be noted that for larger viruses that have an envelope, secondary UV scattering effects may also occur in the nucleocapsid, but these effects are ignored in the current model. All viruses are assumed to be spherical – almost all of the viruses addressed in the model are either spherical or ovoid, with one or two rod-like exceptions.

Little is known about the refractive index of viruses in the UV range. Balch et al. (2000) found the median refractive index of four viruses to be 1.06 in visible light. Water has a refractive index of $n_m = 1.4$ in the ultraviolet range. If we scaled the refractive index of viruses (Balch's value) to that of water (from visible to UV), the estimated real refractive index would be 1.06(1.4/1.33) = 1.12. For the imaginary component we can reasonably assume a value comparable to that of water, $k=1.4$. These values were used as input to a Mie Scattering program (Prahl 2009) to estimate the effects of UV scattering at the wavelength of 253.7 nm, and with negligible concentrations (0.000001 spheres/μm³). The program output was used to generate Fig. 4.14 for water (compare with Fig. 2.16 in Chap. 2 for air) which shows the UV scattering efficiency, the absorption efficiency, and the scattering fraction.

4.11 Prediction of UV Susceptibility

The previous review of relationships between UV susceptibility has highlighted some dead ends, some information gaps, and some promising approaches. Prediction of UV susceptibility by taxonomic methods has produced limited results (Lytle and Sagripanti 2005). The most successful approach to predicting UV susceptibility so far has been the work of the author (Kowalski et al. 2009a, b) and is based largely on the seminal works of those researchers discussed herein. To review, the disruption of normal DNA processes occurs as the result of the formation of photodimers, but not all photoproducts appear with the same frequency. Purines are approximately ten times more resistant to photoreaction than pyrimidines (Smith and Hanawalt 1969). Setlow and Carrier (1966) reports that TT dimers are the most numerous of the cyclobutane pyrimidine dimers at high UV doses. Minor products other than CPD dimers, such as interstrand cross-links, chain breaks, and DNA-protein links occur with much less frequency, typically less than 1/1000 of the number of cyclobutane dimers and hydrates may occur at about 1/10 the frequency of cyclobutane dimers. Although irradiated vegetating cells produce large amounts of cyclobutane pyrimidine dimers, thymine-containing photoproducts isolated from bacterial spores do not include cyclobutane pyrimidine dimers but include spore photoproducts. Spore photoproducts decrease when the spore transforms to a vegetative state and thymine dimers increase. For DNA, the thymine dimers decrease under dry conditions (A-DNA) and the spore photoproduct is formed and can become the dominant photoproduct (Rahn and Hosszu 1969). The rate of spore photoproduct formation is unaffected by high concentrations of thymine dimers but high concentrations of spore photoproduct inhibit dimer formation. The spore photoproduct is favored in dry DNA, in spores, and in dsRNA, which are in A-DNA conformation.

Thymine (or pyrimidine) % content alone is not useful as a predictor of UV rate constants because the exact sequence of DNA bases plays a major role in determining photoreactivity of the thymine and other bases. Wang (1964) first suggested that dimerization is favored when adjacent pyrimidine triplets in ice are suitably oriented and positioned. The effect of base composition can impact the intrinsic sensitivity of DNA to UV irradiation (Smith and Hanawalt 1969). The specific sequence of adjacent base pairs, as well as the frequency of thymines, can be determinants of UV sensitivity. Setlow and Carrier (1966) stated that the probability of photodimerization is approximately proportional to the nearest-neighbor frequencies of the various pyrimidine sequences.

Some 80% of pyrimidines and 45% or purines form UV photoproducts in double-stranded DNA, per studies by Becker and Wang (1989), who also showed that purines only form dimers when adjacent to a pyrimidine doublet. The formation of purine dimers requires transfer of energy in neighboring pyrimidines, and will only occur on the 5′ side of the purine base, hence only 50% of these can form dimers.

Becker and Wang (1985) formulated some simple rules for sequence-dependent DNA photoreactivity as follows:

1. Whenever two or more pyrimidine residues are adjacent to one another, photore-actions are observed at both pyrimidines.
2. Non-adjacent pyrimidines, surrounded on both sides by purines, exhibit little or no photoreactivity.
3. The only purines that readily form UV photoproducts are those that are flanked on their 5' side by two or more contiguous pyrimidine residues.

These rules can be used to extract information from DNA and RNA genomes and will enable computation of the relative probability of photoreactions taking place, a parameter that can be directly compared to UV rate constants as a possible predictor. Table 4.6 summarizes these rules in terms that can be computed numerically.

First we address RNA viruses, which are simpler to model. The dimer proba-bilities are distributed in a circular cross-section of a spherical model of the RNA which is assumed to be subject to a collimated beam of UV irradiation corrected for UV scatter (Kowalski et al. 2009a). Since the physical size of the nucleic acid is directly proportional to the genome size, assuming it is tightly packed in a sphere, the number of base pairs represents the volume, and a cube root of the square is taken to represent the surface area of the cross-section, as illustrated in Fig. 4.15.

The sum of the various potential dimers in the genome can be divided by the cross-sectional area of the DNA sphere to obtain a probability density. Theoretically this density would form a Gaussian distribution as illustrated in Fig. 4.15, but for the purposes of this model it is considered to be distributed evenly across the cross-sectional area. The probability function in the numerator was found by trial and error to yield the best possible fit at an exponent of exactly 0.5, or a square root. Distributing the square root of the probability over the circular cross-section yields a probability density function written as follows:

$$D_v = \frac{\left[\sum tt + F_a \sum \overleftrightarrow{ct} + F_b \sum cc + F_c \sum \overleftrightarrow{YYU}\right]^{0.5}}{\sqrt[3]{BP^2}} \tag{4.2}$$

where

Table 4.6 Potential dimerization sequences

Group	DNA sequence				Dimer
Adjacent pyrimidines	TT	TC	CT	CC	Yes
Purines flanked by doublets	ATT	ACC	ACT	ATC	50% Yes
	GTT	GCC	GCT	GTC	50% Yes
	TTA	CCA	CTA	TCA	50% Yes
	TTG	CCG	CTG	CGT	50% Yes
Surrounded pyrimidines	ATA	ATG	GTA	GTG	No
	ACA	ACG	GCA	GCG	No

D_v = dimerization probability value
tt = thymine doublets (same as TT in the text)
cc = cytosine doublets (same as CC in the text)
$\overleftrightarrow{ct}$= ct and tc (both ways, exclusive, aka CT and TC)
$\overleftrightarrow{YYU}$= purine w/ adjacent pyrimidine doublet (both ways, exclusive)
BP = total base pairs
F_a, F_b, F_c = dimer proportionality constants

The dimer proportionality constant for TT is assumed to be unity in Eq. (4.2). Some evidence is available in the literature to allow estimates of the dimer proportionality constants in DNA. Meistrich et al. (1970) indicate that in *E. coli* DNA, the proportions of TT dimers, CT dimers, and CC dimers are in the ratio 1:0.8:0.2, as did Lamola (1973). Per Setlow and Carrier (1964) the average for three bacteria is 1:0.25:0.13. Patrick (1977) suggests ratios of 1:1:1. Unrau et al. (1973) found the ratio was 1:0.5:0.5.

Figure 4.16 shows a plot of Eq. (4.2) applied to the genomes of RNA viruses (Kowalski et al. 2009a, b). Genomes were obtained from the NCBI database (NCBI 2009). The D_{90} values were computed from the average of all the water-based rate constants (in Appendix B) for which genomes were available – see Kowalski et al. (2009b) for a detailed list of the RNA viruses. No available data sets from Appendix B were excluded if genomes were available except for one extreme outlier (HTLV-1). All of the average D_{90} values were corrected for UV scattering per Table 4.5. These results indicate a fairly definitive relationship across the entire size range. The dimer proportionality constants were adjusted by trial and error to produce the best fit to the data and these values were: 1:0.1:6:4, or F_a=0.1, F_b=6, F_c=4. The UV scattering corrections had little impact (~1%) on these results, as would be expected from their size.

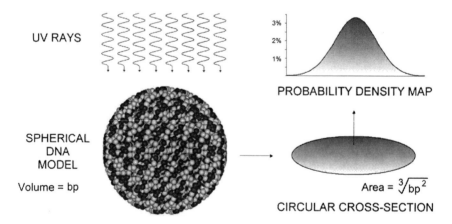

Fig. 4.15 Illustration of the genomic model components, including a figurative probability density map of the cross-sectional area of the spherical DNA model

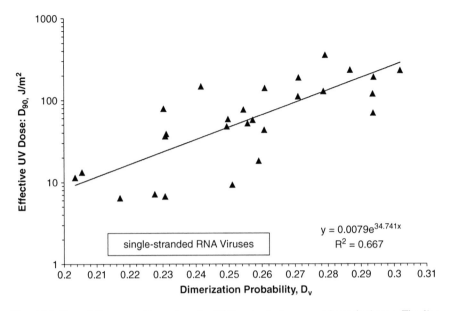

Fig. 4.16 Plot of D_v versus D_{90} values for RNA animal viruses and bacteriophages. The *line* represents a curve fit (equation shown on graph), fit to 27 viruses, representing 70 data sets for UV irradiation tests in water

Application of the model to dsDNA viruses requires some modifications to the equation. DNA is double stranded, with a "template" strand and a complementary strand. The template strand will be accounted for in Eq. (4.2) but the complementary strand is not. However, a TT doublet in the complementary strand will be represented by an AA doublet on the template strand, and so counting base pairs can be done with the template strand alone, by considering the complementary bases. Incorporating the complementary strand bases produces the following equation (Kowalski et al. 2009b):

$$D_v = \frac{\left[\sum (tt + aa) + F_a \sum (\overleftrightarrow{ct} + \overleftrightarrow{ag}) + F_b \sum (cc + gg) + F_c \sum \left(\overleftrightarrow{YYU} + \overleftrightarrow{UUY}\right)\right]^{0.5}}{\sqrt[3]{BP^2}}$$

(4.3)

where

aa = adenine doublets (representing thymine complement)

$\overleftrightarrow{ag}$= ag and ga (both ways, exclusive, aka AG and GA)

gg == guanine doublets (representing cytosine complement)

$\overleftrightarrow{UYY}$= UUY and YUU (both ways, exclusive)

In addition to the doublets and triplets, it was found that the quadruplets onwards also contributed to the model (which they did not for RNA viruses). The effect of the quadruplets, quintuplets, sextuplets, septuplets, and octuplets onwards can be characterized by a factor that accounts for hyperchromicity (see Chap. 2). When multiple pyrimidines, especially thymines, occur sequentially they appear to increase the probability of dimers occurring by a factor of about 2, with the effect leveling off at about 8–10 pyrimidines in a row. Although not enough is known about the hyperchromic effect to quantify it, a factor can be added to Eq. (4.3) to increase the probability of dimerization whenever 3 or more pyrimidines are found in sequence. Since isolated doublets and triplets are already addressed their hyperchromicity factor can be assumed equal to 1. The value of the hyperchromicity factor can then be estimated by curve-fitting the data to obtain the best fit. In the present model the factor linearly increases the probability of dimerization for doublets and triplets based on how many adjacent pyrimidines are present in the genome, up to a value of 8 in a row. Multiples beyond 8 are rare and since they are partly accounted for (every multiple of 9 in a row contains an 8 in a row, etc) they are ignored for the purposes of the current model. Each contribution can be defined as follows:

tt_n = No. of tt doublets within n pyrimidines (template strand)
tc_n = No. of tc doublets within n pyrimidines (template strand)
cc_n = No. of cc doublets within n pyrimidines (template strand)
UYY_n = No. of UYY triplets within n pyrimidines (template strand)
aa_n = No. of aa doublets within n purines (complement strand)
ag_n = No. of ag doublets within n purines (complement strand)
gg_n = No. of gg doublets within n purines (complement strand)
UUY_n = No. of UUY triplets within n purines (complement strand)

The equations for assigning the increase in probability due to hyperchromicity can then be written in terms of the hyperchromicity factor H as follows:

$$tt_h = H \sum_{n=3}^{8} (n \cdot tt_n) \tag{4.4}$$

$$aa_h = H \sum_{n=3}^{8} (n \cdot aa_n) \tag{4.5}$$

$$tc_h = H \sum_{n=3}^{8} (n \cdot tc_n) \tag{4.6}$$

$$ag_h = H \sum_{n=3}^{8} (n \cdot ag_n) \tag{4.7}$$

$$cc_h = H \sum_{n=3}^{8} (n \cdot cc_n) \tag{4.8}$$

$$gg_h = H \sum_{n=3}^{8} (n \cdot gg_n) \tag{4.9}$$

$$UYY_h = H \sum_{n=3}^{8} (n \cdot UYY_n) \tag{4.10}$$

$$UUY_h = H \sum_{n=3}^{8} (n \cdot UUY_n) \tag{4.11}$$

where tt_h = hyperchromic multiplier, or increase in probability of dimerization from all multiple sequences of 3–8 pyrimidines. Similar for all other hyperchromic constants aa_h, tc_h, ag_h, cc_h, gg_h, UYY_h, and UUY_h.

This model of hyperchromicity effects is simplistic. In fact the function is more likely to be logarithmic in nature than linear, but the current model works sufficiently well, at least for DNA viruses – although it provides no added benefit for the RNA model. Equation (4.2) is therefore re-written for DNA as follows:

$$D_v = \frac{\left[\begin{array}{l} \sum (tt + aa + tt_h + aa_h) + F_a \sum (\overleftrightarrow{ct} + \overleftrightarrow{ag} + \overleftrightarrow{ct}_h + \overleftrightarrow{ag}_h) + \\[2mm] F_b \sum (cc + gg + cc_h + gg_h) + F_c \sum \left(\overleftrightarrow{YYU} + \overleftrightarrow{UUY} + \overleftrightarrow{YYU}_h + \overleftrightarrow{UUY}_h \right) \end{array} \right]^{0.5}}{\sqrt[3]{BP^2}} \tag{4.12}$$

where

H = hyperchromicity factor
YY = pyrimidine doublets (both ways exclusive)

The proportionality constants represent the relative proportions of each type of dimer, which differ in RNA and DNA. Applying this model to DNA viruses produces the result shown in Fig. 4.17 (Kowalski et al. 2009a). All DNA animal viruses and bacteriophages with known UV rate constants and genomes were included – see Kowalski (2009b) for a detailed list. All of the D_{90} values were corrected for UV scattering per Table 4.5. The dimer ratios for this curve fit were 1:0.05:40:18 (F_a=0.05, F_b=40, F_c=18), with a hyperchromicity factor $H = 0.67$. It should be noted that since the hyperchromicity factor adds on to the existing doublets and triplets it is the same as multiplying them by a factor of 1.67, which is fairly close to the expected factor of 2 estimated previously. Even though DNA viruses tend to be larger and likely have more protective mechanisms (i.e. protein coats), the pattern of increasing D_{90} with increasing values of D_v seems fairly definitive.

Although the R^2 value is not quite as impressive for DNA as it is for RNA, this may be due to the previously mentioned factor of size – DNA viruses are larger

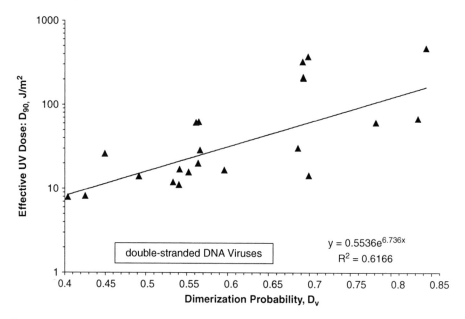

Fig. 4.17 Plot of D_v versus UV rate constant for DNA viruses. The *line* represents a curve fit (equation shown on graph). A total of 77 data sets were used, and these are weighted in the curve fit of the 22 viruses

and may have more innate protection. No available data was omitted from Fig. 4.17 and the only real outliers are two of the four Adenovirus sets at about D_v=0.67. Adenovirus is unusually resistant to UV and may have a chromophore-rich envelope or robust photorepair mechanisms. Adenoviruses also have hemagglutinins on their outer surfaces that may cause them to clump or aggregate which can help protect DNA from UV damage. The aggregation of cells or virions can drastically affect the absorbance through scattering of the incident light (Smith and Hanawalt 1969).

Table 4.7 compares the published estimates of the relative proportions of the various dimer types with the values used in the previous models. The factors shown in the table are the three constants in Eqs. (4.2) and (4.12). The best fit constants are those that were used in the model in the previous figures. The zero values assumed for the constants that were not given by the indicated sources did not have any great influence of the R^2 value. The hyperchromicity factor was zero for all RNA models, and kept at 0.67 for all DNA models. The effects of UV scattering can be seen in the comparisons for the DNA model. Variations of this genomic model are possible, including a version in which the dimer values are multiplied, instead of added as in equations (4.3) and (4.12), and these also provide excellent correlations with UV susceptibility, including high correlations for bacteria on the order of 70% (Kowalski et al. 2009c).

A complete model of microbial inactivation requires consideration of the photoprotective effects afforded by UV absorption in the envelope and nucleocapsid. Photoprotective effects may occur in many bacteria and spores and by incorporating

Table 4.7 Comparison of dimerization proportionality constants

Dimer	Dimer ratio	Factor	Setlow & Carrier (1996)	Meistrich et al. (1970)	Lamola (1973)	Unrau et al. (1973)	Patrick (1977)	Best fit RNA	Best fit DNA
TT		1	1	1	1	1	1	1	1
CT	CT/TT	F_a	0.25	0.8	0.8	0.5	1	0.1	0.05
CC	CC/TT	F_b	0.13	0.2	0.2	0.5	1	6	40
UYY	UYY/TT	F_c	0	0	0	0.5	0	6	18
RNA model R^2 (NS)			61%	60%	60%	64%	62%	66%	–
RNA model R^2			**59%**	**61%**	**61%**	**64%**	**62%**	**67%**	–
Hyperchromicity		H	0.67	0.67	0.67	0.67	0.67	0	0.67
DNA model R^2 (NS)			33%	33%	33%	36%	39%	–	50%
DNA model R^2 (NH)			41%	44%	44%	48%	51%	–	61%
DNA model R^2			**43%**	**46%**	**46%**	**50%**	**53%**	–	**62%**

(NH): No hyperchromicity. (NS): UV scattering not included.

photoprotection factors into a more general model, greater predictive accuracy may
be possible. Since the effect of photoprotection is to reduce the irradiance field, and
thereby reduce the probability of inactivation predicted by the previous Eq. (4.3),
a reduction factor (the overall transmittance) could be added to the equation as
follows:

$$D_v = \tau_P \frac{\left[\sum tt + F_a \sum ct + F_b \sum cc + F_c \sum YYU\right]^{0.5}}{\sqrt{BP^3}} \qquad (4.13)$$

where τ_P = overall transmittance

The value of τ_P will be between zero and one, and represents the reduction of
irradiance between the external UV field and that which strikes the DNA core, or the
transmittance. Although most viruses have such thin capsids that photoprotection is
likely to be nil, this model may apply well to bacteria and spores. The factor can
be applied similarly for Eq. (4.12). The overall photoprotection factor could then
be defined in terms of the various levels of shielding and scattering photoprotection
according to the following relation:

$$\tau_P = S_{uv} \tau_e \tau_n \qquad (4.14)$$

where

S_{uv} = protection factor due to UV scattering
τ_e = protection factor due to shielding by envelope (or cell wall)
τ_n = protection factor due to shielding by nucleocapsid (or cytoplasm)

The protection due to UV scattering Suv has already been addressed and incorporated by correction factors to the UV dose but it forms part of the transmissivity, or
transmittance equation. Additional terms can be added to the previous equations as
necessary, including terms for bacterial photoprotection from the cytoplasm and the
cell walls, and also for possible UV scattering that occurs separately from the envelope (when present) in the nucleocapsid. The equation for transmittance is given by
(Modest 1993):

$$\tau = e^{-(\kappa+\sigma)s} = e^{-\beta s} \qquad (4.15)$$

where

κ = absorption coefficient
σ = scattering coefficient
β = extinction coefficient
s = thickness of material layer

The transmittances τ_e and τ_n could be determined from the capsid and envelope
thicknesses, based on their specific chromophore content or absorptivity, if these are
known. The reduction of transmitted UV irradiance through these components will
be a function of the extinction coefficient of each material layer, which can also be
computed with Eq. (4.15).

References

Aaronson SA. 1970. Effect of ultraviolet irradiation on the survival of simian virus 40 functions in human and mouse cells. J Virol 6(4):393–399.

Abraham G. 1979. The effect of ultraviolet radiation on the primary transcription of Influenza virus messenger RNAs. Virol 97:177–182.

Abshire RL, Dunton H. 1981. Resistance of selected strains of *Pseudomonas aeruginosa* to low-intensity ultraviolet radiation. Appl Envir Microb 41(6):1419–1423.

Ahearn DG, Simmons RB, Switzer KF, Ajello L, Pierson DL. 1991. Colonization by *Cladosporium* spp. of painted metal surfaces associated with heating and air conditioning systems. J Ind Microb 8:277–280.

Ahne W. 1982. Comparative studies of the stability of 4 fish-pathogenic viruses (VHSV, PFR, SVCV, IPNV). Zentbl Vetmed Reihe B 29:457–476.

Albrecht T. 1974. Multiplicity reactivation of human cytomegalovirus inactivated by ultra-violet light. Biochim Biophys Acta 905:227–230.

Allen EG, Bovarnick MR, Snyder JC. 1954. The effect of irradiation with ultraviolet light on various properties of typhus rickettsiae. J Bact 67:718–723.

Alper T, Cramp W, Haig D, Clarke M. 1967. Does the scrapie agent replicate without nucleic acid? Nature (London) 214:764–766.

Antopol SC, Ellner PD. 1979. Susceptibility of *Legionella pneumophila* to ultraviolet radiation. Appl & Environ Microb 38(2):347–348.

Asthana A, Tuveson RW. 1992. Effects of UV and phototoxins on selected fungal pathogens of citrus. Int J Plant Sci 153(3):442–452.

Balch WM, Vaughn J, Novotny J, Drapeau D, Vaillancourt R, Lapierre J, Ashe A. 2000. Light scattering by viral suspensions. Limnol Oceanogr 45(2):492–498.

Ballester N, Malley J. 2004. Sequential disinfection of Adenovirus Type 2 with UV-chlorine-chloramine. J Amer Wat Works Assoc 96(10):97–102.

Barnhart BJ, Cox SH. 1970. Recovery of *Haemophilus influenzae* from ultraviolet and X-ray damage. Photochem Photobiol 11:147–162.

Batch L, Sculz C, Linden K. 2004. Evaluating water quality effects on UV disinfection of MS2 coliphage. J Amer Wat Works Assoc 96(7):75–87.

Battigelli D, Sobsey M, Lobe D. 1993. The inactivation of hepatitis A virus and other model viruses by UV irradiation. Wat Sci Technol 27:339.

Bay PHS, Reichman ME. 1979. UV inactivation of the biological activity of defective interfering particles generated by Vesicular Stomatitis virus. J Virol 32(3):876–884.

Becker MM, Wang Z. 1989. Origin of ultraviolet damage in DNA. J Mol Biol 210:429–438.

Beebe JM. 1959. Stability of disseminated aerosols of *Pastuerella tularensis* subjected to simulated solar radiations at various humidities. J Bacteriol 78:18–24.

Begum M, Hocking A, Miskelly D. 2009. Inactivation of food spoilage fungi by ultraviolet (UVC) irradiation. Int J Food Microbiol 129:74–77.

Bellinger-Kawahara C, Cleaver J, Diener T, Prusiner S. 1987. Purified scrapie prions resist inactivation by UV irradiation. J Virol 61(1):159–166.

Benoit TG, Wilson GR, Bull DL, Aronson AI. 1990. Plasmid-associated sensitivity of *Bacillus thuringensis* to UV light. Appl & Environ Microbiol 56(8):2282–2286.

Benzer S. 1952. Resistance to ultraviolet light as an index to the reproduction of bacteriophage. J Bact 63:59–72.

Bishop JM, Quintrell N, Koch G. 1967. Poliovirus double-stranded RNA: Inactivation by ultraviolet light. J Mol Biol 24:125–128.

Bister K, Varmus HE, Stavnezer E, Hunter E, Vogt PK. 1977. Biological and biochemical studies on the inactivation of Avian Oncoviruses by ultraviolet irradiation. Virology 77(2):689–704.

Bockstahler LE, Lytle CD, Stafford JE, Haynes KF. 1976. Ultraviolet enhanced reactivation of a human virus: Effect of delayed infection. Mutat Res 35:189–198.

Bockstahler LE. 1977. Radiation enhanced reactivation of nuclear replicating mammalian viruses. Photochem Photobiol 25(5):477–482.

Bohrerova Z, Linden KG. 2006. Assessment of DNA damage and repair in *Mycobacterium terrae* after exposure to UV irradiation. J Appl Microbiol 101(5):995–1001.

Bohrerova Z, Shemer H, Lantis R, Impellitteri C, Linden K. 2008. Comparative disinfection efficiency of pulsed and continuous-wave UV irradiation technologies. Wat Res 42:2975–2982.

Boshoff HIM, Reed MB, Barry CE, Mizrahi V. 2003. DnaE2 polymerase contributes to in vivo survival and the emergence of drug resistance in *Mycobacterium tuberculosis*. Cell 113:183–193.

Bossart W, Nuss DL, Paoletti E. 1978. Effect of UV irradiation on the expression of Vaccinia virus gene products synthesized in a cell-free system coupling transcription and translation. J Virol 26(3):673–680.

Bourre F, Benoit A, Sarasin A. 1989. Respective roles of pyrimidine dimer and pyrimidine (6–4) pyrimidone photoproducts in UV mutagenesis of simian virus 40 DNA in mammalian cells. J Virol 63(11):4520–4524.

Bukhari Z, Abrams F, LeChevallier M. 2004. Using ultraviolet light for disinfection of finished water. Wat Sci Technol 50(1):173–178.

Butkus MA, Labare MP, Starke JA, Moon K, Talbot M. 2004. Use of aqueous silver to enhance inactivation of coliphage MS-2 by UV disinfection. Appl Environ Microbiol 70(5):2848–2853.

Butler RC, Lund, V., and Carlson, D. A. 1987. Susceptibility of *Campylobacter jejuni* and *Yersinia enterolitica* to UV radiation. Zbl Vet Med B 29:129–136.

Caballero S, Abad F, Loisy F, LeGuyader F, Cohen J, Pinto R, Bosch A. 2004. Rotavirus virus-like particles as surrogates in environmental persistence and inactivation studies. Appl Environ Microbiol 70(7):3904–3909.

Campbell A, Wallis P. 2002. The effects of UV irradiation on human-derived *Giardia lamblia* cysts. Wat Res 36(4):963–969.

Carlson HJ. 1975. Germicidal lamp inactivation of poliovirus. Am J Publ Health 32:1256–1262.

Casarett AP. 1968. Radiation Biology. Englewood: Prentice-Hall.

Chang JCH, Ossoff SF, Lobe DC, Dorfman MH, Dumais CM, Qualls RG, Johnson JD. 1985. UV inactivation of pathogenic and indicator microorganisms. Appl & Environ Microbiol 49(6):1361–1365.

Chevrefils G, Caron E, Wright H, Sakamoto G, Payment P, Barbeau B, Cairns B. 2006. UV dose required to achieve incremental log inactivation of bacteria, protozoa and viruses. IUVA News 8(1):38–45.

Chick EW, A.B. Hudnell J, Sharp DG. 1963. Ultraviolet sensitivity of fungi associated with mycotic keratitis and other mycoses. Sabouviad 2(4):195–200.

Collier LH, McClean D, Vallet L. 1955. The antigenicity of ultra-violet irradiated vaccinia virus. J Hyg 53(4):513–534.

Collins FM. 1971. Relative susceptibility of acid-fast and non-acid fast bacteria to ultraviolet light. Appl Microbiol 21:411–413.

Cornelis JJ, Su ZZ, Ward DC, Rommelaere J. 1981. Indirect induction of mutagenesis of intact parvovirus H-1 in mammalian cells treated with UV light or with UV-irradiated H-1 or simian virus 40. Proc Natl Acad Sci 78(7):4480–4484.

Cornelis JJ, Rommelaere J. 1982. Direct and indirect effects of ultraviolet light on the mutagenesis of parvovirus H-1 in human cells. EMBO J 1(6):693–699.

Cox CS. 1987. The Aerobiological Pathway of Microorganisms. New York: John Wiley & Sons.

Craik S, Weldon D, Finch G, Bolton J, Belosevic M. 2001. Inactivation of *Cryptosporidium* oocysts using medium- and low-pressure ultraviolet radiation. Wat Res 35:1387–1398.

Crowley DJ, Boubriak I, Berquist BR, Clark M, Richard E, Sullivan L, DasSarma S, McCready S. 2006. The uvrA, uvrB, and uvrC genes are required for repair of ultraviolet light induced DNA photoproducts in *Halobacterium* sp. NRC-1. Saline Systems 2(11):13.

Danner K, Mayr A. 1979. In vitro studies on Borna virus. II. Properties of the virus. Arch Virol 61:261–271.

Darnell MER, Subbarao K, Feinstone SM, Taylor DR. 2004. Inactivation of the coronavirus that induces severe acute respiratory syndrome, SARS-CoV. J Virol Meth 121:85–91.

Das J, Nowak JA, Manilof J. 1977. Host cell and ultraviolet reactivation of ultarviolet-irradiated Mycoplasmaviruses. J Bact 129(3):1424–1427.

David CN. 1964. UV inactivation and thymine dimerization in bacteriophage phiX. Z. Verberbungsl. 95:318–325.

David HL, Jones WD, Newman CM. 1971. Ultraviolet light inactivation and photoreactivation in the mycobacteria. Infect and Immun 4:318–319.

David HL. 1973. Response of mycobacteria to ultraviolet radiation. Am Rev Resp Dis 108: 1175–1184.

Davidovich IA, Kishchenko GP. 1991. The shape of the survival curves in the inactivation of viruses. Mol Gen, Microb & Virol 6:13–16.

Day RS. 1974. Studies on repair of adenovirus 2 by human fibroblasts using normal, xeroderma pigmentosum, and xeroderma pigmentosum heterozygous strains. Cancer Res 34:1965–1970.

de Roda Husman AM, Bijkerk P, Lodder W, Berg Hvd, Pribil W, Cabaj A, Gehringer P, Sommer R, Duizer E. 2004. Calicivirus inactivation by nonionizing (253.7-nanometer-wavelength [UV]) and ionizing (Gamma) radiation. Appl Environ Microbiol 70(9):5089–5093.

DeFendi V, Jensen F. 1967. Oncogenicity by DNA tumor viruses. Science 157:703–705.

Deguchi K, Yamaguchi S, Ishida H. 2005. UV-disinfection reactor validation by computational fluid dynamics and relation to biodosimetry and actinometry. J Wat Technol Environ Technol 3(1):77–84.

Deshmukh D, Pomeroy B. 1968. Ultraviolet inactivation and photoreactivation of avian viruses. Report nr Sci J Minnesota Agricult Exp Station Paper No. 6744.

DiStefano R, Burgio G, Ammatuna P, Sinatra A, Chiarini A. 1976. Thermal and ultraviolet inactivation of plaque purified measles virus clones. G Batteriol Virol Immunol 69:3–11.

Dolman PJ, Dobrogowski MJ. 1989. Contact lens disinfection by ultraviolet light. Am J Opthalmol 108(6):665–669.

Dubunin NP, Zasukhina GD, Nesmashnova VA, Lvova GN. 1975. Spontaneous and induced mutagenesis in western equine encephalomyelitis virus in chick embryo cells with different repair activity. Proc Nat Acad Sci 72(1):386–388.

Duizer E, Bijkerk P, Rockx B, de Groot A, Twisk F, Koopmans M. 2004. Inactivation of caliciviruses. Appl Environ Microbiol 70(8):4538–4543.

Dulbecco R. 1952. A critical test of the recombination theory of multiplicity reactivation. J Bact 63:199–207.

Dulbecco R, Vogt M. 1955. Biological properties of Poliomyelitis viruses as studied by the plaque technique. Ann N Y Acd Sci 61(4):790–800.

Durance CS, Hoffman R, Andrews RC, Brown M. 2005. Applications of Ultraviolet Light for Inactivation of Adenovirus. Toronto, ON: University of Toronto Department of Civil Engineering.

Elasri MO, Miller RV. 1999. Response of a biofilm bacterial community to UV Radiation. Appl & Environ Microbiol 65(5):2025–2031.

EPA. 2006. Biological Inactivation Efficiency by HVAC In-Duct Ultraviolet Light Systems: Steril-Aire, Inc.: U.S. Environmental Protection Agency. Report nr EPA 600/R-06/052.

Fasman, GD 1996. Circular dichroism and the conformational analysis of biomolecules. New York: Plenum Press.

Fletcher LA, Noakes CJ, Beggs CB, Sleigh PA, Kerr KG. 2003. The ultraviolet susceptibility of aerosolised microorganisms and the role of photoreactivation; Vienna. IUVA.

Fletcher L. 2004. The influence of relative humidity on the UV susceptibility of airborne gram negative bacteria. IUVA News 6(1):12–19.

Fluke DJ, Pollard EC. 1949. Ultraviolet action spectrum of T1 bacteriophage. Science 110:274–275.

Fraenkel-Conrat H, Wagner RR. 1981. Comprehensive Virology. New York: Plenum Press.

Freeman AG, Schweikart KM, Larcom LL. 1987. Effect of ultraviolet radiation on the Bacillus subtilis phages SPO2c12, SPP1, and phi 29 and their DNAs. Mut Res 184(3): 187–196.

Frerichs GN, Tweedle A, Starkey WG, Richards RH. 2000. Temperature, pH, and electrolyte sensitivity, and heat, UV and disinfectant inactivation of sea bass (*Dicentrarchus labrax*) neuropathy nodavirus. Aquaculture 185:13–24.

Fulton HR, Coblentz WW. 1929. The fungicidal action of ultraviolet radiation. J Agric Res 38:159.

Furness G. 1977. Differential responses of single cells and aggregates of Mycoplasmas to ultraviolet irradiation. Appl Microbiol 18(3):360–364.

Furuse K, Watanabe I. 1971. Effects of ultraviolet light (UV) irradiation on RNA phage in H2O and in D2O. Virol 46:171–172.

Galasso GJ, Sharp DG. 1965. Effect of particle aggregation on the survival of irradiated Vaccinia virus. J Bact 90(4):1138–1142.

Gates FL. 1929. A study of the bactericidal action of ultraviolet light. J Gen Physiol 13:231–260.

Gates FL. 1934. Results of irradiating *Staphylococcus aureus* bacteriophage with monochromatic ultraviolet light. J Exp Med 60:179–188.

Gerba C, Gramos DM, Nwachuku N. 2002. Comparative inactivation of enteroviruses and adenovirus 2 by UV light. Appl Environ Microbiol 68(10):5167–5169.

Germaine GR, Murrell WG. 1973. Effect of dipicolinic acid on the ultraviolet radiation resistance of *Bacillus cereus* spores. Photochem & Photobiol 17:145–154.

Giese N, Darby J. 2000. Sensitivity of microorganisms to different wavelengths of UV light: Implications on modeling of medium pressure UV systems. Wat Res 34(16):4007–4013.

Gillis HL. 1974. Photoreactivation and ultraviolet inactivation of Mycobacteria in air, MS Thesis. Atlanta, GA: Georgia Technical University.

Gilpin RD, Dillon SB, Keyser P, Androkites A, Berube M, Carpendale N, Skorina J, Hurley J, Kaplan AM. 1985. Disinfection of circulating water systems by ultraviolet light and halogenation. Water Res 19(7):839–848.

Golde A, Latarjet R, Vigier P. 1961. Isotypical interference in vitro by Rous virus inactivated by ultraviolet rays. C R Acad Sci (Paris) 253:2782–2784.

Green CF, Favidson CS, Scarpino PV, Gibbs SG. 2004. Disinfection of selected *Aspergillus* spp. using ultraviolet irradiation. Can J Microbiol 50(3):221–224.

Green CF, Favidson CS, Scarpino PV, Gibbs SG. 2005. Ultraviolet germicidal irradiation disinfection of *Stachybotrys chartarum*. Can J Microbiol 51:801–804.

Griego VM, Spence KD. 1978. Inactivation of *Bacillus thuringiensis* spores by ultraviolet and visible light. Appl Environ Microbiol 35(5):906–910.

Gritz DC, Lee TY, McDonnell PJ, Shih K, Baron N. 1990. Ultraviolet radiation for the sterilization of contact lenses. CLAO J 16(4):294–298.

Guillemain B, Mamoun R, Astier T, Duplan J. 1981. Mechanisms of early and late polykaryocytosis induced by the Bovine Leukaemia virus. J Gen Virol 57:227–231.

Gurzadyan GG, Nikogosyan DN, Kryukov PG, Letokhov VS, Balmukhanov TS, Belogurov AA, Zavilgelskij GB. 1981. Mechanism of high power picosecond laser UV inactivation of viruses and bacterial plasmids. Photochem Photobiol 33:835–838.

Harm W. 1961. Gene-controlled reactivation of ultraviolet-irradiated bacteriophage. J Cell Comp Physiol Suppl 58(1):169.

Harm W. 1968. Effects of dose fractionation on ultraviolet survival of *Escherichia coli*. Photochem & Photobiol 7:73–86.

Harper GJ. 1961. Airborne micro-organisms : Survival tests with four viruses. J Hyg 59:479–486.

Harris GD, Adams VD, Sorenson DL, Curtis MS. 1987. Ultraviolet inactivation of selected bacteria and viruses with photoreactivation of the bacteria. Water Res 21(6):687–692.

Harris MG, Fluss L, Lem A, Leong H. 1993. Ultraviolet disinfection of contact lenses. Optom Vis Sci 70(10):839–842.

Harstad JB, H.M.Decker, A.G.Wedum. 1954. Use of ultraviolet irradiation in a room air conditioner for removal of bacteria. Am Ind Hyg Assoc J 2:148–151.

Havelaar AH. 1987. Virus, bacteriophages and water purification. Vet Q 9(4):356–360.

Hayes SL, White KM, Rodgers MR. 2006. Assessment of the effectiveness of low-pressure UV light for inactivation of *Helicobacter pylori*. Appl Environ Microbiol 72(5):3763–3765.

Hedrick R, McDowell T, Marty G, Mukkatira K, Antonio D, Andree K, Bukhari Z, Clancy T. 2000. Ultraviolet irradiation inactivates the waterborne infective stages of Myxobolus cerebralis: A treatment for hatchery water supplies. Dis Aquatic Org 42(1):53–59.

Helentjaris T, Ehrenfeld E. 1977. Inhibition of host cell protein synthesis by UV-inactivated poliovirus. J Virol 21(1):259–267.

Henderson E, Heston L, Grogan E, Miller G. 1978. Radiobiological inactivation of Epstein-Barr virus. J Virol 25(1):51–59.

Hercik F. 1937. Action of ultraviolet light on spores and vegetative forms of B. megatherium sp. J Gen Physiol 20(4):589–594.

Hijnen WAM, Beerendonk EF, Medema GJ. 2006. Inactivation credit of UV irradiation for viruses, bacteria and protozoan (oo)cysts in water: A review. Wat Res 40:3–22.

Hill WF, Hamblet FE, Benton WH, Akin EW. 1970. Ultraviolet devitalization of eight selected enteric viruses in estuarine water. Appl Microb 19(5):805–812.

Hirai K, Maeda F, Watanabe Y. 1977. Expression of early virus functions in human cytomegalovirus infected HEL cells: Effect of ultraviolet light-irradiation of the virus. J Gen Virol 38:121–133.

Hofemeister J, Bohme H. 1975. DNA repair in Proteus mirabilis. III. Survival, dimer excision, and UV reactivation in comparison with Escherichia coli K12. Mol Gen Genetic 141(2):147–161.

Hollaender A, Oliphant JW. 1944. The inactivating effect of monochromatic ultraviolet radiation on influenza virus. J Bact 48(4):447–454.

Hollaender A. 1955. Radiation Biology, Volume II: Ultraviolet and Related Radiations. New York: McGraw-Hill.

Horneck G, Bucker H, Reitz G. 1985. Bacillus subtilis spores on Spacelab I: Response to solar UV radiation in free space. In: Dring GJ, Ellar DJ, Gould GW, editors. Fundamental and Applied Aspects of Bacterial Spores. London: Academic Press, pp. 241–249.

Hotz G, Mauser R, Walser R. 1971. Infectious DNA from coliphage T1. 3. The occurrence of single-strand breaks in stored, thermally-treated and UV-irradiated molecules. Int J Radiat Biol Relat Stud Phys Chem Med 19:519–536.

Hoyer O. 2000. The status of UV technology in Europe. IUVA News 2:22–27.

Huffman D, Gennccaro A, Rose J, Dussert B. 2002. Low- and medium-pressure UV inactivation of microsporidia Encephalitozoon intestinalis. Wat Res 36(12):3161–3164.

Jagger. 1956. Action spectra of light-restoration in Escherichia coli B/r. Ann Inst Pasteur 91(6):858–873.

Jagger J, Takebe H, Snow JM. 1970. Photoreactivation of killing in Streptomyces: Action spectra and kinetic studies. Photochem Photobiol 12:185–196.

Jensen MM. 1964. Inactivation of airborne viruses by ultraviolet irradiation. Appl. Microbiol. 12(5):418–420.

Jepson JD. 1973. Disinfection of water supplies by ultraviolet radiation. Wat Treat Exam 22:175–193.

Johnson DL, Pearce TA, Esmen NA. 1999. The effect of phosphate buffer on aerosol size distribution of nebulized Bacillus subtilis and Pseudomonas fluorescens bacteria. Aerosol Sci & Technol 30:202–210.

Johnson A, Linden K, Ciociola K, DeLeon R, Widmer G, Rochelle P. 2005. UV inactivation of Cryptosporidium hominis as measured in cell culture. Appl Environ Microbiol 71(5):2800–2802.

Kariwa H, Fujii N, Takashima I. 2004. Inactivation of SARS coronavirus by means of povidone-iodine, physical conditions, and chemical reagents. Jpn J Vet Res 52(3):105–112.

Kassanis B. 1965. Inactivation of a strain of Tobacco necrosis virus and the RNA isolated from it, by ultraviolet radiation of different wavelengths. Photochem Photobiol 4:209–214.

Ke Q, Craik S, El-Din M, Bolton J. 2009. Development of a protocol for the determination of the ultraviolet sensitivity of microorganisms suspended in air. Aerosol Sci Technol 43(4):284–289.

Keller LC, Thompson TL, Macy RB. 1982. UV light-induced survival response in a highly radiation-resistant isolate of the Moraxella-Acinetobacter group. Appl & Environ Microb 43(2):424–429.

Kelloff G, Aaronson SA, Gilden RV. 1970. Inactivation of Murine Sarcoma and Leukemia viruses by ultra-violet irradiation. Virology 42:1133–1135.

Kelner A. 1949. Effect of visible light on the recovery of *Streptomyces griseus* conidia from ultra-violet irradiation injury. Proc Nat Acad Sci 35(2):73–79.

Kethley TW. 1973. Feasibility Study of Germicidal UV Lamps for Air Disinfection in Simulated Patient Care Rooms. San Francisco, CA: American Public Health Association.

Kim T, Silva JL, Chen TC. 2002. Effects of UV irradiation on selected pathogens in peptone water and on stainless steel and chicken meat. J Food Prot 65(7):1142–1145.

Klein B, Filon AR, van Zeeland AA, van der Eb AJ. 1994. Survival of UV-irradiated vaccinia virus in normal and xeroderma pigmentosum fibroblasts; evidence for repair of UV-damaged viral DNA. Mutat Res 307(1):25–32.

Knudson GB. 1985. Photoreactivation of UV-irradiated *Legionella pneumophila* and other *Legionella* species. Appl & Environ Microbiol 49(4):975–980.

Knudson GB. 1986. Photoreactivation of ultraviolet-irradiated, plasmid-bearing, and plasmid-free strains of *Bacillus anthracis*. Appl & Environ Microbiol 52(3):444–449.

Ko G, First MW, Burge HA. 2000. Influence of relative humidity on particle size and UV sensitivity of *Serratia marcescens* and *Mycobacterium bovis* BCG aerosols. Tuber Lung Dis 80(4/5):217–228.

Ko G, Cromenas TL, Sobsey MD. 2005. UV inactivation of adenovirus type 41 measured by cell culture mRNA RT-PCR. Wat Res 39:3643–3649.

Koller LR. 1939. Bactericidal effects of ultraviolet radiation produced by low pressure mercury vapor lamps. J Appl Phys 10:624.

Kowalski WJ. 2001. Design and Optimization of UVGI Air Disinfection Systems [PhD]. State College: The Pennsylvania State University.

Kowalski W, Bahnfleth W, Hernandez M. 2009a. A Genomic Model for the Prediction of Ultraviolet Inactivation Rate Constants for RNA and DNA Viruses; 2009 May 4–5. Boston, MA: International Ultraviolet Association.

Kowalski W, Bahnfleth W, Hernandez M. 2009b. A genomic model for predicting the ultraviolet susceptibility of viruses. IUVA News, 11(2):15–28.

Kowalski W, Bahnfleth W, Hernandez M. 2009c. A genomic model for the prediction of ultraviolet susceptibility of viruses and bacteria. Amsterdam, 2009 Sep 21–22. The Netherlands: International Ultraviolet Association.

Kundsin RB. 1966. Characterization of *Mycoplasma* aerosols as to viability, particle size, and lethality of ultraviolet radiation. J Bacteriol 91(3):942–944.

Lai KM, Burge, H, First, MW. 2004. Size and UV germicidal irradiation susceptibility of *Serratia marcescens* when aerosolized from different suspending media. Appl Environ Microbiol 70(4):2021–2027.

Lamola A. 1973. Photochemistry and structure in nucleic acids. Pure Appl Chem 34(2):281–303.

Latarjet R, Cramer R, Montagnier L. 1967. Inactivation, by UV-, X-, and gamma-radiations, of the infecting and transforming capacities of polyoma virus. Virology 33:104–111.

Lazarova V, Savoye P. 2004. Technical and sanitary aspect of wastewater disinfection by ultraviolet irradiation for landscape irrigation. Wat Sci Technol 50(2):203–209.

Lee JE, Zoh KD, Ko GP. 2008. Inactivation and UV disinfection of Murine Norovirus with TiO2 under various environmental conditions. Appl Environ Microbiol 74(7):2111–2117.

Levinson W, Rubin R. 1966. Radiation studies of avian tumor viruses and of Newcastle disease virus. Virology 28:533–542.

Li D, Craik S, Smith D, Belosevic M. 2007. Comparison of levels of inactivation of two isolates of *Giardia lamblia* cysts by UV light. Appl Environ Microbiol 73(7):2218–2223.

Lidwell OM, Lowbury EJ. 1950. The survival of bacteria in dust. Ann Rev Microbiol 14:38–43.

Liltved H, Hektoen H, Efraimsen H. 1995. Inactivation of bacterial and fish pathogens by ozonation or UV irradiation in water of different salinity. Aquacult Eng 14:107–122.

Liltved H, Landfald B. 1996. Influence of liquid holding recovery and photoreactivation on survival of ultraviolet-irradiated fish pathogenic bacteria. Wat Res 30(5):1109–1114.

Linden KG, Thurston J, Schaefer R, Malley JP. 2007. Enhanced UV inactivation of adenoviruses under polychromatic UV lamps. Appl Environ Microbiol 73(23):7571–7574.

Little JS, Kishimoto RA, Canonico PG. 1980. In vitro studies of interaction of rickettsia and macrophages: Effect of ultraviolet light on *Coxiella burnetti* inactivation and macrophage enzymes. Infect Immun 27(3):837–841.

Lojo MM. 1995. Thymine auxotrophy is associated with increased UV sensitivity in *Escherichia coli* and *Bacillus subtilis*. Mutat Res 347:25–30.

Lovinger GG, Ling HP, Gilden RV, Hatanaka M. 1975. Effect of UV light on RNA directed DNA polymerase activity of murine oncornaviruses. J Virol 15:1273.

Lucio-Forster A, Bowman DD, Lucio-Martinez B, Labare MP, Butkus MA. 2006. Inactivation of the Avian Influenza virus (H5N2) in typical domestic wastewater and drinking water treatment systems. Environ Eng Sci 23(6):897–903.

Luckiesh M. 1946. Applications of Germicidal, Erythemal and Infrared Energy. New York: D. Van Nostrand Co.

Luckiesh M, Taylor AH, Knowles T, Leppelmeier ET. 1949. Inactivation of molds by germicidal ultraviolet energy. J Franklin Inst 248(4):311–325.

Lytle CD. 1971. Host-cell reactivation in mammalian cells. 1. Survival of ultra-violet-irradiated herpes virus in different cell-lines. Int J Radiat Biol Relat Stud Phys Chem Med 19(4):329–337.

Lytle CD, Sagripanti JL. 2005. Predicted inactivation of viruses of relevance to biodefense by solar radiation. J Virol 79(22):14244–14252.

Ma J-F, Straub TM, Pepper IL, Gerba CP. 1994. Cell culture and PCR determination of poliovirus inactivation by disinfectants. Appl Environ Microbiol 60(11):4203–4206.

Malley JP, Ballester NA, Margolin AB, Linden KG, Mofidi A, Bolton JR, Crozes G, Laine JM, Janex ML. 2004. Inactivation of Pathogens with Innovative UV Technologies. Denver, CO: American Research Foundation & American Water Works Association.

Mamane-Gravetz H, Linden KG, Cabaj A, Sommer R. 2005. Spectral sensitivity of *Bacillus subtilis* spores and MS2 coliphage for validation testing of ultraviolet reactors for water disinfection. Environ Sci Technol 39:7845–7852.

Marquenie D, Lammertyn J, Geeraerd A, Soontjens C, VanImpe J, Nicolai B, Michiels C. 2002. Inactivation of conidia of *Botrytis cinerea* and *Monilinia fructigena* using UV-C and heat treatment. Int J Food Microbiol 74:27–35.

Marshall M, Hayes S, Moffett J, Sterling C, Nicholson W. 2003. Comparison of UV inactivation of three *Encephalitozoon* species with that of spores of two DNA repair-deficient *Bacillus subtilis* biodosimetry strains. Appl Environ Microbiol 69(1):683–685.

Martin JP, Aubertin AM, Kirn A. 1982. Expression of Frog Virus 3 early genes after ultraviolet irradiation. Virology 122:402–410.

Martin E, Reinhardt R, Baum L, Becker M, Shaffer J, Kokjohn T. 2000. The effects of ultraviolet radiation on the moderate halophile *Halomonas elongata* and the extreme halophile *Halobacterium salinarum*. Can J Microbiol 46(2):180–187.

Martiny H, Wlodavezyk K, Ruden H. 1988. Use of UV rays for the disinfection of water. II. Microbiological studies of surface water. Zentralbl Bakteriol Mikrobiol Hyg B 186(4):344–359.

Matallana-Surget S, Meador J, Joux F, Douki T. 2008. Effect of the GC content of DNA on the distribution of UVB-induced bipyrimidine photoproducts. Photochem Photobiol Sci 7:794–801.

Maya C, Beltran N, Jimenez B, Bonilla P. 2003. Evaluation of the UV disinfection process in bacteria and amphizoic ameoba inactivation. Wat Sci Technol 3(4):285–291.

McCarthy C, Schaefer J. 1974. Response of *Mycobactrium avium* to ultraviolet irradiation. Appl Environ Microbiol 28(1):151–153.

McClain ME, Spendlove RS. 1966. Multiplicity reactivation of Reovirus particles after exposure to ultraviolet light. J Bact 92(5):1422–1429.

McDevitt JJ, Lai KM, Rudnick SN, Houseman EA, First MW, Milton DK. 2007. Characterization of UVC light sensitivity of vaccinia virus. Appl Environ Microbiol 73(18):5760–5766.

Meistrich M, Lamola AA, Gabbay E. 1970. Sensitized photoinactivation of bacteriophage T4. Photochem Photobiol 11(3):169–178.

Meng Z-D, Birch C, Heath R, Gust I. 1987. Physicochemical stability and inactivation of Human and Simian rotaviruses. Appl Environ Microbiol 53(4):727–730.

Meng QS, Gerba CP. 1996. Comparative inactivation of enteric adenoviruses, Poliovirus and coliphages by ultraviolet irradiation. Wat Res 30(11):2665–2668.

Miller RL, Plageman PGW. 1974. Effect of ultraviolet light on mengovirus: Formation of uracil dimers, instability and degradation of capsid, and covalent linkage of protein to viral RNA. J Virol 13(3):729–739.

Miocevic I, Smith J, Owens L, Speare R. 1993. Ultraviolet sterilization of model viruses important to finfish aquaculture in Australia. Aust Vet J 70(1):25–27.

Mitscherlich E, Marth EH. 1984. Microbial Survival in the Environment. Berlin: Springer-Verlag.

Mongold J. 1992. DNA repair and the evolution of transformation in *Haemophilus influenzae*. Genetics 132:893–898.

Mooers A, Holmes E. 2000. The evolution of base composition and phylogenetic interference. Trends Ecol Evol 15:365–369.

Morita S, Namikoshi A, Hirata T, Oguma K, Katayama H, Ohgaki D, Motoyama N, Fujiwara M. 2002. Efficacy of UV irradiation in inactivating *C. parvum* oocysts. Appl Environ Microbiol 68(11):5387–5393.

Munakata N, Rupert CS. 1972. Genetically controlled removal of "spore photoproduct" from deoxyribonucleic acid of ultraviolet-irradiated *Bacillus subtilis* spores. J Bact 111(1):192–198.

Munakata N, Rupert CS. 1975. Effects of DNA-polymerase-defective and recombination-deficient mutations on the ultraviolet sensitivity of *Bacillus subtilis* spores. Mutat Res 27:157–169.

Nagy R. 1964. Application and measurement of ultraviolet radiation. AIHA J 25:274–281.

Nakamura H. 1987. Sterilization efficacy of ultraviolet irradiation on microbial aerosols under dynamic airflow by experimental air conditioning systems. Bull Tokyo Med Dent Univ 34(2):25–40.

NCBI. 2009. Entrez Genome. Bethesda, MD: National Center for Biotechnology Information.

Neidle S. 1999. Oxford Handbook of Nucleic Acid Structure. New York: Oxford University Press.

Newcombe DA, Schuerger AC, Benardini JM, Dickinson D, Tanner R, Venkateswaran K. 2005. Survival of spacecraft-associated microorganisms under simulated martian UV irradiation. Appl Environ Microbiol 71(12):8147–8156.

Nicholson WL, Setlow B, Setlow P. 1991. Ultraviolet irradiation of DNA complexed with a alph/beta-type small, acid-soluble protein from spores of *Bacillus* or *Clostridium* species makes spore phoptoproduct but not thymine dimers. Proc Nat Acad Sci 88:8288–8292.

Nicholson W, Galeano B. 2003. UV resistance of *Bacillus anthracis* spores revisited: Validation of *Bacillus subtilis* spores as UV surrogates for spores of *B. anthracis* Sterne. Appl Environ Microbiol 69(2):1327–1330.

Nieuwstad T, Havelaar A. 1994. The kinetics of batch ultraviolet inactivation of bacteriophage MS2 and microbiological calibration of an ultraviolet pilot plant. J Environ Sci Health A29(9):1992–2007.

Nomura S, Bassin RH, Turner W, Haapala DK, Fischinger PJ. 1972. Ultraviolet inactivation of Maloney Leukaemia virus: Relative target size required for virus replication and rescue of 'defective' Murine Sarcoma virus. J Gen Virol 14:213–217.

Nozu K, Ohnishi T. 1977. Ultraviolet sensitivity of *Vibrio parahaemolyticus*, a causative bacterium of food poisoning. Photochem Photobiol 26(5):483–486.

Nuanualsuwan S, Mariam T, Himathongkham S, Cliver DO. 2002. Ultraviolet inactivation of feline calicivirus, human enteric viruses, and coliphages. Photochem Photobiol 76(4):406–410.

Nuanualsuwan S, Cliver DO. 2003. Infectivity of RNA from Inactivated Proteins. Appl Environ Microbiol 69(3):1629–1632.

Nwachuku N, Gerba CP, Oswald A, Mashadi FD. 2005. Comparative inactivation of adenovirus serotypes by UV light disinfection. Appl Environ Microbiol 71(9):5633–5636.

O'Hara PJ, Gordon MP. 1980. Ultraviolet inactivation of the midi variant of QBeta RNA: The sites of UV-induced replication inhibition. Photochem Photobiol 31:47–54.

Oguma K, Katayama H, Mitani H, Morita S, Hirata T, Ohgaki S. 2001. Determination of pyrimidine dimers in *Escherichia coli* and *Cryptosporidium parvum* during UV light inactivation, photoreactivation, and dark repair. Appl Environ Microbiol 67(10):4630–4637.

Oguma K, Katayama H, Ohgaki S. 2004. Photoreactivation of *Legionella pneumophila* after inactivation by low- or medium-pressure ultraviolet lamps. Wat Res 38(11):2757–2763.

Oppenheimer J, Hoagland J, Laine J-M, Jacangelo J, Bhamrah A. Microbial inactivation and characterization of toxicity and by-products occurring in reclaimed wastewater disinfected with UV radiation; 1993 May 23–25. Whippany, NJ: Wat Environ Fed.

Oppenheimer JA, Jacangelo JG, Laine J-M, Hoagland JE. 1997. Testing the equivalency of ultraviolet light and chlorine for disinfection of wastewater to reclamation standards. Wat Environ Res 69(1):14–24.

Otaki M, Okuda A, Tajima K, Iwasaki T, Kinoshita S, Ohgaki S. 2003. Inactivation differences of microorganisms by low pressure UV and pulsed xenon lamps. Wat Sci Technol 47(3):185–190.

Owada M, Ihara S, Toyoshima K. 1976. Ultraviolet inactivation of Avian Sarcoma viruses: Biological and biochemical analysis. Virology 69:710–718.

Oye AK, Rimstad R. 2001. Inactivation of infectious salmon anaemia virus, viral hemorrhagic septicaemia virus and infectious pancreatic necrosis virus in water using UVC irradiation. Dis Aquatic Organ 48:1–5.

Patrick MH. 1977. Studies on thymine-derived UV photoproducts in DNA - I. Formation and biological role of pyrimidine adducts in DNA. Photochem Photobiol 25(4):357–372.

Peak MJ, Peak JG. 1978. Action spectra for the ultraviolet and visible light inactivation of phage T7: Effect of host-cell reactivation. Radiat Res 76:325–330.

Peccia J, Hernandez M. 2001. Photoreactivation in Airborne *Mycobacterium parafortuitum*. Appl and Environ Microbiol 67:4225–4232.

Peccia J, Werth HM, Miller S, Hernandez M. 2001a. Effects of relative humidity on the ultraviolet induced inactivation of airborne bacteria. Aerosol Sci & Technol 35:728–740.

Peccia J, Hernandez, M. 2002. UV-induced inactivation rates for airborne *Mycobacterium bovis* BCG. J Occup Environ Hyg 1(7):430–435.

Powell WF. 1959. Radiosensitivity as an index of herpes simplex virus development. Virology: 9:1–19.

Prahl S. 2009. Mie Scattering Calculator. Portland, OR: Oregon Medical Laser Center.

Proctor WR, Cook JS, Tennant RW. 1972. Ultraviolet photobiology of Kilham rat virus and the absolute ultraviolet photosensitivities of other animal viruses: Influence of DNA strandedness, molecular weight, and host-cell repair. Virology 49(2):368–378.

Qualls RG, Johnson JD. 1983. Bioassay and dose measurement in UV disinfection. Appl Microb 45(3):872–877.

Quek PH, Hu J. 2008. Indicators for photoreactivation and dark repair studies following ultraviolet disinfection. J Ind Microbiol Biotechnol 35:533–541.

Rahn RO, Hosszu JL. 1969. Influence of relative humidity on the photochemistry of DNA films. Biochim Biophys Acta 190:126–131.

Rainbow AJ, Mak S. 1970. Functional heterogeneity of virions in human adenovirus types 2 and 12. J Vir 5:188–193.

Rainbow AJ, Mak S. 1973. DNA damage and biological function of human adenovirus after UV irradiation. Int J Radiat Biol 24(1):59–72.

Rastogi VK, Wallace L, Smith LS. 2007. Disinfection of *Acinetobacter baumanii*-contaminated surfaces relevant to medical treatment facilities with ultraviolet C light. Mil Med 172(11):1166–1169.

Rauth AM. 1965. The physical state of viral nucleic acid and the sensitivity of viruses to ultraviolet light. Biophys J 5:257–273.

Rentschler HC, Nagy R, Mouromseff G. 1941. Bactericidal effect of ultraviolet radiation. J Bacteriol 42:745–774.

Rentschler HC, Nagy R. 1942. Bactericidal action of ultraviolet radiation on air-borne microorganisms. J Bacteriol 44:85–94.

Reponen T, Willeke K, Ulevicius V, Reponen A, Grinshpun S. 1996. Effect of relative humidity on the aerodynamic diameter and respiratory deposition of fungal spores. Atmos Environ 30: 3967–3974.

Riley RL, Kaufman JE. 1972. Effect of relative humidity on the inactivation of airborne *Serratia marcescens* by ultraviolet radiation. Appl Microbiol 23(6):1113–1120.

Riley RL, Knight M, Middlebrook G. 1976. Ultraviolet susceptibility of BCG and virulent tubercle bacilli. Am Rev Resp Dis 113:413–418.

Rommelaere J, Vos J-M, Cornelis JJ, Ward DC. 1981. UV-enhanced reactivation of Minute-Virus-of-Mice: Stimulation of a late step in the viral life cycle. Photochem Photobiol 33:845–854.

Ronto G, Gaspar S, Berces A. 1992. Phages T7 in biological UV dose measurement. Photochem Photobiol 12:285–294.

Ross LJN, Wildy P, Cameron KR. 1971. Formation of small plaques by Herpes viruses irradiated with ultraviolet light. Virology 45:808–812.

Rubin H, Temin H. 1959. A radiological study of cell-virus interaction in the Rous sarcoma. Virology 7:75.

Ryan D, Rainbow A. 1986. Comparative studies of host-cell reactivation, cellular capacity and enahnced reactivation of herpes simplex virus in normal, xeroderma pigmentosum and Cockayne syndrome fibroblasts. Mutat Res 166:99–111.

Sako H, Sorimachi, M. 1985. Susceptibility of fish pathogenic viruses, bacteria and fungus to ultraviolet radiation and the disinfectant effect of U.V.-ozone water sterilise on the pathogens in water. Bull Nat Res Inst Aquacult 8:51–58.

Samad SA, Bhattacharya SC, Chatterjee SN. 1987. Ultraviolet inactivation and photoreactivation of the cholera phage 'Kappa'. Radiat Environ Biophjys 26:295–300.

Sanz EN, Davila IS, Balao JAA, Alonso JMQ. 2007. Modelling of reactivation after UV disinfection: Effect of UV-C dose on subsequent photoreactivation and dark repair. Wat Res 41: 3141–3151.

Sarasin AR, Hanawalt PC. 1978. Carcinogens enhance survival of UV-irradiated simian virus 40 in treated monkey kidney cells: Induction of a recovery pathway? Proc Natl Acad Sci 75(1): 356–350.

Scholes CP, Hutchinson F, Hales HB. 1967. Ultraviolet-induced damage to DNA independent of molecular weight. J Mol Biol 24:471–474.

Seemayer NH. 1973. Analysis of minimal functions of Simian virus 40. J Virol 12(6):1265–1271.

Sellers MI, Nakamura R, Tokunaga T. 1970. The effects of ultraviolet irradiation on Mycobacteriophages and their infectious DNAs. J Gen Virol 7(3):233–247.

Selsky C, Weichselbaum R, Little JB. 1978. Defective host-cell reactivation of UV-irradiated Herpes Simplex virus by Bllom's Syndrome skin fibroblasts. In: Hanawalt PC, Friedberg EC, Cox CF, editors. DNA Repair Mechanisms. New York: Academic Press.

Setlow R, Boyce R. 1960. The ultraviolet light inactivation of phiX174 bacteriophage at different wavelengths and ph's. Biophys J 1:29–41.

Setlow JK, Duggan DE. 1964. The resistance of *Micrococcus radiodurans* to ultraviolet radiation. I. Ultraviolet-induced lesions in the cell's DNA. Biochim Biophys Acta 87:664–668.

Setlow RB, Carrier WL. 1966. Pyrimidine dimers in ultraviolet-irradiated DNA's. J Mol Biol 17:237–254.

Setlow J, Boling M. 1972. Bacteriophage of *Haemophilus influenzae* - II. Repair of ultraviolet-irradiated phage DNA and the capacity of irradiated cells to make phage. J Mol Biol 63:349–362.

Severin BF, Suidan MT, Englebrecht RS. 1983. Kinetic modeling of U.V. disinfection of water. Water Res 17(11):1669–1678.

Shafaat H, Ponce A. 2006. Applications of a rapid endospore viability assay for monitoring UV inactivation and characterizing arctic ice cores. Appl Environ Microbiol 72(10):6808–6814.

Shanley JD. 1982. Ultraviolet irradiation of Murine Cytomegalovirus. J Gen Virol 63:251–254.

Sharp G. 1939. The lethal action of short ultraviolet rays on several common pathogenic bacteria. J Bact 37:447–459.

Sharp G. 1940. The effects of ultraviolet light on bacteria suspended in air. J Bact 38:535–547.

Shaw JE, Cox DC. 1973. Early inhibition of cellular DNA synthesis by high multiplicities of infectious and UV-inactivated Reovirus. J Virol 12(4):704–710.

Shimizu A, Shimizu N, Tanaka A, Jinno-Oue A, Roy B, Shinagawa M, Ishikawa O, Hoshino H. 2004. Human T-cell leukaemia virus type 1 is highly sensitive to UV-C light. J Gen Virol 85:2397–2406.

Shin G, Linden K, Arrowood M, Sobsey M. 2001. Low-pressure inactivation and DNA repair potential of *Cryptosporidium parvum* oocysts. Appl Environ Microbiol 67(7):3029–3032.

Shin G, Linden KG, Sobsey MD. 2005. Low pressure ultraviolet inactivation of pathogenic enteric viruses and bacteriophages. J Environ Eng Sci 4(Supp 1):S7-S11.

Shin G. 2008. Inactivation of *Mycobacterium avium* complex by UV irradiation. Appl Environ Microbiol 74(22):7067–7069.

Simonet J, Gantzer C. 2006. Inactivation and genome degradation of poliovirus 1 and F-specific RNA phages and degradation of their genomes by UV irradiation at 254 nanometers. Appl Environ Microbiol 72(12):7671–7677.

Smirnov Y, Kapitulez S, Kaverin N. 1992. Effects of UV-irradiation upon Venezuelan equine encephalomyelitis virus. Virus Res 22(2):151–158.

Smith KC, Hanawalt PC. 1969. Molecular Photobiology: Inactivation and Recovery. New York: Academic Press.

Sommer R, Weber G, Cabaj A, Wekerle J, Keck G, Schauberger G. 1989. UV-inactivation of microorganisms in water. Zentralbl Hyg Umweltmed 189(3):214–224.

Sommer R, Haider T, Cabaj A, Pribil W, Lhotsky M. 1998. Time dose reciprocity in UV disinfection of water. Wat Sci Technol 38(12):145–150.

Sommer R, Cabaj A, Sandu T, Lhotsky M. 1999. Measurement of UV radiation using suspensions of microorganisms. J Photochem Photobiol 53(1–3):1–5.

Sommer R, Pribil W, Appelt S, Gehringer P, Eschweiler H, Leth H, Cabaj A, Haider T. 2001. Inactivation of bacteriophages in water by means of non-ionizing (UV 253.7 nm) and ionizing (gamma) radiation: A comparative approach. Wat Res 35(13):3109–3116.

Stull H, Gazdar A. 1976. Stability of Rauscher Leukemia Virus under certain laboratory conditions. Proc Soc Exp Biol Med 152:554–556.

Sturm E, Gates FL, Murphy JB. 1932. Properties of the causative agent of a chicken tumor. II. The inactivation of the tumor-producing agent by monochromatic ultra-violet light. J Exp Med 55:441–444.

Summer W. 1962. Ultra-Violet and Infra-Red Engineering. New York: Interscience Publishers.

Sussman AF, Halvorson HO. 1966. Spores: Their Dormancy and Germination. New York: Harper & Row.

Templeton MR, Andrews RC, Hofmann R. 2006. Impact of iron particles in groundwater on the UV inactivation of bacteriophages MS2 and T4. J Appl Microbiol 101(3):732–741.

Templeton M, Antonakaki M, Rogers M. 2009. UV dose-response of *Acinetobacter baumanii* in water. Environ Sci Eng 26(3):697–701.

Thompson SS, Jackson JL, Suva-Castillo M, Yanko WA, Jack ZE, Chen CL, Williams FP, Schnurr DP. 2003. Detection of infectious human adenovirus in tertiary-treated and ultraviolet-disinfected wastewater. Water Environ Res 75(2):163–170.

Thurman R, Gerba C. 1989. The molecular mechanisms of copper and silver ion disinfection of bacteria and viruses. CRC Crit Rev Environ Control 18:295–315.

Thurston-Enriquez JA, Haas CN, Jacangelo J, Riley K, Gerba CP. 2003. Inactivation of Feline calicivirus and adenovirus type 40 by UV radiation. Appl Environ Microbiol 69(1):577–582.

Tosa K, Hirata T. 1998. Photoreactivation of *Salmonella* following UV disinfection. Proc IAWQ 19th Biennial Int Conf. 10, Health-Related Water Microbiology.

Tree J, Adams M, Lees D. 1997. Virus inactivation during disinfection of wastewater by chlorination and UV irradiation and the efficacy of F+ bacteriophage as a 'viral indicator'. Wat Sci Technol 35(11–12):227–232.

Tree J, Adams M, Lees D. 2005. Disinfection of Feline calicivirus (a surrogate for Norovirus) in wastewaters. J Appl Microbiol 98:155–162.

Tseng C-C, Li C-S. 2005. Inactivation of virus-containing aerosols by ultraviolet germicidal irradiation. Aerosol Sci Technol 39(1136–1142).

Tyrrell RM, Moss SH, Davies DJG. 1972. The variation in UV sensitivity of four K12 strains of *Escherichia coli* as a function of their stage of growth. Mutat Res 16:1–12.

Unrau P, Wheatcroft R, Cox B, Olive T. 1973. The formation of pyrimidine dimers in the DNA of fungi and bacteria. Biochim Biophys Acta 312:626–632.

UVDI. 2001. Report on Survival Data for *A. niger* and *R. nigricans* Under UVGI exposure. Valencia, CA: Ultraviolet Devices, Inc.

van der Eb AJ, Cohen JA. 1967. The effect of UV-irradiation on the plaque-forming ability of single- and double-stranded polyoma virus DNA. Biochem Biophys Res Comm 28(2):284–293.

VanOsdell D, Foarde K. 2002. Defining the Effectiveness of UV Lamps Installed in Circulating Air Ductwork. Arlington, VA: Air-Conditioning and Refrigeration Technology Institute. Report nr ARTI-21CR/610-40030-01.

VonBrodrotti HS, Mahnel H. 1982. Comparative studies on susceptibility of viruses to ultraviolet rays. Zbl Vet Med B 29:129–136.

Vos JM, Cornelis JJ, Limbosch S, Zampetti-Bosseler F, Rommelare J. 1981. UV-irradiation of related mouse hybrid cells: Similar increase in capacity to replicate intact Minute-Virus-of-Mice but differential enhancement of survival of UV-irradiated virus. Mutat Res 83:171–178.

Walker CM, Ko G. 2007. Effect of ultraviolet germicidal irradiation on viral aerosols. Environ Sci Technol 41(15):5460–5465.

Wang Y, Casadevall A. 1994. Decreased susceptibility of melanized *Cryptococcus neoformans* to UV light. Appl Microb 60(10):3864–3866.

Wang C-H, Tschen S-Y, Flehmig B. 1995. Antigenicity of hepatitis A virus after ultra-violet irradiation. Vaccine 13(9):835–840.

Wang J, Mauser A, Chao SF, Remington K, Treckmann R, Kaiser K, Pifat D, Hotta J. 2004. Virus inactivation and protein recovery in a novel ultraviolet-C reactor. Vox Sang 86(4):230–238.

Wasserman F. 1962. The inactivation of Adenoviruses by ultraviolet irradiation and nitrous acid. Virology 17:335–341.

Webb SJ. 1961. Factors affecting the viability of air-borne bacteria: IV. The inactivation and reactivation of air-borne *Serratia marcescens* by ultraviolet and visible light. Can J Microbiol 7:607–619.

Webb SJ. 1965. Bound Water in Biological Integrity. Springfield, IL: Charles C. Thomas.

Webb SJ, Tai CC. 1970. Differential, lethal and mutagenic action of 254 nm and 320–400 nm radiation on semi-dried bacteria. Photochem Photobiol 12:119–143.

Weidenmann A, Fischer B, Straub U, Wang C-H, Flehmig B, Schoenen D. 1993. Disinfection of hepatitis A virus and MS-2 coliphage in water by ultraviolet irradiation: Comparison of UV-susceptibility. Wat Sci Technol 27(3–4):335–338.

Weigle JJ. 1953. Induction of mutations in a bacterial virus. Proc Natl Acad Sci USA 39:628.

Weinberger S, Evenchick Z, Hertman I. 1984. Transitory UV resistance during germination of UV-sensitive spores produced by a mutant of *Bacillus cereus* 569. Photochem & Photobiol 39(6):775–780.

Weisova H, Vinter V, Starka J. 1966. Heat and UV-resistance of spores of *Bacillus cereus* produced endotrophically in the presence of b-2-thienylalanine. Praha, Acad Sci Bohem 11(5):387–391.

Weiss M, Horzinek MC. 1986. Resistance of Berne virus to physical and chemical treatment. Vet Microbiol 11:41–49.

Wells WF. 1940. Bactericidal irradiation of air, physical factors. J Franklin Inst 229:347–372.

Wetz K, Habermehl K-O. 1982. Specific cross-linking of capsid proteins to virus RNA by ultraviolet irradiation of Poliovirus. J Gen Virol 59:397–401.

Wilson B, Roessler P, vanDellen E, Abbaszadegan M, Gerba C. 1992. Coliphage MS-2 as a UV water disinfection efficacy test surrogate for bacterial and viral pathogens. In: Association AWW, editor. Denver, CO.

Winkler U, Johns HE, Kellenberger E. 1962. Comparative study of some properties of bacteriophage T4D irradiated with monochromatic ultraviolet light. Virology 18:343–358.

Wolff MH, Schneweis KE. 1973. UV inactivation of herpes simplex viruses, types 1 and 2. Zentralbl Bakteriol 223(4):470–477.

Xu P, Peccia J, Fabian P, Martyny JW, Fennelly KP, Hernandez M, Miller SL. 2003. Efficacy of ultraviolet germicidal irradiation of upper-room air in inactivating airborne bacterial spores and mycobacteria in full-scale studies. Atmos Environ 37:405–419.

Yamamoto H, Urakami I, Nakano K, Ikedo M, Yabuuchi E. 1987. Effects of Flonlizer, ultraviolet sterilizer, on *Legionella* species inhabiting cooling tower water. Microbiol Immunol 31(8):745–752.

Yarus M. 1964. The U.V.-resistance of double-stranded phiX174 DNA. J Mol Biol 8:614–615.

Yaun BR, Summer SS. 2002. Efficacy of Ultraviolet Treatments for the Inhibition of Pathogens on the Surface of Fresh Fruits and Vegetables. Blacksburg, VA: Virginia Polytechnic Institute and State University.

Yaun B, Eifert SSJ, Marcy J. 2003. Response of *Salmonella* and *E. coli* O157:H7 to UV energy. J Food Prot 66(6):1071–1073.

Yoshikura H. 1971. Ultraviolet inactivation of murine leukemia and sarcoma viruses. Int J Cancer 7:131–140.

Yoshikura H. 1989. Thermostability of Human Immunodeficiency virus (HIV-1) in a Liquid Matrix is far higher than that of an ecotropic Murine Leukemia virus. Jpn J Cancer Res 80:1–5.

Yoshimizu M, Yoshinaka T, Hatori S, Kasai H. 2005. Survivability of fish pathogenic viruses in environmental water, and inactivation of fish viruses. Bull Fish Res Agen Suppl 2:47–54.

Zahl PA, Koller LR, Haskins CP. 1939. The effects of ultraviolet radiation on spores of the fungus *Aspergillus niger*. J Gen Physiol 16:221–235.

Zavadova Z, Gresland L, Rosenbergova M. 1968. Inactivation of single- and double-stranded ribonucleic acid of encephalomyocarditis virus by ultraviolet light. Acta Virol 12:515–522.

Zavadova Z. 1971. Host-cell repair of vaccinia virus and of double stranded RNA of encephalomyocarditis virus. Nature (London) New Biol 233:123.

Zavadova Z, Libikova H. 1975. Comparison of the sensitivity to ultraviolet irradiation of reovirus 3 and some viruses of the Kemerovo group. Acta Virol 19:88–90.

Zelle MR, Hollaender A. 1955. Radiation Biology Volume II. New York: McGraw-Hill.

Zemke V, Podgorsek L, Schoenen D. 1990. Ultraviolet disinfection of drinking water. 1. Communication: Inactivation of *E. coli* and coliform bacteria. Zentralbl Hyg Umweltmed 190(1–2):51–61.

Zimmer JL, Slawson RM. 2002. Potential repair of *Escherichia coli* following exposure to UV radiation from both medium- and low-pressure UV sources used in drinking water treatment. Appl Environ Microbiol 68(7):3293–3299.

Zimmer J, Slawson R, Huck P. 2003. Inactivation and potential repair of *C. parvum* following low- and medium-pressure ultraviolet radiation. Wat Res 37(14):3517–3523.

Chapter 5

UVGI Lamps and Fixtures

5.1 Introduction

The ultraviolet (UV) lamp is the most critical component of any UVGI air or surface disinfection system. UV lamps come in many shapes, sizes, and UV output power levels, and they may produce different spectral bands of germicidal UV. All germicidal UV lamps will produce light wavelengths of either UVC (100–280 nm), UVB (280–320 nm), or both (UVC/UVB). Wavelengths of UVA (320–400 nm) are not germicidal and UVA lamps (i.e. blacklight and sunlamps) are not considered UVGI lamps and are not addressed in this chapter. UVGI lamps are commonly referred to as UV lamps and these terms are used interchangeably throughout this book. Various types of UV lamps are currently in use for air and surface disinfection applications, including standard medium pressure (MP) UV lamps which produce broadband UVC/UVB wavelengths, low pressure (LP) UV lamps that produce narrow band UVC wavelengths, microwave UV lamps, and light-emitting diodes (LEDs). Technically, UVC lamps are LP lamps that produce UVC wavelengths only, and the term is not necessarily accurate when applied to MP lamps that produce broadband UVC/UVB wavelengths although common usage doesn't always make this distinction.

UV lamps are generally part of fixtures which are installed such as in Heating, Ventilating, and Air Conditioning (HVAC) systems inside ductwork. UV lamp fixtures are installed to control microbial growth on cooling coils. Special fixtures are used in Upper Room and Lower Room UVGI systems where it is necessary to limit UV exposure.

The UV lamp is the main component of any disinfection system and must meet certain standards and performance criteria, and a number of guidelines and standards are currently available (NFPA 2008, ANSI 2005a, b, UL 2004, IESNA 2000). See Chap. 11 for a review of the applicable electrical and mechanical guidelines and standards for UV lamp applications. The lamp ballast must also meet certain standards, as must any lamp fixtures and associated electrical wiring. Standards for these components have been previously defined by other professional societies and

W. Kowalski, *Ultraviolet Germicidal Irradiation Handbook*,
DOI 10.1007/978-3-642-01999-9_5, © Springer-Verlag Berlin Heidelberg 2009

organizations and are summarized in Chap. 11, but are also referenced here (ANSI 2004, NFPA 2008, UL 2004).

5.2 Types and Characteristics of UV Lamps

UV lamps are often called mercury or "amalgam" lamps because they contain solid amalgam "spots" (an amalgam is an alloy of mercury with another element, such as indium or gallium) that controls the mercury vapor pressure. In addition to mercury, UV lamps also contain a starter gas, typically argon. The two most common types of UV lamps are high intensity discharge (HID) lamps (also called high pressure or medium pressure mercury vapor lamps) and low pressure mercury vapor lamps. HID lamps contain mercury gas at pressure of approximately 1000 torr and generate high levels of UV irradiance over a broad range of wavelengths. Low pressure UV lamps contain mercury gas at pressures of about 10 torr or less, and when this gas is stimulated by an electrical charge, it emits UV light in a narrow band of wavelengths centered around 254 nm as shown previously in Fig. 2.1 in Chap. 2, which also shows the broad band UV lamp spectrum.

The different spectra produced by UV lamps can impact germicidal effectiveness in various ways. Due to the narrow spectrum of UVC produced by the low pressure UV lamps, these types of lamps tend to produce much less ozone. Due to the broader spectrum of the medium pressure UV lamps, these types of lamps tend to produce more damage to enzymes and other microbial constituents, limiting or preventing photoreactivation (Masschelein 2002). Studies on *E. coli* shows that photorepair occurred following irradiation by low pressure UV lamps but that no photoreactivation occurred after exposure to medium pressure lamps (Hu et al. 2005). The UV spectra produced may also impact the allowable levels of exposure to actinic radiation. For narrow band UVC, the ACGIH 8-hour exposure limit is 0.002 W/m^2, while for broad band UVC/UVB the ACGIH exposure limit is 0.001 W/m^2. Such effects, although probably insignificant or irrelevant in most cases, may dictate the choice of lamps for certain applications. HID or medium pressure lamps have higher bulb wall temperatures and lower conversion efficiency than low pressure lamps but can achieve higher irradiance levels (CIE 2003). Figure 5.1 shows some typical examples of single-ended lamps.

Low pressure UV lamps have many similarities with fluorescent lamps, which are also low pressure mercury discharge lamps, and components such as ballasts may be used for both. Common pin types allow the use of the same fixtures, which is a convenience in some cases but may be a hazard in others. Other pin types are unique to UV lamps (see Fig. 5.2). Low pressure lamps consist of an envelope made of quartz glass or other UV-transmittant glass, a pair of electrodes, and a mercury amalgam. Ballasts are required to provide the necessary starting voltage across the electrodes and to provide the proper lamp current. The electric current that passes between the electrodes heats up the mercury vapor which stimulates electronic transitions and causes emission of ultraviolet and visible light. In fluorescent lamps,

Fig. 5.1 Typical
single-ended UV lamps. *Top
left* Amway 8G2S, *top right*
Philips PL-S 9 W TUV, next
Virobuster VB101, next
LightTech LTC95WHO, and
bottom Philips PL-L 36 W
TUV

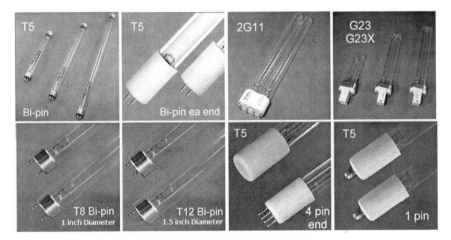

Fig. 5.2 Examples of pin connectors on lamp ends. Images reprinted courtesy of AtlantaLight-Bulbs.com

the glass is coated on the inside with phosphors which absorb UV and re-emit the energy as visible light. In UV lamps, there are no phosphors and the glass, usually quartz glass, is largely transparent to UV wavelengths. The quartz glass can absorb a small fraction of incident light and converts it to heat. It is also possible to use Softglass (sodium-barium glass) which does not transmit at 185 nm and produces no ozone (Schalk et al. 2006)

The glass wall temperature, specifically at the coldest location, dictates the pressure of the mercury vapor inside the lamp, and so determines the total UV output. Cooling of the lamp wall surface, by cold air or by the cooling effects of an airstream, can reduce the UV output of a lamp. In low pressure lamps about 60% of the electrical input power is converted to light, of which about 85% occurs near 254 nm. Overall efficiencies of low pressure lamps tend to be about 30–31%, although these can vary depending on ambient operating conditions and type of ballast. Low pressure UV lamps also emit lower levels of radiation at wavelengths of about 185 nm, 312 nm, 365 nm, and higher. The radiation near 185 nm can produce ozone, while UVA radiation emissions above 400 nm can be visibly perceived as bluish-white light.

UV lamps may have one of two types of cathodes – hot cathodes or cold cathodes. Cold cathodes are instant starting and use a thimble-shaped cylinder of soft iron that allows frequent starting without affecting lamp life (CIE 2003). Cold cathode lamps maintain high output over the life of the lamp. The voltage drop at the cathode is higher than with hot cathode lamps and this results in greater wattage loss and lower efficiency.

Hot cathode lamps use coiled tungsten filaments impregnated with electron-emissive materials at each end of the bulb. They operate at a higher output per unit length and have a higher overall efficiency than cold cathode lamps, for lamps of similar dimensions. Hot cathode lamps are the most common type used in germicidal applications. The lamp life of hot cathode lamps is determined by the rate of loss of the electron emissive coating on the electrodes (IES 1981). Coatings are eroded during starting and evaporate during use. Frequent starting may accelerate lamp aging. The end of lamp life is reached when the coating is completely removed from one or both electrodes.

Lamp starting occurs in two stages. First, a high enough voltage must be applied across the electrodes to initiate ionization of the gas in the lamp. Second, a sufficient voltage must be maintained across the lamp to extend the ionization throughout the lamp and develop an arc. At least three types of starters are available – preheat start, instant start, and rapid start. Preheat-start germicidal lamps use electrical preheating to stimulate electrons and ionize the gas. This lowers the voltage necessary to strike the arc. An automatic starter controls the preheating process, which can take a few seconds. After the cathodes are preheated, a timer shuts off the preheater and the voltage is applied between the cathodes to strike the arc. Once the lamp is operating, the cathode temperature is maintained by the arc.

Instant start lamps, also called cold start lamps, are capable of operating at several current densities, depending on their associated ballasts, which results in these lamps having a range of nominal wattages. Arc initiation in an instant start lamp depends solely on the application of high voltage (400–1000 V) across the lamp (IES 1981). This high voltage ejects electrons from the electrodes by field emission which flow through the gas, ionize it, and initiate an arc discharge. In instant start lamps the ballast provides the voltage necessary to strike the arc and they operate without a preheater or starter. This requires cathodes that will start without enhanced electron emission. Since preheating is unnecessary, only a single pin base is needed at each end of the lamp and these lamps are sometimes called slimline lamps.

Rapid start lamps use low or high resistance electrodes which are heated continuously. Heating is achieved through low voltage windings built into the ballast or by separate low voltage transformers. Starting voltage requirements are similar to those of preheat lamps and lamps usually start in about 1 s. Rapid start ballasts are smaller, less expensive, and have lower power loss than instant start ballasts. High-output (HO) rapid start germicidal lamps operate at higher amperages than most rapid start lamps and produce about 45% more output than similar sized lamps (CIE 2003).

Table 5.1 UV lamp types, dimensions, output, and expected life

Parameter	Preheat tubular miniature	Preheat tubular standard	Preheat compact	Preheat large compact	Instant start (Slimline)	Cold cathode
Nominal lamp wattage, W	4–11	15–115	5–11	18–55	10–75	12–34
UV output, W (after 100 h)	0.5–2.5	4.5–40	1.5–3.5	5.5–17	3.0–25	1.4–11.2
Approx. length range, mm	150–225	450–1215	165–235	225–570	300–1700	300–1200
Nominal diameter, mm	16	28–38	28 max	38 max	16–19	16
Lamp life, hours	6000	5–10K	8000	8000	10000	20000
Base/cap	Min Bipin	Med Bipin	2 pin cap	4 pin cap	Single pin	Single pin
Starter	Yes	Yes	Integrated	No	No	No

NOTE: Nominal diameters are doubled for twin-tube lamps.

The various types of lamp/ballast combinations have general characteristics in common. Table 5.1 summarizes the general categories for various lamp types and their range of dimensions, outputs, and lamp life (adapted from CIE 2003).

Appendix D tabulates many common models of UV lamps along with their geometric specifications, UV output, IESNA ratings, and manufacturers. The tabulation is not inclusive since there are many new models of UV lamps (especially from foreign sources) that are not listed, but Appendix D does provide a representative list of a wide variety of UV lamps which can be modeled with the data provided (see the Chap. 7 section on modeling UV lamps). A new type of UV lamp that is not yet in wide use is the excimer lamp, which is mercury-free but has a much lower efficiency (about 8%) than LP and MP lamps. See Schalk et al. (2006) for a detailed comparison of excimer and LP lamps.

5.3 Lamp Power, Size, and Nomenclature

UV lamp power is specified in terms of nominal total input wattage and nominal total UV output wattage. The nominal UV output power represents a summation of the spectral power distribution, which will tend to differ between low pressure and HID lamps (see Fig. 5.1). The ratio of output power to input power defines the efficiency of the lamp. Most low pressure lamps have efficiencies of about 30–31%. Efficiencies of HID lamps can be lower.

Most UV lamps are identified by model numbers that include coded information about the lamp. Abbreviations for fluorescent lamps, as specified in ANSI_IEC C78.81-2005 and other documents, are generally applicable to UV lamps although these are not always followed. Lamp abbreviations are normally found on the lamp data sheet and they are comprised of six parts:

1. Lamp nominal wattage
2. Lamp nominal length
3. Bulb diameter

4. Lamp shape
5. Lamp base (base code)
6. Circuit or special description

Length is specified in inches, but if the length is given in metric units it will be followed by "mm" immediately after the diameter. Nominal tube diameter is often given as "T-x" where x is in eighths of an inch. The most common nominal diameters for cylindrical lamps are T5 (5/8″ or 16 mm), T6 (6/8″ or 19 mm), T8 (1″ or 26 mm), and T12 (1-1/2″ or 38 mm). The letter G often precedes the wattage in many lamp model numbers and so a G8T5 lamp would be an 8 W lamp of 5/8″ diameter. The lamp base code may be included and is given per ANSI C81.61-2005. The circuit part of the abbreviation follows a slash, which follows the bulb diameter. Typical circuit identifiers and special descriptions include the following:

RS – Rapid start
PH – Preheat start (starter)
IS – Bipin base, instant-start
SP – Single pin base, instant-start
HF – High frequency
HO – High output, 800 mA and 1000 mA, rapid-start
1.5 A – 1500 mA, rapid start
B – Bactericidal lamp
CC – Cold cathode
LP – Low Pressure
HP – High pressure

The following abbreviations are provided as examples:

30 W/36T12/RS	30 W, 36″ T12, rapid-start
37 W/24T12/HO	37 W, 24″ T12, high output, rapid-start
30 W/36T8/PH-B	30 W, 38″ T8, preheat-start, bactericidal

UV lamps generally have the wattage provided separately such that the abbreviation begins with the length, followed by the bulb diameter, followed by the lamp base, and followed by any special description, as in Appendix D. Since the lamp is generally given as a UV lamp, the designation "B" is rarely used. Also, high pressure (HP) lamps are also called medium pressure (MP). Many examples of UV lamp abbreviations are provided in Appendix D, although not all of these follow the traditional conventions described above.

Lamps have different types of connectors at one or both end caps. The connectors can be single-pin, double-pin, or four-pin. The single and double pin connectors are located at both ends of the lamp. The four-pin connectors are single connectors at one end of the lamp, as for u-tube and biaxial lamps.

5.4 Lamp Ballasts

The ballast performs two functions, providing the starting voltage pulse to ionize the gas in the UV lamp tube, and then limiting the current. The ballast provides the high initial voltage required to create the starting arc, and then limits the current to control the gas temperature. Lamp ballasts can be either magnetic or electronic. Some smaller UV lamps have no ballasts.

The similarity of UV lamps to fluorescent lamps allows many tests and parameters to be common to both. Ballast factor (BF) is a term that specifies the percentage of rated lamp output that will be produced when the lamp is operated on a particular commercial ballast (Lindsey 1997). A ballast factor of 95% means the lamp will produce about 95% of its rated output when operated on the ballast.

The efficiencies of different ballasts are compared in terms of the ballast efficiency factor (BEF). BEF is defined as the ratio of ballast factor in percent, to the input power in watts, and is calculated from:

$$ BEF = \frac{BF}{P} \tag{5.1} $$

where

BF = ballast factor
P = input power, W

There are two types of ballasts commonly used to start UV lamps – magnetic ballasts and electronic ballasts. Magnetic ballasts are an old and proven technology but somewhat inefficient. Newer ballasts tend to be electronic, and these are available in several types. There are also iron ballasts, which consist of a core, windings, and other simple components, but these have become uncommon as electronic ballasts have become the norm. Figure 5.3 shows some common examples of UV lamp ballasts.

5.4.1 Magnetic Ballasts

Magnetic ballasts use a core and coil assembly transformer to perform the minimum functions necessary to start the ballast. Magnetic ballasts may be either standard electromagnetic or energy-efficient electromagnetic. The ballast provides a time-delayed inductive kick with enough voltage to ionize the gas mixture in the tube after which the current through the tube keeps the filaments energized. The starter will cycle until the tube lights up. While the lamp is on, a preheat ballast is just an inductor which, at approximately 60 Hz, has the appropriate impedance to limit the current to the UV lamp to the proper value. Ballasts should be fairly closely matched to the lamp in terms of tube wattage, length, and diameter.

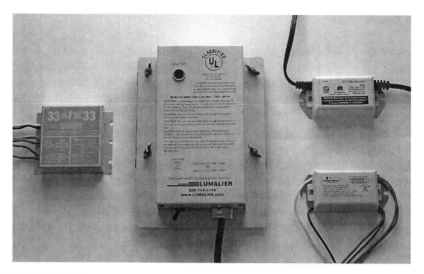

Fig. 5.3 Typical UV lamp ballasts with UL labeling. *Left* and *right* images show ballast component separate from lamp fixture. *Center* is an integrated lamp fixture with internal ballast

5.4.2 Electronic Ballasts

Electronic ballasts are basically switching power supplies. They are basically switching power supplies that eliminate the large, heavy, 'iron' ballast in favor of an integrated high frequency inverter/switcher. Current limiting is then done by a very small inductor, which has sufficient impedance at the high frequency. Properly designed electronic ballasts are very reliable. Whether they actual are reliable in practice depends on the ambient operating temperature and location with respect to the heat produced by the lamps as well as other factors.

There are two basic types of electronic ballasts – hybrid electronic ballasts and high frequency electronic ballasts. Hybrid electronic ballasts are essentially magnetic ballasts that include electronic components to switch off voltage to rapid start lamp coils once the lamp has started. High frequency electronic ballasts operate lamps at frequencies above 20,000 Hz and efficiency is maximized with electronic components matched to optimum lamp characteristics.

5.5 Microwave UV Lamps

UV lamps that are driven by microwave power require no end connectors or pins, and the microwave unit has a separate power supply. No lamp ballast is required. The lamp exterior is made entirely of glass with no openings and no electrodes, the deterioration of which can be a cause of lamp failure in standard UV lamps that employ ballasts. Microwave UV lamps can be used for air and surface disinfection,

Fig. 5.4 Microwave UV
lamp assembly. Photo of
MicroDynamics™ lamp
reprinted courtesy of Severn
Trent Services, UK

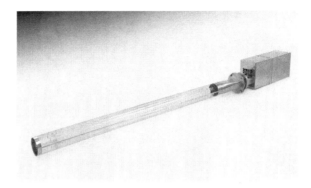

as well as water applications, but are still relatively new (Barkhudarov 2006). They
typically use a small UV lamp to ignite the main lamp, after which the energy level
in the igniter lamp is reduced to prolong life (Fusion 2004). Figure 5.4 shows one
example of a microwave UV lamp assembly.

5.6 UV LEDs

Ultraviolet light emitting diodes (LEDs) are compact light sources that come in a
variety of shapes, including tiny bulb shapes, hemispherical, and flat chips (Bettles
et al. 2007). UV LEDs have certain advantages over UV lamps. They are smaller
and do not require an electronic ballast for starting. UV LEDs may be located where
space does not permit regular UV lamps. Another advantage includes the fact that
UV LEDs do not use the toxic heavy metal mercury. The LED is a p-n junction
semiconductor lamp that emits radiation when biased in a forward direction (IES
1981). Two semiconductor materials are used to create the junction: one having
an excess of electrons (negative or n-type material), and one having a shortage of
electrons (positive or p-type material). Most near-UV (380–400 nm) LEDs (aka
UVB LEDs) employ InGaN quantum well structures with GaN barriers, while deep-
UV LEDs (aka UVC LEDs) require AlGaN alloys with aluminum concentrations
of 50% and higher (Guha and Bojarczuk 1998). Milliwatt power output levels have
been achieved with deep-UV LEDs and research to produce higher power LEDs is
ongoing.

   Although LEDs are relatively low power (i.e. about 100 mW), they can be
installed in larger arrays to produce power levels suitable for airstream disinfection.
LEDs can be modeled as point sources for analytical purposes. One advantage of
UV LEDs is that they can produce UV at the optimum wavelength (about 265 nm)
for germicidal effectiveness. Figure 5.5 shows one example of the spectral output of
a nominal 265 nm UV LED. Experiments on *E. coli* have confirmed that arrays of
UV LEDs producing 270 nm UV can inactivate the bacteria at lower power levels

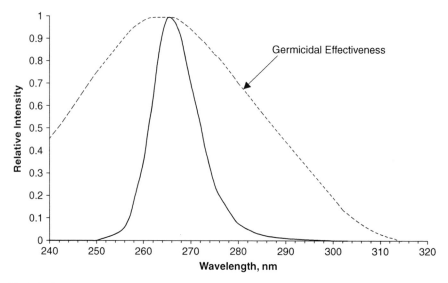

Fig. 5.5 Spectral output of a nominal 265 nm LED superimposed on the germicidal effectiveness curve (compare with Fig. 5.1). Based on data from Seoul Optodevice Co. Ltd., Korea

than typical mercury vapor lamps (Crawford et al. 2005). UV LEDs are also available in nominal 255 nm and other wavelengths.

UV LEDs are available in at least two basic shapes, including the standard bulb type LED (see Fig. 5.6 left) and circular or rectangular flat type LEDs (see Fig. 5.6 right).

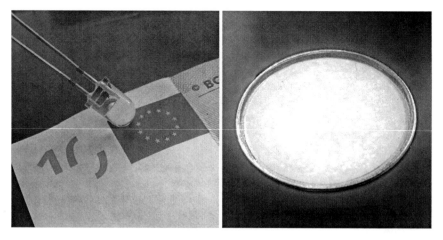

Fig. 5.6 Example of a UV LED (*left*) and a flat visible light LED (*right*). Photos reprinted courtesy of Roithner LaserTechnik, GmbH, Vienna, Austria

5.7 Pulsed UV Lamps

Pulsed Xenon Arc (PXA) lamps or Pulsed UV (PUV) lamps are ac nonpolarized low pressure xenon lamps with two active electrodes (IESNA 2000). These types of lamps generate extremely high UV output for pulses lasting a few milliseconds. A switching reactor in series with the lamp forces a peak of about 50–100 A and pulses about 120 times a second. The spectrum of light produced resembles daylight and has peaks in the UVA, UVB, and UVC ranges. Lamp wattages can range from 300 to 8000 W. Pulsed surface discharge (SD) lamps are also available. For more specific information on Pulsed UV systems, including PUV spectra, see Chap. 16.

5.8 Lamp Ratings

UV lamps are rated by taking photosensor measurements at 1 m from the midpoint of the lamp axis in still air at 25°C (IESNA 2000). Lamp ratings can vary due to the amount of burn-in time and ambient conditions. Typical burn-in times are about 100 h. Appendix D provides lamp ratings where they are available. It is assumed, but may not always be the case, that the measured lamp rating is based upon a lamp that has been burned in for at least 100 h. Lamp testing requirements for assessing lamp ratings are spelled out by the Illuminating Engineering Society of North America (IESNA) in IESNA Guide LM-55 (IESNA 1996). See Chap. 11 for additional guidelines related to lamp testing. Although the standard method for rating UV lamps has been widely used for comparative convenience, it serves little useful purpose otherwise. The fact that measurements are taken in still air makes the rating practically meaningless for airstream applications. However, until a new, more practical, standard is developed the IESNA lamp rating is likely to remain in use. The lamp rating can be estimated analytically from the lamp model equations in Chap. 6.

5.9 Lamp Shapes

UV lamps occur in a variety of shapes and sizes, the most common being cylindrical, U-tube, and biaxial, as shown in Fig. 5.7. Cylindrical lamps may be any length or diameter and are typically characterized by having connectors at both ends, requiring a compatible fixture. Once the most common type of lamp, cylindrical lamps are now being replaced by the more convenient single-ended U-tube and biaxial lamps. U-tube lamps are similar to biaxial lamps except they have a continuously curved bend at the outer end. U-tube and biaxial lamps have single connectors at the base end, a convenience which has expanded their popularity in applications.

UV lamps may be manufactured to almost any specifications or shapes, including multiple coils. Multi-coiled lamps often require connectors at both ends of the UV lamp.

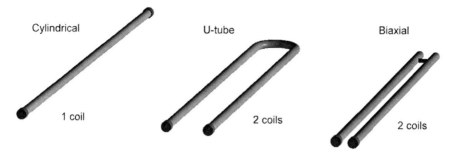

Fig. 5.7 Cylindrical, U-tube, and Biaxial UV lamps

5.10 Lamp Fixtures

UV lamps are generally held in some sort of fixture which contains the ballast when installed. Sometimes the ballast is incorporated into a self-contained fixture that is attached to the side of a duct. UV lamps can also be held in fluorescent lamp fixtures or in fixtures specifically designed for UV lamps (see Fig. 5.8). One of the main differences between fluorescent lamp fixtures and UV lamp fixtures is that fluorescent lamp fixtures are coated with white enamel to enhance visible light, while UV lamp fixtures are often made of aluminum or have aluminum coatings to enhance UV reflectivity. Fixtures that include multiple UV lamps are often installed as a single component, and often have reflective surfaces. Fixtures that are part of prefabricated reflective enclosures are the subject of the following chapter where they are discussed in more detail.

Multiple UV lamp fixtures are often installed in ventilation systems since the UV power requirements can be high. Figure 5.9 shows one example of a multi-lamp fixture consisting of 5 lamps at 45 UV watts each, for a total of 225 UV watts. This particular fixture is installed axially in the duct.

Since many UV lamps have connectors that are identical to those used in fluorescent lamps, many of the fixtures sued for lighting can be used for UV lamps as well. The practice of making UV lamps that have identical connectors and end pins to

Fig. 5.8 Single UV lamp fixture (Model UVS-136) with a biaxial single-ended lamp. Image courtesy of Lumalier, Memphis, TN

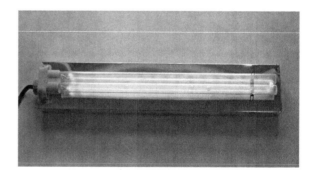

Fig. 5.9 Multiple UV lamp
fixture for installation in
ventilation ductwork. Photo
of Bio-Wall™ provided
courtesy of Sanuvox, St.
Laurent, Quebec

fluorescent lamps is a potential danger that has resulted in injuries due to UV lamps
mistakenly installed in lighting fixtures, but until the existing standards are changed,
ordinary light fixtures will be adaptable for use as UV light fixtures. Most normal
light fixtures use an enameled white background, which is fine for lighting purposes, but since white enamel has a UV reflectivity of only 9% (see Appendix D),
it is a poor choice for UV lamps. Instead, most UV fixtures are made from polished
aluminum, which has a UV reflectivity of about 73%.

5.11 Upper Room Lamp Fixtures

UV lamp fixtures for Upper Room systems, and for Lower Room systems, are
intended to confine the ultraviolet exposure field above or below a particular height,
and limiting the UV exposure in other areas to below ACGIH threshold limit values. This stratification of the UV field is generally accomplished by using a stack
of UV-reflective or UV-absorbent louvers that have a narrow gap (height h) and a
minimal length (distance d) that must be traversed, as shown in Fig. 5.10. The height
and distance of the louvers restrict the angle at which light rays exit and also impact
overall fixture efficiency.

Upper Room systems are usually designed such that no direct UV rays will strike
below a designated height, such as seven feet. In Fig. 5.10, the maximum angle θ at
which unattenuated UV light rays will exit the louvers is given by:

$$\theta_{\mathrm{max}} = a \tan \left(\frac{h}{d} \right) \tag{5.2}$$

where

  h = height between parallel louvers
  d = distance or length of louvers

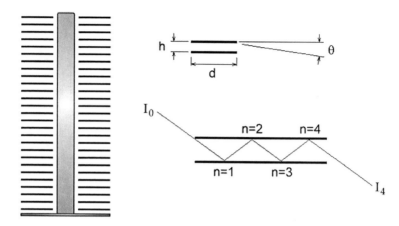

Fig. 5.10 Schematic of Upper Room UV fixture (*left*) and geometry for evaluating reflective attenuation

The angle at which attenuated UV rays exit the louvers is defined by the number of times the rays bounce inside the louvers. In Fig. 5.9, a ray is shown bouncing four times, which will reduce the reflected intensity by an exponent of four, which is a reduction factor (for aluminum) of $(0.73)^4 = 0.28$, or 28%. If the louvers were painted black, the exiting irradiance, I_n, after four bounces would be $(0.07)^4 = 0.00002$, or approximately zero. The general relation for computing the exit irradiance is given by:

$$I_n = I_0 \rho^n \tag{5.3}$$

where

I_0 = inlet irradiance, W/m^2
ρ = reflectivity of louvers
n = number of reflections or bounces

Equations (5.4) and (5.5) may be used to perform computational analysis of complex louver systems to determine the optimum fixture efficiency that can be achieved while still limiting room UV levels to below ACGIH limits. The same principles apply to Lower Room fixtures and barrier systems if they use louvers. See Chaps. 9 and 17 for more information on designing Upper and Lower Room systems

5.12 Lamp Cooling Effects

Most germicidal lamps operate at maximum efficiency in still air at 25°C and produce maximum UV output. Temperatures above or below the optimum value will decrease UV output. HID lamps are less sensitive to ambient temperatures than low pressure lamps but their lifetime can depend on ambient cooling, which makes them suitable for applications in airstreams (CIE 2003). The parameters which most affect the UV output of lamps include (1) lamp age, (2) On/Off cycling rate, and (3) wind chill from ambient temperature and air velocity (Lau et al. 2009).

Lamp UV output is a function of mercury vapor pressure, which is controlled by lamp surface cold spot temperature (the coldest temperature on the lamp surface), and the cold spot temperature depends on several factors, including air temperature, air velocity, lamp type, and lamp orientation (Lau et al. 2008, CIE 2003). Most UV lamps are designed to operate at an air temperature of approximately 21.5°C (70°F) and an air velocity of 2–2.54 m/s (400–500 fpm). Lamp UV output may decrease or increase outside this range. Some lamps can lose 25% or more of their UV output when the air temperature drops from 27°C (80°F) to 16°C (60°F) (Westinghouse 1982). Over ranges of air velocity of 0.51–3.25 m/s (100–640 fpm) and air temperatures of 10–32°C (50–90°F), in-duct UV lamp output can change by as much as 70% (Lau et al. 2009). Figure 5.11 shows a typical example of the cooling effects on the UV output of a model TUV lamp and Fig. 5.12 shows air cooling effects for a typical GPH lamp. Note that lamps can exceed rated output at low air velocities and high air temperatures. Lamps must be sized to account for the cooling effects when operating conditions are not optimum. Lamp manufacturers should be consulted regarding the UV output of lamps operating outside design conditions as they can vary between lamp models.

In order to account for the cooling effects, the operating conditions of the lamp must be known. The lamp wattage must then be 'derated' to account for the cooling effects. The derated wattage is the correct value to use for sizing UV systems and for modeling the lamp computationally.

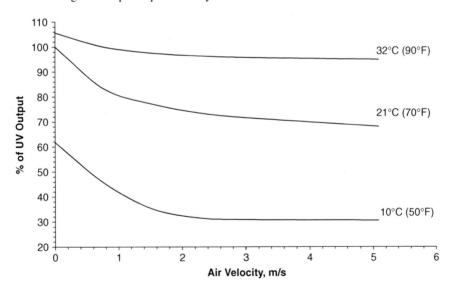

Fig. 5.11 Variation of UV output with air temperature and velocity for a model TUV36W-PLL lamp

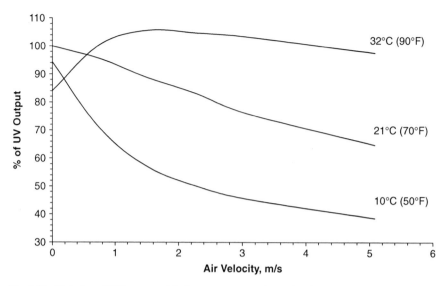

Fig. 5.12 Variation of UV output with air temperature and velocity for a typical model GPH UV lamp

Example 5.1: A GPH model lamp is to be placed in service where the air velocity is 4 m/s and cannot be reduced. The minimum air temperature is expected to be 15.5°C. If the system requires 100 W of UV output for an URV 13 rating, what is the minimum UV wattage that must be installed? Also, what is the total lamp power input required?

Answer: Interpolating between 10–21°C from Fig. 5.12 at 4 m/s, we find the GPH lamp(s) must be derated to about 58% of the nominal rated output. Therefore the minimum required total lamp wattage is (100 W)/0.58 = 172 W.

The minimum wattage is 172 W. At an assumed 31% efficiency, the total lamp power input will be 172/0.31 = 556 W of input power.

In the previous example, the total lamp UV power required is 172 W. If, for example, we chose a model GPH893TP5 lamp (see Appendix D), which is rated at 30 W UV output, we would need at least 6 lamps. These would produce 6 × 30 = 180 W of UV output. Rounding the number of lamps upwards will always produce conservatism in the lamp power estimate. It should be noted, however, that such linear interpolations may not give the most accurate results. Improved accuracy and more general applicability can be obtained by using heat transfer models of the lamps under airflow. Lau et al. (2009) provides details on semi-empirical heat transfer models for three common types of UVGI lamps and demonstrated a predictive accuracy for the UV output to within +/–5%.

Certain UV lamps have controls that will boost UV output in response to the cooling effects of airflow on the plasma temperature (Kowalski and Bahnfleth 2000). Still other lamps may include an infrared blocking UV-transparent shield (i.e., quartz glass or Teflon) to reduce cooling effects. There are well over one hundred different types of UV lamps and their individual performance cannot be generalized. Data should be obtained from the manufacturer of any UV lamp to determine the cooling effects or the limiting design air velocities and temperatures within which the lamps can be efficiently operated.

5.13 Ozone Generation

Ultraviolet radiation at wavelengths of 175–210 nm can convert atmospheric oxygen to ozone. Most medium pressure mercury lamps generate a weak spectral line at 185 nm which produces ozone in combination with oxygen from ambient air (Schalk et al. 2006). Low pressure lamps have no peak in the vicinity of 185 nm but may still produce some detectable ozone. Certain models of low pressure UV lamps block the spectrum at 185 nm and produce no ozone at all.

Although ozone has biocidal properties and may contribute to the overall disinfection rate of UV systems, it is also a powerful oxidizing agent and is hazardous to humans at levels above the ACGIH limit (0.1 ppm for 8 h exposure). It can also damage certain organic materials and some plastics that may be used in ventilation systems. Ozone will react with moisture in the air to produce hydroxyl radicals and other corrosive compounds.

Since ozone also absorbs UV at 253.7 nm, ozone is both created and destroyed by ultraviolet radiation and the UV spectrum of a lamp and operating conditions will dictate the ozone generation rate. Many manufacturers report the ozone generation rate of their lamps. For example, a model G48T6VH/U lamp will produce 5.2 g of ozone per hour under favorable conditions, while a model G48T6L/U lamp will produce no ozone (based on catalog information from manufacturer).

Some UV lamps are used specifically for the purpose of generating ozone and the bulb glass is made of high purity fused quartz which is transparent to ozone-producing wavelengths. The ozone generation rate can be enhanced by adding pure oxygen to the airstream. Other lamps are designed to eliminate ozone production and are designated non-ozone producing lamps. The latter type of lamp have bulbs made of glass or a fused quartz doped to block radiation below 200 nm. In in-duct applications, the level of ozone that may be produced by arrays of UV lamps is greatly diluted by the volume of airflow to the point that the supply air will have non-detectable levels of ozone. This may not be the case for Upper Room systems or local recirculation units, where it may be prudent to use only non-ozone producing lamps, or at least low pressure UV lamps with minimal ozone.

5.14 Lamp Maintenance

UV lamps are typically replaced about once a year or every 9–10 thousand hours of continuous operation. Some lamps, especially small UV lamps, may have shorter lifetimes and the manufacturer's specifications should be consulted for change-out guidelines. Lamps that are operated in ambient conditions that are less than ideal

Fig. 5.13 Example of a
burned-out UV lamp (*bottom*)
compared with a new lamp
(*top*)

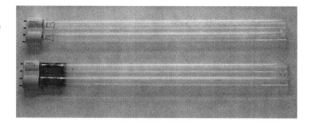

(too hot or too cold) and lamps that are periodically cycled on and off may also have reduced lifetimes. For this reason, the visual inspection of UV lamps should be a regular maintenance item. Ideally, non-UV transmittant portholes (i.e. glass or plastic of appropriate thickness) will be provided so that lamps can be viewed during operation. Alternatively, a radiometer can be used, with appropriate precautions and attire, to verify that UV lamps are operating near their design output. Lamps that have burned out will have dark areas near the electrodes, as in Fig. 5.13.

In-duct UV lamps should be protected from dust by upstream filters and should be regularly inspected for dust accumulation. Upper room systems are likely to accumulate dust, even if room supply air is filtered, and may need periodic cleaning. Lamps can be wiped clean with a cloth dampened with water or a cleaning agent like dilute alcohol, ammonia, or water (Philips 1985). No oils or waxes should be used on the wiping cloth. All maintenance personnel designated to care for UV systems should receive adequate training in both UV system maintenance and UV safety. See also Chap. 11: UVGI Guidelines and Standards for additional information on installation and maintenance of UV lamps.

References

ANSI. 2004. Lamp Ballast – Line Frequency Fluorescent Lamp Ballast. New York: American National Standards Institute. Report nr ANSI C82.1-2004.

ANSI. 2005a. Double-Capped Fluorescent Lamps – Dimensional and Electrical Characteristics. New York: American National Standards Institute. Report nr ANSI_IEC C78.81-2005.

ANSI. 2005b. Electric Lamp Bases. New York: American National Standards Institute. Report nr C81.61-2005.

Barkhudarov EM, Christofi N, Kossyi IA, Misakyan MA, Morton J, Sharp J, Taktakhisvilli IM. 2006. Eradication of bacteria present in water and on surfaces using microwave UV lamps; Zvenigorod, Russia.

Bettles T, Schujman S, Smart J, Liu W, Schowalter L. 2007. UV light emitting diodes – Their applications and benefits. IUVA News 9(2).

CIE. 2003. Ultraviolet Air Disinfection. Vienna, Austria: International Commission on Illumination. Report nr CIE 155:2003.

Crawford MH, Banas MA, Ross MP, Ruby DS, Nelson JS, Boucher R, Allerman A. 2005. Final LDRD Report: Ultraviolet Water Purification Systems for Rural Environments and Mobile Applications. Albuquerque, NM: Sandia National Laboratories. Report nr SAND2005-7245.

Fusion. 2004. F300S/F300SQ Ultraviolet lamp System: Installation, Operation, and Maintenance. Gaithersburg, MD: Fusion UV Systems, Inc.

Guha S, Bojarczuk N. 1998. Ultraviolet and violet GaN light emitting diodes on silicon. Appl Phys Lett 72:415.

Hu JY, Chu SN, Quek PH, Feng YY, Tang, XL. 2005. Repair and regrowth of Escherichia coli after low- and medium-pressure ultraviolet disinfection. Wat Sci Technol 5(5):101–108.

IES. 1981. Lighting Handbook Application Volume: Illumination Engineering Society.

IESNA. 1996. Measurement of Ultraviolet Radiation from Light Sources. New York: Illumination Engineering Society of North America. Report nr IESNA LM-55-96.

IESNA. 2000. IESNA Lighting Handbook HB-9-2000, Reference and Application Chapter 6, Light Sources. New York: Illumination Engineering Society of North America.

IUVA. 2005. General Guideline for UVGI Air and Surface Disinfection Systems. Ayr, Ontario, Canada: International Ultraviolet Association. Report nr IUVA-G01A-2005.

Kowalski WJ, Bahnfleth WP. 2000. UVGI design basics for air and surface disinfection. HPAC 72(1):100–110.

Lau J, Bahnfleth W, Kremer P, Friehaut J. 2008. Investigation of Surface Temperature Distributions of UVC Lamps Under Variable Flow Conditions Using Infrared Camera Measurements; Denmark.

Lau J, Bahnfleth W, Friehaut J. 2009. Estimating the effects of ambient conditions on the performance of UVGI air cleaners. Building and Environment 44:1362–1370.

Lindsey JL. 1997. Applied Illumination Engineering. Lilburn: The Fairmont Press, Inc.

Masschelein WJ. 2002. Ultraviolet Light in Water and Wastewater Sanitation. Boca Raton, FL: Lewis Publishers.

NFPA. 2008. National Electrical Code. Quincy, MA: National Fire Protection Association. Report nr NFPA 70, NFPA 70B.

Philips. 1985. UVGI Catalog and Design Guide. Netherlands: Catalog No. U.D.C. 628.9.

Schalk S, Adam V, Arnold E, Brieden K, Voronov A, Witzke H-D. 2006. UV-lamps for disinfection and advanced oxidation – Lamp types, technologies and applications. IUVA News 8(1):32–37.

UL. 2004. Luminaires. Research Triangle Park, NC: Underwriters Laboratories. Report nr UL 1598.

Westinghouse. 1982. Booklet A-8968, Westinghouse Lighting Handbook. Westinghouse Electric Corp., Lamp Div.

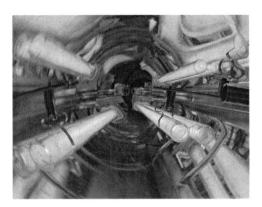

6.1 Introduction

Many types of UVGI systems and equipment are currently available for air and surface disinfection applications. These systems include lamps, fixtures, and ballasts, and often include reflective surfaces to enhance the irradiance field. Various other components may form part of more complete systems, including filters, light baffles, and UV-absorbing surfaces. Disinfection systems are specialized for each application and are either prefabricated or are engineered to adapt to particular applications. The previous chapter reviewed the types of lamps, ballasts, and basic lamp fixtures that are currently available. This chapter reviews complete UV systems, how the components are put together for specific applications, and general characteristics. Table 6.1 classifies the main types of UVGI systems used for air and surface disinfection.

Some of the systems shown in Table 6.1 serve a dual purpose, disinfecting both air and surfaces. Each of these systems has unique performance specifications and testing requirements. Most of these systems are described and analyzed in more detail in the later chapters on applications. The following sections briefly describe these systems and provide examples. The subsequent sections address additional components and auxiliary types of UV equipment.

6.2 Air Disinfection Systems

UV systems for disinfecting air will either use forced air, as with in-duct or recirculation units, or rely on natural air movement for circulation, as in Upper Room or Barrier systems. The degree of effectiveness of these systems varies with local conditions and it cannot be said for certain that one system is better than another. Ultimately, the choice of which system to install in a building will depend on both economics and installation feasibility. The following sections review the basic systems and their components. More detailed information on design and performance is provided in Chap. 8 for airstream disinfection systems and Chap. 9 on Upper Room systems.

W. Kowalski, *Ultraviolet Germicidal Irradiation Handbook*,
DOI 10.1007/978-3-642-01999-9_6, © Springer-Verlag Berlin Heidelberg 2009

Table 6.1 Types of UVGI Systems

Classification	System
Air disinfection	In-duct air disinfection
	Recirculation units
	Upper room systems
	UV barrier systems
	Overhead tank disinfection
Surface disinfection	Equipment & packaging disinfection
	Cooling coil disinfection
	Lower room disinfection
	Overhead surgical site disinfection
	Area/Room disinfection
	Food surface disinfection
	Hand wands
	Mold remediation systems

6.2.1 In-Duct Air Disinfection

In-duct UVGI systems serve the purpose of disinfecting an airstream in a building or zonal ventilation system. They generally consist of UV lamps, fixtures, and ballasts, and rely on existing filters to maintain lamp cleanliness. UV lamp fixtures can be placed almost anywhere in the ductwork or in the air handling unit (AHU). In some systems the lamp fixtures and ballasts are installed internally while in others the fixture and ballasts are installed external to the ductwork, which will result in lower pressure drops. In general, the pressure losses associated with UV lamps in the airstream are minor, especially if the air velocity is within the normal design limits of 2–3 m/s (400–600 fpm). Figure 6.1 shows two types of in-duct systems.

Some modular systems are available for installation in ductwork. Systems that recirculate room air will deliver multiple UV doses to airborne microorganisms and this chronic dosing will enhance system effectiveness. Although the effect is dependent on the room air exchange rate, the result is an increase in overall removal rates in comparison with single-pass system.

Retrofitting UV lamps in an existing ventilation system may require the upgrading or addition of air filters, generally at least a MERV 6–8 filter, but preferably a MERV 10 or higher to maximize air cleaning. Increasing the filter efficiency may reduce the airflow or increase fan motor power draw, depending on fan and motor type. If the existing air velocity in the duct is too high, it may be necessary to expand the duct to reduce air velocity to 2–3 m/s (400–600 fpm) to ensure UV system effectiveness. Some modular systems include filtration. Figure 6.2 shows an example of a UVGI unit with integral fan and filter that can be retrofit in a residential air system.

Auxiliary components and materials may be added to in-duct systems to enhance efficiency or improve safety. Reflective materials such as polished aluminum can

Fig. 6.1 Two types of in-duct systems. *Left*, a single lamp with external ballast that can be attached to ductwork. *Right*, a multi-lamp module that can be retrofit in duct. Images provided courtesy of Light Progress, Snc., Anghiari, Italy

Fig. 6.2 UVGI system designed for retrofitting residential central air units. Photo provided courtesy of Air Clean Assurance Corporation, Houston, TX

be added around the UV lamp fixtures to increase irradiance. Light baffles can be added to reduce downstream irradiance when there is a chance that UV rays may bleed through to air registers into occupied areas. UV-absorbing paint may be used on exit ductwork to reduce bleed-through of UV rays. Access doors may also be necessary for maintenance. Sight-glasses or small windows may be added to allow regular inspection of lamps, although some external fixtures include such options.

In-duct systems that are designed for specific applications and built up from components are known as engineered systems and their performance must be established by analysis or testing. Complete systems that are 'plugged in' to an existing ventilation system will generally have performance specifications that are known in advance. In either case, the actual air disinfection performance will depend on local operating conditions, especially airflow, and building volume, and these can only be determined by analysis or testing. All engineered systems must be built to current electrical standards and that requires that wiring and other components be shielded from UV or be made of UV-resistant materials. Any existing ventilation system components must also be protected from UV degradation or verified as UV-resistant. Retrofitted systems must also be tested to ensure that no stray UV rays exit the duct (i.e. through air supply grilles or gaps) and that no personnel hazards are created.

6.2.2 Recirculation Units

UV recirculation units generally consist of UV lamps and fixtures in a housing containing a blower. They often, but not always, contain a filter to keep the lamps free of dust. If a filter is included, it will generally be at least a MERV 6–8 filter, but units that use a MERV 10 or higher filter will have superior air cleaning performance. Some UV recirculation units include HEPA filters, but this can significantly increase system energy consumption without necessarily improving air quality. The airflow in recirculation units can vary but often they are in the range of 1.4–14 m³/min (50–500 cfm) and are suitable for small rooms or apartments only. Many recirculation units are portable and can be positioned on the floor or on tables. Some units are available for hanging on walls or from ceilings, especially those used in the health care industry. Room recirculation units and upper air systems can be installed to augment in-duct systems or where in-duct installation is not feasible. Figure 6.3 shows two examples of stand-alone units in applications.

Because of their compact size, the internal volume of recirculation units does not allow for extended exposure times and most manufacturers attempt to make up for

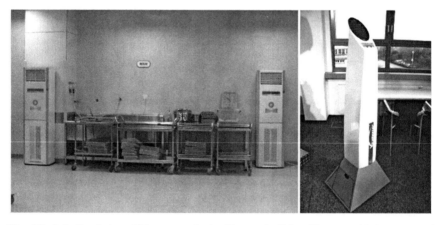

Fig. 6.3 *Left*: Stand-alone UV systems for health care facilities (Photo provided courtesy of Juguang Photoelectric Technology Co., Inc., China). *Right*: Residential unitary UV system (Photo provided courtesy of Virobuster Electronic Air Sterilisation, Enschede, The Netherlands)

this by increasing the total UV wattage. This approach may not guarantee effectiveness, and may not meet the minimum requirements set forth in IUVA guidelines of a 0.25 s exposure time (IUVA 2005). Although UV recirculation units may have a specified UV dose, without this minimum residence time, the actual performance may be less than predicted.

One problem that can occur with UV recirculation units that use high UV wattage is bleed-through of UV rays from the exit, the entrance, or through gaps in the enclosure. Since the distance from the lamps to the exit or entrance grilles is often short, the irradiance levels near the unit may be high and possibly exceed ACGIH TLVs. The best way to deal with this problem is to include light baffles, internal UV-absorbing surfaces, and/or exit filters, which will reduce bleed-through. Stray irradiance can be measured with a radiometer, but a simple way of checking for bleed-through is to place the unit in the dark and observe any visible light leakage. Another problem with compact UV recirculation units is that internal wiring may be exposed to UV and may be subject to deterioration over time. In general, such units must meet specific electrical codes such as those of IESNA, UL, ETL, or other applicable standards, and will have labeling indicating such conformance.

Some recirculation units include auxiliary components such as charcoal filters, PCO (photocatalytic oxidation), or ionization components that contribute to air disinfection or that remove gases and odors. More detailed information on sizing and evaluating UV recirculating units is provided in Chap. 8.

6.2.3 Upper Room Systems

Upper Room systems most often consist of single or multiple Upper Room fixtures installed on the walls or hung from the ceiling. These are relatively simple to install, but performance can depend on a number of factors such as local air movement, ceiling height, location of the fixtures, and local UV reflectivity of the walls and ceiling. There are basically two types of Upper Room fixtures – wall-mounted and ceiling hung varieties with or without louvers (see Fig. 6.4). The choice of which to use depends to some degree on local conditions. In addition to the fixtures, an Upper Room system may include the use of UV-absorbing paints on the walls or ceiling to minimize UV exposure to room occupants. The application of such materials may allow higher UV wattages to be used, which will increase system effectiveness without causing increased hazards. The general design methodology for Upper Room systems, as explained in more detail in Chap. 9, is to provide maximum irradiance in the upper portion of the room while minimizing UV levels below the UV units to less than ACGIH TLVs.

Upper Room systems can provide a cost-effective alternative to forced air systems where budgets are critical or for buildings that rely on natural ventilation. Some Upper Room systems have built-in fans for assisting air movement. See Chap. 9 for more information on Upper Room systems.

Fig. 6.4 Two types of wall-mounted Upper Room fixtures – louvered (*left*) and unlouvered (*right*). Photos reprinted courtesy of American Air & Water, Inc., Hilton Head Island, South Carolina

6.2.4 UV Barrier Systems

UV Barrier systems are generally installed in doorways between rooms and consist of UV lamp fixtures mounted on one or more sides of a doorway (i.e. each side and overhead). The fixtures themselves are often of the Upper Room type with louvers to constrain the UV rays to parallel directions across the narrow doorway. In some cases a barrier system consists only of an overhead UV fixture in a hallway or passage between two rooms. Barrier systems are often considered to be surface disinfection systems that irradiate clothes and materials that pass through the barrier but they are actually intended to irradiate the air that passes between rooms. The UV levels in the barrier can be high and hazardous for anything but a brief passage, and eye protection is essential. Though not in common use, they do provide options in the health care industry and other applications. Figure 6.5 shows an example of a Barrier system fixture.

The ideal barrier system would involve not a doorway, for which the exposure time of the air is brief, but would involve a hallway at least one meter long in which the irradiance levels are balanced to meet the airflow such that an appropriate UV dose would be generated and the air passing through the hallway would be highly disinfected in a single pass. Such a barrier system would necessitate complete body coverage and eye protection for any personnel passing through, and may also require some type of light baffling (i.e. plastic air curtains), protective doorways, or 90° turns.

Fig. 6.5 Louvered barrier system for overhead door or hallway mounting. Photo reprinted courtesy of American Air & Water, Inc., Hilton Head Island, South Carolina

6.2.5 Overhead Tank Disinfection

Overhead tank systems, also called liquid tank conditioners, are a specialized UV application in which the air in the headspace of liquid storage tanks is disinfected. The intent is to control bacteria, yeast, and mold growth which are favored in the warm, moist air space at the top of tanks that store water or foodstuffs like liquid sugar or maple syrup. These systems pressurize the head space with disinfected air that has been filtered and irradiated. Such systems target water-based pathogens as well as food spoilage fungi and bacteria. Figure 6.6 shows an example of a forced air overhead tank UV system with integral filter.

Fig. 6.6 Overhead tank UV system with integral blower and electrostatic filter. Photo provided courtesy of American Air & Water, Inc., Hilton Head Island, South Carolina

6.3 Surface Disinfection Systems

UV surface disinfection systems are used to specifically disinfect surfaces such as those in the food industry and health care industries. Some systems are available for sterilizing bathrooms. Some systems are portable units for remediation. Cooling coil disinfection systems are essentially surface disinfection systems, but they often serve a dual purpose for air disinfection. Area disinfection or decontamination systems can be used to eliminate surface contamination in open areas, either for remediation or for prevention of potential hazards. Portable UVGI systems are available to decontaminate open areas and these are, of course, operated only in unoccupied areas. After Hours UVGI systems are area disinfection systems that are permanently mounted on walls or ceilings and are engaged during times when no occupants are present. After Hours systems are engaged either by timers or disengaged by motion detectors, and provides options for mailrooms, laboratories, or other facilities where contamination potential exists.

6.3.1 Equipment and Packaging Disinfection

Equipment and packaging disinfection systems are used for disinfecting surgical and medical equipment, mail, clothes, pharmaceutical packaging, food packaging, and other types of equipment and materials. Medical equipment disinfection units are in common use today in medical and dental offices and they sterilize everything from reusable needles to scalpels. Systems often consist of a reflective metal enclosure or room lined with reflective aluminum in which one or more UV lamp fixtures are installed. Many systems mount lamp fixture on multiple sides, or else rely on reflectivity to obtain all-around exposure of items. Small units and cabinets may have only a single UV lamp fixture and exposure times of up to 1 h or more are often required. Systems with moving conveyor belts have speeds adjusted such that the exposure time produces an appropriate dose (see Fig. 6.7).

Packaging disinfection systems are used in the pharmaceutical and food industries to sterilize containers that are intended for sterile products. They are often assembly-line type systems with UV lamp fixtures on the sides, bottom, or top. UV can sterilize the external surfaces of packaging components but does not penetrate packaging materials, even transparent plastic or glass. In order to sterilize the internal surfaces the UV rays must shine down inside them, and the degree to which this is effective depends on the packaging design. That is, the packaging components may have to be exposed on the inside for the UV to have any effectiveness. More penetrating forms of radiation, including pulsed light and gamma radiation, are more commonly used for packaging disinfection.

Figure 6.8 shows an example of a laundry disinfection system in which the laundry is passed from an area of higher contamination to a clean area, after irradiation on all sides with the doors closed. In this system, pressurization is maintained to ensure air flows out from the clean side only.

Mail disinfection UVGI Systems are a specialized systems designed just for mail or packages. They may be small systems for disinfecting incoming mail for a business or home, or may be room-sized systems designed to irradiate the surfaces of

Fig. 6.7 Conveyor belt UV system for disinfection of products and packaging. Photo provided courtesy of Light Progress, Snc., Anghiari, Italy

Fig. 6.8 Pass-through
laundry disinfection system.
Laundry trolleys are rolled
into a decontamination
compartment surrounded on
top, bottom, and sides with
UV lamps and reflective
aluminum. Image provided
courtesy of Philips Lighting,
Eindhoven, The Netherlands

packages on palettes. Like other equipment disinfection systems, they are often
capable of producing very high irradiance levels and have specified exposure times,
typically 30–60 min. UV cannot penetrate inside of mail envelopes or packages but
it can disinfect the exterior surfaces. The main application for mail disinfection systems is biodefense, or protection against intentional mailings of bioweapon agents
since such agents can leak through paper and packaging and pose hazards to mail-
room personnel.

6.3.2 Cooling Coil Disinfection

Cooling coil disinfection systems perform surface disinfection as their primary
design intent, but because they are generally installed in the airstream of air handling units, they simultaneously perform some air disinfection. UVGI has enjoyed
much success in this application thanks to favorable economics and energy savings that result from operating clean coils (Shaughnessy et al. 1999, Scheir and
Fencl 1996). Cooling coil systems consist of one or more lamp fixtures mounted in
front, in back, above, below, or on the sides of cooling coils, as shown in Fig. 6.9
(also see the Chap. 8 heading photo – courtesy of Immune Building Systems, Inc.,
New York). These systems operate continuously and therefore can be of relatively
low UV wattage, but often the UV wattage is high so as to ensure that the UV rays
can penetrate the cooling coils and expose shadowed areas. Reflective aluminum is
sometimes added around such systems to remove shadowed areas.

Cooling coil UV systems may serve dual purposes when they irradiate internal
air handling units surfaces to control mold growth, especially in drain pans or places
where moisture may collect. They may also disinfect the air if appropriate wattage
and other design considerations are taken into account.

When cooling coil irradiation systems are retrofit into existing air handling units,
there are often constraints on space, in which case multiple lamp fixtures may have

Fig. 6.9 Cooling coil disinfection system, composed of two reflective UV light fixtures mounted facing a cooling coil (*left*). Photo provided courtesy of Sanuvox, St. Laurent, Quebec

to be installed wherever space permits, boosting power and adding reflectivity as necessary to achieve adequate UV exposure of the surface. Ideally, a coil will have low wattage fixtures placed on both sides, but this is an exception rather than the rule. More often, high wattage systems are placed on one side such that some residual levels of UV will appear on the other side of the coils. Since cooling coils generally sit downstream of filter banks, they may not require the addition of filters. However, if the filters are not rated at least MERV 6–8, the filters may have to be upgraded, and the associated increase in pressure losses may have to be considered. See Chap. 10 for more detailed information on designing cooling coil irradiation systems.

6.3.3 Lower Room Disinfection

Lower Room UVGI systems create a UV field in the lower 1–2 ft of floor space much the same as Upper Room systems for ceilings, except that the intent is generally to disinfect the surface of the floors and air disinfection is a collateral benefit. Fungal spores and bacteria gravitate to the floor and will settle into rugs and carpeting, with the result that airborne concentrations are typically higher near the floor where settled microbes are stirred up by traffic. Irradiating the floor to destroy settled microbes will therefore contribute to reducing airborne contamination levels. The fixtures used for Lower Room systems are much the same as wall-mounted Upper Room systems, and may include reflectors or baffles to direct the irradiance downwards. Exposure hazards are minimal provided personnel use appropriate legwear (unless these systems are only used after-hours). Floor and lower walls surfaces must be non-reflective for UV, which is often the case, so that stray UV rays do not reflect upwards. Though uncommon, such systems could be used in operating rooms, hospital hallways, and in other facilities where the predominant problem is airborne spores and bacteria that tend to settle downwards. Unlike Upper Room

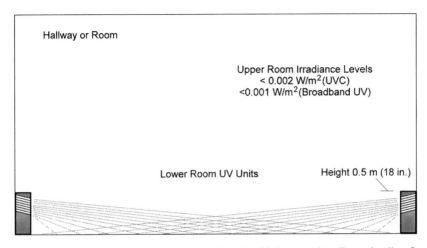

Fig. 6.10 Example placement of Lower Room UV units with downward angling to irradiate floor surface. Recommended ACGIH TLVs above 0.5 m are shown

systems, which often shine UV in parallel rays horizontally across ceiling spaces, Lower Room systems are preferably designed to shine directly across the floor surfaces, angling downwards rather than straight across, as shown in Fig. 6.10. UV exposure above the specified height (0.5 m) have the same TLVs as Upper Room systems.

Lower Room UV systems may be ideal for entrance doorways, such as in hospitals, where foot traffic brings in environment spores. Any such installations, however, should be coupled with motion detectors to prevent hazards to uncovered legs and inquisitive children. It may also be possible to link such systems to automatic doors in double-door entrances.

6.3.4 Overhead Surgical Site Disinfection

UVGI systems designed to expose a surgical site on a patient in an operating room have been used with great success and have become increasingly popular recently as the result of successful demonstration studies (Ritter et al. 2007). The disinfecting capabilities of such systems tend to outweigh the hazards of direct UV exposure of skin or internal organs. Overhead UV systems consist of UV lamp fixtures either hanging over or installed in the ceiling in recessed troffers like normal lighting. Variable controls or ON-OFF controls are provided such that the operating room personnel may adjust the UV intensity before, during, and after procedures. Figure 6.11 shows the installation of UV lamp fixtures in the ceiling directly above the operating table (and above the hanging lighting fixture). Overhead surgical site systems require doctors and nurses to wear protective eyewear, clothing, and skin

Fig. 6.11 Overhead Surgical
UVGI System with UV lamps
in four recessed troffers
above the operating table.
Image provided courtesy of
Dr. Merrill Ritter of the St.
Francis Hospital,
Mooresville, IN

creams. See Chap. 17 for further information on designing Overhead UV Systems
for health care facilities.

6.3.5 Area/Room Disinfection

Entire rooms may be disinfected by continuous or intermittent exposure from UV
lamps. Permanently installed systems, such as those shown in Fig. 6.12, generally
consist of UV lamp fixtures mounted on ceilings or walls. These installations require
periodic cleaning and require personnel protection or shutoff controls during occupation. Portable UVGI Systems are also commonly available that can be moved into
place temporarily to disinfect room surfaces or equipment.

After-Hours UVGI Systems are used in open areas to disinfect surfaces like
walls, floors, tabletops, and equipment during periods when the room is unoccupied.

Fig. 6.12 Two examples of UV lamp fixtures mounted overhead in a bakery (UV fixtures are above
the lighting fixtures) for the control of mold and bacteria. Photos reprinted courtesy of Philips
Lighting, Eindhoven, The Netherlands

Fig. 6.13 Area disinfection unit with integral timer and controls, shown in close-up at right. Image provided courtesy of Lumalier, Memphis, TN

They consist of UV lamp fixtures coupled with a control unit. They can be engaged by timers to operate overnight and they may include controls to disengage when doors are opened or movement is detected (i.e. radar detectors). The controls can also be set to operate continuously and to disengage when anyone enters the area. Figure 6.13 shows an example of an area disinfection unit with integral timers that can be set to both engage and disengage at preset periods (i.e. once occupants have left the room).

6.3.6 Food Surface Disinfection

UV systems are sometimes used in the food industry to disinfect food surfaces but because UV has no penetrating ability it has only limited applications in this regard (Yaun and Summer 2002). The FDA permits the use of UV for control of surface microorganisms provided that there is no ozone production, that no high fat-content foods are irradiated and that irradiance levels are limited to 1 W of UV per 5–10 ft^2 (FDA 2000). UV is, of course, widely used for water disinfection and water may be an ingredient in foodstuffs. The most common use of UV in the food industry is for controlling mold and bacteria in food processing facilities, especially cheesemakers, bakeries, and breweries. By keeping building and equipment surfaces free of fungal spores, yeasts, and spoilage bacteria, UV can enhance product safety, productivity and avoid personnel health hazards.

Figure 6.14 shows an example of a conveyor belt irradiation system for disinfecting food surfaces and packaging. See also Chaps. 10 and 18 for more information on UV applications in the food industry.

Fig. 6.14 UV conveyor belt systems. *Left*: belt with roller chain for transport of food products. *Right*: belt for disinfection of food packaging. Images provided courtesy of Light Progress, Snc., Anghiari, Italy

6.3.7 Mold Remediation Systems

Various types of UV systems can be used to disinfect building or equipment services that have become contaminated with mold, especially as a result of water damage. Most of these systems are the same as those used for area disinfection or Upper Room disinfection. Portable UV units are used in such applications but sometimes systems are installed on a permanent basis to continually control mold growth. One type of system that is currently available involves the use of a flexible UV lamp array that can be slipped through a duct to disinfect internal duct surfaces. The flexible array can be removed after a period of hours or days once the contamination problem has been eliminated, but such systems may also be left in place permanently.

6.3.8 UV Hand Wands

Lightweight battery-operated or rechargeable hand-held UV units are available for surface disinfection applications, such as those shown in Fig. 6.15. These units produce UV light focused downwards, usually with parabolic reflectors, and are suggested for use in health care settings and for home use (i.e. kitchens, toilets, mattresses, clothes, etc.). Typical recommended surface irradiation times are on the order of 10–20 s. Units usually include timers and other controls. Such units must, of course, be used with extreme caution and users should wear eye protection and either wear protective clothing or else ensure that for short duration exposures the UV beam is only directed away from the user and not towards any reflective UV surfaces. Users of such devices should educate themselves regarding the safety hazards of ultraviolet irradiation and the ACGIH threshold limit values (TLVs) before attempting to disinfect surfaces with these devices, and should be aware that invisible UV rays can reflect off many kinds of common surfaces and cause severe eye damage and erythema.

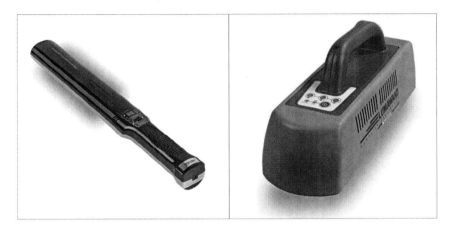

Fig. 6.15 UV Hand Wands, or hand-held portable UV disinfection units. *Left* image provided courtesy of Purelight UK, Ltd, Falconwood, London. *Right* image provided courtesy of Lumalier, Memphis, TN

6.4 UV Reflective Materials

Enhancement of UVGI effectiveness is possible through the use of materials that are highly reflective in the UV spectral range. Polished sheet aluminum is the most common material used in such applications and has a largely specular UV reflectivity of about 75%. Aluminum foil insulation is also currently available and is both inexpensive and easy to install. The most reflective material of all is ePTFE from W.L.Gore, with a diffusive UV reflectivity of approximately 99% (Lash 2000). Surrounding the internal surfaces of an air handling unit with reflective materials may amplify the UV irradiance field and will provide UV exposure to areas that would otherwise be shadowed. Specular materials are most common but diffusive materials will distribute UV rays more evenly. Specular materials with irregular surfaces (like diamond plate) will have increased diffusive components. The technique of peening, or adding numerous indentations to a surface, is used in luminaire design to increase the diffusive component of specular reflectors. Reflective surfaces may require periodic cleaning although upstream filtration will tend to keep both the UV lamps and the reflective materials clean. A list of materials and their UV reflectivities has been provided in Appendix F based on a variety of summaries and sources (NIOSH 1972, Summer 1962, Luckiesh 1945, 1946, Luckiesh and Taylor 1946, Wilcock and Soller 1940, Koller 1952, Toshimasa et al. 1999, Hulburt 1915, and Hagen and Rubens 1902).

6.5 UV Light Baffles

Light baffles may be used to either prevent UV light from escaping a UVGI fixture or to enhance the UV irradiance field. Light baffles that block UV from exiting a system have been used in laboratory studies and can be useful for assuring the safety of UV systems where the possibility of escaping UV rays could cause harm to occupants or furnishings. The basic light baffle will typically consist of a series of blinds, often painted black or using some material that has low UV reflectivity, but which allows unimpaired airflow, such as the one illustrated in Fig. 6.16.

The total attenuation of light by a light baffle depends on the reflectivity of the UV-absorbing surfaces, and the mean number of reflections. The attenuation of light through a light baffle can be computed as follows:

$$A_i = \rho^n \tag{6.1}$$

where

A_i = reduction of irradiance (at exit), fractional
ρ = reflectivity of absorbing surface, fractional
n = mean number of reflections

There are no perfectly absorbing surfaces, and the best absorbers have about 4% reflectivity in the UV range. In the example in Fig. 6.16, assuming a mean of two reflections, the total attenuation would be $(0.04)^2 = 0.0016$, or a 99.84% reduction in the exiting irradiance.

A simple alternative for duct systems is to paint the downstream duct with black or UV-absorbing paint. This will be particularly effective if there is at least one elbow. A $90°$ elbow (a long elbow) at the exit of a UV system through which no UV has a direct path will attenuate at least 96% of the UV rays (at 4% reflectivity). Two elbows in series will reduce the UV by over 99%.

The second type of light baffle is one composed of reflective materials that are designed to enhance and amplify the internal UV irradiance field. Reflective pol-

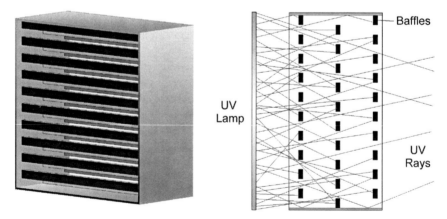

Fig. 6.16 Example of a light baffle (*left*) with UV-absorbing black surfaces facing the UV lamp. *Right* shows a ray-tracing diagram illustrating how the reflected UV rays become attenuated through the baffling

ished aluminum or new materials like ePTFE can be used to increase the irradiance field and thereby reduce the required UV wattage. In theory, it is possible to amplify the effective UV wattage many times over with a well-designed reflective chamber, with consequent energy savings. However, the baffle will increase fan energy consumption due to the pressure drop and is a factor in any energy optimization analysis.

Light baffles will cause an increase in system pressure loss and care should be taken in selecting or designing a baffle to minimize impact on airflow. These pressure losses can be reduced by expanding the ductwork such that the free area through the baffle is at least equal to the area of the duct. That is, if a baffle has only 50% free area (half the duct area is blocked by the baffles) then the duct could be expanded to twice the original size to compensate. This will not completely eliminate the pressure losses through the light baffle, but it may make them manageable. The problem is the flat surface of the baffle plates incurs a pressure loss. Conceivably, the baffle plates could be shaped as airfoils to minimize losses. In any event, the fan power and total airflow must be addressed in any system in which light baffles are installed, as they will likely reduce the airflow, or increase fan power draw, depending on the fan and motor. In general, the power consumption of the UV lamps is almost negligible compared to the power consumption of the fan motor, and therefore it may be difficult to justify the use of light baffles unless there is a personnel hazard.

6.6 UV-Transmittant Materials

There are a limited number of materials that are effectively transparent to UV, including quartz glass, which is used to manufacture the UV lamp bulbs, Softglass, and teflon. Such materials can be used to irradiate materials that must be kept enclosed or separate from the UV lamp system. Some materials are nearly transparent to UV spectra while being simultaneously opaque in the visible light range. One such material, Teflon, has been used in the past as a means of protecting UV lamps from excessive cooling effects in an airstream while simultaneously transmitting UV for air disinfection. Teflon has a low thermal conductivity and acts to keep the lamp hot, potentially providing energy savings. It may also serve to help keep the glass surface of the lamp clean. Lamps which are subject to cooling effects may benefit from the use of such materials especially in cold airstreams or at air velocities that exceed design parameters.

6.7 UV Absorbers

UV absorbing paints and other materials can be used to decrease the reflected irradiance and minimize UV exposure. These materials are useful for controlling the reflected irradiance from Upper Room and Lower Room systems, and can be used to attenuate UV rays at the entrance and exit planes of air disinfection systems. Zinc oxide paints (reflectivity 4–5%) are one example of a useful UV-absorbing material that can be used along with Upper Room systems, or to prevent stray UV rays from exiting ductwork. UV absorbing materials can be used to block UV rays and thereby protect surfaces that may be subject to UV degradation. See Chap. 15 for additional information on UV absorbers and see Appendix F for a list of material reflectivities.

6.8 Flow Straighteners

Flow straighteners have been used in the HVAC industry to deal with lopsided air profiles and to reduce pressure drops through sharp turns. Typically, the airflow on the exit side of the fan will have higher velocities on one side of the duct, and this may produce undesirable effects like excessive pressure losses. In UVGI systems, it is possible that such stratified flow may reduce the overall effectiveness of the air disinfection process. Therefore, the use of flow straighteners may be beneficial in UVGI applications, although there are no studies to justify their use at present. Flow straighteners may increase or reduce the fan pressure losses, depending on the operating conditions of the system, and so some care must be taken to evaluate the relative economics of using flow straighteners for UVGI applications. Traditional flow straighteners consist of a series of parallel plates, or rectangular grilles, or hexagonal grilles of a few centimeters or more in length. Air is forced to divert more evenly across the grilles as a result of the pressure loss, and so the exiting air will be more evenly distributed.

6.9 PCO Systems

Photocatalytic oxidation (PCO) systems have been in use for some time and hold great promise for the removal of odors and gases, something which UV is not effective at. Titanium dioxide, under exposure to light, can react with various organics, including microorganisms, to disinfect air or surfaces. The light source is typically a UV lamp, although visible light systems are possible, and the disinfection is produced by a combination of the PCO and the UV. The substrate on which the titanium dioxide ($TiO2$) is located is often a low-efficiency filter or other material (i.e. glass beads) that basically acts like a filter in order to bring the contaminants in close contact with the activated $TiO2$. The internal components of a PCO system include lamp fixtures and often filters, as shown in Fig. 6.17.

The lamps are positioned to irradiate the surface of the PCO material, whether it is a filter-like cartridge or a glass bead container, and the lamp operates continuously during airflow. Irradiance levels are no higher than in a typical UV air disinfection system. PCO systems apply UV technology, but they are not UV systems per se, and therefore they are not addressed further in this book. There are many good references in the literature that readers may consult for additional information (Kowalski 2006, 2003, Goswami et al. 1997, Jacoby et al. 1996).

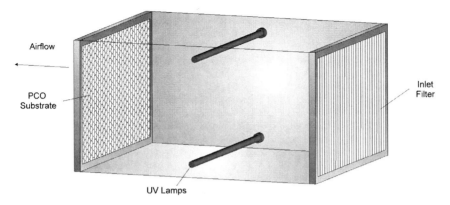

Fig. 6.17 Illustration of a basic PCO system irradiated with UV lamps

6.10 Controls and Integral Sensors

Control units are available that will allow the UV output of lamps to be manually or automatically adjusted. One application is in operating room Overhead UV systems in which the irradiance levels are dialed down during procedures but turned up during unoccupied periods for room disinfection. Integral UV sensors are also available for monitoring UV output of in-duct systems, enabling verification of performance or indicating when UV lamps may need replacement.

References

FDA. 2000. Ultraviolet radiation for the processing and treatment of food. Food and Drug Administration. Report nr 10CFR21, Section 179.39 & 179.41.

Goswami DY, Trivedi DM, Block SS. 1997. Photocatalytic disinfection of indoor air. In: ASME, editor. Transactions of the ASME – Solar Engineering, pp. 92–96.

Hagen E, Rubens H. 1902. The reflection ability of some metals for ultraviolet and infrared radiation. Annalen der Physik 8(5):1–21.

Hulburt E. 1915. The reflecting power of metals in the ultraviolet region of the spectrum. Astrophys J 42:205–230.

IUVA. 2005. General Guideline for UVGI Air and Surface Disinfection Systems. Ayr, Ontario, Canada: International Ultraviolet Association. Report nr IUVA-G01A-2005.

Jacoby WA, Blake DM, Fennell JA, Boulter JE, Vargo LM. 1996. Heterogeneous photocatalysis for control of volatile organic compounds in indoor air. J Air & Waste Mgmt 46:891–898.

Koller LR. 1952. Ultraviolet Radiation. New York: John Wiley & Sons.

Kowalski WJ. 2003. Immune Building Systems Technology. New York: McGraw-Hill.

Kowalski WJ. 2006. Aerobiological Engineering Handbook: A Guide to Airborne Disease Control Technologies. New York: McGraw-Hill.

Lash DJ. 2000. Performance Benefits of Highly Reflective Diffuse Materials in Lighting Fixtures. J Illum Eng Soc Winter:11–16.

Luckiesh M. 1945. Disinfection with germicidal lamps: Control – I. Electrical World Sep. 29:72–73.

Luckiesh M. 1946. Applications of Germicidal, Erythemal and Infrared Energy. New York: D. Van Nostrand.

Luckiesh M, Taylor A. 1946. Transmittance and reflectance of germicidal (l2537) Energy. J Opt Soc Am 36:227–232.

NIOSH. 1972. Occupational Exposure to Ultraviolet Radiation. Cincinnati, OH: National Institute for Occupational Safety and Health. Report nr HSM 73-110009.

Ritter M, Olberding E, Malinzak R. 2007. Ultraviolet lighting during orthopaedic surgery and the rate of infection. J Bone Joint Surg 89:1935–1940.

Scheir R, Fencl FB. 1996. Using UVC Technology to Enhance IAQ. HPAC February.

Shaughnessy R, Levetin E, Rogers C. 1999. The effects of UV-C on biological contamination of AHUs in a commercial office building: Preliminary results. Indoor Environ '99:195–202.

Summer W. 1962. Ultra-Violet and Infra-Red Engineering. New York: Interscience Publishers.

Toshimasa K, Uh J, Norihisa N. 1999. Ultraviolet band reflectance and transmittance for building material. J Arch 525:21–26.

Wilcock D, Soller W. 1940. Paints to reflect ultraviolet light. Ind Eng Chem 32:1446.

Yaun BR, Summer SS. 2002. Efficacy of Ultraviolet Treatments for the Inhibition of Pathogens on the Surface of Fresh Fruits and Vegetables. Blacksburg, VA: Virginia Polytechnic Institute and State University.

Chapter 7

UVGI System Modeling

7.1 Introduction

The modeling of systems for air and surface disinfection can be used to assess the UV dose, which can then be used to determine disinfection rates for specific microbes. The modeling of UV irradiance fields can produce fairly accurate results for the purposes of system sizing. Several components in UV disinfection systems may require modeling – the lamp UV irradiance, the reflective enclosure, if any is used, and the total UV dose imparted by any UV system. The methods presented here can be adapted to evaluating any type of air and surface disinfection system, including cooling coil irradiation systems, Upper Room systems, packaging disinfection, air disinfection, and food disinfection. This chapter presents modeling methods and tools that are available and that have been well-corroborated by empirical data on UV lamps and microbial disinfection rates. These are computational methods and are best applied using software or spreadsheets. Models and methods have been presented in the literature other than those given here, and some of these are accurate while others are either less so or are theoretical models that have not yet been applied (Buttolph and Haynes 1950, Philips 1985, IESNA 2000, Gardner and Shama 1999, Krasnochub 2005). It is not the intent of this chapter to review or compare the various models, but to present the reader with a workable set of tools for designing UV systems for air and surface disinfection. The models detailed herein, while not necessarily simple, represent the most accurate methods currently available. These models are specifically for air and surface disinfection and could be adapted to water disinfection (by incorporating UV attenuation in water) but to the author's knowledge no such adaptations have yet been developed. The modeling methods presented here have been used to develop the tables and charts for system sizing that are presented in Chaps. 8 and 9.

W. Kowalski, *Ultraviolet Germicidal Irradiation Handbook*,
DOI 10.1007/978-3-642-01999-9_7, © Springer-Verlag Berlin Heidelberg 2009

7.2 UV Lamp Modeling

The complete three-dimensional (3D) irradiance field for any enclosed UVGI
system needs to be resolved in order to establish the dose received by any airborne
microbe passing through the field. In theory, the UV dose is the radiative energy
absorbed by the subject microbial population but in fact the actual absorbed dose
is unknown and only the fluence (the incident irradiance) can be measured or calculated, but in keeping with common usage, the term dose is used here. Both the
irradiance field of the lamp and the irradiance field due to a reflective enclosure
can be determined through the use of thermal radiation view factors (Kowalski and
Bahnfleth 2000).

Various models of the irradiance field due to UV lamps have been proposed in the
past, including point source (inverse square law), line source, integrated line source,
and other models (Jacob and Dranoff 1970, Qualls and Johnson 1985, Beggs et al.
2000). The inverse square law simply states that the irradiance decreases as the
inverse of the distance from the source, but although this is adequate for luminaire
design where distances from lights are large, it can be inaccurate for UV design
where distances to the lamp are often much shorter. The model used here is based on
thermal radiation view factors, which define the amount of diffuse radiation transmitted from one surface to another (Modest 1993). Figure 7.1 illustrates a lamp
modeled as a cylinder where the planar area at which the UV irradiance is to be
determined is perpendicular to the axis and is at the edge of the cylinder.

The fraction of radiative irradiance that leaves the cylindrical body and arrives at
a differential area (Modest 1993) is:

$$F = \frac{L}{\pi H} \left[\begin{array}{l} \frac{1}{L} ATAN \left(\frac{L}{\sqrt{H^2 - 1}} \right) - ATAN \left(M \right) \\[2ex] + \frac{X - 2H}{\sqrt{XY}} ATAN \left(M \sqrt{\frac{X}{Y}} \right) \end{array} \right] \qquad (7.1)$$

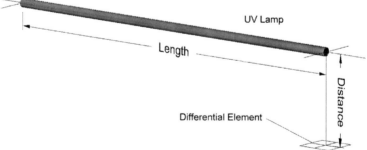

Fig. 7.1 UV lamp modeled as a cylinder some finite distance from a differential element at which
the irradiance is to be computed

The parameters in Eq. (7.1) are defined as follows:

$$H = x/r$$
$$L = l/r$$
$$X = (1 + H)^2 + L^2$$
$$Y = (1 - H)^2 + L^2$$
$$M = \sqrt{\frac{H - 1}{H + 1}}$$

where

$l =$ length of the lamp segment (arclength), cm
$x =$ distance from the lamp, cm
$r =$ radius of the lamp, cm

This equation applies to a differential element located at the edge of the lamp segment. In order to compute the view factor at any point along a lamp it must be divided into two segments. Figure 7.2 shows the predicted results of the irradiance at a series of points that extend from the lamp midpoint. Comparison of the view factor model with lamp photosensor data gives excellent agreement. No other models that have been used, including point source, line source, and integrated line source models, approach the accuracy of the view factor model, especially in the near field (Kowalski and Bahnfleth 2000, Kowalski et al. 2000). Integrated surface models,

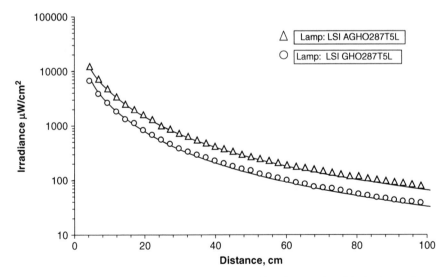

Fig. 7.2 Comparison of view factor model predictions of irradiance at the midpoint for two lamps with photosensor data

which integrate the energy over the cylindrical surface of the lamp, should produce virtually identical results, since that is essentially what the view factor model does. Models which integrate the energy throughout the volume of the gas inside the cylindrical lamp are possible but are unlikely to provide any increased accuracy, since all the light exits through the glass surface.

Equation (7.1) can be used to compute the irradiance at any point beyond the ends of the lamp by applying it twice – once to compute the view factor for an imaginary lamp of the total length (distance between some point and the far end of the lamp) and then subtracting the view factor of the non-existent portion, or ghost portion. This method, known as view factor algebra, is detailed in Kowalski and Bahnfleth (2000), Kowalski (2001), and Modest (1993).

The irradiance field as a function of distance from the lamp axis is simply the product of the surface irradiance and the view factor, where the surface irradiance is computed by dividing the UV power output by the surface area of the lamp:

$$I = \frac{E_{uv}}{2\pi\, rl} F_{total} \tag{7.2}$$

where E_{uv} = UV power output of lamp, μW.

Source code (in C++) is provided for this model in Appendix E which include subroutines for the view factor model. The subroutines will compute the irradiance field for any number of UV lamps and will saves the values in a $50 \times 50 \times 100$ matrix. This software uses view factor algebra to compute the irradiance at points beyond the lamp arclength. It is provided as an example of coding for Eq. (7.1) and requires input and output routines for full implementation. See Kowalski (2001) for the complete source code implementation, which includes input routines, output files, and also reflectivity subroutines.

Implicit in the use of the view factor model is the assumption that microbes are spherical and receive UV rays that pass right through the cell. In this model the cross-sectional area of a sphere is a flat disc that remains perpendicular to a line passing through the lamp axis. Although this may not be a perfect model, since the flat disc only faces the lamp axis and not the lamp length, the error appears to be minor or insignificant (Kowalski et al. 2000).

Several sources may contribute errors to the predictions of the view factor model, including the round off error in the lamp wattage and variations of surface irradiance along the lamp. Manufacturers typically state that the error in the nominal UV wattage is ±1% but this may be optimistic. Blatchley (1997) measured wattage variations in the same model lamp of +3.7%/−2.2%.

The lamp rating for any tubular lamp can be computed by using Eq. (7.1) and modeling the lamp as two equal lamp segments at a distance of 100 cm ($x = 100$). A comparison of predicted ratings using the view factor model for over 90 different UVGI lamps indicates predictive errors within ±9% (Kowalski and Bahnfleth 2000). Biaxial lamps can be modeled as two cylinders while u-tube lamps can be modeled as several cylindrical segments. A convenient simplification for modeling any UV lamp or several lamps is to define a single cylindrical lamp with the same wattage and total surface area – results will often produce nearly the same average irradiance values in an enclosure as a detailed model of the actual lamp(s).

7.3 UV Reflectivity Modeling

Reflectivity of the UVGI enclosure can greatly enhance the irradiance field depending on the enclosed surface area, the reflectivity, and the type of reflectivity. Reflectivity may be diffusive like the reflectivity of clouds or white paper, or it may be specular like the reflectivity of mirrors. Each of these types of reflectivity can be modeled as described in the following sections.

7.3.1 Modeling Specular Enclosures

A model has been specifically developed for specularly reflective, or mirror-like, surfaces that treats the reflections as virtual images, each with their own contribution to the irradiance field (Kowalski et al. 2005). The specular model defines each of the multiple virtual images of the lamp with its own view factor computed from Eq. (7.1) but at the virtual image distance and reduced in intensity by the reflectivity of the surface. That is, if the reflectivity is 75%, and the distance of the lamp to the surface is 10 cm, the first virtual image is at a distance of 20 cm and the computed irradiance (or UV output) is reduced by 75%. All of the virtual images can be treated as if they were separate lamps with their UV power output reduced by the reflectivity of each of the reflective surfaces through which the image passes. Figure 7.3 shows a photograph in which some 8 virtual images of the UV lamp can be seen reflected. The reflectivity is approximately 85% for visible light whereas in the UV spectrum aluminum surfaces will have about 75% reflectivity.

Fig. 7.3 Model UV System with four specular surfaces viewed through a single one-way mirror. At least eight virtual images can be discerned

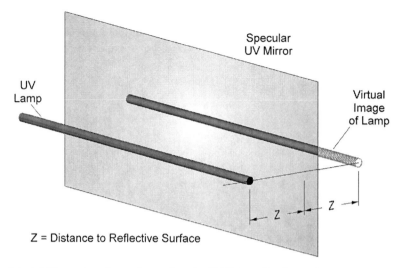

Fig. 7.4 A UV specular reflector (i.e. a mirror for UV light) will reflect a virtual image of the real lamp an equivalent distance (Z) behind the reflector surface

Figure 7.4 illustrates the first reflection of the real lamp in relation to the specular surface. The irradiance contribution of the first specular reflection depends on the distance of its virtual image. Since reflectivity is always less than 100%, the lamp irradiance must be multiplied by the fractional reflectivity. For example, if the surface reflectivity is 75%, then the irradiance of the first virtual lamp image must be multiplied by 0.75, which is the same as multiplying the UV power (E_{UV} in Eq. 7.2) by 0.75. As a result, the entire array of virtual images can be modeled as separate lamps, with the UV power of each lamp reduced by the reflectivity of each real or virtual reflective surface it passes through.

Figure 7.5 illustrates the array of virtual images that are created in a four-sided specular reflective UVGI enclosure. Note that the position of the lamp is mirrored at each reflection. Note also that the first image in the corner is actually a second reflection since the image must pass through two virtual reflective surfaces. The first specular reflection creates 4 virtual images, the second creates 8 images, the third creates 12, and so on. The number of images in each reflection progresses by the following series:

$$(4, 8, 12, 16, 20, 24, \ldots) = 4 \sum_{n=1}^{\infty} n \qquad (7.3)$$

where n = number of reflections

In mathematical terms, the reflected images provide contributions to the irradiance at a point as per the following:

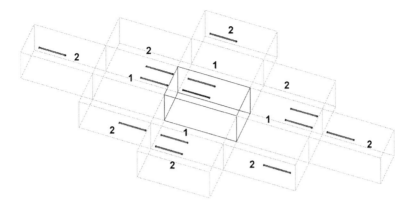

Fig. 7.5 Virtual images of UV lamps in an enclosure with four specular surfaces. The first (1) and second (2) virtual lamp images are shown

$$E_{\text{tot}} = \frac{P_{\text{uv}}}{2\pi r\ell} F_0 + \frac{\rho P_{\text{uv}}}{2\pi r\ell} F_1 + \frac{\rho^2 P_{\text{uv}}}{2\pi r\ell} F_2 + \dots \qquad (7.4)$$

where

F_0 = direct irradiance contribution from UV lamp
F_1 = irradiance contribution from all first reflections
F_2 = irradiance contribution form all second reflections

Equation (7.4) can be simplified as:

$$E_{\text{tot}} = \frac{P_{\text{uv}}}{2\pi r\ell} \left[F_0 + \rho F_1 + \rho^2 F_2 + \dots \right] \qquad (7.5)$$

Since the maximum specular reflectivity likely to be encountered is no more than about 90%, the number of reflections that need be considered in a normal UVGI system is about 5 before the contribution diminishes to insignificant levels. Although 90% reflectivity through 5 virtual surfaces will only reduce the 5th lamp virtual image by about 60% of its irradiance (i.e. $0.90^5 = 0.59$), the drop-off in irradiance due to the distance to the virtual image is far greater. At the same time, the total number of virtual lamps increases per Eq. (7.3).

Analysis of reflections beyond the first six are best handled with a geometric series, especially since the distant specular images will become diffused, and since even the most specular surface has a diffuse component. The geometric series model in the following section on diffuse reflectivity can be applied.

The specular model produces results that are similar in quantity to the diffuse reflective model previously presented but differ in certain qualities. Since specular reflections are more focused there can be differences in the predicted inactivation rates due to the fact that concentrated irradiance fields may be less efficiently distributed in the enclosure. The specular model also requires considerably more

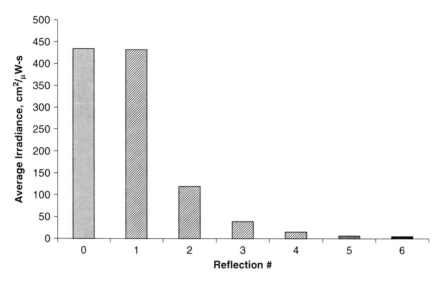

Fig. 7.6 UV irradiance contribution from first six specular reflections inside a rectangular four-sided enclosure with a reflectivity of 90%. Reflection 0 is the direct irradiance contribution from the UV lamp

computation time (IBSI 2003). Figure 7.6 shows the combined results of the irradiance field contributions computed for a typical system. Each of the successive specular reflections 1–6 represents the virtual image from four or more walls of the rectangular enclosure. It should be noted that it is possible for the irradiance due to multiple reflective surfaces to be greater than the initial irradiance at high reflectivities, as was reportedly first observed by Thomas Edison. In fact, it is theoretically possibly to multiply the average irradiance many times over under the right conditions.

In comparison with the diffusive reflectivity model (presented following) the specular model tends to produce higher net values of irradiance and dose, to a degree that depends on each system's geometry. The general explanation of this is that the specular reflections tend to contain the UV light within the enclosure, while the diffusive reflections tend to scatter more light out the entrance and the exit. From a design point of view, it might be more efficient to use specularly reflective materials, but this will also depend on the geometry of the system. On the other hand, the existence of diffusive materials with extremely high reflectivities offers the possibility of creating diffusively reflective systems that are more efficient than specular reflective systems. Combining these two types of reflective surfaces also offers the possibility of maximizing system performance.

7.3.2 Modeling Diffuse Enclosures

Diffuse reflectivity spreads light in all directions, regardless of the incident angle of incoming light rays. Since the reflected radiation spreads in all directions, there is no virtual image, only diffuse light. The entire diffusive surface tends to light up and

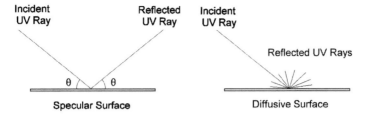

Fig. 7.7 Rays reflected from specular surfaces maintain cohesion while rays reflected from diffusive surfaces are dispersed equally in all directions regardless of the incident angle

the intensity of the reflected irradiance becomes smeared or diffused over the entire surface. Figure 7.7 illustrates this important distinction. Of course, there are no perfect specular surfaces, since there is inevitably some diffusive spreading even in highly specular materials. There are also no perfect diffusive surfaces, since at some incident angle even highly diffusive materials will have some specular component. Clouds may be an exception, but they are not surfaces.

To compute the irradiance field from multiple diffusive surfaces, first determine the direct irradiance at each surface using the lamp model equations (7.1) and (7.2). The resulting matrix of the intensities for each wall, which is actually a surface contour, can be averaged to simplify computations without introducing much error (since diffuse reflectivity tends to even out the irradiance field). The average reflected irradiance for each surface is:

$$I_{vl} = \bar{I}_{Dvl}\rho \; I_{vr} = \bar{I}_{Dvr}\rho \; I_{ht} = \bar{I}_{Dht}\rho \; I_{hb} = \bar{I}_{Dhb}\rho \qquad (7.6)$$

where

$\bar{I}_D$ = average direct irradiance for surface vl, vr, ht, hb
vl = vertical left, vr = vertical right
ht = horizontal top, hb = horizontal bottom

The view factor F_h represents a horizontal element exposed to the diffuse irradiance of a facing parallel wall (view factor 10 per Modest 1993):

$$F_h = \frac{1}{2\pi}\left[\frac{X}{\sqrt{1+X^2}}ATAN\left(\frac{Y}{\sqrt{1+X^2}}\right) + \frac{Y}{\sqrt{1+Y^2}}ATAN\left(\frac{X}{\sqrt{1+Y^2}}\right)\right] \quad (7.7)$$

where

X = Height/x
Y = Length/x
x = distance to the vertical wall.

The view factor F_v represents an element perpendicular to the bottom side of the duct (view factor 11 per Modest 1993):

$$F_v = \frac{1}{2\pi}\left[ATAN\left(\frac{1}{Q}\right) - \frac{Q}{\sqrt{P^2 + Q^2}} ATAN\left(\frac{1}{\sqrt{P^2 + Q^2}}\right)\right] \qquad (7.8)$$

where

$P = $ Height/Width
$Q = Y$/Width
$Y = $ distance to the horizontal wall.

For a rectangular duct use Eq. (7.7) for each of the two vertical surfaces and Eq. (7.8) for each of the two horizontal faces to compute the 3-dimensional irradiance field due to the first reflections. The reflected irradiance I_R at any x, y, z point will be:

$$I_R = I_{vl}F_{vl} + I_{vr}F_{vr} + I_{ht}F_{ht} + I_{hb}F_{hb} \qquad (7.9)$$

where

F_{vl}, $F_{vr} = $ view factor to vertical left and right wall, Eq. (7.7)
F_{ht}, $F_{hb} = $ view factor to horizontal top and bottom wall, Eq. (7.8)

The first reflection is often sufficient to provide a good estimate of the average irradiance. For increased accuracy, multiple diffusive reflections (inter-reflections) can be computed by re-applying Eqs. (7.7) through (7.9). Each surface will absorb some amount of energy from each of the other three surfaces. The amount of energy absorbed from the opposite (parallel) surface can be determined from Eq. (7.7). For surfaces that are adjacent and perpendicular to each other, Eq. (7.8) can be used. These equations can be applied individually on each of the four surfaces to determine the intensity received by each of the other three surfaces. The sum of the three contributions defines the incident energy on the surface. The amount actually reflected will be this total multiplied by the reflectivity, as in the first reflection case. To make accurate predictions with this method it is necessary to subdivide the surfaces into smaller elements and compute the view factors and intensity contributions individually. This integrated sum will account for the variations in intensity across the surfaces. If the variations are not great, the entire surface may be approximated with an average irradiance, as noted previously.

The new surface intensities computed from the first inter-reflection can then be used to determine the second and third inter-reflections in an iterative fashion, simply by repeating this computational process. These computations can be performed for several reflections (i.e. 3–10), at which point there will be little energy left to exchange.

Once the first two or three inter-reflections have been computed, the remaining inter-reflections can be estimated by summing a simple geometric series. Consider the ideal case where the surface irradiance from the first inter-reflection is uniform everywhere, which will be a reasonable approximation for all diffusive reflections. The second inter-reflection intensity, I_{R2}, will be the product of the reflectivity, the total view factor, and the first inter-reflection surface irradiance, I_{R1}, as follows:

$$I_{R2} = I_{R1} \, (\rho F_{tot}) \qquad (7.10)$$

The third inter-reflection irradiance, I_{R3}, will be the reflectivity times the view factor times the second inter-reflection as follows:

$$I_{R3} = I_{R2} \, (\rho F_{tot}) = I_{R1} \, (\rho F_{tot})^2 \qquad (7.11)$$

Each subsequent inter-reflection will be the product of the reflectivity times the view factor, multiplied by the previous inter-reflection, in an infinite series. The total I_{Rtot} for all subsequent inter-reflections can be summed as:

$$I_{Rtot} = I_{R1} \, (\rho F_{tot}) + I_{R1} \, (\rho F_{tot})^2 + I_{R1} \, (\rho F_{tot})^3 + \ldots \qquad (7.12)$$

This common geometric series conveniently sums to:

$$I_{Rtot} = I_{R1} \frac{\rho F_{tot}}{1 - \rho F_{tot}} \qquad (7.13)$$

The first two or three inter-reflections need to be computed before Eq. (7.13) can be applied, after which it will give the total reflected irradiance for all reflections onwards. The reason the geometric series approach works after two or three inter-reflections is because the effect of inter-reflection between surfaces is to average out the irradiance and produce even surface contours.

A convenient simplification is to divide the third inter-reflection intensity by the second. This produces the following equation in which the total inter-reflection irradiance is computed for reflections 3 through infinity:

$$\sum_{i=3}^{\infty} I_R = I_{R3} \frac{\dfrac{I_{R3}}{I_{R2}}}{1 - \dfrac{I_{R3}}{I_{R2}}} = \frac{(I_{R3})^2}{I_{R2} - I_{R3}} \qquad (7.14)$$

The total for all inter-reflections then becomes:

$$I_{Rtot} = I_{R1} + I_{R2} + I_{R3} + \frac{(I_{R3})^2}{I_{R2} - I_{R3}} \qquad (7.15)$$

In the example shown in Fig. 7.8, the first two inter-reflections were used to compute the geometric series total for all the subsequent inter-reflections. The difference between these approaches was found to be only 0.05% after computing

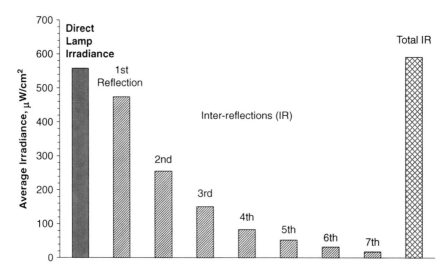

Fig. 7.8 Comparison of the direct and reflected average irradiances predicted for a model test case. The total inter-reflections (Total IR) were computed from a geometric series

30 inter-reflections between rectangular surfaces, each of which was subdivided into 10 × 20 elements (Kowalski 2001).

The above methods can be used to compute the matrix of values defining the inter-reflection irradiance field, which must then be added (point-by-point) to the direct irradiance field matrix and the first reflection irradiance field matrix. The method for the geometric series can also be applied for the specular model (Kowalski et al. 2005). UV dose prediction results from software versions of the diffusive model described above have shown good agreement with laboratory test results (UVDI 2000, 2001). For more detailed information on computing multiple diffusive reflections see Kowalski (2001).

7.3.3 Combined Specular and Diffusive Reflectivity

The reflective surfaces modeled in the previous sections were assumed to be either perfectly specular or perfectly diffusive. In reality many surfaces are a little of both. Specular reflections tend to dissipate after several reflections and become diffuse. Diffuse surfaces always have some incident angle at which light reflects specularly (even concrete will produce specular reflections at extreme angles). The two previous models of reflectivity take such uniquely different approaches that they cannot be combined directly. However, a simple and convenient approximation for estimating the effect of combining specular and diffuse reflectivity is to compute the average irradiance for each type of reflectivity (as described above) and then to add the values proportioned by their estimated contribution as follows:

$$I_{\text{Ravg}} = R \cdot I_{\text{d}} + S \cdot I_{\text{s}} \qquad (7.16)$$

where

I_{Ravg} = average irradiance due to reflectivity
R = fraction of reflectivity that is diffusive
I_d = average irradiance computed via diffusive model
S = fraction of reflectivity that is specular
I_s = average irradiance computed via specular model

In Eq. (7.16) the average irradiance is that of the reflections, not including the direct irradiance from the lamps. If, for example, a semi-rough aluminum surface had a total reflectivity of 60%, of which half is estimated to be specular and half diffuse, then $R = S = 0.5$. Since either I_d or I_s may be higher the effect of combining diffuse and specular reflectivity may be to lower or raise the average irradiance to a degree that depends on system geometry.

7.4 Air Mixing Effects

The velocity profile of a laminar airstream will approach a parabolic shape in long ducts, with the velocity higher towards the center, but fully developed laminar velocity profiles are unlikely to be achieved in real-world installations. The design velocity of a typical UVGI system of is about 2.54 m/s (500 fpm), produces a Reynolds number of approximately 150,000. Turbulent mixing is therefore far more likely to be the norm than laminar flow. Real world operating conditions will lie somewhere between complete mixing and the idealized condition of completely unmixed flow, as shown by Severin et al. (1984) for water-based systems. If complete mixing is assumed, then the average irradiance can be used directly to compute the UV exposure dose regardless of velocity profile.

If incomplete mixing is assumed, for velocity profiles that are not flat and in which the airflow followed parallel streamlines then each streamline would be subject to a dose that depends on the distance from the lamp. The survival rate in this case must be calculated for each streamline segment and summed or integrated to obtain the net survival. The survival S_i for each streamline segment is computed separately and the total survival would be the sum of all streamline segments as follows:

$$S = \sum_{j=1}^{l} \sum_{i=1}^{m} \sum_{k=1}^{n} e^{-kI_{ijk}t} \tag{7.17}$$

where

I_{ijk} = Irradiance at point i, j, k
i = a point defining the x coordinate (width)
j = a point defining the y coordinate (height)

$k =$ a point defining the z coordinate (length)

$t =$ exposure time for each segment defined by point ijk

Equation (7.17) can be used to develop a contour map of the inactivation zones for incomplete mixing. Consider two typical systems as shown in Fig. 7.9a (crossflow), b (axial flow). The inactivation zones developed from the previously described computational methodology are shown in Fig. 7.10a, b for a single lamp system in a reflective chamber for a model microorganism.

In the crossflow condition, the inactivation rate was predicted to be 59–64% (unmixed–mixed). In Fig. 7.9b, using the same UV power as the crossflow case, the range of inactivation rates was predicted to be 53–56%.

In this example, the crossflow configuration could be said to be more efficient due to the wider spread of the highest inactivation zone. However, the length of

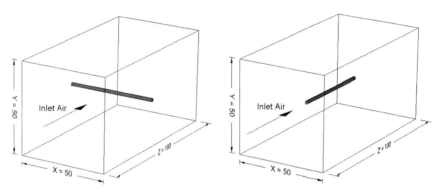

Fig. 7.9   a, b Schematic of crossflow and axial flow configurations

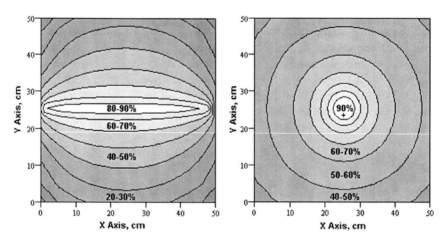

Fig. 7.10   a, b Inactivation zones determined for the crossflow and axial flow configurations, shown as percentage of population inactivated in unmixed air

duct and the reflectivity of the surface also play a major role in determining system efficiency, and the axial flow configuration can also be more efficient under other conditions.

7.5 UV System Optimization

The performance of UVGI systems can be optimized to produce maximum inactivation rates while minimizing energy costs if attention is paid to certain aspects of design. Both reflectivity and duct length, for example, can be used to reduce required UV lamp power. Reflectivity will increase the average irradiance while duct length will increase the average UV dose (by increasing the exposure time). Ignoring, for the moment, fan power considerations, there are design approaches that can increase the performance from a disinfection point of view.

There are several variables that define the design and operation of any UVGI system, and these include the dimensions (WxHxL), the lamp coordinates (x, y, z), airflow (Q), reflectivity (ρ), lamp UV wattage (P), lamp length (l), and lamp radius (r). Some of the variables have optimum values that are fairly well known. The optimum air velocity for most lamps is 2.54 m/s (500 fpm) while the typical operating velocity range of most systems is about 2–3 m/s (400–600 fpm). The optimum operating air temperature is considered to be 20–22°C (68–72°F). The critical design parameters of UVGI systems and their impact on performance has been evaluated using dimensional analysis (Kowalski et al. 2003, 2005). There are eight dimensionless parameters, excluding RH, that define UVGI system performance and these are defined in Table 7.1.

Figure 7.11 compares the effects of specular reflectivity and specific dose (Kowalski et al. 2005). Increasing reflectivity produces an approximately linear increase in inactivation rates but that gains level off as inactivation rates near 100%. It is not too difficult to see from this chart that if the same inactivation rates can be achieved by increasing reflectivity, then the expense of increasing lamp power may be unnecessary, but this is ultimately a matter of economics (i.e. lamp power vs. materials). Similar results are produced by the diffusive model for these two parameters (Kowalski 2001).

For the Z ratio the maximum inactivation rate occurs at a value of 0.5, or exactly centered along the length of the duct, which matches common industry practice. For

Table 7.1 Dimensionless parameters of UVGI systems

Parameter	Definition	Parameter	Definition
Reflectivity	ρ	X ratio	x/W
Aspect ratio	W/H	Y ratio	y/H
Lamp aspect ratio	r/l	Z ratio	z/L
Specific dose	kPL/Q	Height ratio	H/L

Fig. 7.11 Inactivation rate
for specific dose vs.
reflectivity, specular
reflectivity model

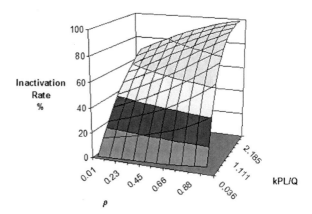

Fig. 7.11 Inactivation rate for specific dose vs. reflectivity, specular reflectivity model

additional information on dimensionless analysis of UVGI systems, see Kowalski
(2001) and Kowalski et al. (2003, 2005).

The previous dimensionless parameters can be used to estimate the system
wattage for any given set of mass flowrate, duct length, and reflectivity. For example, the specific dose can be multiplied by the reflectivity to obtain a dimensionless
parameter that encompasses all the defining parameters of a UVGI system. We call
this dimensionless parameter an F value as follows:

$$F = \rho \left(\frac{kPL}{Q} \right) \tag{7.18}$$

For any given set of design parameters, F will be a constant. We can further
simplify this relation by ignoring the units and assuming that the mass flowrate, Q,
is essentially equivalent to the airflow in CFM. This will be approximately true for
all UVGI systems operated under a normal range of air temperatures. The simplified
relation is now dimensional and we have no need of the rate constant k, which
merely served dimensionless conversion factor. The result is a dimensional F value
equal to some constant, or:

$$F_d = \rho \left(\frac{PL}{CFM} \right) \tag{7.19}$$

The above value of F_d will be approximately constant for all systems of comparable performance or specific dose, and the same reflectivity. Since the dose produced
by the system is also essentially a function of the UVGI Rating Value (URV), we
can define an F constant for each URV. Using the term F_u to represent the constant
F as a function of the URV, we can write:

$$F_u = f(\text{URV}) = \text{Constant} \tag{7.20}$$

References

References175

This allows the writing of a relation between the power and the airflow per unit length of duct as follows:

$$P = F_u \rho \left(\frac{CFM}{L} \right) \tag{7.21}$$

Equation (7.21) implies that the quantity of airflow per unit length is a prime determinant of UVGI system effectiveness, not unlike similar relations used for sizing pumps and fans. Values of F_u may be estimated based upon well-designed UVGI systems of known performance parameters, effectively scaling new systems from proven designs. Although this simplified relationship may provide estimates of UV wattage for any given URV, reflectivity, and airflow per unit duct length, it ignores the remaining dimensionless parameters and errors could exceed $\pm 30\%$. Equation (7.21) will not apply when the dose is so low that the shoulder curve dominates or when the inactivation rate is so high that the second stage of the decay curve dominates.

Economic optimization of UVGI systems is possible by assigning a cost to any system in terms of the defining parameters. The critical cost determinants include the UV lamp power, the duct length, and the reflectivity. However, the operating cost of a UV system may be dwarfed by the fan power consumption if the pressure losses are high, as they might be if a filter is added, or if light baffles are used, and therefore economic optimization of UV systems may require addressing many more factors than just the dimensionless parameters. See Kowalski (2001, 2003) for more information on economic optimization of UV systems.

References

Beggs CB, Kerr KG, Donelly JK, Sleigh PA, Mara DD, Cairns G. 2000. An engineering approach to the control of *Mycobacterium tuberculosis* and other airborne pathogens: A UK hospital based pilot study. Transactions of the Royal Society of Tropical Medicine and Hygiene 94: 141–146.

Blatchley EF. 1997. Numerical modelling of UV intensity: Application to collimated-beam reactors and continuous-flow systems. Wat Res 31(9):2205–2218.

Buttolph LJ, Haynes H. 1950. Ultraviolet Air Sanitation. Cleveland, OH: General Electric. Report nr LD-11.

Gardner DWM, Shama G. 1999. UV Intensity Measurement and Modelling and Disinfection Performance Prediction for Irradiation of Solid Surfaces with UV Light. Food Bioproducts Proc 77(C3):232–242.

IBSI. 2003. UVS: Specular Reflectivity UVGI Analysis Program. New York: Immune Building Systems, Inc.

IESNA. 2000. IESNA Lighting Handbook HB-9-2000, Reference and Application Chapter 6, Light Sources. New York: Illumination Engineering Society of North America.

Jacob SM, Dranoff JS. 1970. Light intensity profiles in a perfectly mixed photoreactor. AIChE J 16(3):359–363.

Kowalski WJ, Bahnfleth WP. 2000. Effective UVGI system design through improved modeling. ASHRAE Trans 106(2):4–15.

Kowalski WJ, Bahnfleth WP, Witham DL, Severin BF, Whittam TS. 2000. Mathematical modeling of UVGI for air disinfection. Quant Microbiol 2(3):249–270.

Kowalski WJ. 2001. Design and Optimization of UVGI Air Disinfection Systems [PhD]. State College: The Pennsylvania State University.

Kowalski WJ, Bahnfleth WP, Rosenberger JL. 2003. Dimensional analysis of UVGI air disinfection systems. Intl J HVAC&R Res 9(3):17.

Kowalski WJ. 2003. Immune Building Systems Technology. New York: McGraw-Hill.

Kowalski WJ, Bahnfleth WP, Mistrick RG. 2005. A specular model for UVGI air disinfection systems. IUVA News 7(1):19–26.

Krasnochub A. 2005. UV disinfection of air, some remarks. IUVA News 7(2):9–13.

Modest MF. 1993. Radiative Heat Transfer. New York: McGraw-Hill.

Philips. 1985. UVGI Catalog and Design Guide. Netherlands: Catalog No. U.D.C. 628.9.

Qualls RG, Johnson JD. 1985. Modeling and efficiency of ultraviolet disinfection systems. Water Res 19(8):1039–1046.

Severin BF, Suidan MT, Englebrecht RS. 1984. Mixing effects in UV disinfection. J Water Pollut Control Fed 56(7):881–888.

UVDI. 2000. Report on Bioassays of *S. marscecens* and *B. subtilis* Exposed to UV Irradiation. Valencia, CA: Ultraviolet Devices, Inc.

UVDI. 2001. UVD: Ultraviolet Air Disinfection Design Program. Version 12. Valencia, CA: Ultraviolet Devices, Inc.

Chapter 8

Airstream Disinfection

8.1 Introduction

UVGI air disinfection systems that operate on airstreams have applications in hospitals, schools, commercial buildings, homes, and in many other facilities. There are two main types forced air UV disinfection systems: In-duct UV systems and stand-alone recirculation units (also called unitary UV systems). A third class of systems exist, barrier systems, which are UV fixtures placed in doorways to disinfect air passing from room-to-room, but they are relatively uncommon. The primary advantage of forced air UV disinfection systems is the ability to control and predict performance. Well designed UV air disinfection systems can produce extremely high levels of clean air.

UV airstream disinfection systems are specifically designed to remove airborne microbes such as bacteria, viruses, and fungi. The removal rates can be predicted based on laboratory measurements of UV rate constants. The methods detailed in Chap. 3 are applied here to real-world systems and examples are presented for computing removal rates of microbes under UV exposure in air.

UV air disinfection invariably requires the use of filters to control dust accumulation on the lamp, and most UV air disinfection systems utilize filters. The benefits of filtration are complementary to UV system operation and combined systems provide optimum removal rates for the entire array of potential airborne pathogens and allergens. Prediction of the removal rates of airborne microbes by filters and the combined effect of filtration and UVGI are addressed here. This chapter then examines general applications of UV air treatment systems that utilize forced air and evaluates a variety of real world installations. Also examined here are field trials of air disinfection systems and summaries of unitary UV system performance.

8.2 UV Air Disinfection Performance

The performance of any UV air disinfection system can be quantified in terms of the UV exposure dose it produces in the airstream. The dose, in turn, is dependent on the mean irradiance in the duct and the exposure time. The exposure time depends

W. Kowalski, *Ultraviolet Germicidal Irradiation Handbook*,
DOI 10.1007/978-3-642-01999-9_8, © Springer-Verlag Berlin Heidelberg 2009

on the airflow and the dimensions of the duct (Width, Height, and Length). The exposure time in seconds, E_t, is computed as follows:

$$E_t = \frac{Vol}{Q} = \frac{WHL}{Q} \tag{8.1}$$

where

Vol = volume of the UV chamber, m^3
Q = airflow, m^3/s
W = width, m
H = height, m
L = length, m

It is implicit in Eq. (8.1) that the airflow is completely mixed in the exposure zone, which is a convenient simplifying assumption for any airstream in which turbulent conditions exist or are approached. At the typical design air velocity of 2.54 m/s (500 fpm) Reynolds numbers will be on the order of 150,000 and complete mixing will be approached (Kowalski et al. 2000). Assuming complete mixing, the survival rate of any airborne microbial population will be predicted by the standard exponential decay model (Eq. 3.2 in Chap. 3) and the UVGI removal rate (RR) can be computed as follows:

$$RR = 1 - e^{-k I_m E_t} \tag{8.2}$$

where

RR = removal rate, fraction or %
k = UV rate constant, m^2/J
I_m = mean irradiance, W/m^2

Determination of the mean irradiance is the key to estimating UV air cleaner performance. Methods for predicting the irradiance around a UV lamp based on a view factor model have been presented in Chap. 7 and this method is used here to tabulate air cleaner system sizes. Table 8.1 shows the predicted mean irradiance for a one square meter duct of various lengths.

Figure 8.1 shows the effect of lamp power on the mean irradiance for a range of duct lengths. Doubling the power roughly doubles the mean irradiance, up to a point. Figure 8.2 shows the mean irradiance vs. duct length, at three reflectivities, for a one square meter duct. It is of some practical importance to note that the effect of increasing duct reflectivity is approximately the same as the effect of increasing UV power.

Before showing the generic dose predictions for the examples in Table 8.2, it is useful to introduce the UVGI Rating Value (URV) scale, which is a convenient way to describe the UV dose produced by any air disinfection system. Table 8.2 shows

Table 8.1 Mean irradiance for a square duct

Duct length		UVP	I_m, W/m^2		
m	ft	W	$\rho = 0\%$	$\rho = 50\%$	$\rho = 75\%$
0.5	1.6		9.12	13.67	17.86
1	3.3		9.01	15.66	21.77
1.5	4.9		6.79	13.00	18.81
2	6.6		5.35	10.79	15.98
2.5	8.2		4.41	9.16	13.75
3	9.8		3.74	7.93	12.01
4	13.1	20	2.85	6.18	9.49
5	16.4		2.34	5.13	7.92
6	19.7		1.93	4.30	6.68
7	23.0		1.71	3.80	5.92
8	26.2		1.47	3.31	5.19
9	29.5		1.29	2.93	4.62
10	32.8		1.24	2.78	4.36
0.5	1.6		18.25	27.33	35.71
1	3.3		18.02	31.32	43.54
1.5	4.9		13.59	26.00	37.63
2	6.6		10.71	21.59	31.96
2.5	8.2		8.81	18.31	27.49
3	9.8		7.49	15.86	24.03
4	13.1	40	5.69	12.36	18.97
5	16.4		4.67	10.25	15.84
6	19.7		3.86	8.59	13.37
7	23.0		3.41	7.60	11.84
8	26.2		2.94	6.62	10.37
9	29.5		2.58	5.87	9.23
10	32.8		2.48	5.57	8.73
0.5	1.6		27.37	41.00	53.57
1	3.3		27.03	46.97	65.32
1.5	4.9		20.38	38.99	56.44
2	6.6		16.06	32.38	47.94
2.5	8.2		13.22	27.47	41.24
3	9.8		11.23	23.78	36.04
4	13.1	60	8.54	18.54	28.46
5	16.4		7.01	15.38	23.76
6	19.7		5.80	12.89	20.05
7	23.0		5.12	11.39	17.76
8	26.2		4.41	9.93	15.56
9	29.5		3.87	8.80	13.85
10	32.8		3.72	8.35	13.09
0.5	1.6		36.49	54.67	71.42
1	3.3		36.04	62.63	87.09
1.5	4.9		27.17	51.99	75.25
2	6.6		21.41	43.18	63.92
2.5	8.2	80	17.62	36.63	54.98
3	9.8		14.97	31.71	48.05
4	13.1		11.39	24.73	37.95
5	16.4		9.35	20.50	31.68
6	19.7		7.73	17.19	26.73

The header row spanning "unitary square-sided duct, Face Area = 1 m^2" appears above the table.

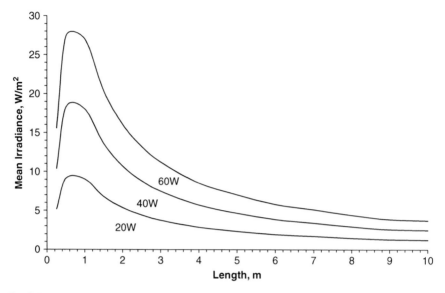

Fig. 8.1 Mean Irradiance vs. Length, at three power levels, for a one square meter duct, 0% reflectivity

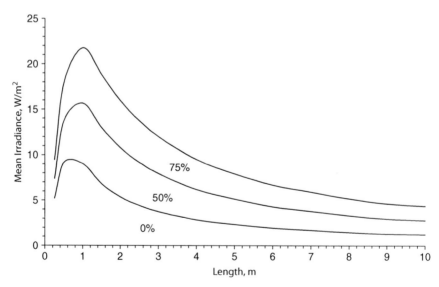

Fig. 8.2 Mean Irradiance vs, Length, at three reflectivities, for a one square meter duct. The UV reflectivity of 50% approximates galvanized steel, while 75% represents polished aluminum

Table 8.2 UVGI rating values (URV) and filter recommendations

URV	Dose J/m^2	Dose µW-s/cm^2	Mean Dose, J/m^2	Filter MERV	% Air disinfection rates from UV alone			
					Adenovirus	Influenza	TB	MRSA
1	0.01	1	0.055	6	0	0	0	1
2	0.10	10	0.15	6	1	1	5	6
3	0.20	20	0.25	6	1	2	9	11
4	0.30	30	0.4	6	2	4	13	16
5	0.50	50	0.63	6	3	6	21	26
6	0.75	75	0.88	6	4	9	30	36
7	1.0	100	1.25	7	5	11	38	45
8	1.5	150	2	8	8	16	51	59
9	2.5	250	3.75	9	13	26	69	77
10	5	500	7.5	10	24	45	91	95
11	10	1000	12.5	11	42	70	99	100
12	15	1500	17.5	12	56	83	100	100
13	20	2000	25	13	66	91	100	100
14	30	3000	35	14	80	97	100	100
15	40	4000	45	15	88	99	100	100
16	50	5000	55	15	93	100	100	100
17	60	6000	70	15	96	100	100	100
18	80	8000	90	15	99	100	100	100
19	100	10000	150	15	100	100	100	100
20	200	20000	250	15	100	100	100	100
21	300	30000	350	15	100	100	100	100
22	400	40000	450	15	100	100	100	100
23	500	50000	750	15	100	100	100	100
24	1000	100000	1500	15	100	100	100	100
25	2000	200000	2500	15	100	100	100	100
UV Rate constants, m^2/J					0.054	0.119	0.4721	0.5957

NOTE: URV 21–25 are newly appended URV definitions. Gray indicates normal design range.

the URV scale, which is based on the indicated UV dose (IUVA 2005). The dose represents the cutoff above which the URV applies. The mean dose is the average from one URV to the next, and is provided for estimation purposes. Most UV systems will normally fall into the URV 10–15 range. When the recommended MERV filter is coupled with the indicated URV system the removal rates of airborne microbes will be approximately the same across the entire array of microbes, by design. Filters below MERV 6 are not recommended for use in protecting UV lamps while MERV 15 filters represent the maximum size filter that would normally be coupled with a UV system. Exceptions to this rule may exist when designing ultraclean operating rooms for hospitals (see Chap. 16). UV removal rates for a few microbes in air are shown as examples – if the filter were accounted for the removal rates would be even higher.

NOTE: URV definitions beyond URV 20 are new and have not been formally approved by IUVA. They have been added due to the fact that unitary UV systems are being developed that go well beyond the original URV scale in an attempt to remove spores from air by UV, sometimes without filters (see Sect. 8.5).

Table 8.3 shows the predicted UV exposure dose for the same data set in Table 8.1, at a design operating air velocity of 2.54 m/s (500 fpm). Shown alongside the UV Dose is the URV to which it corresponds, with the URVs in gray representing the normal recommended design range of UV air cleaning systems (IUVA 2005).

Table 8.3 can be interpolated, but not extrapolated. Table 8.3 can be used to estimate UV power for any ventilation system operating at 2.54 m/s (500 fpm), on a 'per square meter' basis. That is, for a duct of 2 m^2 operating at 500 fpm, the lamp wattage shown in the table would be doubled. Figures 8.3 and 8.4 illustrate the data in Table 8.3, and again it can be observed that the effect of increasing reflectivity is similar to that of increasing UV power. The slight irregularities in Figs. 8.3 and 8.4 at high power and long lengths are artifacts of the computational methods used to generate the discrete data points and all curves will approach horizontal limits asymptotically if they are extended continuously. For the specific relations between power, reflectivity, and airflow illustrated in these curves see the dimensionless analysis in Chap. 7.

Example 1: Sizing a System of Arbitrary Face Area Given a polished aluminum duct with 0.25 m^2 of face area that is 1 m long, determine the required wattage to achieve an URV 14 rating.

In cases where the face area is greater or lesser than 1 m^2, we first size a system of one square meter face area and then proportionate the wattage. From Table 8.3, at 75% reflectivity and URV 14, we get a dose of 34.3 J/m^2.with 80 W. Since the duct face area is one-fourth of a square meter, the required wattage is 0.25(80) = 20 W of UV power.

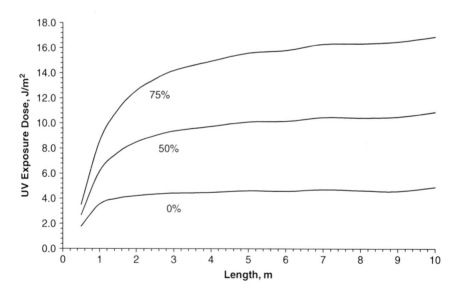

Fig. 8.3 UV dose as a function of duct length, for 20 W UV power and three reflectivities

Table 8.3 UV exposure dose for a square duct

unitary square-sided duct, Face Area = 1 m^2									
Duct length		E$_t$	UVP	D, J/m^2		D, J/m^2		D, J/m^2	
m	ft	sec	W	ρ = 0%	URV	ρ = 50%	URV	ρ = 75%	URV
0.5	1.6	0.20		1.8	8	2.7	9	3.5	9
1	3.3	0.39		3.5	9	6.2	10	8.6	10
1.5	4.9	0.59		4.0	9	7.7	10	11.1	11
2	6.6	0.79		4.2	9	8.5	10	12.6	11
2.5	8.2	0.98		4.3	9	9.0	10	13.5	11
3	9.8	1.18		4.4	9	9.4	10	14.2	11
4	13.1	1.57	20	4.5	9	9.7	10	14.9	11
5	16.4	1.97		4.6	9	10.1	11	15.6	12
6	19.7	2.36		4.6	9	10.1	11	15.8	12
7	23.0	2.76		4.7	9	10.5	11	16.3	12
8	26.2	3.15		4.6	9	10.4	11	16.3	12
9	29.5	3.54		4.6	9	10.5	11	16.5	12
10	32.8	3.94		4.9	9	10.9	11	16.9	12
0.5	1.6	0.20		3.6	9	5.4	10	7.0	10
1	3.3	0.39		7.1	10	12.3	11	17.1	12
1.5	4.9	0.59		8.0	10	15.4	12	22.2	13
2	6.6	0.79		8.4	10	17.0	12	25.2	13
2.5	8.2	0.98		8.7	10	18.0	12	27.1	13
3	9.8	1.18		8.8	10	18.7	12	28.4	13
4	13.1	1.57	40	9.0	10	19.5	12	29.9	13
5	16.4	1.97		9.2	10	20.2	13	31.2	14
6	19.7	2.36		9.1	10	20.3	13	31.6	14
7	23.0	2.76		9.4	10	20.9	13	32.6	14
8	26.2	3.15		9.3	10	20.9	13	32.7	14
9	29.5	3.54		9.3	10	20.8	13	32.7	14
10	32.8	3.94		9.7	10	21.9	13	34.4	14
0.5	1.6	0.20		5.4	10	8.1	10	10.5	11
1	3.3	0.39		10.6	11	18.5	12	25.7	13
1.5	4.9	0.59		12.0	11	23.0	13	33.3	14
2	6.6	0.79		12.6	11	25.5	13	37.7	14
2.5	8.2	0.98		13.0	11	27.0	13	40.6	15
3	9.8	1.18		13.3	11	28.1	13	42.6	15
4	13.1	1.57	60	13.4	11	29.2	13	44.8	15
5	16.4	1.97		13.8	11	30.3	14	46.8	15
6	19.7	2.36		13.8	11	30.4	14	47.4	15
7	23.0	2.76		14.0	11	31.4	14	48.9	15
8	26.2	3.15		13.9	11	31.3	14	49.0	15
9	29.5	3.54		14.0	11	31.2	14	49.1	15
10	32.8	3.94		14.3	11	32.9	14	51.5	16
0.5	1.6	0.20		7.2	10	10.8	11	14.1	11
1	3.3	0.39		14.2	11	24.7	13	34.3	14
1.5	4.9	0.59		16.0	12	30.7	14	44.4	15
2	6.6	0.79		16.9	12	34.0	14	50.3	16
2.5	8.2	0.98	80	17.3	12	36.1	14	54.1	16
3	9.84	1.18		17.7	12	37.5	14	56.8	16
4	13.12	1.57		17.9	12	38.9	14	59.8	16
5	16.4	1.97		18.4	12	40.4	15	62.4	17
6	19.7	2.36		18.3	12	40.6	15	63.1	17

Example 2: Determining the UV Dose

Given 96 W of total lamp power installed in a duct length of 0.75 m with an air velocity of 500 fpm, determine the UV Dose and the URV.

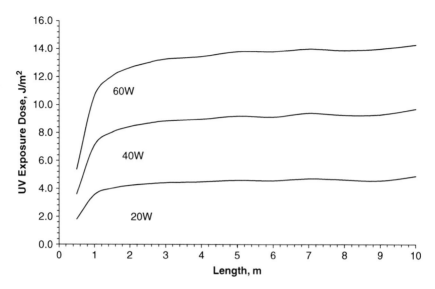

Fig. 8.4 UV dose as a function of duct length, for three UV power levels, at 0% reflectivity

A total input power of 90 W is approximately 30 W of UV output power (at the typical 31% efficiency for UV lamps). If the duct is galvanized steel, the reflectivity can be conservatively assumed to be 50% (see Appendix F).

From Table 8.3 the UV dose at 20 W and ρ=50% is interpolated between 0.5 m and 1 m as (2.7+6.2)/2 = 4.45 J/m². At 40 W, the interpolation is (5.4+12.3)/2 = 8.85 J/m². The UV dose at 30 W is then halfway between these, or (4.45+8.85)/2 = 6.65 J/m². per Table 8.2 this would rate an URV 10.

Example 3: Determining the Required UV Wattage
Given a 1.5 m length of stainless steel duct, with an operating velocity of 500 fpm, determine the approximate UV wattage required to achieve an URV 13 rating.

Stainless steel duct has a reflectivity of about 25%. From Table 8.3 at 60 W UV power, the interpolation of the dose between 0% and 50% reflectivity yields (12+23)/2 = 17.5 J/m². At 80 W the interpolated dose is (16+30.7)/2 = 23.35 J/m².

Since the minimum dose for an URV 13 rating is 20 J/m², the fractional distance between 60 W and 80 W is used to compute the additional watts (above 60 W) as follows:

$$UVP = 60 + \frac{20 - 17.5}{23.35 - 17.5}(80 - 60) = 68.5 \text{ W}$$

Therefore, a minimum of 68.5 W of UV power is required. At a typical 31% efficiency, the minimum total lamp power would be about 68.5/0.31 = 221 W. Once the total wattage of lamps selected for this application was established, the UV dose could be re-estimated per Table 8.3.

Example 4: Sizing a System of Arbitrary Air Velocity

Given a 1 m^2 galvanized duct 1 m long operating at 3.56 m/s (700 fpm) with normal (21°C/70°F) air temperature, in which 100 W of UV lamp power has been installed, determine the UV dose.

Since the air velocity is not the standard 2.54 m/s (500 fpm), two adjustments must be made. First the lamp must be corrected for cooling effects, per manufacturer's curves. Second, the UV dose must be adjusted for the shorter exposure time.

If the lamp is a typical model GPH lamp, then the cooling effect per the manufacturer's curve (see Fig. 5.11 in Chap. 5) results in about 75% of full output. Derating the lamp power then yields 75 W of UV power.

Per Table 8.1, at 50% reflectivity and 1 m length, the mean irradiance is about 47 W/m^2 at 60 W and 62.6 W/m^2 at 80 W. Interpolating the irradiance for 75 W yields 47 + 0.75(62.6–47) = 58.7 W/m^2. The exposure time is 1/3.56 = 0.28 sec. The UV dose is then computed as the exposure time multiplied by the irradiance, or 0.28(58.7) = 16.5 J/m^2. This dose corresponds to an URV 12 rating (see Table 8.2 or Appendix G).

The methods detailed above will provide a conservative estimate of lamp wattage or dose, and this approximation can be refined, if necessary, by more sophisticated techniques, such as computer modeling, photometric testing, or biodosimetric testing. Once the UV dose is established, predictions can be then made on the UV removal rates for any given microbe for which test data exists. The removal rates for the UV system can then be combined with the removal rates for the filter to determine the overall removal rates for any given microbe or for an array of microbes.

8.3 Filtration Performance

UVGI systems use various types of filters, usually in the range of MERV 6–15. The filter performance adds to the removal rates of airborne microbes, especially for large spores that tend to be resistant to UV. Filter performance is described by filter performance curves that depict the filtration efficiency for a range of particle sizes and at a design velocity of 400–500 fpm (ASHRAE 2003). The ASHRAE performance scale known as Minimum Efficiency Reporting Value (MERV) is used to define the performance of medium to high efficiency filters. MERV ratings are determined using ASHRAE Standard 62-2001 to test filter performance in the 0.3–10.0 micron size range (ASHRAE 2001).

For microbial filtration however, filter performance curves do not extend to the range of microbial sizes, or about 0.02–0.3 μm. Therefore, it is often necessary to model filter performance based on MERV curves and extend the curve to the region of interest. The details of filter modeling have been addressed in detail by Kowalski and Bahnfleth (2002 and 2002a), Kowalski et al. (1999), and Fig. 8.5 shows the modeled filter performance curves for MERV 6–15 filters. This figure can be used to estimate the removal rates of any microbe based on the mean diameter (see

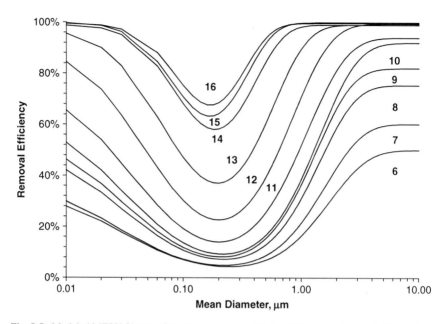

Fig. 8.5 Modeled MERV filter performance curves modeled at 500 fpm, and extended into microbial size ranges

Appendix A, B, & C for microbial logmean diameters). For filtration efficiencies at velocities other than 500 fpm see Kowalski and Bahnfleth 2002)

It can be seen in Fig. 8.5 that removal efficiencies actually increase below about 0.1 micron. The range of lowest filtration efficiencies between about 0.1–0.3 is known as the most penetrating particle (MPP) size range. It is microbes in this MPP range that must be removed at high rates by UV to obtain an improved overall performance curve in combined UV/filtration systems. Microbes in this range can be used as target microorganisms for assessing net removal rates.

Table 8.4 list the main types of filters and compares MERV ratings with their approximate equivalent DSP ratings per ASHRAE (2001). Shown also is the applicable range of filtration that is typical and appropriate for UV air cleaners.

The European Union (EU) uses a standard to rate filters that is based on the same concepts and testing methods as ASHRAE Standard 62 but which differentiates filter types. The EU ratings do not exactly correspond to the MERV standard. Table 8.5 summarizes the European filter rating standards (both old and new) and suggests the approximate MERV equivalent.

8.4 Combined Performance of Filtration and UVGI

Filtration and UVGI are mutually complementary technologies since filtration removes most of the microbes that tend to be resistant to UVGI, and vice-versa.

Table 8.4   Comparison of MERV and DSP filter standards

Filter type	Applicable size range µm	Dust Spot efficiency %	Total arrestance %	MERV rating (est.)	UV appl.
Dust Filters	>10	<20	<65	1	
		<20	65–70	2	
		<20	70–75	3	
		<20	75–80	4	
High Efficiency	3-10	<20	80–85	5	
		<20	85–90	6	
		25–35	>90	7–8	UV System Design Range
	1-3	40–45	>90	9	
		50–55	>95	10	
		60–65	>95	11	
		70–75	>98	12	
	0.3-1	80–90	>98	13	
		90–95	na	14	
		>95	na	15	
HEPA	<0.3	na	na	16–17	
ULPA	<0.3	na	na	18–20	

Table 8.5   European filter classes comparison

Type	New class	Eurovent class	Efficiency %	MERV (Approx.)	Testing	UV application
Coarse dust filter	G1	EU1	<65	1	Dust Arrestance (weight)	
	G2	EU2	65-80	2		
	G3	EU3	80-90	–		
	G4	EU4	>90	–		
Fine dust filter	F5	EU5	40-60	1–2	Dust spot (DSP)	
	F6	EU6	60-80	2–4		
	F7	EU7	80-90	5–6		
	F8	EU8	90-95	7–10		UV System Design Range
	F9	EU9	>95	11–13		
High Efficiency	H10	EU10	85	5–9	NaCl or aerosol	
	H11	EU11	95	10–13		
	H12	EU12	99.5	15		
	H13	EU13	99.95	16		
HEPA	H14	EU14	99.995	17–18		
ULPA	U15	EU15	99.9995	19	Aerosol	
	U16	EU16	99.99995	19–20		
	U17	EU17	99.999995	20		

The larger microbes, like spores, tend to be more UV resistant while most of the microbes in the MPP size range of the filters tend to be susceptible to UVGI.

The standard array of test microbes shown in Table 8.6 can be used to assess overall single-pass performance of a combined UV and filtration system. This table summarizes potentially airborne microbes from the UV rate constant database in

Table 8.6   Standard test array for air disinfection systems

Microbe	Type	UVGI k m²/J	Dia. μm	Filter Removal %							
				MERV 6	MERV 8	MERV 10	MERV 11	MERV 12	MERV 13	MERV 14	MERV 15
Parvovirus H-1	V	0.09200	0.022	21	37	35	52	72	84	95	98
Echovirus 1	V	0.02878	0.024	20	35	34	50	69	83	95	98
Coxsackievirus	V	0.11100	0.027	19	33	31	47	66	80	94	97
Murine Norovirus (MNV)	V	0.03040	0.032	17	30	28	43	62	76	92	95
VEE	V	0.04190	0.065	10	18	17	27	41	56	78	82
Reovirus	V	0.00940	0.075	9	16	15	24	37	52	75	78
Adenovirus	V	0.03900	0.079	9	15	14	23	36	50	73	76
Influenza A virus	V	0.11900	0.098	7	13	12	19	31	44	68	71
Avian Influenza virus	V	0.10600	0.098	7	13	12	19	31	44	68	71
Coronavirus (SARS)	V	0.01000	0.113	6	12	10	18	28	41	64	68
Mycoplasma pneumoniae	B	0.27910	0.177	5	9	8	14	23	37	58	64
Neisseria catarrhalis	B	0.05233	0.177	5	9	8	14	23	37	58	64
Francisella tularensis	B	0.00900	0.2	4	9	8	14	23	37	58	66
Newcastle Disease Virus	V	0.14400	0.212	4	9	8	14	23	37	59	67
Coxiella burnetii	B	0.15350	0.283	4	9	8	15	25	43	64	75
Haemophilus influenzae	B	0.17700	0.285	4	9	8	16	25	43	64	75
Proteus vulgaris	B	0.07675	0.291	4	9	8	16	25	43	65	76
Vaccinia virus	V	0.16040	0.307	5	10	9	16	26	45	66	78
Measles virus	V	0.10510	0.329	5	10	9	17	27	47	69	80
Proteus mirabilis	B	0.28900	0.494	7	15	14	25	39	64	84	94
Pseudomonas aeruginosa	B	0.57210	0.494	7	15	14	25	39	64	84	94
Legionella pneumophila	B	0.19298	0.52	7	16	15	27	41	66	86	95
Rickettsia prowazekii	B	0.17600	0.6	9	19	18	31	47	73	90	97
Serratia marcescens	B	0.28670	0.632	9	20	19	33	49	75	92	98
Mycobacterium tuberculosis	B	0.47210	0.637	9	21	19	33	49	75	92	98
Klebsiella pneumoniae	B	0.05480	0.671	10	22	20	35	52	78	93	99
Corynebacterium diphtheriae	B	0.07010	0.698	10	23	21	37	54	79	94	99
Burkholderia cenocepacia	B	0.03956	0.707	11	24	22	37	54	80	94	99
Listeria monocytogenes	B	0.01480	0.707	11	24	22	37	54	80	94	99
Yersinia enterocolitica	B	0.15351	0.707	11	24	22	37	54	80	94	99
Staphylococcus aureus	B	0.11300	0.866	14	30	28	45	64	87	97	100
Staphylococcus epidermis	B	0.16210	0.866	14	30	28	45	64	87	97	100
Streptococcus pyogenes	B	0.81100	0.894	14	31	29	47	66	88	97	100
Bacillus anthracis spores	BS	0.01988	1.118	19	40	38	57	76	92	99	100
Nocardia asteroides	B	0.00822	1.118	19	40	38	57	76	92	99	100
Bacillus subtilis spores	BS	0.02540	1.12	19	40	38	57	76	93	99	100
Acinetobacter baumannii	B	0.12800	1.225	21	44	42	61	80	94	99	100
Moraxella	B	0.00022	1.225	21	44	42	61	80	94	99	100
Enterobacter cloacae	B	0.03598	1.414	24	51	49	68	85	95	99	100
Aeromonas	B	0.20310	2.098	35	66	68	83	95	96	99	100
Penicillium chrysogenum	FS	0.00434	3.262	44	75	84	91	99	96	99	100
Aspergillus niger	FS	0.00058	3.354	45	76	85	92	99	96	99	100
Candida albicans	F	0.00515	4.899	49	77	91	94	99	96	99	100
Cryptococcus neoformans	FS	0.01670	4.899	49	77	91	94	99	96	99	100
Trichophyton rubrum	FS	0.00411	4.899	49	77	91	94	99	96	99	100
Clostridium tetani	B	0.04699	5	49	77	91	94	99	96	99	100
Stachybotrys chartarum	FS	0.00041	5.623	49	77	92	94	99	96	99	100
Scopulariopsis brevicaulis	FS	0.00344	5.916	50	77	92	94	99	96	99	100
Ustilago zeae	FS	0.06580	5.916	50	77	92	94	99	96	99	100
Rhizopus nigricans	FS	0.00861	6.928	50	77	93	94	99	96	99	100
Mucor mucedo	FS	0.00384	7.071	50	77	93	94	99	96	99	100
Cladosporium herbarum	FS	0.00370	8.062	50	77	93	94	99	96	99	100
Blastomyces dermatitidis	F	0.01645	11.000	50	77	93	94	99	96	99	100
Fusarium oxysporum	FS	0.00886	11.225	50	77	93	94	99	96	99	100

NOTES: B: Bacteria, BS: Bacterial Spore, V: Virus, F: Fungi, FS: Fungal Spore.

Appendices A, B, and C, using average or representative values of rate constants for air, surfaces, or water. The filtration rates shown in Table 8.6 are based on the filter models shown previously in Fig. 8.5. The overall single-pass removal rates

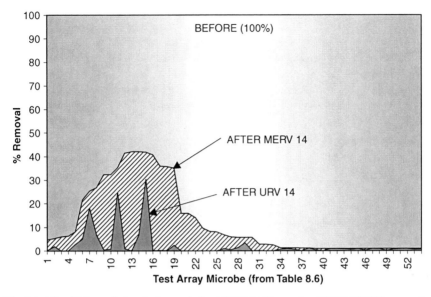

Fig. 8.6 Microbial populations before and after MERV 14 filter and an URV 14 UVGI system (at a UV dose of 35 J/m^2). Numbered microbes are sequential per the Standard Test Array in Table 8.6, ordered from smallest to largest diameters

can be used to compute the CADR for a system, which can then be used to assess in-place performance in a zone of some particular volume. Figure 8.6 shows a graph of the single pass removal of the Standard Test Array of microbes through a MERV 14 filter and an URV 14 UVGI system (at the mean dose of 35 J/m^2). The overall average removal rate for the Standard Test Array is 98%.

Most UV rate constants for microbes remain unknown and predictions can only be made based on existing studies. It can be reasonably assumed, however, that most bacteria and viruses will succumb to UVGI exposure and that most spores will be removed by the filters. In combination, therefore, UV and filtration will provide optimum air cleaning if the components are sized properly.

Combined filtration and UVGI system can be 'tuned' to target the microbes of concern for any particular facility, and to achieve any desired level of disinfection. In the previous example, tuning the system to remove all microbes completely could be accomplished by (1) increasing UV power (or increasing reflectivity), or (2) increasing filter MERV rating, or (3) decreasing the airflow. Economics would likely dictate any such choice, and basic techniques of economic optimization can be used to seek the best solution for any application.

Computing the net removal rate for combined systems is a simple algebraic process. Any microbial population that penetrates the filter is subject to the inactivation rate of the UVGI system. Figure 8.7 illustrates the process, in which the removal rate (filter efficiency) operates the same as the inactivation rate for the UV system.

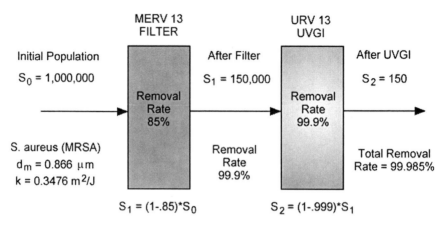

Fig. 8.7 Schematic for computing total combined inactivation rate when a filter is placed in series with a UVGI system

The target microbe here is *Staphylococcus aureus* (or MRSA) which has a known mean diameter and a known UV rate constant as indicated. The term S_0 defines the initial population, while S_n denotes the survivors after each step.

Computations for the intermediate removal rates are shown in Fig. 8.7. The total combined inactivation rate, IR_T, is then:

$$IR_T = (S_0 - S_2)/S_0 = (1,000,000 - 150)/1,000,000 = 99.985\% \qquad (8.3)$$

In mathematical form, we can write the combined total survival for the example in Fig. 8.7 as follows:

$$S_2 = (1 - IR_1)(1 - IR_2)S_0 \qquad (8.4)$$

The process is simply repeated if there are three or more systems in series. This might be the case if there is a prefilter before the primary filter.

8.5 In-duct Air Disinfection Systems

UVGI in-duct air disinfection systems usually consist of lamps, lamp fixtures, or modular lamp arrays placed inside ducts or air handling units. UV lamps are often located in an AHU downstream from the mixing box and air filters, and on the upstream side of the cooling coils. Assuming the UV lamp performs per design specifications and is maintained properly, the primary characteristics that impact airstream UV disinfection are relative humidity (RH), temperature, and air velocity. Increased RH tends to decrease decay rates of bacteria under ultraviolet (UV) exposure (Peccia et al. 2001). Air temperature has a negligible impact on microbial susceptibility to UVGI (Rentschler et al. 1941). However, air temperature can impact the UV output of lamps if design operating values are exceeded. When a UV system operates at air velocities above design UV output will be reduced because of the cooling effect of the air on the lamp.

8.6   EPA Test Results

It is informative to examine test results of actual UV air disinfection systems to see how their design compares with the analytical results presented previously. The EPA tested nine different manufacturers systems and measured their biological inactivation efficiency (EPA 2006a-h). Table 8.7 summarizes the essential operating parameters of eight of these systems. One system used pulsed light and is excluded (see Chap. 16 for a review of this pulsed light system). Another system operated at a much lower airflow and could not be included since system dimensions were not provided. For the system manufacturer see the original EPA test reports (available online). The seven systems are arranged in order of total lamp UV power (UVP). Test protocols involved the aerosolization of microbes and the sampling of downstream air. The test microbes were *Bacillus atrophaeus* spores, *Serratia marcescens* (bacteria), and bacteriophage MS2 (virus). Total reductions of airborne microbes were used as an indicator of system performance.

The mean UV dose was estimated by the EPA based on the survival of *Bacillus atrophaeus* spores ($k = 0.016$ m^2/J per EPA (2006)). Table 8.8 summarizes the test results and includes an additional mean UV dose estimate for bacteriophage MS2, using $k = 0.0448$ m^2/J in air at 50% RH per Walker and Ko (2007). All systems inactivated virtually 100% of *Serratia* and no dose could be computed on this basis. A computer model of the lamps and the reflective enclosure was used to estimate the mean irradiance and the UV dose, and this computed dose is compared with the biodosimetry results in Table 8.7. One problem with this approach is that the reflectivity in the test chambers was not specified, and different materials were used in the tests (i.e. galvanized steel, aluminum sheet, and aluminum insulation, and combinations of these) for which the reflectivity is uncertain. A nominal value of

Table 8.7 Summary of EPA test system operating parameters

System number	EPA Test Report	UVP W	W cm	H cm	L cm	Airflow m³/min	Velocity m/s	Lamp L, cm	Lamp dia cm
50	EPA 600/R-06/050	19	61	61	183	55.78	2.50	53.3	1.9
51	EPA 600/R-06/051	34	61	61	183	55.78	2.50	53.82	1.9
54	EPA 600/R-06/054	81	61	61	183	55.78	2.50	53.4	1.9
55	EPA 600/R-06/055	144	61	61	183	55.78	2.50	61	1.9
52	EPA 600/R-06/052	144	61	61	183	55.78	2.50	61	1.59
49	EPA 600/R-06/049	216	61	61	183	55.78	2.50	38.5	1.9
53	EPA 600/R-06/053	240	61	61	183	55.78	2.50	127	1.9
84	EPA 600/R-06/084	228	30.5	30.5	122	8.40	1.50	79	1.5

Table 8.8 Test results, dosimetry, and computer model predictions

System #	UVP W	Survival Fractions			UV Dose Estimate, J/m²			Model Dose Prediction			
		Serratia	Bacillus	MS2	Bacillus	MS2	Average	Et, sec	I$_m$, W/m²	Dose, J/m²	URV
50	19	0.01	0.96	0.61	2.55	11.03	6.79	0.732	14	10	11
51	34	0.002	0.9999	0.54	0.01	13.75	6.88	0.732	24	18	12
54	81	0.0004	0.91	0.25	5.89	30.94	18.4	0.732	57	42	15
55	144	0.0002	0.6	0.18	31.93	38.28	35	0.732	99	73	18
52	144	0.0004	0.04	0.01	201.18	102.79	152	0.732	101	74	18
49	216	0.0002	0.29	0.02	77.37	87.32	82	0.732	158	115	19
53	240	0.0003	0.07	0.01	166.20	102.79	134	0.732	161	118	19
84	228	0.0006	0.0011	0.0012	425.78	150.12	288	0.811	375	304	21

25% was used for reflectivity in all cases, as this provided the best overall match to test results.

The systems in Table 8.7 are arranged in order of increasing UV power, and this is reflected in the predicted UV dose, but is not reflected in the measured dose, which indicates problems with either the method, the measurements, or the stated power levels. The doses computed on the basis of *Bacillus* and MS2 are shown in Fig. 8.8, compared with predictions of a computer model. Excluded is Test no. 84, which, with a different size and airflow, and a below-design operating velocity, is off the charts in terms of dose, and it likely represents the second stage of the test microbes, and so the predicted model dose may not be valid. Figure 8.8 also shows the estimated URV for each system based on the computed (model) dose.

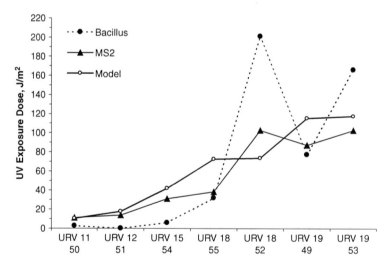

Fig. 8.8 Biodosimetry results for EPA tested systems compared with computer model predictions. Below the horizontal axis is shown the estimated URV based on the model, and the test system number (per Table 8.8)

Table 8.9 System Parameters and Test Results Summary

EPA report #	600/R-06/050	System #	50	EPA report #	600/R-06/051	System #	51

UVP, W	19	Airflow, m³/s	55.78	UVP, W	34	Airflow, m³/s	55.78
Number of lamps	1	Exposure time, s	0.732	Number of lamps	4	Exposure time, s	0.732
Lamp W	58	*Bacillus* survival	0.96	Lamp W	25	*Bacillus* survival	0.9999
Total W	58	MS2 survival	0.61	Total W	100	MS2 survival	0.54
Lamp efficiency	0.33	*Bacillus* dose, J/m²	2.55	Lamp efficiency	0.34	*Bacillus* dose, J/m²	0.01
Lamp length, cm	53.3	MS2 dose, J/m²	11.03	Lamp length, cm	53.82	MS2 dose, J/m²	13.75
Lamp dia, cm	1.9	Average dose, J/m²	6.79	Lamp dia, cm	1.9	Average dose, J/m²	6.88
W, cm	61	Dose URV	10	W, cm	61	Dose URV	10
H, cm	61	Model I_m, W/m²	14	H, cm	61	Model I_m, W/m²	24
L, cm	183	Model dose, J/m²	10	L, cm	183	Model dose, J/m²	18
Velocity, m/s	2.50	Model URV	11	Velocity, m/s	2.50	Model URV	12

EPA report #	600/R-06/054	System #	54	EPA report #	600/R-06/055	System #	55

UVP, W	81	Airflow, m³/s	55.78	UVP, W	144	Airflow, m³/s	55.78
Number of lamps	4	Exposure time, s	0.732	Number of lamps	8	Exposure time, s	0.732
Lamp W	60	*Bacillus* survival	0.91	Lamp W	60	*Bacillus* survival	0.6
Total W	240	MS2 survival	0.25	Total W	480	MS2 survival	0.18
Lamp efficiency	0.34	*Bacillus* dose, J/m²	5.89	Lamp efficiency	0.30	*Bacillus* dose, J/m²	31.93
Lamp length, cm	53.4	MS2 dose, J/m²	30.94	Lamp length, cm	61	MS2 dose, J/m²	38.28
Lamp dia, cm	1.9	Average dose, J/m²	18.42	Lamp dia, cm	1.9	Average dose, J/m²	35.10
W, cm	61	Dose URV	12	W, cm	61	Dose URV	14
H, cm	61	Model I_m, W/m²	57	H, cm	61	Model I_m, W/m²	99
L, cm	183	Model dose, J/m²	42	L, cm	183	Model dose, J/m²	73
Velocity, m/s	2.50	Model URV	15	Velocity, m/s	2.50	Model URV	17

Table 8.10   System parameters and test results summary

EPA report #	600/R-06/052	System #	52	EPA report #	600/R-06/049	System #	49

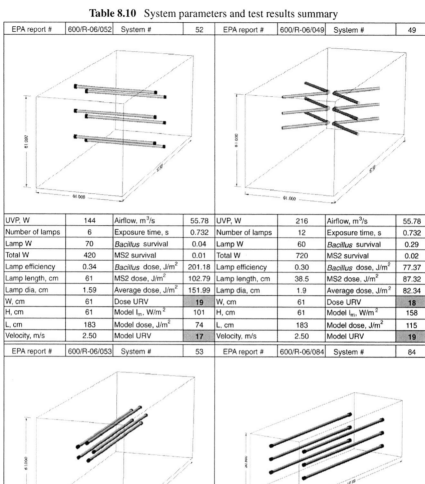

UVP, W	144	Airflow, m³/s	55.78	UVP, W	216	Airflow, m³/s	55.78
Number of lamps	6	Exposure time, s	0.732	Number of lamps	12	Exposure time, s	0.732
Lamp W	70	*Bacillus* survival	0.04	Lamp W	60	*Bacillus* survival	0.29
Total W	420	MS2 survival	0.01	Total W	720	MS2 survival	0.02
Lamp efficiency	0.34	*Bacillus* dose, J/m²	201.18	Lamp efficiency	0.30	*Bacillus* dose, J/m²	77.37
Lamp length, cm	61	MS2 dose, J/m²	102.79	Lamp length, cm	38.5	MS2 dose, J/m²	87.32
Lamp dia, cm	1.59	Average dose, J/m²	151.99	Lamp dia, cm	1.9	Average dose, J/m²	82.34
W, cm	61	Dose URV	19	W, cm	61	Dose URV	18
H, cm	61	Model I_m, W/m²	101	H, cm	61	Model I_m, W/m²	158
L, cm	183	Model dose, J/m²	74	L, cm	183	Model dose, J/m²	115
Velocity, m/s	2.50	Model URV	17	Velocity, m/s	2.50	Model URV	19

EPA report #	600/R-06/053	System #	53	EPA report #	600/R-06/084	System #	84

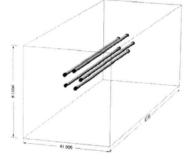 

UVP, W	240	Airflow, m³/s	55.78	UVP, W	228	Airflow, m³/s	8.40
Number of lamps	5	Exposure time, s	0.732	Number of lamps	6	Exposure time, s	0.811
Lamp W	150	*Bacillus* survival	0.07	Lamp W	125	*Bacillus* survival	0.0011
Total W	750	MS2 survival	0.01	Total W	750	MS2 survival	0.0012
Lamp efficiency	0.32	*Bacillus* dose, J/m²	166.20	Lamp efficiency	0.30	*Bacillus* dose, J/m²	425.78
Lamp length, cm	127	MS2 dose, J/m²	102.79	Lamp length, cm	79	MS2 dose, J/m²	150.12
Lamp dia, cm	1.9	Average dose, J/m²	134.50	Lamp dia, cm	1.5	Average dose, J/m²	287.95
W, cm	61	Dose URV	19	W, cm	30.5	Dose URV	20
H, cm	61	Model I_m, W/m²	161	H, cm	30.5	Model I_m, W/m²	375
L, cm	183	Model dose, J/m²	118	L, cm	122	Model dose, J/m²	304
Velocity, m/s	2.50	Model URV	19	Velocity, m/s	1.50	Model URV	21

It can be observed in Fig. 8.8 that MS2 provides a better overall match to the
model than the *Bacillus* spores. This may be due to the fact that the URV 11 through
15 systems do not produce enough dose to shift the *Bacillus* spore response out of
the shoulder region, and so the model, based on a single stage decay curve, over-
predicts the inactivation rate. The survival fractions for *Bacillus* spores (91–99.99%)
are so high that the spores are likely in a shoulder region. In a typical UVGI air
cleaning system the spores would be largely removed by the filters, and so clearly
the use of spores is a poor choice for measuring the effectiveness of the UV component of a normal size air cleaning system. A summary of all the EPA tested systems,
their critical operating data, and the test and modeling results is presented in Tables
8.8 and 8.9. Some of the data which was not provided in the original test reports was
estimated or obtained from the system manufacturers, such as some duct lengths and
the lamp diameters. A correction to the total input watts (not UV watts) for system
53 was provided by the manufacturer.

8.7 Unitary UV Systems

Stand alone recirculation units employing ultraviolet lamps come in all shapes and
sizes and are suited for various applications, including health care, schools, residential environments, and commercial buildings. Figure 8.9 shows a high-volume
unitary UV system for use in hospitals. This particular unit draws air from near the
floor, where microbial concentrations are highest, and delivers clean air at breathing
height. It is suitable for areas of about 400 ft^2 and multiple units can be placed in
larger rooms.

Figure 8.10 shows a unitary UV system with a detachable fan module that can
be used as a stand-alone unit. The UV component can also be installed as a retrofit
in ductwork with or without the fan (see the duct retrofit example in Chap. 18,
Fig. 18.1).

Fig. 8.9 Example of a UV
recirculation unit for health
care applications. Photo
courtesy of Juguang
Photoelectric Technology Co.
Inc., China

Fig. 8.10 Unitary UV
module with detachable fan
assembly (*left*). Image
courtesy of Virobuster
Electronic Air Sterilisation,
Enschede, The Netherlands

Unitary systems operate much the same as in-duct UV systems except the airflow is rarely controlled to operate near 2.54 m/s (500 fpm) and many systems have multiple operating speeds. Often, the tendency to make these units compact results in short exposure times. In order to make up for the low doses that result from short exposure times, manufacturers will often boost the UV wattage and assume reciprocity – that is, they assume the dose is a linear function of exposure time and that more wattage will make up for less exposure. This may not be the case if the exposure time is so short that the exposed airborne microbes remain in the 'shoulder' region of their decay curves. Microbial decay in the shoulder region is not linear on a log scale, and reciprocity is an invalid assumption in this region.

Another major difference between unitary systems and in-duct systems is that they need to be compared in terms of the clean air delivery rate (CADR) rather than the UV dose. For in-duct systems the airflow is pre-defined, and if designed per ASHRAE guidelines, in-duct systems will deliver an appropriate amount of air to each occupied building zone. For unitary systems, the performance will vary with the zone volume (or floor area) in which the unit is placed. Therefore, each unit's performance must ultimately be evaluated in terms of the range of zone volumes for which it is appropriate.

In the previously reviewed EPA test reports, only one unitary system was tested, system no. 84 in Table 8.9. This unit operates at 300 cfm (low speed) but has very high power. The biodosimetric test results indicate the average dose is 288 J/m^2 (URV 20), while the modeled dose is 304 J/m^2 (URV 21). Table 8.10 provides a summary of basic operating parameters for a variety of unitary air disinfection systems that are available from the indicated manufacturers. Most of the information is as reported by the manufacturer, except for the UV dose which is rarely reported. The UV dose was computed analytically for most of the systems except where it was supplied or where laboratory biodosimetric test results were available. The URV is based on the indicated dose, per Table 8.2.

Methodologies and test methods have been proposed for measuring the single pass efficiencies and effectiveness of room air cleaners (Foarde et al. 1999 and 1999a, Hanley et al. 1995, Janney et al. 2000). The only standards currently used for testing or measuring the effectiveness of unitary UV systems are ANSI (2002) and the updated version ANSI (2006) which are general methods for testing portable air cleaners.

Table 8.11 Examples of unitary UV systems and specifications

Manufacturer	Model	Airflow/CADR cfm	Airflow/CADR m³/min	Prefilter	Primary Filter	UVP W	Dose J/m²	URV	Notes
sterilAir AG	LSK2018	11.8	0.33	none	none	7	173	19	
sterilAir AG	LSK2036-U	29.2	0.83	none	none	30	315	21	
Amcor	AM-45	30	0.85	none	none	1.96	26.9	13	
Amcor	AM-45C	31	0.88	Yes	none	25	23.2	13	PCO, carbon
Sanuvox	P-900	35	0.99	Yes	none	4.76	48	15	multispeed
Amcor	AM-45C	40	1.13	Yes	none	25	17.8	12	PCO, carbon
Virobuster	Steritube	44	1.25	G4	none	57	592	23	multispeed
Amcor	AM-45C	45	1.27	Yes	none	25	15.9	12	PCO, carbon
Virobuster	Steritube	59	1.67	G4	none	57	443	22	multispeed
Amcor	AM-45	60	1.70	none	none	1.96	13.5	11	
Virobuster	Steritube	74	2.083	G4	none	57	355	21	multispeed
sterilAir AG	UVR2250-1	88	2.50	optional	none	27	190	19	
Holmes Group	BAP920-U	100	2.83	none	MERV15	22	69.2	17	PCO
Air Clean Assurance	HDU	110	3.11	Yes	MERV14	12	83.6	18	MERV8 prefilt.
BARO GmbH & Co	Air Wetech L	118	3.33	screen	none	128	274	20	
Holmes Group	BAP920-U	125	3.54	none	MERV15	22	55.3	16	PCO
UV Superstore	UV-SC-2AB/10SS	150	4.25	Yes	HEPA	30	86.5	18	tank top unit
Holmes Group	BAP920-U	150	4.25	none	MERV15	22	46.1	15	PCO
Sterilite	Arianne 250-N	153	4.33	Yes	none	144	94	18	multispeed
Holmes Group	BAP920-U	175	4.96	none	MERV15	22	39.5	14	PCO
BARO GmbH & Co	AirTube C	177	5.00	optional	none	64	137	19	
Sterilite	Arianne 250-N	177	5.00	Yes	none	144	100	19	multispeed
Sterilite	Arianne 250-N	188	5.33	Yes	none	144	115	19	multispeed
UVC LLC	Airwave	200	5.66	none	MERV13	16	46.7	15	
Eco-Rx	RX-400	210	5.95	screen	none	66	41	15	lab test dose
sterilAir AG	UVR2250-2	265	7.50	optional	none	46	324	21	
sterilAir AG	UVR2250-4	265	7.50	optional	none	92	649	23	
Novatron	Bioprotector BP114i	300	8.49	none	MERV13	228	288	20	lab test dose
NQ Industries	NQ Clarifier Medical	350	9.91	Yes	HEPA	11.16	16.3	12	carbon filter
Calutech	ADU	400	11.33	Yes	optional	62	74	17	carbon filter
NQ Industries	NQ500	500	14.16	Yes (2)	HEPA	117	328	21	multispeed

NOTE: CADR is approximately equal to the airflow for all systems.

8.8 Barrier Systems

Ultraviolet light barrier systems are used to provide isolation of areas subject to microbiological contamination. Barrier systems have been used in hospitals to create a pathway by which air transferring from one patient area to another is disinfected as it passes through a curtain of UV irradiation (Koller 1965, Buttolph and Haynes 1950). The barriers typically consist of an archway equipped with UV lamps, usually on both vertical sides and sometimes overhead as well. The UV lamp fixtures usually include baffles or louvers to direct the UV across the barrier and confine the UV rays to a narrow zone, much like Upper Room systems, and so minimize the irradiance in the room. Irradiance levels within the barrier must be high enough to achieve a high disinfection rate in the brief period it takes for air to pass through. Experiments have shown that barrier systems can greatly reduce airborne concentrations of bacteria and that they can reduce rates of respiratory illness (Koller 1965,

Wheeler et al. 1945). Wells (1938) found that an air barrier system was effective in preventing the spread of chickenpox in an isolation ward. Robertson et al. (1939) demonstrated that ultraviolet curtains could reduce airborne bacteria by 95%. Del-Mundo and McKhann (1941) used barrier systems across the cubicles of individual patients to reduce cross-infections from 12.5% to 2.7%. Sommer and Stokes (1942) found that UV barrier systems were effective at reducing the airborne bacterial counts in a hospital ward and had some impact on cross-infections. Sauer et al. (1942) showed that a UV barrier system effectively controlled cross-infections in a nursery. Robertson et al. (1943) conducted a number of tests in a children's hospital and was able to show that UV curtains could cut cross-infections in half or better in most cases.

Performance of barrier systems is dependent on the airflow through the barrier and the irradiance levels. Since irradiance levels are often high, there is a manifest hazard to personnel. Personnel who pass through a UV barrier may be subject to hazardous levels of UV, and may be required to wear protective clothing, skin creams, and goggles, etc.

8.9 Zonal Modeling

The ultimate performance indicator is how well a UV air cleaning system performs in actual service. In any installation, a specific volume of room air must be targeted for disinfection. The degree of aerobiological cleanliness that can be achieved depends on the clean air delivery rate (CADR), the zone volume, the zone supply air (if separate), the infiltration, and the contamination sources (or release rate of microbes). Figure 8.11 illustrates how an air cleaning system interacts with the zone volume, indoor sources, and infiltration. A steady state condition will always be approached in which the removal rate for the air cleaner will balance out with the sources to produce some ambient level of airborne microbial concentration.

Various approaches can be taken to solve for the ambient concentration of microbes in indoor air, including steady state solutions, transient solutions, and multizone modeling. Most of these models assume complete mixing, which is a reasonable approximation for real-world conditions of occupancy and forced air ventilation.

8.9.1 Steady State Single Zone Model

A steady state model assumes the airborne concentrations are controlled to remain constant over time, and the model is independent of the zone volume. One simple steady state model, called a system model, that accounts for filtration and outdoor air ventilation has been provided by Grimm and Rosaler (1990). This model can be adapted for predicting concentrations in a building modeled as a single zone of

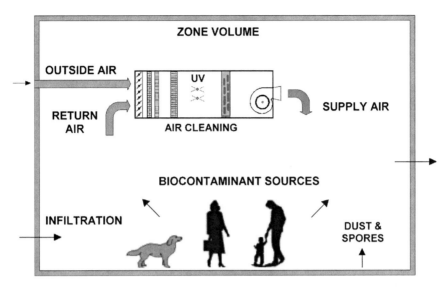

Fig. 8.11 The interaction of an air cleaning system with a volume and its contaminant sources determines the aerobiological quality of the indoor environment

uniform concentration. The steady state (SS) concentration due to a contaminant release rate R_R is:

$$C_{ss} = \frac{R_R}{CADR} \qquad (8.5)$$

here

C_{ss} = steady state airborne concentration, cfu/m^3
R_R = source release rate, cfu/min
CADR = clean air delivery rate, cfm

Release rates can vary greatly depending on indoor bioburden, number of occupants, and the release rate of microbes from each occupant. Release rates of bacteria on the order of 1000 cfu/min per person might not be uncommon. The CADR is the total clean air delivered to the zone and is defined as the removal efficiency times the airflow.

$$CADR = RE \cdot Q \qquad (8.6)$$

where

RE = removal efficiency, fractional
Q = total airflow, m^3/min

In general, any well-designed UV system combined with a filter will have removal efficiencies near 99% for a broad range of pathogens and allergens (such as the Standard Test Array), and the CADR will be approximately equal to the total airflow.

As an example of single zone steady state model prediction of microbial concentrations in a building, the system model presented in Grimm and Rosaler (1999) is restated here in aerobiological terms and assumes that only a single filter and a UV system is present. In this model the outside air is assumed to contain biocontaminants.

$$C_i = \frac{R_R + Q_{oa}P_fP_uC_0}{Q_{oa} + Q(1 - P_fP_u)} \tag{8.7}$$

where

  C_i = steady state indoor concentration, cfu/m^3
  C_0 = outdoor airborne concentration, cfu/m^3
  P_f = filter penetration, fractional
  P_u = UV penetration, fractional

Note that in Eq. (8.7) the product P_fP_u is the total penetration through the UV air cleaner, including the filter, and RE= $(1-P_fP_u)$. Equation (8.7) can be used to evaluate the impact of changing outside air flowrates in conjunction with UV system removal efficiencies. If there is no outside air, $Q_{oa}=Q_{ex}=0$, Eq. (8.7) reverts to Eq. (8.5).

The removal efficiency is relative to a specific microbe, but Eq. (8.7) could be applied repeatedly for an array of microbes to obtain the net average indoor airborne concentration. Alternatively, a representative microbe could be chosen from the most penetrating particle size range (for the combined filter and UV system) to produce a conservative estimate.

The decrease in concentration of airborne microbes in zone with a UV system can be used as an indicator of system performance. The rate of decrease could, for example, be compared to that achieved by ventilation with clean air alone, as opposed to recirculation air cleaning. The result can be interpreted in terms of equivalent air exchange rates, which is a method used to evaluate UV systems that target TB (Riley and Kaufman 1972, Ko et al. 2002).

8.9.2 Transient Modeling

When it is desired to know the airborne concentrations in a zone over time, a transient model must be used. Calculus can be used to estimate indoor contaminant concentrations in a perfectly mixed single zone model, as can computational methods. The classic model for complete mixing in a single zone is a linear first order differential equation (Boyce and DiPrima 1997). The airborne concentration over time will equal the amount of contaminants entering from the outside air minus the amount of contaminants exiting through the exhaust air plus the release rate of the source, as follows:

$$\frac{d\,(VC)}{dt} = Q\,(C_{oa} - C) + R_R \tag{8.8}$$

where

V = volume, m^3
C = airborne concentration, cfu/m^3
C_{oa} = airborne concentration in outside air, cfu/m^3
R_R = release rate of contaminant, cfu/min

Taking the outside air concentration to be zero, or $C_a = 0$, produces the following:

$$V\frac{dC}{dt} = -QC + R_R \tag{8.9}$$

Rearranging Eq. (8.9) and integrating over time, t, results in the solution for the concentration as a function of time, C(t), as follows:

$$C(t) = \frac{R_R}{Q}\left[1 - \exp\left(-\frac{Q}{V}t\right)\right] \tag{8.10}$$

In Eq. (8.10), R_R/Q represents the steady state concentration. It can be rewritten in a slightly more convenient form by defining the number of room air changes per hour, ACH, as the airflow, Q, divided by the room volume, V, as follows:

$$ACH = \frac{Q}{V} \tag{8.11}$$

Rewriting Eq. (8.10) in its simplified form produces:

$$C(t) = \frac{R_R}{Q}\left[1 - \exp\left(-ACHt\right)\right] \tag{8.12}$$

If the initial concentration is some value C_0, the decay of the zone concentration can be derived in a similar fashion and written as the following:

$$C(t) = C_0\left[\exp\left(-ACHt\right)\right] \tag{8.13}$$

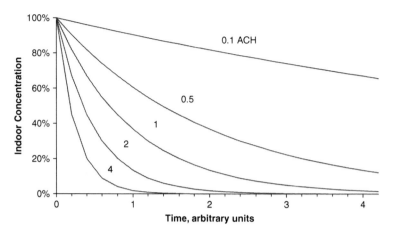

Fig. 8.12 Decay of zone concentrations from an initial concentration for different air change rates

The equivalent air exchange rate (EAC) for any given microbe can be substituted in Eq. (8.13). Figure 8.12 shows the results of Eq. (8.13) for a range of air exchange rates.

Computational methods can be used to integrate the airborne concentration over time by using a finite number of time steps, minute-by-minute being the most convenient. For each minute a finite volume of disinfected air replaces an equal volume of mixed room air. The removal rate of microorganisms is given by the indoor concentration, $C(t)$ at any given time t, multiplied by the airflow rate, Q_o, which is the volume of disinfected air displaced into and out of the zone volume (i.e. the CADR or total airflow, which are typically approximately the same). For each minute of the analysis, the number of microorganisms removed is:

$$N_{out} = C(t) \cdot Q_o \qquad (8.14)$$

It is assumed that any outdoor air entering passes through the air cleaner, for an in-duct system. For a room recirculation unit the microbes added to the zone will have to be accounted for. The rates of microbes generated internally are specified per minute, and are designated B_{in} and V_{in} respectively. The total population of microbes, $N(t)$, that will exist in the building, for any given minute t, will then be the previous minute's population plus the current minute's additions, minus the number exhausted to the outside air:

$$N(t) = N(t-1) + N_{in} - N_{out} + B_{in} + V_{in} \qquad (8.15)$$

The zone microbial concentration will be defined as the building microbial concentration divided by the building volume, for any given minute, or:

$$C(t) = \frac{N(t)}{V} \qquad (8.16)$$

This incremental calculation process, carried out for, say, 8–12 h, provides a close approximation of continuous flow, with the result being an exponential decrease in the building microbial concentration over time, much the same as that depicted in Fig. 8.10. As a check on the validity of any incremental spreadsheet model, the results can be compared with the steady state concentration per Eq. (8.5), which represents the condition that will be approached after sufficient time. For more information on developing spreadsheet transient models, including the condition when the outside air is unfiltered and contains separate sources, see Kowalski and Bahnfleth (2003) and Kowalski (2006).

8.9.3 Multizone Modeling

For buildings that are composed of multiple zones with separate airflows, the best approach is to use multizone modeling software. Software tools are available that perform multizone modeling with internal generation of biocontaminants, and which incorporate a variety of factors that can mimic real-world effects such as interior leakage, exterior leakage, wind pressure effects, stack effects, and plate-out effects. Several multizone modeling software packages are currently available including CONTAMW, DOE-2, ESP-r, Risk V1.0, and others (Dols et al. 2000, Sparks 1995, Axley 1987, Demokritou 2001).

The CONTAMW program is publicly available (NIST 2002) and simulates the release and distribution of contaminants in a multizone building with a ventilation similar to the previously described numerical integration or spreadsheet model. CONTAMW has the ability to model complex building systems and the contaminant sources can be varied between zones. Extensive information is available in the literature on multizone modeling methods from NIST and other sources (Musser et al. 2001, Musser 2000, Yaghoubi et al. 1995, Walton 1989), especially through the NIST website (NIST 2002). For evaluation of multizone zone models in which partial mixing occurs within zones, it is possible to combine these models with computational fluids dynamics (CFD) methods (Musser 2001, Musser et al. 2001).

8.10 Zonal Protection Factors

The previous modeling methods will predict airborne concentrations in any zone in which a UV air cleaner is operating. Though these results are useful for making comparisons of air cleaning rates, they provide no absolute comparative basis. That is, the airborne concentrations alone do not tell us whether or not the environment is healthy, or whether the transmission of airborne disease will be inhibited. Certainly, the kinds of major reductions of airborne concentrations of pathogens and allergens in indoor air that are possible with UV air cleaning systems will provide for a healthier environment, but the question of whether or not these systems will interdict airborne disease transmission or prevent secondary infections (or allergic reactions to allergens) remains unanswered. However, prediction of the airborne concentrations is a first step towards determining how well a UV system will protect occupants against disease, because, once concentration are known, it is only necessary to compute the inhaled dose and estimate the percentage of infections that may result.

For the steady state condition addressed previously, it is a simple matter to estimate the inhaled dose of microbes. A normal breathing rate for sedentary office workers is about $B_r = 0.01$ m^3/min (0.353 cfm) per Heinsohn (1991). the 8-hour inhaled dose (D_8) from the steady state condition is the breathing rate multiplied by the concentration as follows:

$$D_8 = 8(60)B_r \cdot C_{ss} = 4.8 \left(\frac{R_R}{Q_{oa} + RE \cdot (Q - Q_{oa})} \right) \tag{8.17}$$

When no air filtration is present (i.e. the baseline condition), Eq. (8.17) reduces to:

$$D_8 = 4.8 \left(\frac{R_R}{Q_{oa}} \right) \tag{8.18}$$

For transient models, or multizone models, the inhaled dose can be computed in the same manner, typically minute-by-minute, and the total inhaled dose can be summed for the time period under consideration (i.e. 8 h). The inhaled dose can also be computed directly by integration of Eq. (8.13), after multiplying by the breathing rate, as follows.

$$D_8 = \int_0^8 B_r C_0 \left[e^{-ACHt} \right] dt \tag{8.19}$$

This integral is resolved for t = 8 h as follows:

$$D_8 = \frac{B_r C_0}{ACH} \left(1 - e^{-ACHt} \right) = \frac{B_r C_0}{ACH} \left(1 - e^{-8ACH} \right) \tag{8.20}$$

Once the inhaled dose is established, for any given pathogen, the infection rate can be computed based on the ID$_{50}$, which is the 50% infection dose or the dose that will cause infections in 50% of an exposed population. A simplified mathematical model for approximating the infection rate has been proposed by Kowalski (2006) and is given as follows:

$$y = 0.5^{0.1} {}^{\left(\frac{D - ID_{50}}{ID_{50}} \right)} \tag{8.21}$$

where

y = fraction of occupants infected
D = inhaled dose, cfu

Equation (8.21) can be used for an 8 h inhaled dose, or any other time period under consideration, if the ID_{50} for a particular pathogen is known. Very few ID_{50} values are actually known, but those that are have been tabulated or estimated in Kowalski (2006). In fact, typical airborne concentrations of specific microbes are equally unknown, and it becomes convenient and almost necessary to simply assume a generic value for the ID_{50} (i.e. 1.0). Once this assumption is made, it becomes possible to compare UV air cleaner performance on the basis of how well the system protects occupants from infection. That is, the removal rates and CADR or any air cleaner, when placed into operation in a zone or building, can be defined in how well it reduces theoretical infections. The computed value of y in Eq. (8.21), say for an 8 h day with the air cleaner operating, can be compared with the same value when the UV system is not operating. The complement of y (or 1-y) in Eq. (8.21) is called the Zonal Protection Factor (ZPF) since it defines the degree of protection offered by the UV system (or any other type of air cleaner) to the occupants. It forms a more complete basis for describing the effectiveness of a UV air disinfection system since it accounts for the interaction of the air cleaner with the zone and places the comparative result on a more absolute basis (as opposed to disinfection efficiency or equivalent air exchange rates). Furthermore, it is generic for all pathogens and depends directly on the removal rate, the CADR, and the zone volume. For 8 h of exposure to airborne microbes the ZPF can be written as follows:

$$ZPF = 1 - 0.5^{0.1^{\left(\frac{D - ID_{50}}{ID_{50}}\right)}} \qquad (8.22)$$

For the purposes of comparing the performance of any air cleaning system in a zone volume, the ID_{50} can be assumed to be unity. The ZPF method can be applied to any building that can be reasonably modeled as a single zone with complete mixing, such as those illustrated in Fig. 8.13a,b, provided the airflow is equally distributed on a unit area basis. This approach may not necessarily apply to multizone buildings with varying air distribution or that has separate ventilation systems, such as shown in Fig. 8.13c. When applied to whole buildings, it may be called a BPF, for Building Protection Factor (Kowalski and Bahnfleth 2004). The BPF may also represent the overall protection factor for a building composed of zones which each have a their own ZPF value. The BPF may, therefore, be considered the integrated or weighted sum of the ZPFs for individual zones in a building.

When a building is composed of zones 1 through n with different ZPF values for each zone, the BPF may be estimated as follows:

$$BPF = ZPF_1 FA_1 + ZPF_2 FA_2 + \ldots = \sum_{1}^{n} ZPF_n FA_n \qquad (8.23)$$

where FA_n = floor area of zone n

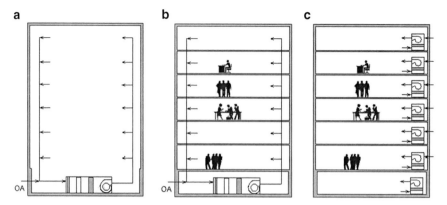

Fig. 8.13 *Left* (**a**) shows a model single zone building. *Center* (**b**) shows a multizone model with a single AHU. *Right* (**c**) shows a multizone model with separate ventilation systems

In buildings in which occupancy is unevenly distributed, it may be more appropriate to weight the ZPF values in Eq. (8.23) by occupancy rather than by floor area. In complex, multistory buildings, however, is it more accurate to use a multizone model as described previously.

8.11 UV Air Disinfection Field Trials

Although published studies on the ability of UV in-duct air disinfection to reduce indoor microbial populations have been few in number, they can easily demonstrate major reductions in indoor airborne concentrations of pathogens and allergens. Robertson et al. (1939) was among the first to test an installed in-duct air disinfection system and showed that up to 99% of the bacteria passing through air ducts could be removed with UV, although no measurements were taken on the indoor concentrations. Allegra et al. (1997) used a unitary UV system without a filter to reduce airborne concentrations of bacteria in a room, and obtained a 93–99% reduction of airborne counts within a matter of minutes.

Evidence that UVGI is effective at reducing disease incidence has accumulated over the decades but most of these studies involved upper room air disinfection. Limited epidemiological data is available on UV air disinfection systems but most of it confirms that reductions in disease transmission are possible. In 1937 the first application of UVGI to a school ventilation system dramatically reduced the incidence of measles, and subsequent similar applications enjoying comparable success (Wells 1955). Rosenstern (1948) showed that UV in air conditioning systems could reduce cross-infections by 71% or more in a nursery, although the system also employed UV barrier systems. Riley and O'Grady (1961) demonstrated the complete elimination of tuberculosis (TB) bacilli from hospital ward exhaust air using in-duct UV. Schneider et al. (1969) applied UV both in-duct and in corridors and effectively controlled pathogens in an isolation unit. One recent study, by Menzies et al. (2003), studied the effectiveness of UVGI in a forced air ventilation system. In this study respiratory symptoms in commercial office buildings were reduced from 3.9% to 2.9% after the installation of an in-duct UVGI system, representing a net decrease of 27% in respiratory symptoms. Another study, on mold-sensitized children afflicted with asthma who lived in residential homes equipped with in-duct UV systems, showed that UV air disinfection was effective in decreasing respiratory symptoms (Bernstein et al. 2006).

A study by Dionne (1993) on the incidence of respiratory infections at a day care center showed a negligible reduction in illness, but did demonstrate a reduction in airborne concentrations of microorganisms. It is often the case in epidemiological studies that natural disease incidence is so low (i.e. about 1–5% on any given day) that to achieve statistical significance requires either a large population or a trial period lasting years. In this study the population was 122 children and the trial period was 12 weeks. Airborne concentrations of bacteria were reduced from about 500 cfu/m^3 to about 160 cfu/m^3 after 3 weeks. The airflow or CADR through the UV units was low by current standards and ranged between 1.1 and 1.8 ACH.

References

Allegra L, Blasi F, Tarsia P, Arosio C, Fagetti L, Gazzano M. 1997. A novel device for the prevention of airborne infections. J Clinical Microb 35(7):1918–1919.

ANSI. 2002. Standard method for Measuring Performance of Portable Household Electric Cord-connected Room Air Cleaner. Washington, DC: American National Standards Institute. Report nr ANSI/AHAM AC-1-2002.

ANSI. 2006. Method for Measuring Performance of Portable Household Electric Room Air Cleaner. Washington, DC: American National Standards Institute, Association of Home Appliance manufacturers. Report nr ANSI/AHAM AC-1-2006.

ASHRAE. 2001. Standard 62: Ventilation for acceptable indoor air quality. Atlanta: ASHRAE.

ASHRAE. 2003. Handbook of Applications. Atlanta: American Society of Heating, Refrigerating, and Air-Conditioning Engineers.

Axley JW. 1987. Indoor Air Quality Modeling. Gaithersburg, MD: NBS. Report nr NBSIR 87–3661.

Bernstein J, Bobbitt R, Levin L, Floyd R, Crandall M, Shalwitz R, Seth A, Glazman M. 2006. Health effects of ultraviolet irradiation in asthmatic children's homes. J Asthma 43:255–262.

Boyce WE, DiPrima RC. 1997. Elementary Differential Equations and Boundary Value Problems. New York: John Wiley & Sons, Inc.

Buttolph LJ, Haynes H. 1950. Ultraviolet Air Sanitation. Cleveland, OH: General Electric. Report nr LD-11.

DelMundo F, McKhann CF. 1941. Effect of ultra-violet irradiation of air on incidence of infections in an infant's hospital. Am J Dis Child 61:213–225.

Demokritou P. 2001. Modeling IAQ and Building Dynamics. In: Spengler JD, Samet JM, McCarthy JF, editors. Indoor Air Quality Handbook. New York: McGraw-Hill.

Dionne J-C. 1993. Assessment of an ultraviolet air sterilizer on the incidence of childhood respiratory tract infections and day care centre indoor air quality. Indoor Environ 2:307–311.

Dols WS, Walton GN, Denton KR. 2000. CONTAMW 1.0 Users Manual. Springfield, VA: NTIS.

EPA. 2006a. Biological Inactivation Efficiency by HVAC In-Duct Ultraviolet Light Systems: Steril-Aire, Inc.: U.S. Environmental Protection Agency. Report nr EPA 600/R-06/052.

EPA. 2006b. Biological Inactivation Efficiency by HVAC In-Duct Ultraviolet Light Systems: Ultra-vIolet Devices, Inc.: U.S. Environmental Protection Agency. Report nr EPA 600/R-06/049.

EPA. 2006c. Biological Inactivation Efficiency by HVAC In-Duct Ultraviolet Light Systems: Lumalier.: U.S. Environmental Protection Agency. Report nr EPA 600/R-06/055.

EPA. 2006d. Biological Inactivation Efficiency by HVAC In-Duct Ultraviolet Light Systems: American Ultraviolet Corp.: U.S. Environmental Protection Agency. Report nr EPA 600/R-06/054.

EPA. 2006e. Biological Inactivation Efficiency by HVAC In-Duct Ultraviolet Light Systems: Atlantic Ultraviolet Corp.: U.S. Environmental Protection Agency. Report nr EPA 600/R-06/051.

EPA. 2006 f. Biological Inactivation Efficiency by HVAC In-Duct Ultraviolet Light Systems: Dust Free.: U.S. Environmental Protection Agency. Report nr EPA 600/R-06/050.

EPA. 2006 g. Biological Inactivation Efficiency by HVAC In-Duct Ultraviolet Light Systems: Novatron, Inc.: U.S. Environmental Protection Agency. Report nr EPA 600/R-06/084.

EPA. 2006 h. Biological Inactivation Efficiency by HVAC In-Duct Ultraviolet Light Systems: Sanuvox Technologies, Inc.: U.S. Environmental Protection Agency. Report nr EPA 600/R-06/053.

Foarde KK, Hanley JT, Ensor DS, Roessler P. 1999. Development of a method for measuring single-pass bioaerosol removal efficiencies of a room air cleaner. Aerosol Sci Technol 30:223–234.

Foarde KK, Myers EA, Hanley JT, Ensor DS, Roessler PF. 1999a. Methodology to perform clean air delivery rate type determinations with microbiological aerosols. Aerosol Sci Technol 30:235–245.

Grimm NR, Rosaler RC. 1990. Handbook of HVAC Design. New York: McGraw-Hill.

Hanley JT, D.D.Smith and D.S.Ensor. 1995. A fractional aerosol filtration efficiency test method for ventilation air cleaners. ASHRAE Transactions 101(1):97.

Heinsohn RJ. 1991. Industrial Ventilation: Principles and Practice. New York: John Wiley & Sons, Inc.

IUVA. 2005. General Guideline for UVGI Air and Surface Disinfection Systems. Ayr, Ontario, Canada: International Ultraviolet Association. Report nr IUVA-G01A-2005.

Janney C, Janus M, Saubier LF, Widder J. 2000. Test Report: System Effectiveness Test of Home/Commercial Portable Room Air Cleaners.: U.S. Army Soldier, Biological Chemical Command. Report nr Contract N. SPO900-94-D-0002, Task No. 491.

Ko G, First MW, Burge HA. 2002. The characterization of upper-room ultraviolet germicidal irradiation in inactivating airborne microorganisms. Environ Health Perspect 110(1):95–101.

Koller LR. 1965. Ultraviolet Radiation. New York: John Wiley & Sons.

Kowalski WJ, Bahnfleth WP, Whittam TS. 1999. Filtration of Airborne Microorganisms: Modeling and prediction. ASHRAE Trans 105(2):4–17.

Kowalski WJ, Bahnfleth WP, Witham DL, Severin BF, Whittam TS. 2000. Mathematical modeling of UVGI for air disinfection. Quant Microbiol 2(3):249–270.

Kowalski WJ, Bahnfleth WP. 2002. MERV filter models for aerobiological applications. Air Media Summer:13–17.

Kowalski WJ, Bahnfleth WP. 2002a. Airborne-Microbe Filtration in Indoor Environments. HPAC Eng 74(1):57–69.

Kowalski WJ, Bahnfleth WP. 2003. Immune-Building Technology and Bioterrorism Defense. HPAC Eng 75 (Jan.)(1):57–62.

Kowalski WJ, Bahnfleth WP. 2004. Proposed Standards and Guidelines for UVGI Air Disinfection. IUVA News 6(1):20–25.

Kowalski WJ. 2006. Aerobiological Engineering Handbook: A Guide to Airborne Disease Control Technologies. New York: McGraw-Hill.

Kowalski WJ. 2007. Air-Treatment Systems for Controlling Hospital-Acquired Infections. HPAC Eng 79(1):28–48.

Menzies D, Popa J, Hanley JA, Rand T, Milton DK. 2003. Effect of ultraviolet germicidal lights installed in office ventilation systems on workers' health and wellbeing: double-blind multiple crossover trials. The Lancet 362(November 29):1785–1791.

Musser A. 2000. Multizone modeling as an indoor air quality design tool. Finland: Espoo.

Musser A, Palmer J, McGrattan K. 2001. Evaluation of a fast, simplified computational fluid dynamics model for solving room airflow problems. Gaithersburg, MD: NIST.

Musser A. 2001. An analysis of combined CFD and multizone IAQ model assembly issues. ASHRAE Trans 106(1):371–382.

NIST. 2002. CONTAMW: Multizone Airflow and Contaminant Transport Analysis Software. Gaithersburg, MD: National Institute of Standards and Technology

Peccia J, Werth HM, Miller S, Hernandez M. 2001. Effects of relative humidity on the ultraviolet induced inactivation of airborne bacteria. Aerosol Sci Technol 35:728–740.

Rentschler HC, Nagy R, Mouromseff G. 1941. Bactericidal effect of ultraviolet radiation. J Bacteriol 42:745–774.

Riley RL, O'Grady F. 1961. Airborne Infection. New York: The Macmillan Company.

Riley RL, Kaufman JE. 1972. Effect of relative humidity on the inactivation of airborne *Serratia marcescens* by ultraviolet radiation. Appl Microbiol 23(6):1113–1120.

Robertson EC, Doyle ME, Tisdall FF, Koller LR, Ward FS. 1939. Air contamination and air sterilization. Am J Dis Child 58:1023–1037.

Robertson EC, Doyle ME, Tisdall FF, Koller LR, Ward FS. 1943. Use of ultra-violet radiation in reduction of respiratory cross-infections in a children's hospital. JAMA 121:908–914.

Rosenstern I. 1948. Control of air-borne infections in a nursery for young infants. Am J Dis Child 75:193–202.

Sauer LW, Minsk LD, Rosenstern I. 1942. Control of cross infections of respiratory tract in nursery for young infants. JAMA 118:1271–1274.

Schneider M, Schwartenberg L, Amiel JL, Cattan A, Schlumberger JR, Hayat M, deVassal F, Jasmin CL, Rosenfeld CL, Mathe G. 1969. Pathogen-free Isolation Unit – Three Years' Experience. Brit Med J 29 March:836–839.

Sommer HE, Stokes J. 1942. Studies on air-borne infection in a hospital ward. J Pediat 21:569–576.

Sparks LE. 1995. IAQ Model for Windows: Risk Version 1.0 User Manual.: US Environmental Protection Agency. Report nr EPA-600/R-96–037.

Walker CM, Ko G. 2007. Effect of ultraviolet germicidal irradiation on viral aerosols. Environ Sci Technol 41(15):5460–5465.

Walton GN. 1989. Airflow network models for element-based building airflow modeling. ASHRAE Transactions 1989:611–620.

Wells WF. 1938. Air-borne infections. Mod Hosp 51:66–69.

Wells WF. 1955. Airborne Contagion and Air Hygiene. Cambridge, MA: Harvard University Press.

Wheeler SM, Ingraham HS, Hollaender A, Lill ND, Gershon-Cohen J, Brown EW. 1945. Ultraviolet light control of airborne infections in a naval training center. Am J Pub Health 35:457–468.

Yaghoubi MA, Knappmiller K, Kirkpatrick A. 1995. Numerical prediction of contaminant transport and indoor air quality in a ventilated office space. Particul Sci Technol 13:117–131.

Chapter 9

Upper Room UV Systems

9.1 Introduction

Upper Room UV systems, sometimes called Upper Air systems, create a germicidal zone of UV rays that are confined to the upper portion of a room, known as the UV zone or stratum. Air that enters into this field is disinfected, along with any exposed surfaces. UV exposure levels in the lower room are maintained below the ACGIH 8-hour exposure limit of 30 J/m^2 (broadband UV) or 60 J/m^2 (254 nm). Upper Room systems operate continuously in occupied areas and if properly designed and installed are safe (see photo above, image courtesy of Lumalier, Memphis, TN). They can be a relatively simple and effective means of controlling airborne infection and can be cost-effective for many types of facilities, including hospital applications, waiting rooms, prisons, and homeless shelters. Upper Room UV may be preferred over other devices since they are passive (no moving parts), have modest costs, and low energy consumption. One of the stated advantages of Upper Rooms systems is that they intercept microbes inside the room where they may be generated by occupants, thereby controlling infection at the source (First et al. 1999). Continuous upper air UV disinfection is considered to be the most practical method in resource-limited countries where mechanical systems are often lacking and in cold climates where energy losses from high air turnover rates are prohibitive (WHO 1999).

Upper Room systems have been in use for decades and data that has accrued from installations in hospitals, schools, army barracks, and other indoor environments has shown them to be predictably effective provided the systems are properly designed and appropriate safety precautions are taken. Upper room systems have been shown to be effective against a wide array of airborne viruses and bacteria, including chickenpox, measles, mumps, varicella, TB, and cold viruses. Applications have shown them to reduce respiratory infections and decrease absenteeism. This chapter cites some twenty studies in which disease incidence rates were decreased and in only a few of these were the results inconclusive. Issues of safety have been largely addressed through good design practices and the history of its use shows very few cases of accidental eye and skin burns.

W. Kowalski, *Ultraviolet Germicidal Irradiation Handbook*,
DOI 10.1007/978-3-642-01999-9_9, © Springer-Verlag Berlin Heidelberg 2009

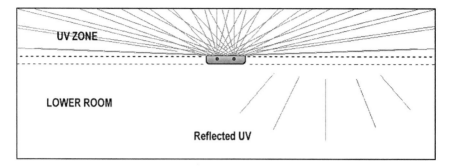

Fig. 9.1 Upper Room UV system with inverted fluorescent light fixture

Some Upper Room systems use naked UV lamps or fluorescent fixtures turned upside down as shown in Fig. 9.1. Many modern fixtures employ baffles or louvers, and sometimes parabolic mirrors, to obtain stratified layers of UV irradiance overhead while minimizing UV exposure in occupied areas below.

A summary of epidemiological performance data is provided in this chapter. Safety considerations are addressed here as they apply to design considerations, but they are addressed in more detail in Chap. 12.

9.2 Types and Design of Fixtures

Upper Room UVGI fixtures are available in a variety of sizes and configurations, including rectangular wall-mounted units and circular ceiling-mounted units (Dumyahn and First 1999). Most of these are passive devices but some include fans to enhance air movement. A circular fixture can be placed in the center of a room, such as a classroom. A rectangular fixture can be placed along the walls, usually in multiple locations. The fixture contains one or more UV lamps and often has a series of stacked parallel louvers to direct the UV rays. Some rectangular fixtures contain parabolic reflectors to direct the UV in parallel exit beams. Typical lamp input wattages range from about 18 to 36 W, with UV lamp output wattages of about 5–10 W, although with fixture efficiencies of about 5% or lower, actual UV output to the upper room zone is much less. Figure 9.2 shows two examples of Upper Room lamp fixtures, one rectangular and one circular, each with louvers.

Fig. 9.2 Rectangular Upper Room UV fixture (*left*), and circular model (*right*). Photos courtesy of Lumalier, Inc., Memphis, TN

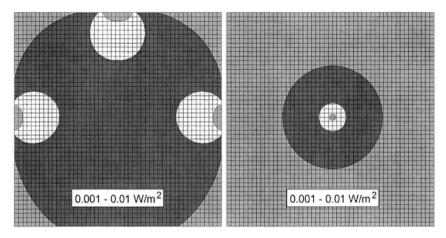

Fig. 9.3 Irradiance contours for three rectangular fixtures (*left*) and a circular fixture (*right*) of the same total wattage in a 400 ft^2 room. Contours increase on a log scale

Figure 9.3 illustrates the irradiance contours produced by rectangular and circular fixtures of the same total power, as generated by computer modeling. The left contour shows the contours for three 20 W (input) lamps, a system that produces a mean irradiance of 0.009 W/m^2 (9 μW/cm^2) in a 400 ft^2 room. The right contour shows a single 60 W circular lamp fixture that produces a mean irradiance of 0.012 W/m^2 (12 μW/cm^2) The circular fixture is more efficient in this application because it has less internal and external blockage of light.

Circular, ceiling-hung, fixtures will produce the highest UV levels in the center of the upper room, as shown in Fig. 9.3, and will tend to produce a higher mean irradiance due to higher luminaire efficiency (although it can depend on the specific design of the fixture). One advantage of circular fixtures is that often only a single fixture may suffice, whereas with rectangular units two or more may be needed to provide full room coverage.

The irradiance field is mostly confined to the upper zone, where levels are highest, but lower levels of UV will inevitably spread to the room below. Irradiance levels in the stratum are typically about 0.005–0.1 W/m^2 (0.5–10 μW/cm^2) but can be higher depending on ceiling height and proximity to the lamp fixture. Using a strict interpretation of the ACGIH TLV for 8-hour exposure 30 J/m^2 (broadband) or 60 J/m^2 (254 nm), irradiance values in the lower room would be less than 0.002 W/m^2 for UVC (254 nm) and 0.001 W/m^2 for broadband UV (UVC/UVB or 200–320 nm). However, measured values of irradiance at eye level for installed systems show that actual UV dose at eye level was far lower than the TLV, typically less than a few percent, and it is suggested a maximum eye-level UVC irradiance of 0.002–0.004 W/m^2 (0.2–0.4 μW/cm^2) will allow improved germicidal effectiveness in the Upper Room without causing undue hazards (First et al. 2005, Nardell et al. 2008, CIE 2003).

Since the light is blocked from the lower room, a stratified layer of irradiance will exist above the lower plane of the upper room area, as shown in Fig. 9.4. Depending

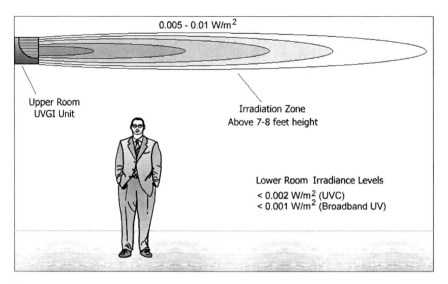

Fig. 9.4 Upper Room system illustrating irradiation contours in the upper room above 7–8 ft

on available space, the cutoff height of the UV zone may be 7–8 ft or higher, well above head height for most people.

The use of louvers in Upper Room units is intended to keep UV levels below ACGIH limits in the occupied areas of the room. The louvers function to keep rays parallel and horizontal. Sometimes these are reflective aluminum or else they are painted black. Reflectors inside the fixtures, though not common, serve the same function – to direct the outgoing rays in a parallel and horizontal direction. The ideal reflector would be a parabolic specular mirror (not a circle, ellipse, or hyperbola) so that the reflected light rays exit as parallel beams (Elmer 1989).

Figure 9.5 illustrates the effect of parabolic reflectors and louvers on the attenuation of UV rays. A ray that travels directly through the louvers, unattenuated, will have a factor of 1.0 (100% transmission). A ray that reflects off the specular parabolic surface is reduced by $3/4$ (75% reflectivity). A ray that reflects n times will be reduced per the following equation:

$$R_i = I_D \rho^n \qquad\qquad (9.1)$$

where

R_i = reflected irradiance, W/m^2
I_D = direct (unattenuated) irradiance, W/m^2
ρ = reflectivity
n = number of reflections

If black louvers are used, the reflectivity will be on the order of about 4–5%, and the reduction in exiting UV rays will be considerable. Such an approach can be very

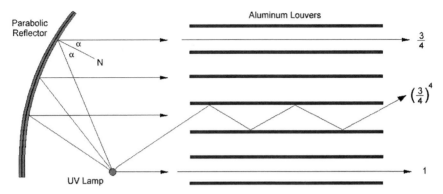

Fig. 9.5 Cross-section schematic of a UV lamp, a parabolic reflector, and an array of aluminum louvers (UV reflectivity = 0.75), showing the attenuation of UV rays as they bounce at incident angles (α) from the normal (N) to the surface

effective at confining the irradiance to the upper zone, but also consumes a lot of energy that will convert mostly to heat. The loss in radiated UV through such fixtures can be described in terms of the luminaire efficiency, which is the ratio of light output from a fixture to that of a naked lamp (Lindsey 1997, IESNA 2000). Luminaire efficiencies of Upper Room fixtures appear to be on the order of 0.5–3% per Dumyahn and First (1999), but the luminaire efficiency could be increased with parabolic reflectors. Luminaire efficiency is also a function of the louver spacing, with greater spacing producing higher efficiency, but less stratification in the upper zone.

Figure 9.5 illustrates the internal dynamics of light rays in an Upper Room fixture that employs both louvers and a parabolic reflector. All rays are assumed to bounce off reflective surfaces at the incident angle α, relative to the normal (perpendicular to the surface). Each reflection will reduce the irradiance, at that point, by 1–ρ, where ρ is the reflectivity.

Incident UV rays that reflect off surfaces are partly absorbed at the fractional rate (1–ρ), which invariably converts almost entirely to heat (for metal and glass). Approximately one half (180° span) of the lamp energy will reflect off any internal parabolic mirror and be reduced by about 75%. Since very few of the UV rays will travel directly from the lamp through the louvers, a great deal of attenuation goes on in the louvers due to multiple reflections, regardless of reflectivity. For example, six reflections off an aluminum surface will reduce irradiance by 90%. Also, the leading edge of the louvers will tend to block a small fraction of UV rays. The result is that the irradiance field created by the Upper Room unit will produce only a fraction of the irradiance of a naked UV lamp, and the luminaire efficiency will be low. In fact, most of the irradiation that escapes the louvers will likely be the half that reflects off the parabolic surface, and without a reflector these units can be very inefficient. The luminaire efficiency of an Upper Room unit will always be increased by the addition of parabolic reflectors. Black-painted louvers will decrease the luminaire efficiency, compared to aluminum louvers, but will also reduce reflectance into the lower room.

Computer models can be developed based on UV lamp irradiance models (see Chap. 7) combined with the geometric and reflectivity considerations described above. In fact, a circular UV lamp fixture can be approximated by a bare lamp with the power output reduced by the luminaire efficiency. Reflecting plates in rectangular fixtures can be approximated by modeling the reflection as a second UV lamp reduced by the amount of the reflectivity, and positioned accordingly. Another feasible approach to design Upper Room fixtures is to use lighting software of the kind used in illumination engineering to design luminaires. Such software often uses ray-tracing methods to establish reflected lighting intensities and are perfectly adaptable to UV lighting and reflectivity modeling in rooms.

9.3 Upper Room System Sizing

Various methods have been used or suggested for the design or Upper Room systems, including empirical methods, rule-of-thumb methods, and ACH or airflow-based methods. The size of an Upper Room system is usually specified in terms of total watts of lamp input. The number and wattage of lamp fixtures can be based on mean irradiance levels, room floor area and ceiling height, or even total number of occupants. All of these approaches can provide adequate results.

The basic design principle is to maximize the UV dose or fluence in the Upper Room while minimizing irradiance in the lower room, preferably to levels below the ACGIH TLV for 8 h exposure to actinic UV (200–320 nm). This latter criteria tends to impose limits on the irradiance achievable in the upper UV zone and hence the design of Upper Room systems is both constrained and simplified. Experience has shown that the total lamp wattage can be related to the total floor area, for any given room height. Koller (1939) proposed that the number of 15 W or 30 W UV lamps could be distributed per floor area based on ceiling height, as shown in Table 9.1.

The criteria of using one 30 W lamp for every 200 ft^2, as per the 8 ft ceiling in Table 9.1, is widely quoted and has been adopted in guidelines (First et al. 1999, Macher 1993, Riley et al. 1976, CIE 2003, ASHRAE 2008). As will be shown later in the section on performance, these rules-of-thumb will generally provide adequate results.

Other early design methods were developed by Luckiesh and Holladay (1942), who developed tables and charts which could be used with any room height, width,

Table 9.1 Upper Room Design Guidelines per Koller (1939)

Ceiling Height, ft	8	10	12	14	16
Floor area per each 30 W lamp, ft^2	200	250	300	350	400
Ceiling Height, m	2.44	3.05	3.66	4.27	4.88
Floor area per each 30 W lamp, m^2	18.58	23.23	27.87	32.52	37.16

and length to estimate the average flux density in the upper portion of a room, based on the flux emitted by the lamp fixture. Much of their information was developed based on field investigations of installed systems. The Upper Room systems they studied produced an average irradiance in the upper irradiation zone of about 0.002–0.014 W/m^2 (2–14 μW/cm^2). Exposure times in the irradiation zones due to room air movement rates were estimated to be on the order of 20–30 s, yielding dose estimates of up to 0.41 J/m^2. Modern Upper Room systems produce up to one hundred times the mean irradiance of these earlier studies.

The average irradiance can be modeled by computer software or can be measured with a series of photosensor readings. An example of such irradiance measurements in the three dimensional upper room space is given by Xu et al. (2003), who determined the average irradiance in an 87 m^3 room to be about 0.42 W/m^2 (42 μW/cm^2) with a total lamp input of 216 W, and using spherical actinometers. Miller and Macher (2000) measured the irradiance of a 35.3 m^3 room stratum to be 0.25 W/m^2 with 5 W of UV output and 0.52 W/m^2 with 10 W of UV output. Dumyahn and First (1999) used a weighted model of the irradiance field to calculate the effective UV output of Upper Room fixtures, W (watts) according to the following:

$$W = R \sum_{j=1}^{m} \sum_{i=1}^{n} E_{ij} \Delta z_i \Delta \theta_j \qquad (9.2)$$

where

R = radius (distance) from fixture
E_{ij} = UV irradiance for area ij
Δz_i = vertical offset
$\Delta \theta_j$ = lateral offset (in radians)

The UV dose in the upper zone is equivalent to the mean or average irradiance in the UV zone multiplied by the residence time, which depends on the airflow through the zone. Room air mixing is a complex function of many factors and this simple model can only provide estimates of the mean which may be subject to wide variations that depend on specifics of the actual room (Rudnick and First 2007). If air is supplied directly to the zone at a rate of Q m^3/s, then the system may be modeled as shown in Fig. 9.6, where the entire upper UV zone is a control volume with complete mixing, and the dose is simply the product of the mean irradiance and the residence time. In this simplistic model, the residence time will be a function of the airflow traversing the distance L (normal to the airflow plane). The UV exposure dose can be written in terms of the airflow as follows:

$$D = I_m t_{res} = I_m \frac{V}{Q} \qquad (9.3)$$

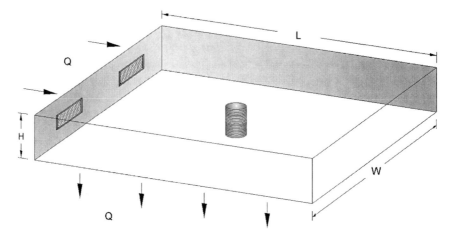

Fig. 9.6 Modeling the upper room UV zone as a control volume when air is supplied directly through ventilation grilles

where

I_m = mean irradiance, W/m^2
t_{res} = residence time, s
V = volume of UV zone, m^3
Q = airflow, m^3/s

The dose computed with Eq. (9.3) can be applied regardless of whether air is supplied or exhausted from the UV zone, or both. It can also be applied to conditions of no forced airflow or natural ventilation provided the airflow Q can be estimated. For an in-duct system (operating at 400–500 fpm), Eq. (9.3) would completely define system performance, but for Upper Room systems the degree of air mixing and other factors will contribute to overall system performance, as addressed later. For the moment we have a means of sizing systems and comparing UV Zone performance, separate from room performance. It also allows direct comparison of Upper Room systems with in-duct systems in term of the UVGI Rating Value, or URV, which provides perspective on the range of UV doses produced.

Table 9.2 summarizes data from Upper Room installations and test configurations from which the UV exposure dose in the UV zone can be computed or deduced. None of the studies provided the UV dose directly and it had to be estimated based on irradiance and residence time (where provided) or else estimated from biodosimetric data. The URV is shown for each of the estimated doses. The mean irradiance data for Rahn et al. (1999) is based on actinometric readings. A typical design rating for in-duct UVGI systems would be about URV 10–15. Although this set of data may not be entirely representative of modern design practice, it would appear that Upper Room systems fall within the general dose range of in-duct systems. One source cites a suggested recommendation that the target dose for mean irradiance of

Table 9.2 Comparison of Upper Room System UV Doses

Lamp	UV power	Zone volume	Mean irradiance	Residence time	UV dose	URV	Data ref
W	W	m^3	W/m^2	s	J/m^2		
72	(22)	9.6	0.44	–	0.7	5	Rahn et al. (1999)
60	(19)	90.00	1.0000	–	(0.89)	6	Macher et al. (1992)
59	15	45.20	–	–	(2.29)	8	Ko et al. (2002)
216	(67)	13.90	0.4200	–	(3.8)	9	Xu et al. (2003)
120	33	81.55	0.7632	20	15.26	12	Luckiesh and Holladay (1942)
90	44	244.66	0.5587	30	16.76	12	Luckiesh and Holladay (1942)
30	(10)	36.00	0.4700	–	(17)	12	Miller (2000)
120	44	163.11	0.7180	30	21.54	13	Luckiesh and Holladay (1942)
120	44	127.43	0.7029	30	21.09	13	Luckiesh and Holladay (1942)
120	44	81.55	1.0162	30	30.49	14	Luckiesh and Holladay (1942)
120	44	32.56	1.5716	30	47.15	15	Luckiesh and Holladay (1942)
148	46	64.40	–	–	(76)	17	Riley et al. (1976)

Note: Values in parentheses are author's estimates computed from Reference data.

Upper Room systems be about 0.5 J/m^2 (URV 5), but based on Table 9.2 and the URV scale, this value would seem low (First et al. 1999).

Based on the estimated URV ratings in Table 9.2, Upper Room systems would seem to fall within normal ranges for in-duct systems, as one might expect from their similarity. Of course, many Upper Room systems operate at low or no airflow and the doses these systems produce would be proportionately higher. In in-duct system design, URV 13 often represents a plateau above which there are diminishing returns for increased power. It is possible this is a natural plateau for Upper Room systems also, and that approximately 50 W of input power for every 200 ft^2 would be a reasonable upper limit for designing Upper Room systems, and that 30 W for every 200 ft^2 is a reasonable lower limit. However, much depends on room airflow mixing effects and so it is difficult to establish a general requisite URV for any Upper Room system, but the approach can provide some confidence that a given system design is within normal parameters.

Given the UV dose, the decay of microbial populations in air is described by the classic exponential decay equation (see Chap. 2). In an upper room the decay may be due to additional factors such as room air exchange rate, deposition on surfaces, and natural decay. For a room model (as opposed to an in-duct UV chamber) it is appropriate to define the decay rate constant as due to all possible removal rates or inactivation factors, and the equivalent rate constant, k^{eq}, can be used to evaluate the room effectiveness of Upper Room systems as follows:

$$C_t = C_0 e^{-k_{eq}t} \qquad (9.4)$$

where

C_t = airborne concentration at time t, cfu/m^3
C_0 = initial airborne concentration, cfu/m^3

In Eq. (9.4), the value of k_{eq} can be determined as the slope of the exponential decay curve of the room concentration of airborne microorganisms measured over time in the absence of significant room airflow. In a room that has continuous generation of contaminants over time, Eq. (9.4) can be written as:

$$C_t = \frac{q_c}{Vk_{eq}} + \left(C_0 - \frac{q_c}{Vk_{eq}}\right)e^{-k_{eq}t} \tag{9.5}$$

where

q_c = the rate at which contaminants enter the room, cfu/s
V = volume of the room, m^3

9.4 Air Mixing Effects

An essential factor in the effectiveness of an Upper Room system is the air turnover rate between the upper and lower room. This turnover, or air exchange rate, defines the duration of a microbe's exposure. Too rapid a passage of lower room air through the irradiation stratum may result in insufficient exposure time, but that the greatest reduction in infectious microorganisms may occur when the greatest rate of air exchange occurs between the upper and lower room areas (CIE 2003).

Riley et al. (1976) presented the disappearance of *Mycobacterium bovis* (BCG) from a room with an upper air UVGI system in terms of the equivalent air changes per hour, defined as the slope of the plot of the natural logarithm of the airborne BCG count versus time in hours. First et al. (1999) define the equivalent air exchange (EAC) more generally as the number of air exchanges in a well-mixed room that would be required to reduce the number of viable airborne bacteria to the same degree as UV alone, and present the following equation:

$$EAC = -\ln\left(\frac{N_s}{N_0}\right) = kIt = kD \tag{9.6}$$

where

N_s = microbial population at time t (based on airborne counts)
N_0 = microbial population at time t-0
k = microbe rate constant, m^2/J
I = irradiance, W/m^2
t = time, seconds
D = UV exposure dose, J/m^2

Equation (9.7) will provide the EAC for any given microbe. For example, if *Serratia marcescens* (k= 0.221 m^2/J) were exposed to a UV dose of 54.3 J/m^2 ($I \times t$), the EAC would be (0.221) × (54.3) = 12 air changes per hour (ACH). For *Streptococcus pneumoniae* (k= 0.055 m^2/J) the same dose would produce an EAC = 3 ACH. The dependence of EAC on microbial species is not a serious drawback provided some average or representative microbe (i.e. *S. marcescens*) is used as an indicator. Studies on reduction rates of *Mycobacterium tuberculosis* have shown equivalent air exchange rates of 10–25 ACH can be achieved (CDC 2005).

The primary obstacle to modeling the overall performance or Upper Room systems is that the movement of air into and out of the upper room is subject to much unpredictability. Room air currents rise with heat (i.e. body heat or from equipment) and can be greatly influenced by any airflow through the room from existing ventilation supply and exhaust ducts. If the room has no forced air ventilation, then the effectiveness of the system depends on natural convection currents to carry microorganisms into the irradiation zone. Figure 9.5 illustrates these processes for an operating room equipped with an Upper Room UV system. It is not uncommon for hospital and other facilities to have supply air enter near the ceiling and exit near the floor, and so in this example the entire airflow Q will pass through the UV zone as in an in-duct UVGI system. Biocontaminants, shown as q_c in Fig. 9.7, can hail from room occupants, including the patient, as well as from other sources like the floor and equipment.

Riley and Permutt (1971) developed a framework for modeling these systems in which the microbial inactivation rate in the upper zone could be predicted if the air velocity across the upper-lower interface was known. However, such air velocities can vary widely and the results become highly dependent on assumptions or average measurements. A multizone model was developed by Nicas and Miller (1999) for Upper Room systems based on the following equation:

$$Q_{ex} = 0.5(l \times w)v_{int} \tag{9.7}$$

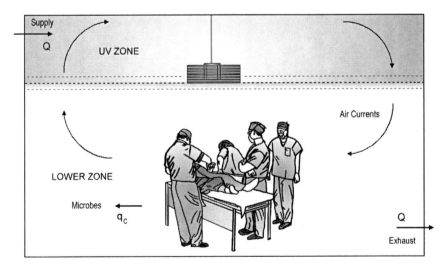

Fig. 9.7 Air will circulate into and out of the Upper Room UV zone while contaminants will be released continuously by occupants or other sources

where

Q_{ex} = volume flow rate across the zone interface, m^3/s
l= length of room, m
w = width of room, m
v_{int} = average velocity across the interface, m/s

A theoretical model developed by Beggs and Sleigh (2002), examined various scenarios involving room airflow rates and found that interzonal air exchange rates were somewhat independent of room air exchange rates. For UV zones that don't have forced ventilation, Eq. (9.7) can be used to estimate the airflow Q in Eq. (9.3) to compute the UV dose.

In the case for complete mixing of room air, which is not an unreasonable assumption for real world applications, even with natural ventilation, the modeling of system performance can be based on the average residence time of particles in the upper irradiation zone. The residence time, t_{res}, of any particle in a room is as follows:

$$t_{res} = \frac{1}{ACH} \tag{9.8}$$

where ACH = air change rate, air changes per hour

The average residence time, t_{uv}, of particles in the upper UV irradiance zone is defined as follows:

$$t_{uv} = t_{res}\frac{h_{uv}}{h_r} \tag{9.9}$$

where

h_{uv} = height of the UV zone, m
h_r = height of the room, m

Rudnick and First (2007) present an alternate model for assessing UVGI effectiveness of Upper Room systems that employs dimensionless parameters in a one-box or two-box model. They define a dimensionless *Index of Effectiveness* as follows:

$$I_e = \frac{1}{\frac{1}{N_I} + \frac{1}{N_M}} \tag{9.10}$$

where

N_I = Irradiation Number, dimensionless
N_M = Vertical Mixing Number, dimensionless

The *Irradiation Number* defines the effectiveness of the air disinfection, and is defined as:

$$N_I = \frac{k \sum \overline{d}_i P_i}{V \cdot \mathrm{ACH_o}} \tag{9.11}$$

where

$k =$ UV rate constant, m^2/J
$d_i =$ mean length of the UV rays, m
$P_i =$ Power output of lamp or fixture, W
$V =$ Room volume, m^3
$\mathrm{ACH_o} =$ Outside air exchange rate

In Eq. (9.11) the product of the mean ray length and the UV lamp power is summed across the room volume. The outside air exchange rate is the outside air flowrate divided by the room volume. The *Vertical Mixing Number* defines the relationship between the air flowing from the lower to the upper room and the outside air flowrate, as follows:

$$N_M = \frac{q}{Q} \tag{9.12}$$

where

$q =$ airflow from the lower zone to the upper room $Q =$ outside airflow

For more details on computing the Index of Effectiveness, computational methods, examples, and ranges of typical values, see Rudnick and First (2007).

9.5 Room Design

The planning an Upper Room installation should consider the physical measurements of the space, room occupancy, and air movement. Rooms may have to be modified to make the best and safest use of Upper Room UV systems. Ventilation systems should maximize air mixing to optimize system effectiveness. Good air mixing is necessary for such systems to be effective (WHO 1999). When rooms lack adequate air movement, the addition of mixing fans is a satisfactory solution (CIE 2003). Relative humidity should be less than about 75%, or typical indoor conditions, but an RH exceeding 80% may reduce system effectiveness due to microbial protective responses.

UV-absorbing paints containing zinc oxide can be used on ceilings and walls to minimize reflected UV (see Chap. 15). Usually, painting with flat white or dark paints is sufficient to reduce reflected UV to about 4–5%. If UV levels exceed safe limits, all highly UV-reflecting surfaces should be removed or altered.

Retrofitting a room with Upper Room UVGI can be limited by ceiling height. For floor to ceiling heights less than 2.3 m (7.5 ft) eye exposure hazards may increase. If vertical air circulation is insufficient to elevate infectious particles into the irradiation zone, system effectiveness is reduced.

9.6 Performance

The effectiveness of Upper Room air disinfection systems depends on UV power, fixture location, local reflective surfaces, degree of air mixing between the stratum and the lower room volume, microbial susceptibility, and relative humidity (First et al. 2007, Rudnick and First 2007). In-duct air disinfection systems can be defined by a single-pass efficiency and/or a clean air delivery rate. For Upper Room systems the overall room effectiveness can be defined by the following relation (Beggs and Sleigh 2002):

$$E = 1 - \left(\frac{C_{\text{on}}}{C_{\text{off}}} \right) \tag{9.13}$$

where

E = room effectiveness, fraction or %
C_{on} = airborne concentration of microbes with system On, cfu/m^3
C_{off} = airborne concentration with system Off, cfu/m^3

In effect, Eq. (9.10) treats the entire room as a single large air disinfection unit in the same fashion as the UV zone was treated to establish the dose and the URV.

Performance of Upper Room systems can be measured via microbiological testing in which the airborne concentrations of microorganisms are sampled. In such tests the Before condition (UV Off) is compared with the After condition (UV On) and the percentage reduction in airborne bacteria is computed. It is possible to use such results for biodosimetry (i.e. to estimate UV dose). An EAC can also be determined from such test results, for each microbe tested. See Chap. 14 for more detailed information on microbiological testing and biodosimetry of Upper Room systems.

In one test of an Upper Room system by Xu et al. (2003), an 87 m^3 room was equipped with four 36 W units mounted in the corners and one 72 W unit mounted in the center. These units created a 30 cm wide upper zone and the average irradiance or fluence rate was measured with actinometers to be about 0.44 W/m^2 (44 μW/cm^2). The removal rates of several microbes from the room air was used to determine the effective clean air delivery rate (CADR), which proved to be about 1392 m^3/h (819 cfm). It was found in a second study that operating the Upper Room system in combination with other air cleaners resulted in an additive effect in terms of microbial removal rates and CADR (Kujundzic et al. 2006).

It has been noted in the literature that increasing the air change rate, or airflow through the UV zone, can decrease the effectiveness of the system (Miller and

Macher 2000, Beggs and Sleigh 2002, Xu et al. 2003). Clearly, airflow through the zone and UV dose are related. This situation, in fact, is hardly any different from an enclosed in-duct UVGI system, since both employ the passage of air through a UV irradiation zone. Per Eq. (9.3), increasing the ACH decreases the UV dose, but of more importance is how much the effectiveness of the system is impacted by increasing ACH, and the effectiveness depends ultimately on the CADR, the clean air delivery rate. The CADR is the flowrate multiplied by the UV efficiency (or kill rate, KR). For in-duct systems this efficiency may approach 99% and so the CADR is approximately equal to the flowrate, Q. This may not be the case for Upper Room systems, and we write the CADR as follows:

$$\text{CADR} = Q_z KR = Q_z \left(1 - e^{-kI_m t_{res}}\right) \qquad (9.14)$$

where $KR =$ kill rate or inactivation rate, fraction

Plotting equating (9.12) for the general case where $k = 1 \ m^2/J$ and $I_m = 1 \ W/m^2$, and $V_z = 100 \ m^3$, and for increasing airflow, we obtain Fig. 9.8.

Clearly, from Fig. 9.8, the CADR will always increase with increasing airflow, up to about 6 ACH, which corresponds to a CADR of 0.167 m^3/s in this example. Though the UV dose decreases in the UV zone, increasing airflow will always increase the CADR, and hence will increase the net effectiveness of the system. However, above about 6 ACH (for this example) further increases may be negligible. As for in-duct systems, this can only be true up to a point because eventually the increased airflow will produce velocities across the UV lamp of greater than about 500 fpm, and lamp output may decrease as a result. The reports in the literature of

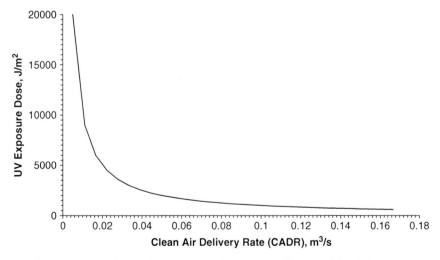

Fig. 9.8 Generic example of a plot of CADR vs. UV Exposure Dose, as airflow is increased from 0 to 6 ACH

decreasing effectiveness with increasing ACH are likely due to both experimental error and the fact that at low doses microbes (especially spores) will have a shouldered response such as in the data from Xu et al. (2003) for *Bacillus subtilis* spores.

There are many more in-place studies on Upper Room systems than there are for any other type of UV system. In an epidemiological study on a Naval Training Center by Miller et al. (1948), UV lamps totaling 750 W input were placed in the upper room and 840 W were placed in the lower room. With a floor area of 4000 ft^2, or the equivalent of about 42 W per 200 ft^2 upper and 37.5 W per

Table 9.3   Results of Field Trials of Upper Room Systems

Location	Infection	Infection Cases		Decrease	References
		Before	UVGI	Net (Percentage)	
Germantown Friends School, PA	Mumps	11	2	–	82	Wells (1938)
The Cradle, Evanston	Respiratory infection	14.5	4.6	9.9	68	Sauer et al. (1942)
Germantown Friends School, PA	Measles	–	–	–	20	Wells (1943)
Combined results for 4 PA schools	Measles	75%	43%	32	43	Wells (1943)
Germantown Friends School, PA	Cold viruses	2122	1738	–	18	Wells (1943)
National Training School for Boys, DC	Respiratory infection	–	–	–	0	Schneiter et al. (1944)
Camp Sampson Naval Training Station, NY	Respiratory infection	–	–	–	20	Wheeler et al. (1945)
St. Luke's Hospital, NY	Respiratory infection	–	–	–	33	Higgons and Hyde (1947)
New York state 3-school study	Measles epidemic	–	–	–	0	Perkins et al. (1947)
Great Lakes Naval Training Station, IL	Respiratory infection	–	–	–	19	Miller et al. (1948)
Mexico, Cato-Meridian and Port Byron schools	Chickenpox	38.5%	28.9%	9.6	25	Bahlke et al. (1949)
Southall Elementary schools, England	Measles	12.97%	11.10%	1.9	14	MRC (1954)
Southall Elementary schools, England	Mumps	9.97%	3.92%	6.1	61	MRC (1954)
Southall Elementary schools, England	Chickenpox	7.49%	6.24%	1.3	17	MRC (1954)
Home for Hebrew Infants, NY	Varicella epidemic	97%	0%	97	100	Wells (1955)
Mexico and Cato-Meridian schools	Mumps epidemic	235	59	–	75	Wells (1955)
Port Byron School	Mumps epidemic	49%	45.90%	3.1	6	Wells (1955)
Pleasantville and Mt. Kisco	Measles	227	217	–	4	Wells (1955)
Pleasantville and Mt. Kisco	Chickenpox	297	104	–	65	Wells (1955)
Livermore CA veteran's Hospital	Influenza	19%	2%	17	89	McLean (1961)
Boston Homeless Shelter	Tuberculosis	–	–	–	78	Nardell (1988)
North Central Bronx Hospital	TB conversions in staff	2.5%	1.0%	2.0	60	EPRI (1997)
National Homeless Shelters	Tuberculosis	–	–	–	7.3	National TB Coalition (2001) (preliminary)
Average net decrease				39%	

200 ft^2 lower. In the Germantown school study Wells et al. (1942) provided about 0.0134 W/m^2 in the classrooms. In the classroom study by Perkins et al. (1947), the upper zone was irradiated with 0.11–0.22 W/m^2, while the lower room area was kept below 0.0002–0.0005 W/m^2, typically with three 30 W lamps in each classroom. In a study on classrooms by Gelperin et al. (1951), three 30 W lamps placed on the walls produced a minimum of 0.093 W/m^2 (93 μW/cm^2) in the upper zone and a maximum of about 0.002 W/m^2 (2 μW/cm^2) in the lower zone (below 6.5 ft). These results are shown in Table 9.3, except for the latter since they were mixed and inconclusive, due at least partly to an unusually low incidence of measles, mumps, and chickenpox in the 1949–1950 school year.

In a study by Escombe et al. (2009), in which guinea pigs were exposed to the exhaust air from a TB ward, 35% of the controls developed TB infections while only 9.5% developed infections when Upper Room systems were in use, yielding a net decrease of 74%.

9.7 Testing

Invariably, photosensor measurements will be taken to establish that no hazardous levels of UV exist below the UV zone in the habitable areas. Testing may also be performed to define the irradiance field in the UV zone.

Figure 9.9 shows how the irradiance levels decreases with distance from a rectangular and circular UV fixtures, taken from the example in Fig. 9.3 for a 400 ft^2 room with 60 W total input power.

Most Upper Room irradiance fields will fall within about 0.001–10 W/m^2, or a four log range, although irradiance may be much higher close to the fixture. It would seem convenient and appropriate for measurement purposes to define the contour

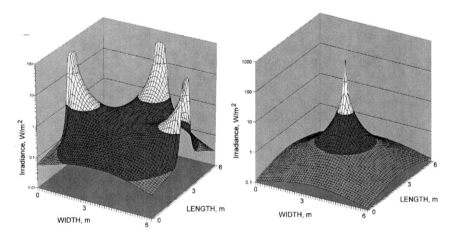

Fig. 9.9 Computed irradiance levels for wall-mounted rectangular fixtures (*left*) and a circular fixture (*right*)

lines that divide each log difference in irradiance. For circular fixtures, photosensor readings values could be taken in concentric circles around the fixture. In each case the contour divide between each log (i.e. 0.001, 0.01, 0.1, 1, 10, etc) should be sought and identified. For rectangular fixtures measurements can be taken in concentric semicircles around the fixtures up to the point that the contours intersect.

Figure 9.10 shows some actual measurements from a rectangular fixture. Photosensor readings were taken in a grid in the front of the unit and towards the sides. Any grid of sufficient resolution should be able to provide an estimate of the mean irradiance field. Since the irradiance field is more concentrated close to the lamp fixtures, more measurements may be taken closer, and fewer farther, to provide increased resolution of the irradiance field.

When multiple fixtures exist in a room and face each other on opposite walls, the question may arise as to how to handle the measurements. It is appropriate in such cases to take two measurements facing opposite and add them directly, provided there are no additional fixtures contributing to the local irradiance field. The result of this approach will approximate the same result obtained through the use of spherical sensors, such as are used in spherical actinometry. Actinometry can be used effectively to measure the irradiance field in a room equipped with an Upper Room system (Rahn et al. 1999, Rahn 2004, Xu et al. 2003). For more on photosensor measurement of Upper Room system irradiance levels, see Chap. 13.

Microbiological testing can also be conducted by means of air sampling. The ability of the system to reduce airborne counts of bacteria or fungi is a prime indicator of system performance and the results can be used directly with Eq. (9.11)

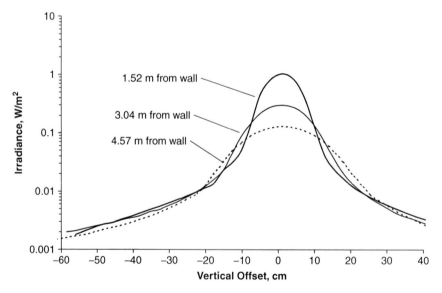

Fig. 9.10 Irradiance measurements for a wall-mounted Upper Room unit. Based on data from Dumyahn and First (1999)

to determine overall room effectiveness. Dosimetry can also be used to estimate the UV exposure dose. For more information on microbiological testing see Chap. 14.

9.8 Upper Room UV Safety

The primary safety hazard of Upper Room systems is that occupants may be exposed to direct or reflected UV rays for prolonged time periods. A fraction of UV will be reflected into the lower room but this will typically be insignificant since most room materials will tend to have UV reflectivities less than about 5%. Measurements should be made to verify that unsafe UV levels do not exist. Technically, levels of UV irradiance in the lower room should be below those that produce the ACGIH TLV dose after 8 h of exposure (0.002 W/m^2 for UVC and 0.001 W/m^2 for broadband UV) although, as noted previously, there may be a wider range of acceptable irradiance levels and it is suggested by Nardell et al. (2008) and First et al. (2005) that a maximum eye-level UVC irradiance of 0.002–0.004 W/m^2 (0.2–$0.4 \mu\text{W/cm}^2$) would be a practical irradiance limit. The ACGIH TLVs for UV can depend on the type of UV lamp used in the fixtures since narrow-band UVC lamps (low pressure mercury lamps) emit most of their UV at 254 nm, while broadband UV lamps (medium pressure mercury lamps) emit a broader spectrum of UV between 200 and 400 nm. Since the erythemal curve varies with wavelength, the limits are different depending on the UV spectra produced, with the ACGIH TLV being 30 J/m^2 for broadband UV and 60 J/m^2 for narrow band UVC (254 nm). Scientists with the Tuberculosis Ultraviolet Shelter Study (TUSS) are now using an irradiance limit of 0.004 W/m^2 at eye level as their design criteria, which takes into account the body's designed defenses: shading of upper lids and brows, and other factors that limit actual exposure (CIE 2003).

If irradiance levels are too high in the lower room, the first option is to paint the ceiling and UV zone wall areas. The use of paints with low reflectivity or the coating of surface with UV absorbers are options for reducing levels in the lower room without impacting the efficacy of the Upper Room system. If this fails, reducing lamp wattage, or blocking portions of the fixture with metal plates will work, although it may also decrease system effectiveness. For information on monitoring and maintaining safe levels of UV see First et al. (2005) and Nardell et al. (2008).

Fading, bleaching, and photodegradation of certain materials is possible even at safe levels of UV, due to the continuous exposure. Figure 9.11 shows an installation in which a house plant located too close to a fixture withered away in a matter of weeks. For more information on what materials are safe to use under continuous UV exposure, see Chap. 15.

Low levels of ozone may be generated by UV lamps, especially broadband or medium pressure UV lamps. If ozone is detected, measurements should be taken to ensure ozone concentrations do not exceed 0.1 ppm for any 8 h period, per ACGIH TLVs. Upper air UVGI systems should have safety switches to disengage the units for servicing. Warning signs may also be appropriate. For more detailed information of the safety hazards of ultraviolet radiation see Chap. 12.

Fig. 9.11 Plants placed in
the vicinity of Upper Room
systems may not survive
continuous exposure

References

ASHRAE. 2008. Handbook of Applications: Chapter 16: Ultraviolet Lamp Systems. Atlanta, GA: American Society of Heating, Refrigerating, and Air-Conditioning Engineers.

Bahlke AM, Silverman HF, Ingraham HS. 1949. Effect of ultra-violet irradiation of classrooms on spread of mumps and chickenpox in large rural central schools. Am J Pub Health 41: 1321–1330.

Beggs CB, Sleigh PA. 2002. A quantitative method for evaluating the germicidal effects of upper room UV lights. J Aerosol Sci 33:1681–1699.

CIE. 2003. Ultraviolet Air Disinfection. Vienna, Austria: International Commission on Illumination. Report nr CIE 155:2003.

Dumyahn T, First M. 1999. Characterization of ultraviolet upper room air disinfection devices. Am Ind Hyg Assoc J 60(2):219–227.

Elmer WB. 1989. The optical design of reflectors. Salem, MA: TLA Lighting Consultants, Inc.

EPRI. 1997. UVGI for TB Infection Control in a Hospital. Palo Alto, CA: Electric Power Research Institute. Report nr TA-107885.

Escombe A, Moore D, Gilman R, Navincopa M, Ticona E, Mitchell B, Noakes C, Martinez C, Sheen P, Ramirez R, et al. 2009. Upper-Room ultraviolet light and negative air ionization to prevent tuberculosis transmission. PLoS Med 6(3):312–322.

First MW, Nardell EA, Chaisson W, Riley R. 1999. Guidelines for the application of upper-room ultraviolet germicidal irradiation for preventing transmission of airborne contagion – Part II: Design and operational guidance. ASHRAE J 105:869–876.

First MW, Weker RA, Yasui S, Nardell EA. 2005. Monitoring human exposures to upper-room germicidal ultraviolet irradiation. J Occup Environ Hyg 2(5):285–292.

First M, Rudnick SN, Banahan KF, Vincent RL, Brickner PW. 2007. Fundamental factors affecting upper-room ultraviolet germicidal irradiation – part I. Experimental. J Occup Environ Hyg 4(5):321–331.

Gelperin A, Granoff MA, Linde JI. 1951. The Effect of Ultraviolet Light upon Absenteeism from Upper Respiratory Infections in New Haven Schools. Am J Pub Health 41:796–805.

Higgons RA, Hyde GM. 1947. Effect of ultra-violet air sterilization upon incidence of respiratory infections in a children's institution. New York State J Med 47(7).

IESNA. 2000. IESNA Lighting Handbook HB-9-2000, Reference and Application Chapter 6, Light Sources. New York: Illumination Engineering Society of North America.

Ko G, First MW, Burge HA. 2002. The characterization of upper-room ultraviolet germicidal irradiation in inactivating airborne microorganisms. Environ Health Perspectives 110(1):95–101.

Koller LR. 1939. Bactericidal effects of ultraviolet radiation produced by low pressure mercury vapor lamps. J Appl Phys 10:624.

Kujundzic E, Matalkah F, Howard CJ, Hernandez M, Miller SL. 2006. UV Air Cleaners and Upper-Room Air Ultraviolet Germicidal Irradiation for Controlling Airborne Bacteria and Fungal Spores. J Occup Environ Hyg 3:536–546.

Lindsey JL. 1997. Applied Illumination Engineering. Lilburn: The Fairmont Press, Inc.

Luckiesh M, Holladay LL. 1942a. Designing installations of germicidal lamps for occupied rooms. Gen Electric Rev 45(6):343–349.

Macher JM, Alevantis LE, Chang YL, Liu KS. 1992. Effect of ultraviolet germicidal lamps on airborne microorganisms in an outpatient waiting room. Appl Occup Environ Hyg 7(8): 505–513.

Macher JM. 1993. The use of germicidal lamps to control tuberculosis in healthcare facilities. Infect Contr Hosp Epidem 14:723–729.

McLean R. 1961. The effect of ultraviolet radiation upon the transmission of epidemic influenza in long-term hospital patients. Am Rev Resp Dis 83:36–38.

Miller WR, Jarrett ET, Willmon TL, Hollaender A, Brown EW, Lewandowski T, Stone RS. 1948. Evaluation of ultra-violet radiation and dust control measures in control of respiratory disease at a naval training center. J Infect Dis 82:86–100.

Miller SL, Macher JM. 2000. Evaluation of a methodology for quantifying the effect of room air ultraviolet germicidal irradiation on airborne bacteria. Aerosol Sci & Tech 33:274–295.

MRC. 1954. Air Disinfection with Ultra-violet Irradiation; Its Effect on Illness among School-children by the Air Hygiene Committee. London: Medical Research Council, Her Majesty's Stationary Office. Report nr 283.

Nardell EA. 1988. Chapter 12: Ultraviolet air disinfection to control tuberculosis. In: Kundsin RB, editor. Architectural Design and Indoor Microbial Pollution. New York: Oxford University Press, pp. 296–308.

Nardell EA, Bucher SJ, Brickner PW, Wang C, Vincent RL, Began-McBride K, James MA, Michael M, Wright JD. 2008. Safety of upper room ultraviolet germicidal air disinfection for room occupants: Results from the tuberculosis ultraviolet shelter study. Pub Health Rep 123:52–60.

Nicas M, Miller SL. 1999. A multi-zone model evaluation of the efficacy of upper-room air ultraviolet germicidal irradiation. Appl & Environ Occup Hyg J 14:317–328.

Perkins JE, Bahlke AM, Silverman HF. 1947. Effect of ultra-violet irradiation of classrooms on the spread of measles in large rural central schools. Am J Pub Health 37:529–537.

Rahn RO, Xu P, Miller SL. 1999. Dosimetry of room-air germicidal (254 nm) radiation using spherical actinometry. Photochem Photobiol 70(3):314–318.

Rahn RO. 2004. Spatial distribution of upper-room germicidal UV radiation as measured with tubular actinometry as compared with spherical actinometry. Photochem Photobiol 80(2): 346–350.

Riley RL, Permutt S. 1971. Room air disinfection by ultraviolet irradiation of upper air: Air mixing and germicidal effectiveness. Arch Environ Health 22:201–219.

Riley RL, Knight M, Middlebrook G. 1976. Ultraviolet susceptibility of BCG and virulent tubercle bacilli. Am Rev Resp Dis 113:413–418.

Rudnick SN, First MW. 2007. Fundamental factors affecting upper-room ultraviolet germicidal irradiation – Part II. Predicting effectiveness. J Occup Environ Hyg 4(5):352–362.

Sauer LW, Minsk LD, Rosenstern I. 1942. Control of cross infections of respiratory tract in nursery for young infants. JAMA 118:1271–1274.

Schneiter R, Hollaender A, Caminita BH, Kolb RW, Fraser HF, duBuy HG, Neal PA, Rosenblum HG. 1944. Effectiveness of ultra-violet irradiation of upper air for the control of bacterial air contamination in sleeping quarters. Am J Hyg 40:136.

Wells WF. 1938. Air-borne infections. Mod Hosp 51:66–69.

Wells WF, Wells MW, Wilder TS. 1942. The environmental control of epidemic contagion; I – An epidemiologic study of radiant disinfection of air in day schools. Am J Hyg 35:97–121.

Wells WF. 1943. Air disinfection in day schools. Am J Pub Health 33:1436–1443.

Wells WF. 1955. Airborne Contagion. Sciences AotNAo, editor. New York: New York Academy of Sciences.

Wheeler SM, Ingraham HS, Hollaender A, Lill ND, Gershon-Cohen J, Brown EW. 1945. Ultraviolet light control of airborne infections in a naval training center. Am J Pub Health 35: 457–468.

WHO. 1999. Guidelines for the Prevention of Tuberculosis in Health Care Facilities in Resource Limited Settings. Geneva: World Health Organization. Report nr WHO/CDS/TB/99.269.

Xu P, Peccia J, Fabian P, Martyny JW, Fennelly KP, Hernandez M, Miller SL. 2003. Efficacy of ultraviolet germicidal irradiation of upper-room air in inactivating airborne bacterial spores and mycobacteria in full-scale studies. Atmos Environ 37:405–419.

Chapter 10

UV Surface Disinfection

10.1 Introduction

The disinfection of surfaces is perhaps the simplest and most predictable application of ultraviolet germicidal radiation. UV is highly effective at controlling microbial growth and at achieving sterilization of most types of surfaces. Early applications included equipment sterilization in the medical industry. Modern applications include pharmaceutical product disinfection, area disinfection, cooling coil and drain pan disinfection, and overhead UV systems for surgical suites. Such applications often involve using bare UV lamps and as such there may be UV hazards associated with them. Cooling coil disinfection with ultraviolet light has proven so effective that such installations often pay for themselves in short order. The use of UV is fairly common in the packaging industry and in the food processing industry where it is sometimes used for irradiating the surfaces of foodstuffs. Lower room UV systems are not common although they have been used in hospitals in the past. This chapter provides basic design information for each type of surface disinfection system based on theoretical analysis and field testing results. Good design practices are discussed and general guidelines are provided.

10.2 Microbial Growth Control

Direct UVGI exposure can sterilize any surface given enough time, and in many surface disinfection applications exposure is continuous. Theoretically, low levels of irradiance could control any microbial growth because the exposure time is extended indefinitely. In practical applications, however, there could be crevices, shadowed areas, thick insulation, and murky stagnant water where UV rays may not completely penetrate. Therefore it is important to either use a calculated level of irradiance as well as reflective surfaces or to assure complete exposure of all surfaces and at all angles. Both air and surface disinfection systems may be necessary to reduce microbial loads and ensure healthy air in any building. One advantage

W. Kowalski, *Ultraviolet Germicidal Irradiation Handbook*,   233
DOI 10.1007/978-3-642-01999-9_10, © Springer-Verlag Berlin Heidelberg 2009

of surface disinfection systems is that it attacks microbial growth at the source. In ventilation systems the source may be cooling coils, drain pans, dust in ductwork, and sometimes filters themselves.

10.3 Modeling Surface Disinfection

Modeling the disinfection of surfaces with ultraviolet radiation is relatively straight-forward and has two components – (1) modeling the UV irradiance at the surface, and (2) modeling the microbial response. The surface irradiance can be modeled using the techniques previously described for lamp modeling in Chap. 7. Given a UV lamp with known power it is only necessary to know the lamp dimensions (arclength and radius), and its position relative to the surface (see Fig. 10.1) in order to determine the irradiance at any point on the surface. The mean irradiance on a surface is the most useful single parameter for predictive purposes.

When a single lamp is positioned in front of the coil surface, the irradiance at any point on the coil surface can be determined using the cylindrical view factor model of the lamp, as detailed by Kowalski et al. (2000). Computer algorithms for this view factor model have been provided in Appendix E per Kowalski (2001, 2006). Figure 10.2 shows an example of an irradiance contour of a UV lamp located approximately 1 ft (30 cm) from a square surface 1 m on each side, with no reflectivity. This configuration produces a mean irradiance of 3.17 W/m^2 based on a view factor model of the lamp.

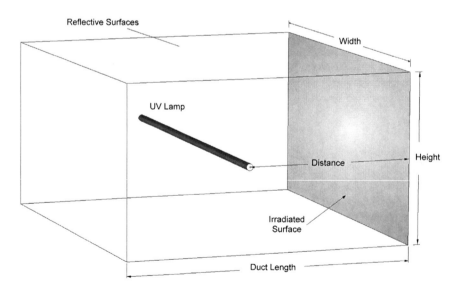

Fig. 10.1 Surface irradiance depends primarily on the lamp dimensions, UV power, reflectivity, and distance of the lamp from the surface (Length)

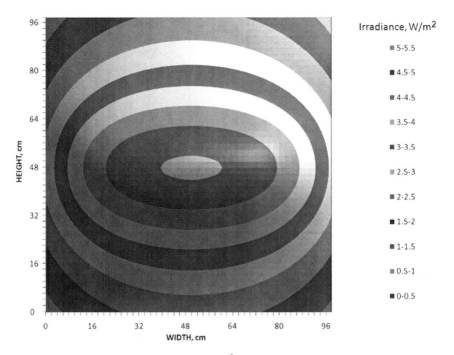

Fig. 10.2   Predicted irradiance contours on a 1 m² surface with a 10 UV W lamp centered 30 cm from surface, 0% reflectivity. Mean irradiance = 3.17 W/m²

A microbial population on a surface exposed to UV will decay exponentially over time, but given enough time the dose-response curve will often have two stages. The microbial population behaves like two separate populations – one that succumbs rapidly to UV and another that resists UV. The single stage and two-stage curves have been addressed in Chap. 3 (see Eqs. (3.2 and 3.3). The use of single stage rate constants is acceptable provided the dose limits of the original studies are not exceeded. Beyond the limits of published test results, two stage decay curves should be used. Unfortunately, data on two stage decay curves for UV inactivation of spores is limited. It should be recognized that published data on UV rate constants only applies to the dose range in the experimental results, and that attempting to predict inactivation rates beyond those achieved in the experiment amounts to extrapolation. Such extrapolation may be valid if the population displays only a single stage, but the prevailing evidence suggests that most microbial populations will display a second stage if irradiated long enough.

Prediction errors can result when two-stage decay is modeled as a single stage and extrapolated beyond the D_{90} value or applicable upper limit (UL). Figure 10.3 shows an example of a two stage decay curve of *Penicillium italicum*. Curve is based on laboratory data from Asthana and Tuveson (1992) fitted to a two stage model (see Table 3.1 in Chap. 3 for the two stage parameters). For the two stage curve, k_1 represents what would normally be called a single stage rate

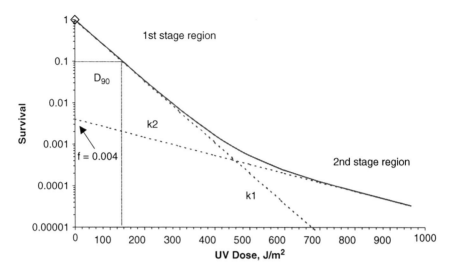

Fig. 10.3 The two stage decay curve for *Penicillium italicum* under extended UV exposure. The single stage model (k_1) will underestimate the required dose for sterilization (assuming sterilization is six logs of reduction)

constant, while k_2 represents the second stage (the resistant fraction). Extrapolating the second stage curve all the way to the axis, as shown in the figure, yields the population fraction *f*. The single stage rate constant is accurate at the D$_{90}$ (90% disinfection) mark, but if it were used to predict disinfection rates greater than about 99.9% (3 logs of reduction), they would be in error.

It is clear from this example that after extended exposure the single stage model will overpredict the survival rate of the spores. This two stage behavior under prolonged exposure is typical for most microbes and indicates the potential need to use the two stage model when evaluating cooling coil surface disinfection. This approach is hampered, however, by a general lack of two stage decay data.

10.4 UV Cabinets

Biological safety cabinets are used in hospitals and laboratories to disinfect medical instruments, surgical tools, vials, containers, etc. Items are placed inside these cabinets and exposed to UV for measured periods of time, typically about 30 min. UV lamps are typically located inside the cabinets above the work surfaces. Some cabinets are equipped with an interlocking switch to deactivate the lamp but it is up to personnel to verify that the UV lamp is off prior to performing work in a cabinet. One safety option is the use of fluorescent labels inside the cabinet that will glow if the UV lamp is on. Levels of UV irradiance in these systems varies widely but a typical cabinet might produce 0.25–10 W/m^2. These levels of irradiance will typically sterilize virtually every microbe in Appendix C within 30–60 min. Figure 10.4 shows three examples of UV cabinets.

Fig. 10.4   Three examples of UV Biological safety cabinets. *Left* and *Middle*: Stainless steel UV cabinets. *Right*: Cabinet for disinfecting knives. Images provided courtesy of Light Progress, Snc., Anghiari, Italy

The use of ultraviolet light in biological safety cabinets has never received formal approval from the CDC, NIH, or the NSF (National Safety Foundation) and there are no performance criteria or testing standards for such applications (NSF 2004, NIH 1995). The problem here is that UV only sterilizes surfaces and has no penetrating power, and its usefulness will be proportionately limited when microbial contamination exists within dust, dirt, grease, or under shadowed areas of work surfaces (Burgener 2004). As a result, UV is never recommended as the sole means of disinfection but is used in conjunction with cleaning by chemical means. A study by Birch (2000) in which ultraviolet disinfection cabinets were compared against traditional means of disinfecting stored instruments found that although microbial contamination levels were greatly reduced by UV cabinets, there was typically some residual level of contamination that resisted sterilization.

UV cabinets can be used effectively provided care is taken to clean instruments of dirt and grease prior to irradiation. Instruments can be rotated during the disinfection process to help ensure there are no shadowed areas left unexposed. The most effective cabinets generally employ internal reflective surfaces to both increase the irradiance and eliminate shadows. Cabinets with four reflective sides can more than double the mean irradiance. Cabinets with six reflective surfaces can greatly amplify the mean irradiance at any angle and are the most efficient. Polished aluminum is normally used for reflective surfaces but diffusive reflective materials like ePTFE can maximize efficiency and create the most even irradiance fields. UV irradiation of polymerase chain reaction (PCR) supplies and equipment has used to decontaminate DNA with irradiation times of up to 8 h. Cone and Fairfax (2009) present a protocol for verifying the disinfection of DNA in PCR products.

Mail disinfection systems are similar to medical equipment disinfection cabinets except that some are as large as rooms. Typically these require about 20–30 min to sterilize the surface of envelopes and packages. As with biological safety cabinets, UV does not penetrate and can only sterilize the surfaces of envelopes and packages.

10.5 Area Disinfection Systems

Area disinfection systems are of two types, portable and permanent. Portable UV area disinfection units are available in many sizes and types and are typically placed in contaminated areas to disinfect whole rooms (see Fig. 10.5). These units either include timers or are manually engaged and disengaged for specified periods, which may be hours or days. Such units can be used in hospitals or sometimes for remediation of mold or bioweapon agents.

Permanently installed area disinfection systems can be used to eliminate surface contamination in open areas, either for remediation or for prevention of potential hazards, and they are generally only used in unoccupied areas. One type of permanent fixture, called an After Hours UVGI system, is designed to be mounted on walls or ceilings. It is engaged during times when no occupants are present and can be disengaged either by timers, switches, or by motion detectors. They provide options for hospitals, laboratories, mailrooms, and in the food industry where contamination potential exists or where there is a need to keep all surfaces sterilized.

A common characteristic of area disinfection systems is the high levels of UV irradiance produced, which are typically far above levels that can be tolerated safely by occupants even for short periods of time. Attention must be paid to their safe use and procedural controls are necessary to ensure no hazards result. It is possible to design area disinfection systems that operate at low levels of UV irradiance that are safe for limited periods of room occupancy (i.e. 1–8 h) since levels below ACGIH TLVs are still capable of disinfecting microbial contamination. But, other than Upper Room systems which produce UV levels of up to 0.002 W/m^2 in the lower room as a byproduct of operation, no systems of this type are currently being marketed.

Fig. 10.5 Portable area disinfection unit using multiple UV lamps. Images provided courtesy of Lumalier, Inc., Memphis, TN

Anderson et al. (2006) compared UV area disinfection with chemical cleaning in a hospital environment. Using four ceiling and nine wall mounted units in three areas, a patient room, a bathroom, and an anteroom, mean irradiance levels of 2.2 W/m^2, 2 W/m^2, and 1.4 W/m^2, respectively, were measured at the floor and at other surfaces. Irradiance levels varied from 0.08 W/m^2 under a shelf to 6.82 W/m^2 at 0.9 m above the floor. Microbial counts on surfaces decreased by 93% with UV exposure after about 40 min. Cleaning and disinfection with chloramines were used in combination and the results are depicted in Fig. 10.6. It can be observed that significant disinfection was obtained with UV alone but the best results were obtained via a combination of cleaning, UV, and chloramines.

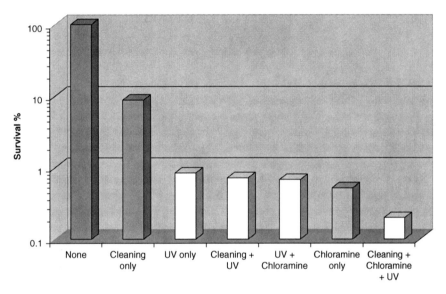

Fig. 10.6 Comparison of area disinfection methods and combinations. Based on data from Anderson et al. (2006)

Schneider et al. (1969) describes a pathogen-free isolation unit in a hospital in which the surrounding corridors, by which the rooms are entered, were irradiated for 12 h a day by five UV lamps. This approach, in combination with rigorous disinfection protocols, succeeded in reducing infections among chemotherapy patients with reduced immunity.

10.6 Cooling Coil Irradiation

Cooling coil irradiation systems have been among the most successful applications of UV, largely due to the fact that they can save energy. UV can sterilize fungal spores and bacteria on the surfaces of cooling coils and reduce energy consumption while simultaneously reducing airborne infections that may be caused by

Fig. 10.7 Cooling coil
irradiation system using UV
lamp fixtures and reflective
aluminum panels. Photo
provided courtesy of
Lumalier, Memphis, TN

Fig. 10.7 Cooling coil irradiation system using UV lamp fixtures and reflective aluminum panels. Photo provided courtesy of Lumalier, Memphis, TN

contaminated ventilation equipment. Cooling coil irradiation can improve indoor air by attacking microbial growth at the source when it occurs inside air handling units and ventilation ductwork. Figure 10.7 shows a typical installation designed for irradiating cooling coils while simultaneously irradiating the airstream.

Placement of UV lamps in air conditioners and between filters and cooling coils to control microbial growth had been suggested by several sources (Buttolph and Haynes 1950, Harstad et al. 1954, GE 1950, Luciano 1977, Grun and Pitz (1974). One manufacturer recommends placing a 15 W lamp 1 m from the surface of cooling coils (Philips 1985). Currently, the use of UV for controlling microbial growth on cooling coils enjoys considerable popularity due to the fact that it has been demonstrated to save energy (Shaughnessy et al. 1999, Scheir and Fencl 1996, Levetin et al. 2001, ELP 2000). Since a biofilm on a cooling coil reduces the space between the coils fins (increasing pressure loss and reducing airflow) and decreases the heat transfer coefficient, keeping these surfaces clean is necessary to maintain design performance.

For cooling coil applications, the irradiance on the coil surface need be only a fraction of the average irradiance used in air disinfection applications, because the exposure is typically continuous, or for extended time periods (8 h a day). Levels of irradiance in cooling coil systems vary from about 0.5 W/m^2 to as high as 1000 W/m^2. The question of what is the appropriate level of irradiance is somewhat arbitrary due to the fact that continuous exposure of a surface at almost any level of UV irradiance will eventually result in sterilization of the surface.

In typical cooling coil disinfection systems, UV lamps are located so as to irradiate either the upstream or downstream surface of the coil. In the example shown in Fig. 10.8 UV lamps are positioned on both the upstream and downstream sides of a cooling coil. Often, it is not possible to position lamps on both sides of a coil like this, or the designer considers it unnecessary and only one side is irradiated.

Often, the target microorganisms for cooling coil disinfection are spores, including fungal spores and bacterial spores. Common fungal spores like *Aspergillus,*

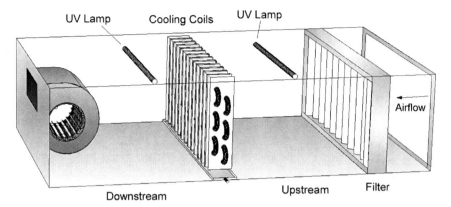

Fig. 10.8 Air handling unit with UV lamps irradiating both upstream and downstream sides of a cooling coil

Penicillium, and *Cladosporium*, and bacterial spores like *Bacillus subtilis*, may be used as indicator microorganisms. Any vegetative bacteria or viruses are likely to be eradicated by UV doses suitable for sterilizing spores.

Consider *Cladosporium* spores with a UV rate constant of 0.00384 m²/J (Luckiesh 1946). Under constant exposure at 1 W/m² (100 μW/cm²), the survival rate will be about one in a million (0.0001%) after one hour, per the classic single stage exponential decay equation. If mathematical sterilization is taken to be six logs of reduction it can be seen that even at 0.10 W/m² (10 μW/cm²) the surface will be sterilized within about 10 h. This is a simplistic example since the population decay is more likely to involve two stages, and since this only addresses the surface of the coil facing the UV lamp. The opposite side of the coil will see considerably lower levels of UV and therefore it is sometimes prudent to use higher levels on the irradiated side to facilitate UV penetration.

Irradiating both sides of a cooling coil may be an ideal approach, but it is not always possible due to space constraints. When only one side of a coil is irradiated, the opposite side will take longer to sterilize, but eventually it should become clean. There is a self-accelerating cleaning effect under UV when aluminum fins are irradiated, since the cleaner they become the higher their reflectivity becomes. In the event the fins are dense (i.e. more than 15 fpi) or damaged, UV penetration may be limited and it may be prudent to oversize the upstream side to assure sufficient irradiation of the downstream side.

The mean irradiance on the surface of a cooling coil is often the design basis of any system, but the minimum irradiance on the surface of the coil may dictate higher lamp wattage since the corners or sides may have much lower irradiance levels than the average. This is illustrated in Fig. 10.9, which shows the irradiance contours on the face of a 120 cm by 124 cm cooling coil over which a 10 W lamp has been located 31 cm (12) away from the face of the coil. With a reflectivity of 50%, the mean irradiance is 6.3 W/m² (630 μW/cm²), based on computer modeling. The minimum irradiance is 1.5 W/m² (150 μW/cm²) and occurs in the corner. The

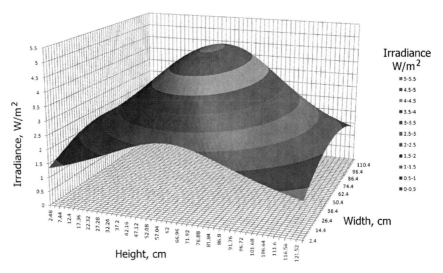

Fig. 10.9 Irradiance contours on the surface of a cooling coil with one 10 W (UV output) lamp 31 cm (12") from the face of the coil

minimum irradiance can be increased with the use of reflective panels on two or more sides of the enclosing duct.

The lamp wattage required for any cooling coil application is basically dependent on the surface area of the coil, the desired mean irradiance, and/or the minimum irradiance in any corner of on the sides. In general, a mean irradiance of approximately $1–10 \, \text{W/m}^2$ ($100–1000 \, \mu\text{W/cm}^2$) on the upstream side should provide eventual sterilization on both sides of the cooling coil. The minimum permissible irradiance can be on the order of $0.5–1 \, \text{W/m}^2$ ($50–100 \, \mu\text{W/cm}^2$) but this minimum must be considered to apply to a specific area of the coil. The higher the irradiance on the upstream side, the more UV will penetrate to the opposite face in a fixed period of time. A minimum area of about $10–100 \, \text{cm}^2$ should be appropriate for measuring the minimum irradiance.

The mean irradiance will vary with distance, and can be increased by reflectivity. Figure 10.10 shows the effect of distance from a 10 W (UV output) lamp on the irradiance on a $1 \, \text{m}^2$ surface for various reflectivities. The reflectivity of 50% represents galvanized steel, and the 75% reflectivity represents polished aluminum. Reflectivity inside the duct will not only significantly increase the mean irradiance, but will help create a more even irradiance contour and raise the minimum irradiance in the corners and on the sides.

Table 10.1 gives the Mean Irradiance values on a $1 \, \text{m}^2$ surface for various distances, wattages, and reflectivities when the lamp is centered over the surface, based on computer modeling. Gray areas indicate the normal design range of cooling coil disinfection systems ($1–10 \, \text{W/m}^2$). Although these predicted values apply to a $1 \, \text{m}^2$ surface, they can be used to conservatively estimate the required UV wattage for any given coil area since multiple lamps will have overlapping irradiance contours

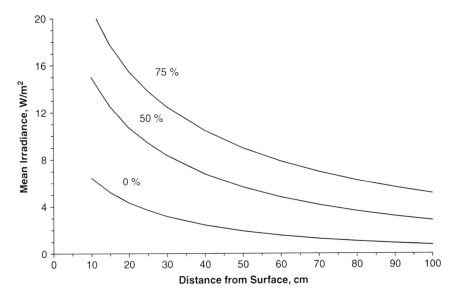

Fig. 10.10 Variation of Mean Irradiance with distance for a 1 m² surface using a 10 W lamp (UV output), at 0%, 50%, and 75% reflectivity

that will increase the overall mean irradiance. For lamps that are not centered, the estimated wattage from Table 10.1 may not be conservative.

Table 10.1 can be used to estimate the mean irradiance by interpolation and since it represents 1 m², it can conservatively approximate larger cooling coil surfaces. Multiple interpolations can be used between distance, wattage, and reflectivity. To size a cooling coil system using Table 10.1, simply look up the distance between the

Table 10.1: Mean Irradiance on a 1 Square Meter Surface

Lamp distance (cm)	Mean irradiance, W/m²								
	0% Reflectivity			50% Reflectivity			75% Reflectivity		
	10 W	20 W	30 W	10 W	20 W	30 W	10 W	20 W	30 W
10	6.45	12.89	19.34	15	30.07	45.11	20.91	41.83	62.74
15	5.19	10.38	15.57	12.5	25.00	37.50	17.73	35.46	53.19
20	4.32	8.65	12.97	10.7	21.47	32.21	15.51	31.02	46.54
25	3.68	7.35	11.03	9.41	18.81	28.22	13.83	27.65	41.48
30	3.17	6.34	9.51	8.35	16.70	25.05	12.48	24.97	37.45
40	2.43	4.86	7.29	6.77	13.54	20.31	10.45	20.91	31.36
50	1.92	3.84	5.76	5.64	11.28	16.92	8.97	17.95	26.92
60	1.55	3.10	4.64	4.79	9.58	14.37	7.84	15.69	23.53
70	1.27	2.54	3.81	4.13	8.26	12.38	6.95	13.90	20.85
80	1.06	2.12	3.18	3.6	7.20	10.81	6.23	12.45	18.68
90	0.89	1.79	2.68	3.18	6.35	9.53	5.63	11.26	16.89
100	0.76	1.53	2.29	2.83	5.65	8.48	5.13	10.25	15.38
120	0.57	1.14	1.72	2.28	4.57	6.85	4.33	8.66	13.00
140	0.44	0.89	1.33	1.89	3.78	5.67	3.73	7.46	11.19
180	0.29	0.57	0.86	1.36	2.73	4.09	2.89	5.77	8.66
200	0.24	0.47	0.71	1.18	2.36	3.54	2.58	5.15	7.73

lamp and coil, and then find the desired range of irradiance. The lamp wattage may then be found by interpolation.

Example: Determine the minimum UV output wattage and the lamp input power required to provide 5 W/m^2 (500 μW/cm^2) irradiance on a 2 m^2 coil surface that is 30 cm from the UV lamp fixture.

Answer: From Table 10.1 at 0% reflectivity, a 10 W lamp produces 3.17 W/m^2 and a 20 W lamp produces 6.34 W/m^2. Interpolation gives us the fraction (5–3.17)/(6.34–3.17) = 0.58. The lamp wattage will be 10 + 0.58(10 W) = 15.8 W, for each square meter. For a 2 m^2 area the lamp UV wattage is simply doubled to 2(15.8) = 31.6 W (UV output). Lamp total wattage will then be the UV watts divided by the lamp efficiency. At a typical efficiency of 31%, the total lamp wattage required for the installation will be about 102 W (total lamp input power).

The use of Table 10.1 will be conservative when multiple lamps are used since each lamp will contribute some irradiance to the (1 m^2) surface area irradiated by the other lamp. Additional conservatism will occur from reflectivity (which was ignored) and when the lamps are selected, unless they can be selected for exactly 31.6 UV W.

Figure 10.11 shows an example of multiple lamps placed over a 6 m^2 coil surface. The six lamps have 10 W UV output each, have a 60 cm arclength, are located 50 cm from the coil face, and the reflectivity is 50%. Computer modeling predicts the mean irradiance to be 6.3 W/m^2. Compared with the estimate from Table 10.1 for a single lamp, which is 5.6 W/m^2, it can be seen the Table 10.1 estimate is roughly 12% conservative.

Placing UV lamps in front of a cooling coil entrance or exit plane will produce an irradiance contour on the leading edges or exit edges of the coil fins. Photometer measurements can be used to confirm irradiance levels at the upstream and downstream coil face. A good indicator of the effectiveness of UVGI is cooling system performance, since the elimination of surface contamination should theoretically restore heat exchange efficiency and airflow to original design values. Parameters that can be measured before and after a UV installation to verify system performance include pressure drop across the coil, coil total airflow (as measured by traverse readings), coil entering wet bulb temperature, coil leaving dry bulb temperature, entering and exiting chilled water temperature, chilled water flowrate, fan rpm, fan motor amperage. Annual hours of cooling system operation may be useful also, and maintenance costs and manhours can also be recorded.

Under UV exposure, the disinfection of cooling coil surfaces follows the basic mathematical decay models detailed previously, but because of the extended exposure a two stage decay curve may be the best model. Since the second stage becomes dominant in the long run, it is a better predictor than the single stage rate constant, as noted previously.

Reports in the literature and from sources in the industry indicate that the disinfection of cooling coils with UV is capable of providing payback periods of about 2–4 years (Steril-Aire 2000). The cleaning action of UV on fouled coils reduces

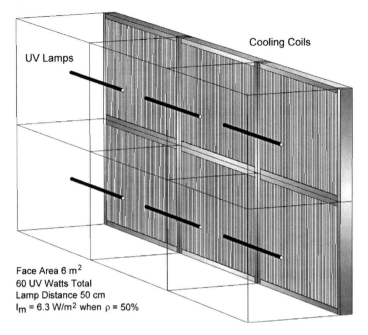

UV Lamps

Cooling Coils

Face Area 6 m^2
60 UV Watts Total
Lamp Distance 50 cm
I_m = 6.3 W/m^2 when ρ = 50%

Fig. 10.11 Example of multiple lamps irradiating a cooling coil surface, showing design parameters and computed mean irradiance

maintenance costs and improves energy efficiency so much that the retrofit of a UV cooling coil cleaning system typically pays for itself in about 2–4 years. The energy savings results from two factors – the first being the reduced pressure loss through the coils after the fouling is removed, and the second being the increased rate of heat transfer in the coils when the fouling film is gone.

Certain basic design guidelines for cooling coil disinfection systems can be summarized from IUVA (2005), from common practices, and the author's recommendations as follows:

- Filtration: MERV 6 (minimum), MERV 8–11 (recommended)
- Air velocity 400–500 fpm
- Air temperature 40°F–110°F
- Maximum ballast operating temperature of 40°C or 50°C (104°F or 122°F) depending on ballast type
- Ballast placement: external, or shielded from heat if internal.
- Lamp placement: upstream, downstream, or both sides of coils
- Lamp distance from coil face: 1–4 ft (30–120 cm)
- Recommended coil average irradiance: 1–10 W/m^2
- Minimum coil irradiance per 10 cm^2: 0.50 W/m^2

The above recommendations should not be considered to be strict requirements as these matters are still under study. All electrical wiring should be in accordance with UL/ETL requirements. Alarms or disconnect switches should be provided to disengage the UV lamps if an access door is opened. Warning signs should be located in the vicinity and proper training should be provided to maintenance personnel.

For verification of coil disinfection system performance, surface sampling for fungi and/or bacteria could be performed before installation, and then follow-up testing could be performed after about two weeks of operation. Major reductions in coil contaminants would suggest effective disinfection while the absence of all fungal contamination would indicate complete sterilization.

10.7 Overhead Surgical Site Systems

Overhead UVGI systems are used in hospitals to directly expose surgical sites during surgery in operating rooms. Such systems have been successfully used for decades to reduce surgical site infections (Kraisl et al. 1940, Overholt and Betts 1940, DelMundo and McKhann 1941, Lowell et al. 1980). Overhead systems involve using one or more UV lamp fixtures suspended over the operating table or across the entire ceiling. They typically produce a UV irradiance of about 0.25–0.3 W/m^2 (25–30 μW/cm^2) at table height. They require full body coverage on the part of surgeons and personnel, and this includes UV-proof clothing, gloves, goggles or visors, and protective skin creams (Young 1991). A study by Ritter et al. (2007) on an overhead system installed in a hospital operating room demonstrated significant reductions in surgical site infections over a ten year period. The overhead system consisted of eight UV lamp fixtures suspended overhead in the OR that produced 25 μW/cm^2 at table height. This system was able to reduce the surgical site infection rate from 1.77% to 0.5%.

The level of irradiance produced by overhead systems is capable of inactivating all major surgical site infection (SSI) bacteria. Table 10.2 summarizes the inactivation rates predicted for this system at table height for all SSI bacteria for which UV rate constants are known, except *Acinetobacter*. *Acinetobacter* has an estimated UV rate constant of 0.00021 m^2/J, making it one of the most UV resistant microbes known, but since it is also one of the largest bacteria and can be easily filtered (Kowalski 2006). For UV rate constants references see Appendix A.

Table 10.2 also shows the survival rates at 0.0083 W/m^2 (the NIOSH two-hour TLV for UVC) and also at 0.002 W/m^2, (the ACGIH TLV for 8 h exposure to UVC). It can be observed that even at the TLV significant reductions are possible for most SSI microbes. Under this latter condition, no special attire would be required for occupancy periods less than the specified limit.

Figure 10.12 illustrates the inactivation rates produced by 25 μW/cm^2 based on Table 10.2. It is clear that most bacteria, and particularly those of greatest concern

Table 10.2:   Predicted Survival of SSI Microbes with Overhead UV

% Survival at UV irradiance						0.25 W/m²		25 µW/cm²	
k, m²/J	0.01	0.02093	0.03598	0.0548	0.07675	0.085	0.096	0.105	0.105
t, hr	Candida	Streptococcus	Enterobacter	Klebsiella	P. mirabilis	Staphylococcus	E. coli	Serratia	Pseudomonas
0	100	100	100	100	100	100	100	100	100
0.25	10.54	0.9012	0.0305	0.0004	3.16E-06	4.9E-07	4.2E-08	5.5E-09	5.5E-09
0.5	1.11	0.0081	9.3E-06	2.0E-09	1.0E-13	2.4E-15	1.7E-17	3.0E-19	3.0E-19
0.75	0.1171	7.32E-05	2.8E-09	8.6E-15	3.2E-21	1.2E-23	7.2E-27	1.7E-29	1.7E-29
1	0.0123	6.6E-07	8.6E-13	3.8E-20	1.0E-28	6.0E-32	3.0E-36	9.1E-40	9.1E-40
2	1.5E-06	4.3E-15	7.5E-27	1.4E-41	1.0E-58	3.6E-65	9.0E-74	8.3E-81	

% Survival at NIOSH TLV for UVC, 2 h limit						0.0083 W/m²		0.83 µW/cm²	
0	100	100	100	100	100	100	100	100	100
0.25	92.80	85.53	76.43	66.41	56.36	53.00	48.82	45.64	45.64
0.5	86.12	73.15	58.42	44.10	31.77	28.09	23.83	20.83	20.8316
0.75	79.92	62.56	44.65	29.29	17.91	14.88	11.63	9.51	9.50787294
1	74.17	53.51	34.13	19.45	10.09	7.89	5.68	4.3395	4.33954796
2	55.01	28.63	11.65	3.78	1.02	0.622239748	3.22E-01	1.88E-01	1.88E-01
4	30.26	8.20	1.36	0.14	0.01	0.003871823	1.04E-03	3.55E-04	3.55E-04
8	9.16	0.67	0.02	0.0002	1.08E-06	1.4991E-07	1.08E-08	1.26E-09	1.26E-09
24	0.08	3.03E-05	6.23E-10	8.57E-16	1.25E-22	3.36894E-25	1.26E-28	1.99E-31	1.99E-31

% Survival rates at NIOSH TLV for UVC, 8 h limit						0.002 W/m²		0.2 µW/cm²	
0	100	100	100	100	100	100	100	100	100
1	93.05	86.01	77.18	67.40	57.55	54.23	50.10	46.95	46.95
2	86.59	73.98	59.56	45.42	33.11	29.41	25.10	22.05	22.0469
4	74.98	54.73	35.48	20.63	10.97	8.65	6.30	4.8606	4.86064034
8	56.21	29.95	12.59	4.26	1.20	0.7476	0.3968	2.36E-01	0.23625824
12	42.15	16.39	4.47	0.88	0.13	0.06464595	2.50E-02	1.15E-02	1.15E-02
24	17.76	2.69	0.20	0.01	0.0002	4.1791E-05	6.25E-06	1.32E-06	1.32E-06
48	3.16	0.07	0.0004	5.96E-07	3.02E-10	1.74649E-11	3.90E-13	1.74E-14	1.74E-14

Note: Gray areas indicate six or more logs of reduction.

(*Streptococcus* and *Staphylococcus*) are reduced to below 1% within the first 20 min. Most surgical procedures take about 1–2 h, and it can be seen that a six log reduction is achieved within about 45 min. In an actual OR, since levels would rarely exceed a few thousand cfu/m³, this level of inactivation should provide sterile operating conditions.

Computer analysis of the irradiance field for a model overhead system with a single lamp of 10 UV watts output will produce about 0.25–0.30 W/m² at table height in the center of the OR. The model dimensions are 731 cm (24 ft) L × 610 cm (20 ft) W × 305 cm (10 ft) H. The contour diagram for the UV irradiance is shown graphically in Fig. 10.13. In this system only one UV lamp is used and is centered in the room, in contrast to the Ritter et al. (2007) system which uses 8 UV lamps distributed across the ceiling. The Ritter system obtains 0.25 W/m² all across the room while the single lamp model produces 0.25 W/m² only above the operating table.

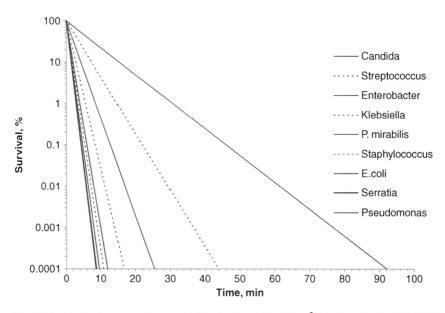

Fig. 10.12   Inactivation rates of common SSI microbes at 25 μW/cm², based on data in Table 10.2

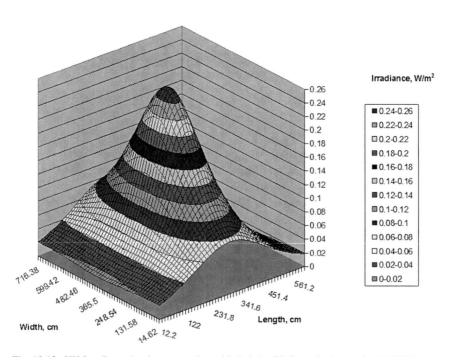

Fig. 10.13   UV Irradiance levels at operating table height (3') for a single overhead 10 UV watt lamp in a 24' × 20' room with a 10' ceiling, based on computer lamp modeling results

Two Operating Modes can be incorporated into the Overhead UV System – Continuous UV Mode and Decontamination Mode. The continuous mode would be used during surgery and this system would be under internal by the Circulator, the Surgeon, or other OR personnel. Decontamination Mode serves as an area disinfection system and would operate when the OR is unoccupied and would sterilize all internal surfaces. It would disengage automatically when anyone entered the room. Decontamination Mode can be controlled by a timer such that it operates for a maximum of perhaps 30–60 min and may re-engage automatically after personnel leave the room. Control can be linked to a door switch or a motion detector such that if no motion is detected for several minutes it will engage, and if any motion is detected at all (i.e. an opening door) it will shut off automatically.

10.8 Lower Room Systems

Lower Room systems have much in common with Upper Room systems but they are treated separately in this chapter because they play a more important role in disinfecting the floor, which is a primary source of dust and airborne microbial contamination. Fungal spores and bacteria tend to gravitate to the floor, can survive in dust, and become stirred up by activity (Lidwell and Lowbury 1950). Airborne bacterial and fungal spore levels tend to be higher near the floor, and lower room systems that sterilize floor surfaces should, in theory, reduce airborne concentrations, although there is no published data at present that directly supports this view.

Wilmon et al. (1948) applied lower room UV systems in conjunction with upper room systems in a Naval barracks. They found that suppressing dust with oil coatings alone could reduce the infection rate, indicating a contribution to respiratory infections from floor dust, but that infections rates dropped the most when UV was used in conjunction. Both Wheeler et al. (1945) and Miller et al. (1948) used lower room UV systems at a Naval training center to reduce respiratory infections, but they were used in conjunction with upper room systems and the effectiveness of the lower room systems could not be isolated.

Lower room UV systems create a zone of irradiance in the lower room, typically about 18 inches (~45 cm) from the floor. Levels of irradiance are similar to those of upper room systems, or up to about 1 W/m². The safety hazards associated with lower room systems are lessened by the fact that eye exposure is unlikely, except perhaps for infants. However, personnel in areas equipped with lower UV must have full leg coverings, and nylon stockings will not completely block UV. Lower room systems can be improved, and even rendered safe, by the use of levels of irradiance below the ACGIH TLV, or 0.001 W/m² (for broadband UV). Levels below the ACGIH limit will still have an effect on microorganisms since continuous exposure will eventually sterilize them. Another improvement is to direct the UV rays downwards at an angle so as to give maximum exposure to the floor surface, as shown in Fig. 10.14.

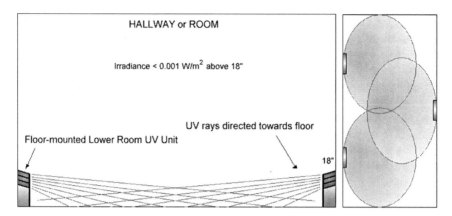

Fig. 10.14 Illustration of an area equipped with lower room units in which UV rays are angled downwards towards the floor. Image on right shows overlapping floor contours when multiple units are used

10.9 Food and Packaging Irradiation

UV irradiation has been used to disinfect the surfaces of foodstuffs during various stages of manufacturing, packing, storage, and distribution, as well as being used to disinfect areas of food production plants (Hui 2006). Applications include decontamination of surfaces of equipment in bakeries, cheese plants, meat processors, and packaging disinfection for containers such as boxes, caps, bottles, cartons, tubes, films and foils (Koutchma and Stewart 2005). At sufficient doses, UV can inactivate all known food spoilage microorganisms, including bacteria, yeasts, and mold spores, as well as food pathogens such as viruses. Unlike chemical treatments for foods, UV does not introduce any toxins or residues

Direct UV irradiation of foods being processed on assembly lines and in cold storage, including meats, cheeses, and baked goods (Philips 1985). The primary objective is usually to control mold by eradicating fungal spores and spoilage microbes that may settle on food surfaces, since UV will generally not penetrate below the surfaces. Direct irradiation of fruits and vegetables can cause discoloration. High levels of UV may also impair the flavor of products containing fats and oils. FDA allows the use of UV for control of surface microorganisms provided that no ozone is produced, that no high fat-content foods are irradiated and that irradiance levels are limited to 1 W of UV per 5–10 ft^2 (FDA 2000). The latter limits equate to an irradiance of 1.08–2.15 W/m^2. The FDA does not specify exposure times of UV dose limits. Table 10.3 provides a summary of the predicted disinfection rates for a variety of common foodborne pathogens and food spoilage microbes. The top two sections of Table 10.3 show the disinfection rates of foodborne pathogens over time at the upper and lower FDA limits. The bottom section shows disinfection rates of food spoilage microbes at the FDA upper limit. UV rate constants are shown in the table (see Chap. 4 for UV rate constant references, all of

Table 10.3: Predicted Survival of Foodborne Pathogens and Spoilage Microbes

Foodborne pathogens	% survival at FDA max irradiance				2.15	W/m²	215	µW/cm²
k, m²/J	0.015	0.017	0.02	0.0767	0.08	0.085	0.096	0.126	0.135
t, sec	Listeria	Clostridium	Bacillus	Salmonella	Campylobacter	Staphylococcus	E. coli	Vibrio	Shigella
0	100	100	100	100	100	100	100	100	100
10	72.43	69.38	65.05	19.22	17.91	16.08	12.69	6.66	5.49
20	52.47	48.14	42.32	3.70	3.21	2.59	1.61	0.44	0.30
30	38.00	33.40	27.53	0.71	0.57	0.42	0.20	0.03	0.02
60	14.44	11.16	7.58	0.0050	0.0033	0.0017	0.0004	8.73E-06	2.73E-06
120	2.09	1.25	0.57	2.55E-07	1.09E-07	2.99E-08	1.75E-09	7.62E-13	7.47E-14
300	0.01	0.0017	0.0002	3.27E-20	3.89E-21	1.55E-22	1.28E-25	5.07E-34	1.53E-36
600	3.95E-07	2.99E-08	6.24E-10	1.07E-41	1.52E-43	2.40E-46	1.65E-52	2.57E-69	2.33E-74

Foodborne pathogens	% survival at FDA lower limit				1.08	W/m²	108	µW/cm²
0	100	100	100	100	100	100	100	100	100
10	85.04	83.23	80.57	43.68	42.15	39.93	35.46	25.65	23.27
20	72.33	69.27	64.92	19.08	17.76	15.95	12.57	6.58	5.41
30	61.51	57.65	52.31	8.33	7.49	6.37	4.46	1.69	1.26
60	37.83	33.23	27.36	0.69	0.56	0.41	0.20	0.0284	0.0159
120	14.31	11.04	7.49	0.0048	0.0031	0.0016	3.95E-04	8.09E-06	2.52E-06
300	0.78	0.41	0.15	1.61E-09	5.53E-10	1.10E-10	3.10E-12	1.86E-16	1.01E-17
600	0.01	0.0016	0.0002	2.60E-20	3.06E-21	1.20E-22	9.63E-26	3.47E-34	1.02E-36
1200	3.61E-07	2.70E-08	5.53E-10	6.76E-42	9.38E-44	1.44E-46	9.26E-53	1.21E-69	1.04E-74

Spoilage microbes	% survival rates at FDA max irradiance				2.15	W/m²	215	µW/cm²
k, m²/J	0.00077	0.0021	0.0023	0.0041	0.00718	0.0128	0.0219	0.0701	0.1046
t, sec	Rhizopus	Cladosporium	Aspergillus	Fusarium	Penicillium	Mucor	Serratia	Corynebacterium	Pseudomonas
0	100	100	100	100	100	100	100	100	100
60	90.54	76.27	74.33	58.93	39.60	19.18	5.93	0.01	0.00
120	81.98	58.17	55.24	34.72	15.69	3.68	0.35	1.40E-06	1.90E-10
240	67.21	33.84	30.52	12.06	2.46	0.14	0.0012	1.95E-14	3.63E-22
480	45.17	11.45	9.31	1.45	0.06	0.0002	1.53E-08	3.82E-30	1.32E-45
960	20.41	1.31	0.87	0.02	3.66E-05	3.36E-10	2.34E-18	1.46E-61	1.73E-92
1920	4.16	0.0172	0.0075	4.46E-06	1.34E-11	1.13E-21	5.48E-38	2.12E-124	3.00E-186
3840	0.17	2.954E-06	5.67E-07	1.99E-13	1.8024E-24	1.27394E-44	3E-77	4.508E-250	

Note: Gray areas indicate six or more logs of reduction.

which are surface rate constants). Gray areas indicate where better than six logs of reduction are obtained. Only common pathogens and spoilage microbes for which UV rate constants are known have been included.

Kim et al. (2002) tested the effect of UV irradiation on inactivating *Listeria*, *E. coli*, and *Salmonella* on chicken meat. With an irradiance of 5 W/m², pathogens were reduced by up to 1.28 logs after 3 min, which was considerably less than reductions obtained on stainless steel (approximately 8 logs). Differences were attributed to the rough surfaces of the meat which provide protection of bacteria from UV. Stermer et al. (1987) reported UV could reduce bacteria on the surface of round steak by about 2 logs. Huang and Toledo (1982) found that 3 W/m² could reduce microbial counts on the surface of mackerel by 2–3 logs and prolonged the shelf life. Yaun and Summer (2002) studied the effects of UV on surface pathogens on tomatoes, lettuce, and apples. Reductions of *E. coli* and *Salmonella* of about 2–3 logs were obtained using obtained irradiance levels of 15–240 W/m². Shama and

Anderson (2005) report that UV exposure of fruits can induce hormesis, a process in which compounds produced by low level UV exposure can promote resistance to microbial attack.

One of the problems with irradiating food surfaces is obtaining constant irradiance all across the surface of the food. On food assembly lines, UV lamps are typically placed in tunnels around and above the moving belt. Reflective surfaces in the tunnel can enhance both the irradiance levels and the coverage. Gardner and Shama (1999) have presented a model that can be used to estimate the irradiance on the surfaces of three-dimensional solids.

One application of UV in the food industry involves tank head space disinfection (Philips 1985). UV systems can be used to disinfect displacement air used for pressurizing tanks that hold perishable fluids. Such storage tanks are highly susceptible to contamination and colonization by bacteria and mold. For more detailed information on food pathogens and spoilage microbes in the food industry see Chap. 18.

10.10 Summary

UV surface disinfection systems can be effective against viruses, bacteria, and fungal spores when properly designed and operated. Exposure hazards can be limited by operational protocols as well as by good design practices. Table 10.4 offers a general perspective on UV surface disinfection systems and shows typical mean irradiance levels as well as some estimated UV doses and the target microbes for each type of system. For additional information on applications of the surface disinfection systems discussed above, see Chap. 16.

Table 10.4: Summary of Operating Parameters for Surface Disinfection Systems

System	Typical irradiance W/m^2	Typical exposure Time, hr	Typical dose J/m^2	Target microbes
UV cabinets	0.25–10	0.5–1	500–36,000	Viruses, bacteria, spores
PCR disinfection	0.25–10	4–8	3600–72,000	DNA
Area disinfection	1–10	1–24	3000–20,000	Viruses, bacteria, spores
Cooling coil irradiation	1–10	continuous	–	Bacteria and fungal spores
Overhead UV	25–30	1–2	3000	SSI bacteria and viruses
Lower room UV	0.1–1	continuous	–	Bacteria and fungal spores
Food surface irradiation	1.08–2.15	< 1	60–1000	Food pathogens and spoilage microbes

References

Anderson BM, Banrud H, Boe E, Bjordal O, Drangsholt F. 2006. Comparison of UV C light and chemicals for disinfection of surfaces in hospital isolation units. Infect Contr Hosp Epidemiol 27(7):729–734.

Asthana A, Tuveson RW. 1992. Effects of UV and phototoxins on selected fungal pathogens of citrus. Int J Plant Sci 153(3):442–452.

Birch R. 2000. A study into the efficacy of ultraviolet disinfection cabinets for storage of auto-claved podiatric instruments prior to use, in comparison with current practices. Northampton: University College Northampton.

Burgener J. 2004. Position paper on the use of ultraviolet lights in biological safety cabinets. Appl Biosafety 11(4):228–230.

Buttolph LJ, Haynes H. 1950. Ultraviolet Air Sanitation. Cleveland, OH: General Electric. Report nr LD-11.

Cone R, Fairfax M. 2003. Protocol for ultraviolet irradiation of surfaces to redcue PCR contamination. PCR Methods and Appl 3:S15–S17.

DelMundo F, McKhann CF. 1941. Effect of ultra-violet irradiation of air on incidence of infections in an infant's hospital. Am J Dis Child 61:213–225.

ELP. 2000. How UV-C lamps saved one company $58,000. Electric Light & Power 78(2).

FDA. 2000. Ultraviolet radiation for the processing and treatment of food. Sliver Spring, MD: Food and Drug Administration. Report nr 10CFR21, Section 179.39 & 179.41.

Gardner DWM, Shama G. 1999. UV intensity measurement and modelling and disinfection performance prediction for irradiation of solid surfaces with UV light. Food Bioproducts Proc 77(C3):232–242.

GE. 1950. Germicidal Lamps and Applications. USA: General Electric. Report nr SMA TAB: VIII-B.

Grun L, Pitz N. 1974. U.V. radiators in humidifying units and air channels of air conditioning systems in hospitals. Zbl Bakt Hyg B159:50–60.

Huang YW, Toledo RE. 1982. Effect of high doses of high and low intensity UV irradiation on surface microbiological counts and storage-life of fish. J Food Sci 47:1667–1669.

Hui YH, editor. 2006. Handbook of Food Science, Technology, and Engineering. Boca Raton, FL: CRC Press.

IUVA. 2005. General Guideline for UVGI Air and Surface Disinfection Systems. Ayr, Ontario, Canada: International Ultraviolet Association. Report nr IUVA-G01A-2005.

Kim T, Silva JL, Chen TC. 2002. Effects of UV irradiation on selected pathogens in peptone water and on stainless steel and chicken meat. J Food Prot 65(7):1142–1145.

Koutchma T, Stewart C. 2005. Applications and efficacy of UV light for foods. IUVA News 7(4).

Kowalski WJ, Bahnfleth WP, Witham DL, Severin BF, Whittam TS. 2000. Mathematical modeling of UVGI for air disinfection. Quant Microbiol 2(3):249–270.

Kowalski WJ. 2001. Design and optimization of UVGI air disinfection systems [PhD]. State College: The Pennsylvania State University.

Kowalski WJ. 2006. Aerobiological Engineering Handbook: A Guide to Airborne Disease Control Technologies. New York: McGraw-Hill.

Kraissl CJ, Cimiotti JG, Meleney FL. 1940. Considerations in the use of ultra-violet radiation in operating rooms. Ann Surg 111:161–185.

Levetin E, Shaughnessy R, Rogers CA, Scheir R. 2001. Effectiveness of germicidal UV radiation for reducing fungal contamination within air-handling units. Applied & Environ Microbiol 67(8):3712–3715.

Lidwell OM, Lowbury EJ. 1950. The survival of bacteria in dust. Ann Rev Microbiol 14:38–43.

Lowell JD, Kundsin RB, Schwartz CM, Pozin D. 1980. Ultraviolet radiation and reduction of deep wound infection following hip and knee arthroplasty. In: Kundsin RB, editor. Airborne Contagion, Annals of the New York Academy of Sciences. New York: NYAS, pp.285–293.

Luciano JR. 1977. Air Contamination Control in Hospitals. New York: Plenum Press.

Luckiesh M. 1946. Applications of Germicidal, Erythemal and Infrared Energy. New York: D. Van Nostrand Co.

Miller WR, Jarrett ET, Willmon TL, Hollaender A, Brown EW, Lewandowski T, Stone RS. 1948. Evaluation of ultra-violet radiation and dust control measures in control of respiratory disease at a naval training center. J Infect Dis 82:86–100.

NIH. 1995. Primary Containment for Biohazards: Selection, Installation and Use of Biological Safety Cabinets. National Institute of Health C, U.S. Department of Health and Human Services, editor. Washington, DC: U.S. Government Printing Office.

NSF. 2004. International Standard, American National Standard: Class II (Laminar Flow) Biohazard Cabinetry. Michigan: National Sanitation Foundation. Report nr NSF/ANSI 49-2004A.

Overholt RH, Betts RH. 1940. A comparative report on infection of thoracoplasty wounds. J Thoracic Surg 9:520–529.

Philips. 1985. UVGI Catalog and Design Guide. Netherlands: Catalog No. U.D.C. 628.9.

Ritter M, Olberding E, Malinzak R. 2007. Ultraviolet lighting during orthopaedic surgery and the rate of infection. J Bone Joint Surg 89:1935–1940.

Scheir R, Fencl FB. 1996. Using UVC Technology to Enhance IAQ. HPAC February.

Schneider M, Schwartenberg L, Amiel JL, Cattan A, Schlumberger JR, Hayat M, deVassal F, Jasmin CL, Rosenfeld CL, Mathe G. 1969. Pathogen-free Isolation Unit – Three years' experience. Brit Med J 29 March:836–839.

Shama G, Anderson P. 2005. UV hormesis in fruits: A concept ripe for commercialization. Trends Food Sci Technol 16:128–136.

Shaughnessy R, Levetin E, Rogers C. 1999. The effects of UV-C on biological contamination of AHUs in a commercial office building: Preliminary results. Indoor Environ '99:195–202.

Steril-Aire I. 2000. UVC lights save on energy while cleaning coils. HPAC Eng 72(1):131–132.

Stermer RA, Lasater-Smith M, Brasington CF. 1987. Ultraviolet radiation – an effective bactericide for fresh meat. J Food Sci 50:108–111.

Wheeler SM, Ingraham HS, Hollaender A, Lill ND, Gershon-Cohen J, Brown EW. 1945. Ultraviolet light control of airborne infections in a naval training center. Am J Pub Health 35:457–468.

Willmon TL, Hollaender A, Langmuir AD. 1948. Studies of the control of acute respiratory diseases among naval recruits. Am J Hyg 48:227–232.

Yaun BR, Summer SS. 2002. Efficacy of Ultraviolet Treatments for the Inhibition of Pathogens on the Surface of Fresh Fruits and Vegetables. Blacksburg, VA: Virginia Polytechnic Institute and State University.

Young DP. 1991. Ultraviolet Lights for Surgery Suites. Mooresville: St. Francis Hospital.

Chapter 11

UVGI Guidelines and Standards

11.1 Introduction

A variety of guidelines and standards are available or under development that provide information on engineering design, applications, and safety considerations for air and surface disinfection UVGI systems. Guideline documents have been issued by government institutions, professional organizations, and by some private sources. Some of them merely address UVGI in passing while others provide detailed information for installations. There are currently no comprehensive standards for UVGI systems although several organizations are in the process of working towards such goals as they relate to each organizations interests. There are four major categories for UVGI guidelines and standards:

(1) Electrical Standards
(2) Measurement and Testing Standards
(3) Applications Guidelines
(4) UV Safety Guidelines

These topics are addressed in the above order in the following sections. Guidelines focusing primarily on the last category, UVGI Safety, are covered in more detail the following chapter. A fifth category is also possible: UV Photodegradation of Materials – this subject has no guidelines but is covered separately in Chap. 15.

This chapter provides an overview of the various documents that currently provide guidance for designing and installing UVGI air and surface disinfection systems for specific applications, or that provide critical information regarding the testing of UVGI systems. It is not the intent of this chapter to reproduce the content of the subject documents but merely to identify the focus and summarize general or key points. Readers should consult the source documents listed in the references for additional information.

W. Kowalski, *Ultraviolet Germicidal Irradiation Handbook*,
DOI 10.1007/978-3-642-01999-9_11, © Springer-Verlag Berlin Heidelberg 2009

11.2 UVGI Electrical Guidelines

UVGI lamps, ballasts, and fixtures are primarily electrical components, as are any associated controls and switches, and are subject to existing electrical guidelines and standards. Hot cathode germicidal lamps are similar in physical dimensions and electrical characteristics to standard preheat fluorescent lamps and they operate on the same auxiliaries. Cold cathode germicidal lamps are instant-start lamps with the cylindrical cold cathode type of electrode. Slimline germicidal lamps are instant-start lamps that operate at current densities that depend on the ballast used. UVGI electrical components have long been subject to guidelines for illumination equipment, particularly those provided by the American National Standards Institute (ANSI), the National Fire Protection Association (NFPA), and the Illumination Engineering Society of North America (IESNA). The safety of these electrical components is generally certified by laboratories such as Underwriter's Laboratories (UL), the Electrical Testing Laboratory (ETL), and the Canadian Standards Association (CSA). In addition, any electronic components of UVGI systems (i.e. electronic ballasts) are expected to conform to Federal Communications Commission (FCC) Class A specifications for electromagnetic interference and radio frequency interference (EMI/RFI), but EMI/RFI is not a covered in this chapter. Electronics may also meet ISO 9002 requirements but ISO is not a topic of this chapter.

Several of the guidelines covered in the Applications section of this chapter address some electrical aspects of UVGI applications but these are generally redundant to the documents addressed below. Table 11.1 summarizes the documents addressed in this section.

Title: **National Electrical Code**
Identifier: **NFPA 70, NFPA 70B**
Source: **National Fire Protection Association (NFPA)**
Synopsis:
The National Electrical Code sets standards and defines recommended practices for all aspects of electrical wiring and equipment installation in commercial buildings, residential housing, and industrial facilities, and assures that safe installations

Table 11.1 Electrical standards and guidelines summary

Document	Identifier	Source
National Electrical Code	NFPA 70, NFPA 70B	NFPA
Double-Capped Fluorescent Lamps – Dimensional and Electrical Characteristics	ANSI_IEC C78.81-2005	ANSI
Lamp Ballast – Line Frequency Fluorescent Lamp Ballast	ANSI C82.1-2004	ANSI
Luminaires	UL 1598	UL
Lighting Handbook: Reference and Application	IESNA HB-9-2000	IESNA

do not pose fire hazards. Section 90.2 of this Code covers the installation of electrical conductors, equipment, and raceways; signaling and communications conductors, equipment, and raceways; and optical fiber cables and raceways for public and private premises, including buildings, structures, mobile homes, recreational vehicles, and floating buildings. This code implicitly applies to all electrical conduit and connectors for UVGI systems. NFPA 70B, Recommended Practice for Electrical Equipment Maintenance, provides recommended practices for preventive maintenance of electrical and electronic equipment and is not intended to duplicate or supersede instructions from manufacturers. In addition, related sections of other NFPA documents may apply, including, NFPA 90A (Standard for the Installation of Air-Conditioning and Ventilating Systems), NFPA 90B (Standard for the Installation of Warm Air Heating and Air-Conditioning Systems), and NFPA 91 (Standard for Exhaust Systems for Air Conveying of Vapors, Gases, Mists, and Noncombustible Particulate Solids). Reference NFPA (2008).

Title: **Double-Capped Fluorescent Lamps – Dimensional and Electrical Characteristics**
Identifier: **ANSI_IEC C78.81-2005**
Source: **American National Standards Institute (ANSI)**
Synopsis:
This standard sets forth the physical and electrical characteristics of the principal types of fluorescent lamps intended for application on conventional line frequency circuits and high frequency electronic circuits. Specifications are given for both the lamp and the interactive features of the lamp and ballast, for double-based lamps of linear shape. Physical dimensions of the lamp models are specified, along with dimensioning principles and abbreviations for lamp model numbers. Lamp operating characteristics, including voltage, current, and wattage are specified based on operation with a reference ballast. Lamp starting requirements are given, including upper temperature limits and voltage between terminals. Single-based lamps and U-tube lamps are addressed in ANSI C78.901-2005. Performance specifications are addressed in ANSI C78.5-2003 and in IEC 60081-1997. Reference ANSI (2005).

Title: **Lamp Ballast – Line Frequency Fluorescent Lamp Ballast**
Identifier: **ANSI C82.1-2004**
Source: **American National Standards Institute (ANSI)**
Synopsis:
This standard is intended to cover ballasts that have rated open-circuit voltages of 2000 V or less and are intended to operate lamps at a frequency of 50–60 Hz. This comprises ballasts for hot-cathode fluorescent lamps, modified rapid start (cathode cutout), or instant start, and also ballasts for cold-cathode fluorescent lamps which come within this voltage range. The ballast and lamp combinations covered in this document are normally intended for use in rooms with ambient temperatures of 10–40°C. See also ANSI C82.3-2002 (Reference Ballasts for Fluorescent Lamps), ANSI C82.11-2002 (High Frequency Fluorescent Lamp Ballasts, ANSI C78.180-2003 (Specifications for Fluorescent Lamp Starters). Reference ANSI (2004).

Title: Luminaires
Identifier: UL 1598
Source: Underwriters Laboratories Incorporated
Synopsis:
This standard applies to luminaries used in non-hazardous locations that are
intended for installation on branch circuits of nominal 600 V or less between conductors in accordance with the US National Electrical Code (NEC) ANSI/NFPA
70, the Canadian Electrical Code, Part 1 (CEC), CSA C22.1, and with the
Mexican National Electrical Code, NOM-001-SEDE. This standard addresses
mechanical construction, electrical construction, fluorescent luminaires, ballasts,
temperature testing, mechanical testing, electrical testing, and marking. Ballasts and
starters are also addressed in standards UL 542, UL 935, and UL 1029. A number of
additional UL standards are cited which are relevant to ultraviolet lamp components,
switches, wiring, electronics, and fixtures, and readers should consult UL 1598 for
relevant document citations. Reference UL (2004).

Title: Lighting Handbook: Reference and Application
Identifier: IESNA HB-9-2000
Source: Illumination Engineering Society of North America (IESNA)
Synopsis:
The sections of the IESNA Handbook related to UVGI include fundamentals
regarding germicidal ultraviolet light, UVGI application information for air disinfection and sanitization (surface disinfection), UV safety information, principles of photometry, and fundamentals of lamps, ballasts, and lamp fixtures that
are generally applicable to both lighting and UVGI. Lamp specifications are provided in tables in Chap. 8 of the reference volume for many models of medium
and high pressure UV lamps, including rated wattage, rated UV output, and
rated average life, as well as pulsed light sources and UV lamps used for non-
germicidal applications (i.e. curing). Also provided is information on warm-up
times, starting voltages, operating voltages, and starting amperes. The reference
volume also provides design information for fixtures and reflectors (Chap. 6) and
methods for performing lighting calculations (Chap. 9) which may be applied to
UVGI. The Applications volume provides information on the germicidal effects
of ultraviolet light and some general information on applications and safety,
including upper room systems and in-duct air disinfection. Reference IESNA
(2000).

11.3 Testing Guidelines and Standards

This section provides an overview of the current state of testing guidelines and standards for UVGI systems. Testing of lamps and ballasts is covered in greater detail
in Chap. 13. Much of the material on testing relates to lamps, ballasts, and switches.
Table 11.2 summarizes the documents covered in this section.

Table 11.2   Testing and measurement standards and guidelines summary

Document	Identifier	Source
Measurement of Ultraviolet Radiation from Light Sources	IESNA LM-55-96	IESNA
Fluorescent Lamps - Guide for Electrical Measurements	ANSI C78.375-1997	ANSI
Performance Testing for Lighting Controls and Switching Devices with Electronic Fluorescent ballasts	NEMA 410-2004	NEMA
IES Approved Method for the Electrical and Photometric Measurements of Fluorescent Lamps	LM-9-88	IESNA
IESNA Approved Method for Photometric Testing of Reflector-Type Lamps	LM-20-1994	IESNA
IESNA Guide to Spectroradiometric Measurements	LM-58-1991	IESNA
IESNA Lighting Handbook, Reference and Application Chapter 6, Light Sources	IESNA HB-9-2000	IESNA
Method for Measuring Performance of Portable Household Electric Room Air Cleaners	ANSI/AHAM AC-1-2006	ANSI AHAM

Title: **Measurement of Ultraviolet Radiation from Light Sources**
Identifier: **IESNA LM-55-96**
Source: **Illumination Engineering Society of North America (IESNA)**
Synopsis:

The measurement method described in this guide applies to incoherent light source emissions in the 200–400 nm region, excluding solar radiation and emissions from welding arcs and lasers. This document is intended to promote uniformity in the measurement of UV. It provides basic definitions and terminology, describes test equipment, and summarizes test requirements. Test requirements include the burn-in period (called seasoning) for UV lamps, control of the ambient temperature at 25°C (77°F), keeping ambient air velocity below 4.6 m/min (15 fpm), control of voltage variations and stray radiation that may impact photosensors. The spectroradiometer requirements are discussed in some detail, including input optics, detectors, amplifiers, and corrections for the appropriate action spectrum. The use of a small integrating sphere as the input optics is recommended for accuracy. The use of radiometric standards and proper calibration is discussed. No specific procedures or tests are addressed in this document but a list of items to be included in the test report are provided. Reference IESNA (1996).

Title: **Fluorescent Lamps – Guide for Electrical Measurements**
Identifier: **ANSI C78.375-1997**
Source: **American National Standards Institute (ANSI)**
Synopsis:
This standard describes the procedures to be followed and the precautions to be taken in obtaining uniform and reproducible measurements of electrical character-istics of fluorescent lamps under standard conditions when operated on alternating current (AC) circuits. These methods are applicable both to lamps having hot cathodes and to lamps of the cold cathode variety. The electrical characteristics measured include lamp current, lamp voltage, and lamp power. The methods noted in this standard apply to lamps operated at common power-line frequencies, 50–60 Hz. For ballast measurements, refer to ANSI C82.2-2002, For Lamp Ballasts – Method of Measurement of Fluorescent Lamp Ballasts. Reference ANSI (1997).

Title: **Performance Testing for Lighting Controls and Switching Devices with Electronic Fluorescent Ballasts**
Identifier: **NEMA 410-2004**
Source: **National Electrical Manufacturers Association (NEMA)**
Synopsis:
This document provides guidance for the design and testing of lighting controls and switching devices to be used with electronic fluorescent ballasts. It covers the def-inition, measurement, and test of characteristics relevant to the use and application of lighting controls and electronic fluorescent ballasts. This standard covers devices rated 120 and 277 VAC, intended to control electronic fluorescent ballast loads up to 16 amps of steady current. The worst case ballast inrush current expected to be encountered in field applications is defined, and uniform test criteria for compatibil-ity is established. Peak currents and pulse widths are defined, and typical waveforms are provided. Reference NEMA (2004).

Title: **IES Approved Method for the Electrical and Photometric Measurements of Fluorescent Lamps**
Identifier: **LM-9-88**
Source: **Illumination Engineering Society of North America (IESNA)**
Synopsis:
This document describes methods for electrical and photometric measurements of fluorescent lamps that are also applicable to ultraviolet lamps, especially the elec-trical portions. Although the measurement of visible light is not of any practical use for ultraviolet lamps, IESNA LM-55-96 can be consulted for the applicable UV spectrum. Reference IES (1988).

Title: **IESNA Approved Method for Photometric Testing of Reflector-Type Lamps**
Identifier: **LM-20-1994**
Source: **Illumination Engineering Society of North America (IESNA)**
Synopsis:
This document describes methods for electrical and photometric measurements of fluorescent lamps that use reflectors. These methods are also applicable to

ultraviolet lamps that have reflectors, except that measurement of visible light is of no importance for ultraviolet lamps. For measurement of lamp output in the ultraviolet spectrum, refer to IESNA LM-55-96. Reference IESNA (1994).

Title: **IESNA Guide to Spectroradiometric Measurements**
Identifier: **LM-58-1991**
Source: **Illumination Engineering Society of North America (IESNA)**
Synopsis:
This document offers general guidelines and methods for spectroradiometric measurements of lamps and although intended primarily for luminaries, it can be useful for the measurement of the ultraviolet output of UV lamps. Reference IESNA (1991).

Title: **IESNA Lighting Handbook, Reference and Application**
                 Chapter 6, Light Sources
Identifier: **IESNA HB-9-2000**
Source: **Illumination Engineering Society of North America (IESNA)**
Synopsis:
This document was reviewed in the previous section on electrical guidelines but is listed here due to its general applicability and importance as a source document for the measurement of ultraviolet lamp electrical characteristics and ultraviolet output. Reference IESNA (2000).

Title: **Method for Measuring Performance of Portable Household**
                 Electric Room Air Cleaners
Identifier: **ANSI/AHAM AC-1-2006**
Source: **American National Standards Institute, Association of Home**
                 Appliance Manufacturers
Synopsis:
This standard is designed to evaluate portable household electric room air cleaners regardless of the particle removable technology utilized. The primary criteria evaluated is the clean air delivery rate (CADR), which is defined here as the rate of contaminant reduction in a test chamber when the air cleaning unit is turned on, minus the natural rate of decay when the unit is off, multiplied by the volume of the test chamber. The contaminants are limited to smoke, dust, and pollen in the 0.10–11 μm size range. This standard employs a 28.54 m^3 (1008 ft^3) test chamber in which the maximum CADR measurement is 450 cfm for pollen and smoke, and 400 cfm for dust. The air cleaner is placed on a table or the floor during operation. The standard provides for a 'Clean Air Delivery Rate' seal to be placed on the packaging of certified air cleaners, which states the CADR for smoke, dust, and pollen, and states the suggested floor area of the room for which the unit is suitable. The ability of the air cleaners to remove other microbiological contaminants are considered outside the scope of this standard. However, in spite of this exclusion the test methods detailed can be adapted (unofficially) for use in estimating the CADR for UV systems using microbiological testing techniques, and it could be expected that ANSI, AHAM, or ARI may eventually adapt this standard for such purposes or

produce a separate standard based on the ANSI/AHAM approach. Reference ASI (2002, 2006).

11.4 UVGI Application Guidelines

Guidelines covered in this chapter include those that refer to UVGI systems specifically or that address them incidentally as they relate to the guideline topic. Only official guidelines issued or drafted by government agencies or professional organizations are addressed here. Many of these are either Health Care guidelines and standards, and guidelines for bioterrorism defense. A variety of state and local guidelines have been issued that are largely culled from the documents discussed below, and these are not discussed here due to their redundancy. Some contradictions and inconsistencies may exist within and between these documents but little or no comment is made in this regard since they tend to be minor, and readers are advised to seek source information (or information in other chapters of his book), and because these guidelines are subject to ongoing revision. Most of these guidelines also address UVGI safety issues but this subject is addressed only incidentally here since it is covered comprehensively in the following chapter. Table 11.3 summarizes the documents addressed in this section. See also (FJNTC 2000).

A limited number of guidelines, catalog methods, and articles issued by companies or private sources and universities are available for assisting designers with various aspects of implementing specific types of UVGI systems and these are cited in the references but are not specifically addressed in this chapter due to the fact that they are equipment-specific, based on limited data, or may be of an unverified or outdated nature (Westinghouse 1982, Sylvania 1981, Philips 1985, Luciano 1977, Luckiesh and Holladay 1942, Luckiesh 1945, 1946, GE 1950, Buttolph and Haynes 1950).

Title:	**Guidelines for preventing the transmission of *Mycobacterium tuberculosis* in health-care facilities**
Identifier:	**MMWR 2005 (54) RR-17**
Source:	**Centers for Disease Control and Prevention (CDC)**
Synopsis:	

The CDC, Department of Health and Human Services, first addressed the use of UVGI for the purpose of preventing tuberculosis transmission in 1994. The latest version of these guidelines, addressed here, describes UVGI as an air cleaning method that can be used to increase the equivalent air changes per hour (ACH), but cannot be used to replace air filtration. The stated CDC approach to defining UVGI effectiveness, in terms of equivalent air changes (EAC), is based upon the method described by First et al. (1999) and Riley et al. (1976). In this method the EAC is defined as the negative logarithm of the survival of UV-irradiated microbes. In effect, this is the slope of the UV survival curve or the purge rate, both of which obey simple logarithmic decay. See Chap. 7 for a more detailed description of this design method.

Table 11.3 UVGI applications guideline and standards summary

Document	Identifier	Source
Guidelines for preventing the transmission of *Mycobacterium tuberculosis* in health-care facilities	MMWR 2005 (54) RR-17	CDC
Guidelines for Environmental Infection Control in Health-Care Facilities	MMWR 2003 (52) RR-10	CDC
Prevention and Control of Tuberculosis in Correctional and Detention facilities	MMWR 2006 (55) RR-9	CDC
Environmental Control for Tuberculosis: Basic Upper-Room Ultraviolet Germicidal Irradiation Guidelines for Healthcare Settings	NIOSH 2009-105	NIOSH
Guidelines on the Design and Operation of HVAC Systems in Disease Isolation Areas	TG 252	USACE
Guidelines for the Prevention of Tuberculosis in Health Care Facilities in Resource Limited Settings	WHO/CDS/TB/99.269	WHO
CIE Technical Report: Ultraviolet Air Disinfection	CIE 155:2003	CIE
Facilities Standards for the Public Buildings Service	PBS-P100	GSA
ASHRAE Handbook Chapter: UV Systems and Equipment	(draft chapter)	ASHRAE
Advanced HVAC Systems for Improving Indoor Environmental Quality and Energy Performance of California K-12 Schools	(none)	CEC
Building Retrofits for Increased Protection Against Airborne Chemical and Biological Releases	EPA 600/R-071/157	EPA
Reference Manual to Mitigate Potential Terrorist Attacks Against Buildings: Providing Protection to People and Buildings	FEMA 426	FEMA
General Guideline for UVGI Air and Surface Disinfection Systems	IUVA-G01A-2005	IUVA
Guideline for Design and Installation of UVGI Air Disinfection Systems in New Building Construction	IUVA-G02A-2005	IUVA
Guideline for the Design and Installation of UVGI In-Duct Air Disinfection Systems	IUVA-G03A-2005	IUVA
Ultraviolet Radiation Systems for Use in the Ventilation Control of Commercial Cooking Operations	UL Standard 710C (outline)	UL
Ultraviolet Radiation for the Processing and Treatment of Food	10CFR21, Section 179.39 & 179.41	FDA

Only three types of UVGI systems are addressed: Upper Room UVGI, In-Duct UVGI, and portable or room UVGI Recirculation units. These guidelines recommend that experts be consulted to ensure proper design and safe operation of the system, and that appropriate measures be taken to protect personnel for UV hazards.

Duct irradiation applications are described as including systems that recirculate room air, systems that irradiate return air from patient rooms, waiting rooms, emergency departments (EDs), and general use areas. The intent of the latter is to prevent undiagnosed TB from contaminating recirculating air.

Upper-air irradiation, or Upper Room UVGI systems, are described which can be used to control TB. It is suggested that a minimum ceiling height of 8 ft is necessary for a substantial volume of upper air to be irradiated without over-exposing occupants. Upper Room UV systems can be used in all rooms in which aerosols might be generated, patient rooms, waiting rooms, EDs, corridors, central areas, etc. They may also be used in operating rooms and adjacent corridors where procedures are performed on TB patients, and medical settings in correctional facilities. Studies are cited in which equivalent air exchange rates of 10–25 ACH were obtained, and it is noted that inlet ceiling diffusers and ceiling fans may enhance the air mixing between upper and lower portions of the room. It is suggested that ceiling fans be used to continuously mix the air unless good mixing has otherwise been demonstrated. Ventilation rates up to 6 ACH do not appear to adversely affect the performance of upper-air UVGI in a well-mixed room. Designers should consider room geometry, UV output of fixtures, room ventilation, and desired EAC in determining the numbers, types, and location of UV fixtures. UVGI fixtures should be spaced to reduce overlap while maintaining an even irradiance zone in the upper air. Wall-mounted, corner-mounted, and ceiling-mounted upper-air systems should have louvers or baffles that block downward radiation and/or to block radiation below the horizontal plane of the fixtures.

UVGI-containing-portable room air cleaners are described as air recirculation units containing UVGI lamps and filters. These units can be used in Airborne Infection Isolation (AII) rooms as an adjunct method of air cleaning. They can also be used in waiting rooms, EDs, corridors, central areas, and other areas where patients with undiagnosed TB may contaminate the air. It is noted that combining portable room-air recirculation units with Upper Room systems is possible and that the performance of combined systems, in terms of the EAC, is approximately additive.

Labeling, Maintenance and Monitoring issues are also addressed in the CDC TB guidelines. Warning signs need to be posted in any areas subject to UV irradiation. Maintenance issues include the replacement of UV lamps per manufacturer's recommendations (typical 10,000 h or about once a year) and the cleaning of lamps that accumulate dust. Lamps should be checked once per month for dust accumulation. Lamps should be replaced if they fail or flicker, or if the UV output drops below the minimum levels required per the design specifications. Maintenance personnel must turn UVGI systems off before performing maintenance or entering the upper room area. Protective clothing such as goggles, gloves, face shields, and sunscreen should be used if potential UV exposure can exceed NIOSH limits.

In-duct UVGI systems require proper tagging and maintenance. Lock-out protocols should be used on access doors and these doors should have viewports or inspection windows by which the UV lamps can be observed in operation. Access doors should also have appropriate warning signs. Access doors should have automatic electric switches or other devices that turn off the UV lamps when the door is opened.

UV lamps and UVGI equipment (i.e. ballasts) should be listed with the Underwriters Laboratories (UL) or Electrical Testing Laboratories (ETL) for their specific application, and they should be installed in accordance with the National Electrical Code. Since UV lamps must ultimately be disposed, consideration should be given to models that have relatively low amounts of mercury (i.e. 5 mg or less).

These TB guidelines also address safety issues associated with Upper Room systems and cite the CDC/NIOSH REL of 0.006 J/cm^2, or 60 J/m^2 (at 254 nm) for 8 h of exposure. Monitoring requirements include the measurement of UV levels in the lower room at eye level to verify NIOSH limits are not exceeded at the time of installation, or whenever changes are made to the area that could affect UV levels. Monitoring results should be recorded along with all maintenance records. UV irradiance levels should be measured in the upper air to verify design requirements are met. Reference CDC (2005).

Title: **Guidelines for Environmental Infection Control in Health-Care Facilities: Recommendations of CDC and the Healthcare Infection Control Practices Advisory Committee (HICPAC)**
Identifier: **MMWR 2003 (52) RR-10**
Source: **Centers for Disease Control and Prevention (CDC)**
Synopsis:

This CDC guideline addresses the use of UVGI for infection control in health care facilities. Intended to address a broader range of health threats in hospitals than just tuberculosis, this guideline addresses UVGI as a supplemental engineering control method, and only addresses upper air UVGI and in-duct UVGI systems. Specific recommendations are provided for air-handling systems in health-care facilities. For UVGI, the recommendations are classified as Category II, 'Suggested for implementation and supported by suggestive clinical or epidemiological studies, or a theoretic rationale.'

UVGI fixtures can be installed on the wall near the ceiling or suspended from the ceiling from the ceiling as an upper air unit. UVGI can be installed in the air-return duct of an Airborne Infection Isolation area. UVGI can be installed in designated enclosed areas of booths for sputum induction, as per CDC TB Guidelines (CDC 2005).

For infection-control and ventilation requirements for Airborne Infection Isolation rooms, where supplemental engineering controls for air cleaning are indicated from a risk assessment of the area, install UVGI units in the exhaust air ducts of the HVAC system to supplement HEPA filtration or install UVGI fixtures on or near the ceiling to irradiate upper room air, per CDC TB Guidelines (CDC 2005).

Curiously, the CDC provides a Category IB recommendation (Strongly recommended for implementation and supported by certain experimental, clinical, or epidemiological studies and a strong theoretic rationale) *against* the use of overhead UVGI systems, saying, 'Do not use ultraviolet (UV) lights to prevent surgical-site infections.' The relevant citations given by the CDC, however, seem to provide strong support for the use of such systems (Lidwell et al. 1982, Taylor et al. 1995) and other uncited publications indicate such an approach can be effective (Ritter et al. 2007, Sanzen et al. 1989, Lowell et al. 1980, Hart 1968, Goldner and Allen 1973, Carlsson et al. 1986, Berg-Perier et al. 1992). It is likely this negative recommendation will be altered in the future. Reference CDC (2003).

Title: **Prevention and Control of Tuberculosis in Correctional and Detention Facilities: Recommendations from CDC**
Identifier: **MMWR 2006 (55) RR-9**
Source: **Centers for Disease Control and Prevention (CDC)**
Synopsis:
In 2006 the CDC issued an update to this report which refers to the use of UVGI as a secondary environmental control (CDC 2006). Secondary environmental controls consist of controlling the airflow to prevent contamination of air in areas adjacent to the source (Airborne Infection Isolation rooms) and cleaning the air using HEPA filtration and/or UVGI. The application of UVGI is stated as increasing the number of equivalent air changes per hour (ACH) and detailed recommendations are provided for the ACH in various correctional facility areas. Supplemental methods such as UVGI may be combined with mechanical filtration in areas that do not have single-pass ventilation to increase effective air cleaning. Such controls are to be used and maintained properly and their strengths and limitations should be recognized. The report states that the engineering design and operational efficacy parameters for UVGI as a secondary control measure (i.e. portable UVGI units, upper-room UVGI, and in-duct UVGI) continue to evolve and require special attention in their design, selection, and maintenance. The application of these environmental controls is detailed in the CDC Guidelines for Preventing TB Transmission (CDC 2005).

When UVGI is used, it should be applied inside the existing HVAC ductwork or in the upper room of the area to be treated to ensure that organisms are inactivated. Upper-air systems should be designed, installed, and monitored to ensure both sufficient irradiation in the upper room to inactivate *M. tuberculosis* and to ensure safe levels of UVGI in the occupied space. Reference CDC (2006).

Title: **Environmental Control for Tuberculosis: Basic Upper-Room Ultraviolet Germicidal Irradiation Guidelines for Healthcare Settings**
Identifier: **NIOSH 2009-105**
Source: **Department of Health and Human Services(DHHS), Centers for Disease Control and Prevention(CDC), National Institute of Occupational Safety and Health (NIOSH)**
Synopsis:
In 2009, The Department of Health and Human Services, CDC/NIOSH, issued this document, which contains detailed information for dealing with tuberculosis in healthcare settings using Upper Room UVGI systems. This document reports

on the results of a study which addressed a number of parameters including (1) the irradiance level in the upper room that provides a UVGI dose over time that kills or inactivates an airborne surrogate of *Mycobacterium tuberculosis*, (2) how to best measure UVGI fluence levels, (3) the effect of air mixing on UVGI performance, (4) the relationship between mechanical ventilation and UVGI systems, (5) the effects of humidity and photoreactivation (PR), and (6) the optimum placement of UVGI fixtures. The results of the study indicates that appropriately designed and maintained Upper Room UVGI systems may kill or inactivate airborne TB bacteria and increase the protection afforded to healthcare workers while maintaining a safe level of UVGI in the occupied lower portion of the room.

This document is intended to provide information to healthcare managers, facility designers, engineers, and industrial hygienists regarding the parameters necessary to install and maintain effective Upper Room UVGI systems. These guidelines reference numerous studies showing the effectiveness of UVGI in controlling mycobacteria but point out the lack of present knowledge concerning the parameters necessary to ensure effective system design. The suggested range of UV irradiance in the upper room is tentatively specified as '30–50 μW/cm^2'. Tables are provided correlating ACH with removal rates in a perfectly mixed room for the purpose of selecting the equivalent ACH of UVGI systems. The importance of air mixing in rooms using upper-room UVGI is noted (see Sect. 9.2.1). Relative humidity issues are addressed in detail and it is suggested that RH be controlled to 60% or less, in agreement with the CDC TB guidelines. Photoreactivation is noted as one concern of high relative humidity. It is recommended that the design temperature for operating Upper Room UVGI systems be consistent with standard practice for hospitals and outpatient facilities, in which temperatures range from 68°F to 75°F (20°C to 24°C). Safety issues are addressed and the CDC/NIOSH RELs for UV exposure are recapitulated. They also cite studies that suggest limits may be too low and that increasing the limits may result in higher effectiveness without undue hazards (Nardell et al. 2008, First et al. 2005).

A variety of specific recommendations are provided regarding UV lamps, fixtures, power levels and placement, based on prior studies. A general 'rule-of-thumb' is quoted with caution from sources stating that one 30 W (input) lamp fixture be installed for every 200 ft^2 of floor area or for every seven people in the room, whichever is greater (First et al. 1999, Macher 1993). It is suggested that the use of louvered upper-room fixtures in rooms with 8-foot ceilings that provide 0.17 W/ft^2 (1.83 W/m^2) should be effective in inactivating mycobacteria, and that for ceilings 9 ft or greater, 0.085 W/ft^2 (0.91 W/m^2) should be effective.

Installation and lamp maintenance recommendations, and safety protocols, are provided that are in accordance with standard recommendations provided elsewhere (i.e. CDC 2005, CIE 2003). Similarly, measurement guidelines for upper-air systems are discussed in detail and provided based on other sources (i.e. Miller et al. 2002, Rudnick 2001, Dumyahn and First 1999) and the limitations of such measurements are noted. Reference NIOSH (2009).

Title: **TB in Homeless Shelters: Reducing the Risk through Ventilation, Filters, and UV**
Identifier: **(none)**
Source: **Francis J. Curry National Tuberculosis Center, Institutional Consultation Services, California Department of Health Services**
Synopsis:

This guideline offers an all-around approach to controlling TB in homeless shelters and address aspects of ventilation, filtration, In-duct UVGI, and Upper-Room UVGI system design. Safety issues are addressed and NIOSH limits on UV exposure are recapitulated. It recommends Upper Room UVGI be used only where the relative humidity is below 70%. In regard to filters, it only discusses ASHRAE 25% filters (MERV 7) and HEPA filters. It states that UVGI can be installed in conjunction with filters to disinfect recirculated air, and implies that an appropriately designed, installed, and maintained UVGI system should provide near 100% removal of TB bacilli (in approximate equivalence to the use of 100% outside air). It recommends that a UVGI in-duct system be installed by professionals but provides no specific details on sizing or selection of equipment.

Upper Room UVGI systems are addressed as a mean of reducing the risk of TB infections. Pointers are offered on safe application of Upper Room systems and the use of radiometers to verify exposure levels is recommended. Detailed suggestions are made regarding finding a contractor, planning the installation, and ensuring safety, including the use of non-reflective UV materials and paints. Design recommendations given are in accordance with those provided in traditional sources (CDC 2005, CIE 2003). Warning signs and staff training are advised, as well as the appointment of a responsible person to periodically check UV levels and lamp operation.

Title: **Guidelines on the Design and Operation of HVAC Systems in Disease Isolation Areas**
Identifier: **TG 252**
Source: **U.S. Army Center for Health Promotion and Preventive Medicine (USACHPPM), U.S. Army Corps of Engineers (USACE)**
Synopsis:

This guideline, issued in 2000 by the, provided some detailed recommendations for the use of UVGI systems in areas used to house and treat disease isolation patients. According to the guideline, UVGI can be used as a method of air disinfection to supplement other engineering controls for disease isolation areas. UVGI cannot be used alone but can be used in conjunction with an engineered HVAC system that has been designed in accordance with other relevant sections of this guideline. UVGI systems shall not be used in areas subject to relative humidity greater than 70%. Circuits connecting UVGI units shall be arranged for either delayed automatic or manual connection to the equipment branch of the emergency power system when facilities are provided with emergency power.

Two types of UVGI systems are addressed, duct irradiation and Upper Room air irradiation systems. Duct irradiation is recommended for isolation and treatment rooms where air is recirculated solely within the room where the system is not equipped with HEPA filtration, and for recirculation systems in patient rooms, waiting rooms, emergency rooms, and other general use areas where there may be unrecognized infectious patients. UVGI may not be used as a substitute for HEPA filter requirements. Ducted systems involve UV lamps mounted perpendicular to the airflow in exhaust ductwork to decontaminate air prior to recirculation. Other requirements for system design and maintenance are in accordance with common practice and typical guidelines such as CDC (2005).

Upper Room irradiation systems are recommended for isolation and treatment rooms as a supplemental method of air cleaning and can also be used in patient rooms, waiting rooms, and emergency rooms. Upper Room systems may be used to supplement the existing ventilation if the HVAC system is incapable of producing the required number of air changes per hour required for existing facilities (as specified in other sections of these guidelines). Upper Room systems are not to be used for this purpose in new or renovated facilities. At a minimum, TB isolation rooms shall be provided with 6 ACH and a direct exhaust system, when equipped with this method of UVGI protection. Upper-room systems are not to be used when a room is connected to a recirculating HVAC system, except for return air systems where direct exhaust is not possible. Upper Room systems shall use louvers (shields) to direct irradiation to the upper room area and minimize exposure to patients and staff. Supply air shall be drawn through the radiation field from air registers and this air shall pass down into the room, over the patient, and out through the exhaust register.

UVGI operation, maintenance, use of warning signs, and measurements are specified in terms typical of other documents (such as the CDC TB Guidelines) or as specified by manufacturers. Monitoring of rooms with UV shall verify that the NIOSH relative exposure limit (REL) is not exceeded and rooms exceeding the REL shall be promptly serviced. UVGI shall not be used as a substitute for negative pressure rooms, and shall not be installed in series with HEPA filters, since they provide no significant additional benefit. Reference USACE (2000).

Title: **Guidelines for the Prevention of Tuberculosis in Health Care Facilities in Resource Limited Settings**
Identifier: **WHO/CDS/TB/99.269**
Source: **World Health Organization (WHO)**
Synopsis:
WHO issued this document in 1999. These guidelines are intended to provide Member States with limited resources with inexpensive and effective control strategies for the prevention of tuberculosis transmission in patients and health care workers. Environmental controls are a secondary priority (after administrative control measures) intended to reduce the concentration of droplet nuclei in the air in high-risk areas, and are not to be used in place of administrative controls.

The guidelines state that in some climates, or in certain high-risk areas of a facility, where the use of natural and mechanical ventilation may not be feasible, UVGI may provide a less expensive alternative to environmental control measures that require structural alterations to a facility. This approach may be useful in larger wards, TB clinic waiting areas, or inpatient areas where TB patients congregate. UVGI may be applied in sputum collection booths using bare bulbs to irradiate the entire booth when it is not occupied.

Continuous upper air irradiation (Upper Room UVGI) may be used when health care workers or patients are in the room if shielding (louvers) are used to prevent UV exposure hazards to occupants. Continuous upper air irradiation is stated to be the most applicable method in resource-limited countries. Good air mixing is necessary for such systems to be effective.

Portable UVGI units may be used, as well as in-duct UVGI in closed mechanical systems. If portable UVGI floor units are used, attention must be paid to lamp placement, since corners may receive inadequate radiation. Reference WHO (1999).

Title: **CIE Technical Report: Ultraviolet Air Disinfection**
Identifier: **CIE 155:2003**
Source: **Commission Internationale de L'Eclairage (CIE)**
Synopsis:

The CIE or International Commission on Illumination issued this Technical Report in 2003, which summarizes the present state of knowledge of ultraviolet radiation (UVR) air disinfection and provides recommendations for future work in research, standardization and testing procedures. This report provides fundamental information related to UVGI applications including measurement, dosimetry, action spectra of germicidal UV, survival curves, decay rate constants, and technical information on lamps and ballasts.

Upper Room air disinfection is stated to have an effectiveness that depends on (1) germicidal fluence, (2) microbial susceptibility, and (3) relative humidity. Another essential factor is air turnover rate between the upper and lower room, which defines the duration of a microbe's exposure. Retrofitting of an existing building can be limited by ceiling height. For floor to ceiling heights less than 2.3 m (7.5 ft) eye exposure hazards may increase. If vertical air circulation is insufficient to elevate infectious particles into the irradiation zone, efficacy diminishes. New construction should integrate these factors in the beginning of the design process. The design methodology suggested for Upper Room systems is the equivalent air changes, or the calculation of the additional air changes needed to produce the same clearance rate of viable microbes irradiated with UVGI. It is stated that for a room with 6 ACH, an installed upper-room UVGI system can achieve 10–20 additional ACH equivalent.

The report suggests that the Upper Room UVGI design criteria regarding the fluence necessary to inactivate the range of microorganisms to be controlled should be based on a risk assessment. Planning a UVGI installation should consider the physical measurements of the space and room occupancy. Referencing Riley et al. (1976,

1989), it is suggested that one 30 W UV lamp is needed for every seven occupants, or else one 30 W (input) lamp per every 18.6 m^2 (200 ft^2). For rooms in excess of 4.7 m the volume of upper room air should be considered. Upper Room fixtures should be louvered to prevent direct radiation in the lower room. Some fraction of UV will be scattered into the lower room but will commonly be insignificant due to the fact that most room materials will tend to have UV reflectivities less than about 5%. Measurements should be made to verify that unsafe UV levels do not exist. The interim guidelines from First et al. (1999) are deferred to, as well as those provided by WHO (1999) and Coker et al. (2001).

The report suggests that Upper Room UVGI may be preferred over other devices since they are passive (no moving parts), have modest costs, low energy consumption, and kill microbes directly in the space in which they are released. It is acknowledged that rapid passage of lower room air through the upper room irradiation zone may result in insufficient exposure time, but that the greatest reduction in infectious microorganisms may occur when the greatest rate of air exchange occurs between the upper and lower room areas. For rooms that lack adequate air movement, the use of mixing fans is a satisfactory solution.

The report states that portable self-contained UVGI systems are applications that are limited by the number of complete room air changes because fan capacity is limited to avoid excessive noise, vibrations, and drafts. It is also stated that UVGI air duct irradiation may be important when it is necessary to recirculate air, but suggests that in-duct air disinfection does little to protect occupants that are in the same room with an infectious source. Reference CIE (2003).

Title: **Facilities Standards for the Public Buildings Service**
Identifier: **PBS-P100**
Source: **General Services Administration (GSA)**
Synopsis:
The GSA publishes this document, which establishes design standards and criteria for new buildings, major and minor alterations, and historic structures for the Public Buildings Service (GSA 2003). The standard includes specific requirements for HVAC components and includes a requirement for the use of ultraviolet lamps (UVC emitters/lamps). It is stated that ultraviolet lamps shall be incorporated downstream of all cooling coils in air-handling units, and above all drain pans, to control airborne and surface microbial growth. Applied lamp fixtures must be those specifically manufactured for this purpose. Safety interlocks and features shall be provided to limit hazards to operating staff. Reference GSA (2003).

Title: **ASHRAE Handbook Chapter 16: Ultraviolet Lamp Systems**
Identifier: **ASHRAE 2008**
Source: **American Society of Heating, Refrigerating, and Air Conditioning Engineers (ASHRAE)**
Synopsis:
ASHRAE has incorporated this chapter in the 2008 ASHRAE Handbook of Systems and Equipment. The chapter provides an overview of UVGI fundamentals, UV lamp

and ballast fundamentals, application guidelines, maintenance guidelines, and safety information. The application guidelines address three types of UVGI systems, In-duct airstream disinfection, air handler surface disinfection, and Upper-air (Upper Room) systems. Considerable background information is provided on UV lamps and ballasts, including cooling and heating effects, and lamp aging.

Per the guidelines, in-duct UVGI systems are generally engineered to achieve a required level of air disinfection and are often unique to each installation. Variables that must be factored in to properly size a UVGI system include duct dimensions, air velocity, lamp cooling effects due to air velocity and air temperature, lamp fouling, age of the lamps, type of power supply, reflectivity, relative humidity, and loca-tion or orientation of the lamps. Mounting the lamps in the supply air plenum will ensure that the return and make-up air are treated. Performance can be improved by increasing the UV reflectivity of the duct walls. The number and location of the UVGI fixtures is dictated by the average percentage reduction desired of the targeted bio-contaminants, and taking into account the above-mentioned variables.

Air handler component surface disinfection systems include those designed to disinfect cooling coils and drain pans. Cooling coils can collect and retain partic-ulates, including microbes, over time. With relative humidities ranging from 30 to 100%, damp coils and drain pans can facilitate the growth of bacteria and mold. UVGI can be applied to HVAC systems to clean components and maintain them free of microbial contamination, thereby maintaining heat exchange efficiency and design airflow. Indoor air quality can be improve by reducing microbial growth on system components, and significant energy savings are possible. It is recommended that coils be cleaned prior to UVGI installation. UV lamps should be mounted in proximity to the coils and spaced to allow even distribution of irradiation over the surface. UVGI fixtures must be designed to withstand moisture and condensate and to operate over the full range of system operating temperatures. UVGI sys-tems should operate continuously and electrical interlocks should be provided to de-energize the system when it is accessed for maintenance. UVGI systems can be located upstream or downstream of the coils but each approach can affect system performance due to the operating conditions. The lamp chamber shall be provided with viewports to allow an operating system to be safely observed.

Upper Room UVGI systems should contain baffles or louvers to direct UV irradi-ation to the upper air space, and safety switches to disengage the units for servicing. Occupied spaces with ceilings lower than 10 ft should have fixtures mounted at least 7 ft high. The basic criterion of installing one 30 W (input power) for every 200 ft^2 is cited per Riley et al. (1976) to produce a uniform distribution of 30–50 μW/cm^2. Ventilation systems should maximize air mixing to receive the greatest benefit. Relative humidity should be less than 60%, and an RH exceeding 80% may reduce effectiveness. If UV levels exceed safe limits, all highly UV-reflecting sur-faces should be removed or altered. UV-absorbing paints containing zinc oxide can be used on ceilings and walls to minimize reflected UV.

All UV lamp chambers should be equipped with electrical disconnect devices, and positive disconnection devices are preferred over switches. Disconnection devices shall be located outside of the lamp chamber and adjacent to the primary access panel or door. Switches shall be wired in series such that opening any access door will de-energize the system. On/Off switches shall not be located in the same location as general room lighting. Switches must be positioned such that only authorized persons have access to them and should be locked to ensure that they are not accidentally turned on or off.

Other recommendations regarding lamp maintenance, disposal, measurements, warning labels, and safety are largely in accordance with other guidelines (CDC 2005, CIE 2003). UV radiation measurements should be taken at the time of installation, lamp replacement, and whenever modifications are made to the UV system or the room. Personnel should be adequately trained in all relevant aspects of UV system operation, safety, maintenance, and lamp disposal. Reference ASHRAE (2008).

Title: **Advanced HVAC Systems for Improving Indoor Environmental Quality and Energy Performance of California K-12 Schools**
Identifier: **(none)**
Source: **California Energy Commission CEC**
Synopsis:
The CEC released this Consultant Report in 2006, which focuses on recommended Code/Guideline actions for the use of UV systems in schools. The report offered recommendations for three types of UVGI systems: in-duct UVGI, Upper Room UVGI, and cooling coil disinfection systems. Limited details are offered regarding installation but guideline references are provided including studies demonstrating both health benefits for students and building energy savings. It is recommended that filtration be used in conjunction with in-duct systems and that the level of filtration can reduce lamp maintenance. Lamps are to be replaced annually or if UV output drops below 70%. Lamps can be cleaned with a soft lint-free cloth moistened with glass cleaning agents. Manual cleaning of cooling coils is recommended prior to lamp installation. It is suggested that UVGI systems be operated when the school buildings are unoccupied (i.e. summer vacation) to deactivate microorganisms and flush them from the buildings prior to occupancy. Lamps should be oversized initially to account for decreased output with age, in accordance with manufacturers' design practices.

For cooling coil disinfection systems, lamp placement should provide good coverage of the coil face. The travel path of the UV rays should be directly through the gaps in the coil fins. The manufacturer has the responsibility, in general, of adequately sizing the product to meet the conditions required by the application. Quoting a manufacturer, it is suggested that 24 in. of UV lamp length be used for every 4 ft^2 of coil face area, and that the ideal distance between the fixture and the coil is half the distance between the rows or half the height of a one row coil if it is less than 24 in. IUVA draft guidelines (IUVA 2005) are cited regarding the avoidance of corrugated fins and limiting fin spacing to 8–12 fins per inch. Combining coil disinfection with air disinfection is recommended for maximum effectiveness. Reference CEC (2006).

Title: **Building   Retrofits   for   Increased   Protection   Against**
                 Airborne Chemical and Biological Releases
Identifier: **EPA 600/R-071/157**
Source: **Environmental Protection Agency (EPA)**
Synopsis:
This report from the EPA addresses various retrofit options, including UVGI, that
building owners can consider to protect buildings from chemical and biological
releases in or near buildings. UVGI is discussed in terms of its general application to
air cleaning in conjunction with filtration, and its economics, but no specific guidelines are provided. However, references are made to the CDC (2005) guidelines, the
ASHRAE (2008) Handbook chapter, and the draft IUVA (2005) guidelines. Reference EPA (2007).

Title: **Reference Manual to Mitigate Potential Terrorist**
                 Attacks Against Buildings
Identifier: **FEMA 426**
Source: **Federal Emergency Management Agency (FEMA)**
Synopsis:
This document was issued by FEMA as part of their Risk Management Series.
UVGI is addressed as one of the air filtration and pressurization systems that can
be used to mitigate biological terrorist threats. Minimal information is provide for
design guidance but references are provided. UVGI systems are noted to be installed
in conjunction with high efficiency filtration systems. It is noted that retrofitting
UVGI systems for in-duct applications can be relatively simple if sufficient space
is available. Attention must be paid to maintaining design air velocity and the
temperature of the UV lamps since cooling of the lamps can significantly affect
UV output. Polished aluminum reflective panels can be used to increase the intensity of the UVGI field. The design velocity of a typical system is similar that for
particulate filters, or about 400 fpm. Designs combining UVGI with filtration can
be very effective against biological agents since smaller microbes, which are difficult to filter out, tend to be more susceptible to UVGI. UVGI safety measures, such
as duct access interlocks that turn off the lamps when the duct housing is opened,
should be used. Reference FEMA (2003).

Title: **General Guideline for Air and Surface Disinfection Systems**
Identifier: **IUVA-G01A-2005 (Draft)**
Source: **International Ultraviolet Association (IUVA)**
Synopsis:
The IUVA has prepared a draft guideline for public comment as part of a series of
planned guidelines for UVGI systems. This guideline provides general information
on UVGI system design and many specific recommendations. Details and examples
of lamp cooling effects are provided to assist designers but for specific lamps the
manufacturer specifications should be used. Lamp ballast location should be considered due to heating effects, including radiative heating, that can cause malfunc-

tions. For in-duct applications, a rating system is provided (UVGI Rating Value or URV) that categorizes UVGI systems based on the UV exposure dose and that can simplify equipment selection. In general, systems rated URV 10–15 are typical for air disinfection (see the URV Table in Appendix G). MERV filters can be directly matched to the URV of a UVGI system for optimum performance, simplifying filter selection. Filters must be constructed of UV-resistant materials – glass media are acceptable but plastic fiber filters and plastic structural components and binders may be subject to UV degradation. General information on maintenance, safety, etc. are provided for all types of UVGI systems and these are similar to those found in other documents (i.e. CIE 2003, CDC 2005, ASHRAE 2008). Specific guidelines for various types of UVGI systems are given in the associated documents in the IUVA series, though most of these have not been released for public review yet. Reference IUVA (2005).

Title: **Guideline for Design and Installation of UVGI Air Disinfection Systems in New Building Construction**
Identifier: **IUVA-G02A-2005 (Draft)**
Source: **International Ultraviolet Association (IUVA)**
Synopsis:
The IUVA has issued a draft guideline for public comment, which provides more specific information than the General Guideline. The subject of new building construction is treated separately due to the fact that it is simpler and more cost-effective to design air treatment systems for new buildings than to retrofit them in existing buildings. It is expected that the design of new buildings will meet minimum requirements for airflow and air exchange rates as specified in ASHRAE 62.1 and that outside air supply will be filtered for dust.

It is recommended that for recirculated air, a filter in the range of MERV 11–15 be included upstream of the UVGI system, and that filters be constructed of UV-proof materials. At least 2 ft of space should be provided downstream of the filters, or upstream and downstream, of the cooling coils, but avoiding the area downstream of the heating coils. The UVGI system should preferably be designed in the range of URV 11–15. In areas that cannot be served by the central air handling unit, it is suggested that local recirculation UVGI units be installed and that they have between 1 and 6 ACH.

Cooling coil UVGI systems are recommended to reduce energy costs, and can be combined with air disinfection systems. Corrugated fins should not be used on the cooling coils since they inhibit penetration by UV rays. Cooling coil fin spacing should not exceed 8–12 fins per inch. Cooling coils should be irradiated on both the upstream and downstream sides if possible. Performance goals for new buildings equipped with UVGI air disinfection are suggested in terms of the concentrations of airborne bacteria and fungi, with an upper limit of 500 cfu/m^3 for bacteria, 150 cfu/m^3 for fungi, 10 gr/m^3 for pollen, and an Indoor/Outdoor ratio of 0.1 (or 10% of the outdoor levels) subject to the above concentration limits. Other guideline requirements (i.e. maintenance, safety, etc.) are to be as specified as per IUVA-G01A-2005. Reference IUVA (2005a).

Title: **Guideline for the Design and Installation of UVGI In-Duct Air Disinfection Systems**
Identifier: **IUVA-G03A-2005 (Draft)**
Source: **International Ultraviolet Association (IUVA)**
Synopsis:

The IUVA has issued a draft guideline for public comment, which is intended for general purpose applications of UVGI for air disinfection. This document has been issued for comment in draft form and is subject to revision, but provides detailed recommendations for UVGI system installation. Any commercial building to be retrofitted must meet minimum requirements for ventilation as specified in ASHRAE Standard 62.1, and these include 20 cfm per occupant or 15% outside air.

Recommendations include a minimum exposure time of 0.25 s and a minimum open duct length of at least 2 ft for lamp installation. A UVGI Rating Value (URV) of 11–15 is recommended, and return air filters rated MERV 10–15. Lamps may be located around cooling coils to serve a dual function as a cooling coil disinfection system. Preferred locations for located UVGI systems include downstream of filter banks, upstream and downstream of the cooling coils, upstream of heating coils, in supply ductwork, in the fan housing, and in the mixing plenum. Lamp ballasts should be located so as to avoid heating effects. Design air velocity should be about 500 fpm, with variations of ±100 fpm being possibly acceptable. UV lamps should not be located in mixing plenums where air temperatures could drop below the normal lamp operating limit of about 45–50°F, unless manufacturer's lamps can adjust to such conditions. Lamps and ballasts should not be located in areas where water condensation may cause problems, except that lamps suited for water applications may be used in such circumstances.

UVGI systems should be operated full-time when the HVAC operates full time. UVGI systems may be disengaged during unoccupied periods. Shutoff switches should be included and located near the UVGI system, and are preferably mounted externally. Access door switches should be included to disengage the UVGI system upon opening the access doors. Motion detectors may also serve this purpose. High temperature cutouts should be used to disengage the UVGI system if the equipment exceeds design operating temperature. Reflective surfaces inside the duct are suggested as a means of increasing UV irradiance and of delivering UV rays to areas that would otherwise receive limited exposure. Reflective surfaces may require periodic cleaning. Reference IUVA (2005b).

Title: **Ultraviolet Radiation Systems for Use in the Ventilation Control of Commercial Cooking Operations**
Identifier: **UL Standard 710C (outline)**
Source: **Underwriters Laboratories Incorporated**
Synopsis:

These requirements cover UV lamp systems used for the reduction of grease laden vapors from commercial cooking equipment when installed within Listed Exhaust Hoods for Commercial Cooking Equipment. These requirements do not address the effectiveness of UV to reduce grease-laden vapors. This outline applies to UV lamp assemblies for use in locations that are intended for installation on branch circuits of nominal 600 V or less between conductors in accordance with the National Electrical Code, ANSI/NFPA 70. This standard is not yet available except in outline form and is tentatively scheduled for issuance on 3-22-2009. Reference UL (2009).

Title: **Ultraviolet Radiation for the Processing and Treatment of Food**
Identifier: **10CFR21, Section 179.39 and Section 179.41**
Source: **Food and Drug Administration (FDA)**
Synopsis:
The Code of Federal Regulations (CFR), Title 21 (Food and Drugs), Chap. 1 (Food and Drug Administration, Department of Health and Human Services), Subchapter B (Food for Human Consumption), Part 179 (Irradiation in the production, processing, and handling of food), Subpart B (Radiation and radiation sources), Section 179.39, states that ultraviolet radiation for the processing and treatment of food may be safely used under the following conditions:

(a) The radiation sources consist of low pressure mercury lamps with 90% emission at a wavelength of 253.7 nm.
(b) The ultraviolet radiation is used or intended for use on food products without ozone production: high fat-content food irradiated in vacuum or in an inert atmosphere; intensity of radiation, 1 W per 5–10 ft^2, for surface microorganism control.

Section 179.41 states that pulsed light may be safely used for the treatment of food under the following conditions:

(a) The radiation sources consist of xenon flashlamps designed to emit broadband radiation consisting of wavelengths between 200–1100 nm and operate so that the pulse duration is no longer than 2 ms.
(b) The treatment is used for surface microorganism control.
(c) Foods treated with pulsed light shall receive the minimum treatment reasonably required to accomplish the intended technical effect.
(d) The total cumulative treatment shall not exceed 12 J/cm^2 (120,000 J/m^2). Reference FDA (2000).

11.5 UVGI Safety Guidelines

Ultraviolet light poses health hazards to humans and a variety of documents have been issued to address UVGI safety. Many of the guidelines covered in the previous section on UVGI Applications have addressed aspects of UVGI safety and these documents are not revisited here. Only primary and source documents are reviewed in this section. A more comprehensive and detailed treatment of UVGI safety, based upon the following documents, is presented in Chap. 12. Documents related to mercury vapor lamps, which can release UV when damaged, are not addressed here, not are documents related to solar or sun lamp hazards. For an

Table 11.4 UV safety guidelines and standards summary

Document	Identifier	Source
Threshold Limit Values for Chemical Substances and Physical Agents and Biological Exposure Indices	ACGIH TLVs	ACGIH
Occupational Exposure to Ultraviolet Radiation (Criteria for a Recommended Standard)	HSM 73-11009	NIOSH
Ultraviolet Radiation Guide	NEHC-TM92-5	NEHC
Nonionizing Radiation Guide Series, Ultraviolet Radiation	ISBN 0-932627-08-8	AIHA
Ultraviolet Radiation Related Exposures: Broad-spectrum ultraviolet (UV) radiation, UVA, UVB, UVC, Solar Radiation, and Exposure to Sunlamps and Sunbeds	10th Report on Carcinogens	USDHHS
Working Safely with Ultraviolet Irradiation: Policy and Procedures	(none)	CUHSD
IRPA Guidelines on Protection Against Non-ionizing Radiation	Chapter 3	IRPA
Environmental Health Criteria 160: Ultraviolet Radiation	EHC 160	UNEP/WHO
Eye and Face Protectors	CSA Standard Z94.3-02	CSA
Health Hazard Evaluation Reports	HETA 91-148-2236, 91-0187-2544, 92-171-2255, 2001-0483-2884	NIOSH
Guidelines on Limits of Exposure to Ultraviolet Radiation of Wavelengths Between 180 nm and 400 nm (Incoherent Optical Radiation)	ICNIRP 2004	ICNIRP

informative review of solar UV hazards see Juchem et al. (1998). For information on ultraviolet hazards from fluorescent lamps, see NEMA document LSD 7-1999 (NEMA 1999). Curiously, the Code of Federal Regulations (CFR) does not explicitly address employee exposure to ultraviolet radiation in 29CFR 1910.97, Nonionizing Radiation, so this CFR section is not addressed here. See also UNEP (1994) and McKinley and Repacholi (1994)

Title: **Threshold Limit Values for Chemical Substances and Physical Agents & Biological Exposure Indices**
Identifier: **ACGIH TLVs**
Source: **American Conference of Governmental Industrial Hygienists (ACGIH)**
Synopsis:
This document establishes allowable threshold limit values (TLVs) for direct ocular and skin exposure to ultraviolet radiation. TLVs for occupational exposure are based

upon parameters for exposure dose (exposure time and irradiance levels). Limits are established in terms of the maximum time periods that may be spent under specific irradiance levels. Exposures to broad-band sources are to be determined using weighting formulas, however, it is noted that no regulations exist that specify exposure limits at the specific wavelengths of UVA, UVB, or UVC, and all regulations are generalized under broad-spectrum UV. The permissible broadband UV exposure for unprotected eye and skin exposure may range from 0.1 $\mu W/cm^2$ (for 8 h a day) to 30,000 $\mu W/cm^2$ (for 0.1 s). Reference ACGIH (2004).

Title: **Occupational Exposure to Ultraviolet Radiation (Criteria for a Recommended Standard)**
Identifier: **HSM 73-11009**
Source: **National Institute for Occupational Safety and Health (NIOSH)**
Synopsis:
This document was issued in 1972 and presents a series of recommendations as a basis for developing a standard for occupational exposure to ultraviolet light. It provides extensive citations on the health effects of UV. The recommended exposure limits for ultraviolet radiation are provided based on ACGIH TLVs, and are essentially identical to the limits provided in later NIOSH documents. Sources of UV are identified and the particular types of skin and eye damage that can result from UV exposure are explained in detail. Results are presented from various studies on UV exposure to demonstrate the adequacy of the recommended TLVs. The etiology of UV damage to skin and eyes is explained and graphs of erythemal effectiveness, or the action spectra for human skin, are presented. Protection and control measures are suggested and UV measurement methods are discussed. Reference NIOSH (1972).

Title: **Ultraviolet Radiation Guide**
Identifier: **Technical Manual NEHC-TM92-5**
Source: **Navy Environmental Health Center (NEHC)**
Synopsis:
This document serves as a guide for industrial hygienists who are responsible for making recommendations for the protection of workers from potential health effects of ultraviolet radiation. Acute biological effects of UV radiation on the eyes and the skin are discussed, including erythema, skin photosensitization, keratoconjunctivitis. Chronic effects of UV such as skin aging and certain types of skin cancer are also addressed. Sources of UV radiation and hazard assessments are presented. Relative exposure limits (RELs) are quoted from the National Institute for Occupational Safety and Health and the proper evaluation of these RELs in dose assessment are summarized. Control measures are discussed, including administrative controls, personal protective equipment (PPE), eyewear, topical screening, and warning signs. Reference NEHC (1992).

Title: **Nonionizing Radiation Guide Series, Ultraviolet Radiation**
Identifier: **ISBN 0-932627-08-8**

Source: **American Industrial Hygiene Association (AIHA)**
Synopsis:
This guide provides detailed information on sources of ultraviolet radiation, biological and health effects, exposure guidelines, measurement instrumentation, engineering controls, administrative controls, and personal protection. Biological effects addressed include erythema, photosensitivity, skin aging, immune systems effects, skin cancer, photokeratoconjunctivitis, cataracts, lens fluorescence, and other eye effects. Reference AIHA (2001).

Title: **Ultraviolet Radiation Related Exposures: Broad-spectrum Ultraviolet (UV) radiation, UVA, UVB, UVC, Solar Radiation, and Exposure to Sunlamps and Sunbeds**
Identifier: **Tenth Report on Carcinogens**
Source: **U.S. Department of Health and Human Safety (USDHHS)**
Synopsis:
This document is part of an ongoing series of reports on carcinogens and it focuses on the relation between UV and skin tumors or skin cancer, based on the latest epidemiological and clinical studies. According to this report, radiation in the UVA, UVB, and UVC spectrum 'is reasonably anticipated to be a human carcinogen,' based on limited evidence of carcinogenicity from human mechanistic studies and sufficient evidence from animal studies. It is reported, however, that no epidemiological studies in humans have adequately evaluated UVC carcinogenicity in humans. Some detailed results are presented from studies that provide a causative basis for possible mechanisms of UV carcinogenicity in humans. Reference USDHHS (2002).

Title: **Working Safely with Ultraviolet Irradiation: Policy and Procedures**
Identifier: **(none)**
Source: **Columbia University Health Sciences Division (CUHSD)**
Synopsis:
This document provides information and guidelines for the safe use of ultraviolet radiation to protect the health of personnel using or potentially exposed during their work activities. This policy covers are research laboratories as well as other areas at the Columbia University Medical Center where ultraviolet radiation is used, including research or germicidal purposes. UV hazards are addressed as well as exposure limits (referenced from NIOSH and ACGIH), and control measures. Specifically covered are germicidal lamps used in biosafety cabinets, clinical areas, and laboratories. Reference CUHSD (2005).

Title: **IRPA Guidelines on Protection Against Non-ionizing Radiation**
Identifier: **Chapter 3**
Source: **International Radiation Protection Association (IRPA)**
Synopsis:
Chapter 3 of this publication addresses 'Guidelines on Limits of Exposure to Ultraviolet Radiation of Wavelengths Between 180 nm and 400 nm (Incoherent Optical

Radiation).' The purpose of these guidelines is to deal with the basic principles of protection against ultraviolet light so that they may serve as guidance to the various international and national bodies or experts who are responsible for the development of regulations., recommendations, or codes of practice to protect the workers and the general public from the potential adverse effects of ultraviolet radiation. Exposure Limits (ELs) are provided for the eyes and for skin that are similar to those published by NIOSH and ACGIH. These ELs are specified by spectral wavelength in some 56 increments and each increment is provided with a value for relative spectral effectiveness. Supplemental information is provided on protective measures and UV measurements, as well as the action spectrum for human skin. See also IRPA (1985). Reference IRPA (1991).

Title: **Environmental Health Criteria 160: Ultraviolet Radiation**
Identifier: **EHC 160**
Source: **United Nations Environment Programme (UNEP) and the World Health Organization (WHO)**
Synopsis:
This document reviews the biological health effects reported from exposure to ultraviolet light and serves as the scientific rationale for the development of guidelines for UV safety. It cites the seminal scientific publications that provide a basis for understanding the biological health effects of ultraviolet light. An update of the 1979 Environmental Health Criteria 14. Reference WHO (1994).

Title: **Eye and Face Protectors**
Identifier: **CSA Standard Z94.3-02**
Source: **Canadian Standards Association**
Synopsis:
This standard relates general and specific requirements for eye and face protectors for industrial and educational processes that include potential UV exposure. Also outlined are the performance requirements tests. Luminous transmittance is determined using Standard Illuminant A as described in CIE 2003 and testing is based upon the test method ASTM D 1003. Transmittance requirements calculations are provided for general purpose filters for protection against ultraviolet light. Reference CSA (2000).

Title: **Health Hazard Evaluation Reports**
Identifier: **HETA 91-148-2236, HETA 91-0187-2544, HETA 92-171-2255, HETA 2001-0483-2884**
Source: **National Institute of Occupational Safety and Health (NIOSH)**
Synopsis:
The Hazard Evaluations and Technical Assistance Branch (HETAB) of NIOSH conducted these field investigations of possible health hazards related to occupational exposure to ultraviolet light. In HETA 91-148-2236, a survey was conducted to measure UV levels in rooms equipped with Upper Room UV devices in a hospital. It was

found that levels in the work areas were below 0.1 μW/cm^2 except at distances very close to the UV lamps. The ventilation was also evaluated and recommendations were made regarding labeling, worker training, maintenance issues, and procedures for minimizing time spent in rooms with UV.

In HETA 91-0187-2544, an evaluation was conducted at a hospital in response to employee requests regarding TB transmission and potential exposure to UV. In this case a UV system had been installed but was not placed into service until this evaluation was completed. When the UV system was activated by investigators it was found that levels exceeded NIOSH limits for 8 h of exposure. Recommendations included installation of UV shields, employee training, warning labels, the deactivation of lamps that produce levels exceeding NIOSH limits until the problem is resolved, and preventive maintenance programs. It was also noted that the lamp fixture geometry and louver design created unique scattering angles that were the major causes of the unacceptably high intensities produced.

In HETA 92-171-2255, an investigation was conducted into a county medical examiner's office in which a high rate of TB exposure among employees had occurred. Measurements were taken of UV intensities in areas where UV lamps were used to control TB transmission. Six UV lamps suspended from the ceiling were found to produce levels of UV irradiation (5–60 μW/cm^2) that limited occupancy of the area to 20 min. In some areas the UV intensities were even higher. It was recommended that personal protective equipment (PPE) be used more diligently and fitted more tightly, since gaps in the fit had resulted in skin erythema. In-duct UV was also used in this facility for air disinfection of recirculated air. Although the in-duct UV lamps produced no hazards, it was noted that the duct return register was open to the ceiling cavity and that if the ceiling panels were removed it would produce an exposure hazard. Various control interlocks and training were recommended, but it was found that the TB transmission problem was likely related to low air exchange rates. Other recommendations included improved monitoring of UV levels, worker training, restricted access and the use of indirect instead of direct UV irradiation. It was also suggested that reviews of workers' past medical histories might identify conditions that could be exacerbated by UV exposure or that may result in UV hypersensitivity.

In HETA 2001-0483-2884, reports of skin and eye problems were investigated among immigrations inspectors at an airport. It was found that UV lamps used by inspectors for validating documents were producing UV shine on anyone in close proximity that exceeded NIOSH limits. The lamps may have been left on continuously, although they were only needed for brief periods. Since the UV devices only required UVA lamps (black light lamps), not UVC, lamps, it was recommended the lamps be replaced, that shielding plates be used, and that training be provided to workers. References NIOSH (1991a, b, 1992, 2001).

Title: **Guidelines on Limits of Exposure to Ultraviolet Radiation of Wavelengths Between 180 nm and 400 nm (Incoherent Optical Radiation)**

Identifier: **ICNIRP 2004**
Source: **The International Commission on Non-Ionizing Radiation Protection**
Synopsis:
This document summarizes UV exposure limits and spectral weighting functions (as per NIOSH, ACGIH, and CIE), and provides supportive information on the effects of UV on skin and eyes. Source information is provided as part of the rationale for exposure limits, including types of skin cancer, skin phototypes, action spectra for melanoma and other chronic effects, ocular action spectra. Calculation details are provided for interpolating the spectral weighting functions at wavelengths other than those specified in the standard tables. Reference ICNIRP (2004).

References

ACGIH. 2004. Threshold Limit Values and Biological Exposure Indices. Cincinnati, OH: American Conference of Governmental Industrial Hygienists.

AIHA. 2001. Nonionizing Radiation Guide Series, Ultraviolet Radiation. Akron, OH: American Industrial Hygiene Association. Report nr ISBN 0-932627-08-8.

ANSI. 2003. American National Standard Practice for Occupational and Educational Personal Eye and Face Protective Devices. New York: American National Standards Institute. Report nr ANSI Z87.1-2003.

ANSI. 2005. Safety in Welding, Cutting, and Allied Processes. New York: American National Standards Institute. Report nr ANSI Z49.1-2005.

ASHRAE. 2008. Handbook of Applications: Chapter 16: Ultraviolet Lamp Systems. Atlanta, GA: American Society of Heating, Refrigerating, and Air-Conditioning Engineers.

Berg-Perier M, Cederblad A, Persson U. 1992. Ultraviolet radiation and ultra-clean air enclosures in operating rooms. J Arthroplasty 7(4):457–463.

Buttolph LJ, Haynes H. 1950. Ultraviolet Air Sanitation. Cleveland, OH: General Electric. Report nr LD-11.

Carlsson AS, Nilsson B, Walder MH, Osterberg K. 1986. Ultraviolet radiation and air contamination during total hip replacement. J Hosp Infect 7:176–184.

CDC. 2003. Guidelines for Environmental Infection Control in Health-Care Facilities. MMWR 52(RR-10).

CDC. 2005. Guidelines for preventing the transmission of *Mycobacterium tuberculosis* in health-care facilities. In: CDC, editor. Federal Register. Washington: US Govt. Printing Office.

CIE. 2003. Ultraviolet Air Disinfection. Vienna, Austria: International Commission on Illumination. Report nr CIE 155:2003.

Coker I, Nardell EA, Fourie B, Brickner PW, Parsons S, Bhagwandin N, Onyebujoh P. 2001. Guidelines for the utilisation of ultraviolet germicidal irradiation (UVGI) technology in controlling the transmission of Tuberculosis in health care facilities in South Africa. Pretoria, South Africa: Medical Research Council.

CSA. 2007. Eye and Face Protectors. Mississauga, Ontario: Canadian Standards Association. Report nr CSA Standard Z94.3-07.

CUHSD. 2005. Working Safely with Ultraviolet Irradiation: Policy and Procedures. Columbia University Health Sciences Division.

Dumyahn T, First M. 1999. Characterization of ultraviolet upper room air disinfection devices. Am Ind Hyg Assoc J 60(2):219–227.

EPA. 2007. Building Retrofits for Increased Protection Against Airborne Chemical and Biological Releases. Environmental Protection Agency. Report nr EPA 600/R-071/157.

FDA. 2000. Ultraviolet radiation for the processing and treatment of food. Food and Drug Administration. Report nr 10CFR21, Section 179.39 & 179.41.

FEMA. 2003. Reference Manual to Mitigate Potential Terrorist Attacks Against Buildings. Federal Emergency Management Agency. Report nr FEMA 426.

First MW, Nardell EA, Chaisson W, Riley R. 1999. Guidelines for the application of upper-room ultraviolet germicidal irradiation for preventing transmission of airborne contagion – Part II: Design and operational guidance. ASHRAE J 105:869–876.

First MW, Weker RA, Yasui S, Nardell EA. 2005. Monitoring human exposures to upper-room germicidal ultraviolet irradiation. J Occup Environ Hyg 2(5):285–292.

FJCNTC. 2000. TB in Homeless Shelters: Reducing the Risk through Ventilation, Filters, and UV. Francis J. Curry National Tuberculosis Center, Institutional Consultation Services, California Department of Health Services.

GE. 1950. Germicidal Lamps and Applications. USA: General Electric. Report nr SMA TAB: VIII-B.

Goldner JL, Allen BL. 1973. Ultraviolet light in orthopedic operating rooms at Duke University. Clin Ortho 96:195–205.

Hart D, Postelthwait R. 1968. Postoperative wound infections: a further report on ultraviolet irradiation with comments on the recent 1964 National Research Council Cooperative report. Ann Surg 167(5):728–743.

IARC. 1992. Solar and Ultraviolet Radiation, Volume 55. Cancer IAfRo, editor. Lyon, France: IARC.

ICNIRP. 2004. Guidelines on limits of exposure to ultraviolet radiation of wavelengths between 180 nm and 400 nm (Incoherent Optical Radiation). Health Phys 87(2):171–186.

IESNA. 1996. Measurement of Ultraviolet Radiation from Light Sources. New York: Illumination Engineering Society of North America. Report nr IESNA LM-55-96.

IESNA. 2000. IESNA Lighting Handbook HB-9-2000, Reference and Application Chapter 6, Light Sources. New York: Illumination Engineering Society of North America.

IRPA. 1985. IRPA guidelines on limits of exposure to ultraviolet radiation of wavelengths between 180 nm and 400 nm (Incoherent Optical Radiation). Health Phys 49:331–340.

IRPA. 1991. IRPA Guidelines on Protection Against Non-ionizing Radiation. Argentina: International Radiation Protection Association. Report nr Chapter 3.

IUVA. 2005. General Guideline for UVGI Air and Surface Disinfection Systems. Ayr, Ontario, Canada: International Ultraviolet Association. Report nr IUVA-G01A-2005.

IUVA. 2005a. Guideline for Design and Installation of UVGI Air Disinfection Systems in New Building Construction. Ayr, Ontario, Canada: International Ultraviolet Association. Report nr IUVA-G02A-2005.

IUVA. 2005b. Guideline for Design and Installation of UVGI In-Duct Air Disinfection Systems. Ayr, Ontario, Canada: International Ultraviolet Association. Report nr IUVA-G03A-2005.

Juchem PP, Hochberg J, Winogron A, Ardenghy M, English R. 1998. Health Risks of Ultraviolet Radiation. SBCP 13(2).

Lidwell OM, Lowbury EJ, Whyte W, Blowers R, et al. 1982. The effect of ultraclean air in operating rooms on deep sepsis in the joint after total hip and total knee replacement: a randomized study. Br Med J (Clin Red Ed) 285(6334):10–14.

Lowell J, Kundsin R. 1977. Ultraviolet Radiation: Its Beneficial Effect on the Operating Room Environment and the Incidence of Deep Wound Infcetion Following Total Hip and Total Knee Arthroplasty. Murray Hill, NJ: American Ultraviolet Company. Report nr A-810.

Lowell JD, Kundsin RB, Schwartz CM, Pozin D. 1980. Ultraviolet radiation and reduction of deep wound infection following hip and knee arthroplasty. In: Kundsin RB, editor. Airborne Contagion, Annals of the New York Academy of Sciences. New York: NYAS, pp. 285–293.

Luciano JR. 1977. Air Contamination Control in Hospitals. New York: Plenum Press.

Luckiesh M, Holladay LL. 1942. Designing installations of germicidal lamps for occupied rooms. Gen Electric Rev 45(6):343–349.

Luckiesh M. 1945. Disinfection with germicidal lamps: Air-II. Electrical World Oct.13:109–111.

Luckiesh M. 1946. Applications of Germicidal, Erythemal and Infrared Energy. New York: D. Van Nostrand Co.

Macher JM. 1993. The use of germicidal lamps to control tuberculosis in healthcare facilities. Infect Contr Hosp Epidem 14:723–729.

McKinlay AF, Repacholi MH. 1999. Ultraviolet Radiation Exposure, Measurement and Protection. Chilton, UK: Nuclear Technology Publishing.

Miller SL, Hernandez M, Fennelly K, Martyny J, Macher J. 2002. Efficacy of ultraviolet irradiation in controlling the spread of tuberculosis. NIOSH. Report nr NTIS PB2003-103816.

Nardell EA, Bucher SJ, Brickner PW, Wang C, Vincent RL, Began-McBride K, James MA, Michael M, Wright JD. 2008. Safety of Upper Room Ultraviolet Germicidal Air Disinfection for Room Occupants: Results from the Tuberculosis Ultraviolet Shelter Study. Pub Health Rep 123:52–60.

NEHC. 1992. Ultraviolet Radiation Guide. Norfolk, VA: Navy Environmental Health Center, Bureau of Medicine and Surgery. Report nr Technical Manual NEHC-TM92-5.

NIOSH. 1972. Occupational Exposure to Ultraviolet Radiation. Cincinnati, OH: National Institute for Occupational Safety and Health. Report nr HSM 73-110009.

NIOSH. 1991a. Health Hazard Evaluation Report. Cincinnati, OH: National Institute of Occupational Safety and Health. Report nr HETA 91-148-2236.

NIOSH. 1991b. Health Hazard Evaluation Report. Cincinnati, OH: National Institute of Occupational Safety and Health. Report nr HETA 91-0187-2544.

NIOSH. 1992. Health Hazard Evaluation Report. Cincinnati, OH: National Institute of Occupational Safety and Health. Report nr HETA 92-171-2255.

NIOSH. 2001. Health Hazard Evaluation Report. Cincinnati, OH: National Institute of Occupational Safety and Health. Report nr HETA 2001-0483-2884.

NIOSH. 2008. Engineering Controls for Tuberculosis: Upper-Room Ultraviolet Germicidal Irradiation Guidelines. Atlanta, GA: Centers for Disease Control.

Philips. 1985. UVGI Catalog and Design Guide. Netherlands: Catalog No. U.D.C. 628.9.

Ritter M, Olberding E, Malinzak R. 2007. Ultraviolet lighting during orthopaedic surgery and the rate of infection. J Bone Joint Surg 89:1935–1940.

Rudnick S. 2001. Predicting the ultraviolet radiation distribution in a room with multilouvered germicidal fixtures. AIHA J 62(4):434–445.

Sanzen L, Carlsson AS, Walder M. 1989. Occlusive clothing and ultraviolet radiation in hip surgery. Acta Orthop Scand 60(6):664–667.

Sylvania. 1981. Sylvania Engineering Bulletin 0-342, Germicidal and Short-Wave Ultraviolet Radiation: GTE Products Corp.

Taylor GJS, Bannister GC, Leeming JP. 1995. Wound disinfection with ultraviolet radiation. J Hosp Infect 30:85–93.

UL. 2004. Luminaires. Research Triangle Park, NC: Underwriters Laboratories. Report nr UL 1598.

UL. 2009. Ultraviolet Radiation Systems for Use in the Ventilation Control of Commercial Cooking Operations. Research Triangle Park, NC: Underwriters Laboratories. Report nr UL Standard 710C (outline).

UNEP. 1994. Environmental Health Criteria 14: Ultraviolet Radiation. New York: United Nations Environment Programme. Report nr UN79.

USACE. 2000. Guidelines on the Design and Operation of HVAC Systems in Disease Isolation Areas. Aberdeen, MD: U.S. Army Center for Health Promotion and Preventive Medicine. Report nr TG252.

USDHHS. 2002. Ultraviolet Radiation Related Exposures: Broad-spectrum ultraviolet (UV) radiation, UVA, UVB, UVC, Solar Radiation, and Exposure to Sunlamps and Sunbeds. Washington, DC: U.S. Department of Health and Human Services. Report nr 10th Report on Carcinogens.

USEPA. 1997. Volume V: Health Effects of Mercury and Mercury Compounds. Mercury Study Report to Congress. Cincinnati, OH: Office of Health and Environmental Criteria, and

Assessment Office, Office of Air Quality Planning and Standards, Research Triangle Park, NC. Report nr EPA 452/R-97-007.

USEPA. 2002. National primary drinking water regulations: Announcement of the result of EPA's review of existing drinking water standards and request for public comment. Federal Regis 67(74).

Wester U. 2000. Analytic expressions to represent the hazard ultraviolet spectrum of ICNIRP and ACGIH. Radiat Protect Dosim 91:231–232.

Westinghouse. 1982. Booklet A-8968, Westinghouse Lighting Handbook: Westinghouse Electric Corp., Lamp Div.

WHO. 1999. Guidelines for the Prevention of Tuberculosis in Health Care Facilities in Resource Limited Settings. Geneva: World Health Organization. Report nr WHO/CDS/TB/99.269.

Chapter 12

UVGI Safety

12.1 Introduction

Ultraviolet germicidal irradiation (UVGI) used for the disinfection of air and surfaces is biocidal to microorganisms but presents a variety of potential health hazards to humans as well, including eye damage, skin burns, and even has the potential to cause skin cancer. Because germicidal UV rays are invisible to the eye, humans may be subject to hazardous doses of UV long before they realize it. Good engineering design, proper procedures, automatic controls, and common sense can all aid in minimizing the hazards from UV exposure. This chapter reviews the nature of the health hazards posed by UVGI, cites the seminal studies that support our knowledge of these hazards, and presents a detailed summary of the existing safety guidelines, standards, and exposure limits that directly or indirectly relate to UVGI. A variety of related issues, including practical engineering approaches to safety for specific UVGI applications are also addressed. Although the health effects discussed in this chapter apply to virtually all UV sources, the focus of this chapter is almost strictly on UVGI sources, and the health effects of other sources of germicidal UV (i.e. sunlight, arc welding, etc.) are not addressed here but such information may be found in the various references cited. Much supportive information on UVGI safety issues can be found in the various safety guidelines and standards and readers should refer to Chap. 11 for a complete list and review of these documents, from which much of the information in this chapter is culled.

12.2 The Germicidal UV Spectrum

Ultraviolet radiation is a subset of that spectrum of electromagnetic radiation known as nonionizing radiation, a category that includes microwaves and radio frequencies. The UV radiation spectrum has been subdivided into three primary bands, UVA (320–400 nm), UVB (280–320), and UVC (100–280). In many current standards, the cutoff between UVA and UVB is at 315 nm, but because this UVB cutoff is currently under debate, because actinic radiation drops off sharply between 315 and 320 nm, and because the 280–320 nm band is also the erythemal region, the newer convention is adopted here in full expectation that it will become the universal standard. Wavelengths above 320 nm (UVA) are not considered very hazardous and these commonly occur in sunlight and are used in tanning beds (Duchene et al. 1991). Wavelengths below 180 nm do not transmit through air (they are absorbed) and are known as vacuum UV (VUV). Therefore, the spectrum of concern for biological safety and health is the range 180–320 nm, which includes UVB and UVC (above 180 nm). Wavelengths below 320 nm produce the most significant adverse health effects and these wavelengths are collectively known as actinic ultraviolet radiation. It should be recognized that these spectrum limits are not necessarily hard and fast limits since there is a certain dose dependency that varies with the light spectrum. That is, some small amounts of UVB can actually be beneficial to health (i.e. from sunlight), and very low levels of UVC may have no long-term effects since human cells can recover from low-level damage.

Units and terminology for health effects of UV on animals and humans are the same as for germicidal effects of UV on microbes. UV exposure is quantified in terms of the irradiance, preferably measured in units of W/m^2. Cumulative UV exposure is quantified in terms of the exposure dose, preferably measured in units of J/m^2. A variety of similar units have also been used to quantify exposure and dose, and these may be readily converted (see Chap. 1 for conversion factors).

Different frequencies of UV have a greater or lesser effect on skin and eyes, relative to some peak value, and the spectral effectiveness can be normalized in the same way that the germicidal effectiveness is normalized (see Fig. 3.1). The entire spectrum of actinic ultraviolet radiation is quantified in terms of the relative spectral effectiveness as shown in Fig. 12.1. The relative spectral effectiveness reaches a peak at about 270 nm.

12.3 Biological Effects of UV Exposure

Numerous investigations have been conducted into the biological effects of exposure to UV radiation. The sun is the principal source of natural UV radiation while artificial sources include all types of UV lamps and certain industrial processes that produce UV as a by-product. Artificial sources of UV radiation can occur in hospitals, laboratories, in industry, and in any building in which UVGI is used to disinfect the air or to sterilize surfaces such as cooling coils or medical equipment. Insignificant levels of UV can be produced by fluorescent lamps but certain incandescent lamps can produce high levels of UV when damaged. The main goal of this section is to address the various biological effects of UV within the context of UVGI applications, and therefore the health issues related to UV exposure from the sun, from sunlamps, and from incidental non-UVGI sources are not specifically addressed here (for non-UVGI sources of UV see, for example, NEHC 1992). However, most of the health issues are the same for UVGI as for non-UVGI sources since they all present hazards to the skin and the eyes, the two main targets of UV

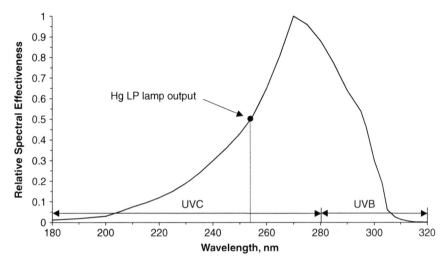

Fig. 12.1 Relative spectral effectiveness (spectral weighting function) for actinic ultraviolet radiation (IRPA 1991). Also shown is the spectral output of a low pressure mercury lamp (254 nm) and the band limits for UVC and UVB

damage. In addition to skin and eye damage, it is possible that tissue damage could result from Overhead UV systems that expose internal tissues and organs during surgery, but in such cases the unpigmented tissues will suffer much the same damage as skin.

12.3.1 UV Effects on Eyes

The most serious effects of UV are those which occur on the eyes. Most ultraviolet light is not detectable by the visual receptors in mammals and ocular damage may occur without the subject being aware of it until long after the exposure when it manifests itself as pain. Excessive exposure to UV can cause photokeratitis and conjunctivitis (or keratoconjunctivitis), and other corneal injuries, potentially including cataracts in the eye lens.

Actinic UV is strongly absorbed by the cornea and conjunctiva. UV-induced photokeratitis is typically followed by conjunctivitis, the combined damage being known as keratoconjunctivitis. Acute keratoconjunctivitis is an inflammation of the cornea and conjunctiva after excessive exposure to actinic ultraviolet radiation (UVB and UVC). This is also known as snow-blindness or welder's flash. The injury can be extremely painful but is usually temporary because of the recuperative powers of the epithelial layer of the eye. The latent period is usually 4–12 h from the time of UV exposure, and the damage is dependent on both the UV dose and the UV spectrum (see Fig. 12.1). It takes at least 8 h for visual incapacitation to become evident and the individual may be visually incapacitated for 48 h (Pitts and Tredici 1971). The latent period varies inversely with the intensity of exposure. Symptoms can include blurred vision, photophobia or sensitization to light, lacrimation or tears, blepharospasm (painful uncontrolled blinking), and a sensation of sand in the eyes. Symptoms, including severe pain, may last from 6 to 24 h and recovery may take up to 48 h (NEHC 1992). Conjunctivitis develops more slowly than photokeratitis and may be accompanied by erythema of the facial skin around the eyelids.

The threshold dose to cause keratoconjunctivitis is about 40 J/m^2 (4 mJ/cm^2) for actinic UV. The peak of the action spectrum for keratoconjunctivitis occurs in the range 265–275 nm with maximum sensitivity occurring at 270 nm. The ocular system, unlike the skin, does not develop tolerance or resistance to repeated ultraviolet exposure. According to IRPA (1991) guidelines, corneal injury from UVA exposure requires levels exceeding 10,000 J/m^2 (1 J/cm^2).

Cataract formation and photodegradation of the eye lens by UV exposure has not been demonstrated in humans and has only been observed in animal studies at extremely high doses (NEHC 1992). Cataract may be any opacification or loss of transparency of the crystalline lens (Cullen 2005). The lens of the eye has approximately the same sensitivity to UV as the cornea, but the cornea tends to filter out UVC wavelengths, thereby reducing the risk of cataracts. Wavelengths above 295 nm can be transmitted through the cornea and are absorbed by the lens (Duchene et al. 1991). Transient and permanent opacities have been produced in animal experiments by exposure to UV in the 295–320 nm range. It is unknown whether chronic exposure can produce lenticular opacity in other than animal models However, the action spectra indicate that UVA and UVB are potentially cataractogenic (Cullen 2005).

Additional UV effects on the eyes include lens fluorescence, pterygium, and climatic droplet keratopathy or CDK (AIHA 2001). UVA penetrates beyond the cornea but its absorption in the aqueous humour and the lens itself produces a harmless transient fluorescence during exposure. Absorption of UVA in the lens likely contributes to progressive yellowing with age and may be a factor in causing cataracts (NRL 1999). Unlike the skin, the ocular system does not develop tolerance to repeated ultraviolet exposure (NIOSH 1972).

Pterygium is a benign fleshy, triangular growth on the conjunctiva that can result from chronic actinic radiation exposure, the source often being solar UVA and UVB. It typically appears at the 3 o'clock or 9 o'clock position. It may spread across and distort the cornea, induce astigmatism, and change the refractive power of the eye. In a two-stage process, conjunctivalisation of the cornea occurs in which tissue is characterized by extensive chronic inflammation, cellular proliferation, connective tissue re-modeling, and angiogenesis (Cullen 2005). Symptoms may include decreased vision, and the sensation of a foreign body in the eye (Taylor 1981). Pterygium is an active, invasive inflammatory process, of which focal limb failure is a key feature (Cullen 2005).

Climatic droplet keratopathy (CDK), also known as spheroid degeneration of the cornea, is a degenerative condition considered to result from exposure to ultraviolet light, and a strong link exists between this disease and reflected solar UV radiation such as from snow or water. CDK was previously described as hyaline degeneration. CDK is caused by an aggregation of proteins modified by advanced glycation end products (AGEs). Etiological findings suggest that ultraviolet radiation and aging, both of which accelerate AGE formation, are closely related to the development of CDK (Kaji et al. 2007). CDK develops slowly, involving the interpalpebral cornea and the conjunctiva, and may produce debris that resembles dots or droplets. These are particles of disintegrated surface cells and whole cells shed from the surface, and they indicate an acceleration of the loss of surface cells (Cullen 2005). CDK is asymptomatic in its early stages, but vision progressively worsens as the changes spread centrally from the periphery. Additional symptoms include mild to severe discomfort or corneal anesthesia.

12.3.2 UV Effects on Skin

Ultraviolet radiation exposure can produce various effects on skin including erythema, photosensitivity, skin aging, immune system damage, and skin cancer. Much of the literature on skin damage from UV exposure relates to solar exposure or tanning equipment. This section attempts to focus primarily on skin damage from germicidal UV lamps.

Erythema is the reddening of skin (i.e. sunburn) after exposure to actinic UV. The threshold dose at which erythema occurs depends on wavelength, anatomical site, and the time since exposure (Duchene et al. 1991). Erythema is a photochemical process in which the skin reddens due to overexposure in the UVB and UVC regions (i.e. about 30 J/m^2 at 270 nm). It can also be produced by UVA but only at very high doses (i.e. greater than $100,000 \text{ J/m}^2$). It has been shown that erythema from UVB is more severe and persists longer than for shorter wavelengths. The increased severity and persistence of the erythema likely results from the deeper penetration of UVB wavelengths into the epidermis. The threshold spectra for the maximum sensitivity of skin to erythema varies from 250 to 297 nm (IRPA 1985). The degree of erythema is most commonly expressed by the Minimum Erythemal Dose (MED).

Erythema occurs in different grades from light to severe, and the respective action spectra can be quite different. The maximum sensitivity for the most severe type of erythema occurs between 290 and 300 nm. Skin pigmentation also impacts the degree of erythema, with high levels of pigmentation offering more protection against UV damage. For untanned, lightly pigmented skin, the MED is about $60-300 \text{ J/cm}^2$ (Parrish et al. 1982). Skin pigmentation and tanning may result in an increase in the minimum erythema dose by an order of magnitude. Figure 12.2 shows the variation of the skin erythema action spectrum based on early studies, or what is known as the 'standard erythemal curve.' It has since been concluded that a true action spectrum for vasodilation cannot be presented in such simplistic fashion and some alternative curves have been developed (NIOSH 1972).

Melanin, the pigment responsible for skin coloration, is present in varying amounts in the epidermis of the skin and acts like a light-blocking filter. The higher the density of melanin, the more ultraviolet wavelengths are absorbed in the outer layers and the greater protection is offered to the deeper layers of skin. The threshold for mild sunburn is about 8 times greater for dark pigmented Negro skin as for

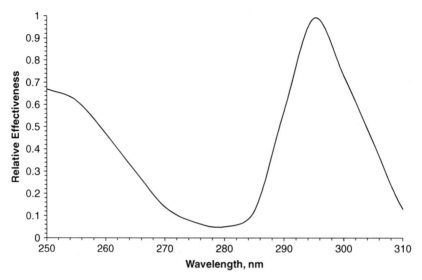

Fig. 12.2 Standard curve for relative erythemal effectiveness. Based on data from Coblentz et al. (1932)

Caucasian skin, and Caucasians are about 120 times more susceptible to severe sunburn (NIOSH 1972).

Mild doses of ultraviolet radiation cause slight thickening of the epidermis, initially due to inter- and intra-cellular edema. After about 72 h, the mitotic rate accelerates and increased cellular production contributes to the thickening of the epidermis. All layers of the epidermis except the basal are thickened and remain so with further stimulation, providing protection against potential damage by ultraviolet radiation. The palms of the hand and the soles of the feet are perfect examples of how thickening of the epidermis renders skin resistant to sunburn and erythema.

Topical or oral application of certain drugs and chemicals can cause the skin to become hypersensitive to ultraviolet light, and lists of such sensitizing agents are available in the literature (Pathak 1969).

The effects of chronic or repeated exposure to ultraviolet light in individuals who have little or no pigmentation or other skin protection include basophilic degeneration of the connective tissue, fragmentation of the elastic tissue, (senile elastosis), and carcinogenesis (NIOSH 1972). Epidemiological and clinical studies have implicated solar ultraviolet radiation as an etiological factor in skin cancer and it is believed that chronic exposure to man-made sources of UV can also produce carcinogenesis (USDHHS 2002). Light-skinned Caucasians are inherently more susceptible to skin cancer and this may prove true in the case of accidental or chronic exposure to UV sources other than sunlight.

Broad-spectrum ultraviolet radiation is known to be a human carcinogen based on sufficient evidence from epidemiological and clinical studies. UVA, UVB, and UVC radiation can reasonably be anticipated to be carcinogenic based on limited evidence of carcinogenicity from studies in humans and sufficient evidence from studies in laboratory animals (USDHHS 2002). Broad-spectrum UV exposure induces skin tumors (papilloma and squamous cell carcinoma) and eye tumors (spindle-cell sarcoma) in albino rats and skin tumors (fibrosarcoma and/or squamous cell carcinoma) in mice, hamsters, and opossums. Broad-spectrum UV is absorbed by DNA and causes direct and indirect DNA damage with the potential to induce mutations, as demonstrated in mechanistic studies on human tissue (AIHA 2001). Broad-spectrum UV causes skin cancers via DNA damage, immunosuppression, tumor promotion, and mutations in the p53 tumor suppression gene.

Molecules that absorb UV contain segments that react with light called chromophores. The main chromophores of the epidermis include DNA, urocanic acid, tryptophan, tyrosine, and the melanins. DNA absorbs broad-spectrum UV (mainly UVB and UVC) and yields intermediates that react with free radicals and various photoproducts with varying mutagenic potential. DNA photoproducts include pyrimidine dimers, pyrimidine-pyrimidone (6-4) photoproducts, thymine glycols, and DNA exhibiting cytosine and purine damage and other damage, such as DNA strand breaks, cross-links, and DNA protein cross-links (IARC 1992).

UVB is considered to be the major cause of skin cancer, even though it does not penetrate as deeply as UVA or react with the outer skin layer as vigorously as UVC. UVB causes the following four major base modifications in human DNA: cyclobutane-type pyrimidine dimers, (6-4) photoproducts and corresponding Dewar isomers, and thymine glycols. Both UVA and UVB induce 8-hydroxydeoxyguanosine production from guanosine by the action of singlet oxygen (Griffiths et al. 1998).

UVC wavelengths are absorbed by DNA and induce mutagenic photoproducts that cause damage similar to UVB, but so far no epidemiological studies have adequately evaluated UVC carcinogenicity in humans. Exposure to high doses of UV radiation from devices emitting UVC have caused skin tumors in rats (keratoacanthoma-like skin tumors) and in mice (squamous-cell carcinoma and fibrosarcoma) (IARC 1992). UVC has the strongest genotoxic effects on DNA and directly damages DNA through base modifications. Mutations of the p53 tumor suppressor gene have been found in over 90% of human squamous-cell carcinomas, mostly from sun-exposed skin. The observed p53 mutations most frequently involved C to T or CC to TT transitions at pyrimidine-pyrimidine sequences (Wikonkal and Brash 1999).

12.4   UV Exposure Limits

Exposure limits (ELs) for UV have been established and are published by various professional organizations and agencies, although the form in which these limits are presented may vary slightly (CIE 2003, IRPA 1991, NIOSH 1972, ACGIH 2004, AIHA 2001). Table 12.1 summarizes the UV Exposure Limits (ELs), the UV wavelength bands, and the Relative Spectral Effectiveness weighting functions. The Relative Spectral Effectiveness is also known as the Spectral Weighting Factor. This table adopts the new convention in which the UVA spectrum begins at 320 nm. See Fig. 12.1 for a plot of the relative spectral effectiveness.

The European Union (EU) regulations for Health and Safety in the workplace have adopted the ICNIRP (2004) limit values, following the general Directive EU-89/391 for the protection of workers (EC 1996). The EU directive covers the whole optical spectrum (ultraviolet, visible, and infrared) and is intended to protect welders, laser operators, steel workers, glass workers, and outdoor workers.

The exposure limits in Table 12.1 apply to both skin and eyes and for general or occupational exposure to UV incident upon the skin or eye within an 8 h period. These EL values may be used to evaluate potentially hazardous exposure from UV from any incoherent sources, which excludes lasers. Most incoherent UV

Table 12.1 UV Exposure Limits and Spectral Weighting Functions

Band	Wavelength nm	Relative spectral effectiveness	Exposure limit, J/m^2	Band	Wavelength	Relative spectral effectiveness	Exposure limit, J/m^2
UVC	180	0.012	2500	UVA	320	0.001	29000
	190	0.019	1600		322	0.00067	45000
	200	0.03	1000		323	0.00054	56000
	205	0.051	590		325	0.0005	60000
	210	0.075	400		328	0.00044	68000
	215	0.095	320		330	0.00041	73000
	220	0.12	250		333	0.00037	81000
	225	0.15	200		335	0.00034	88000
	230	0.19	160		340	0.00028	110000
	235	0.24	130		345	0.00024	130000
	240	0.3	100		350	0.0002	150000
	245	0.36	83		355	0.00016	190000
	250	0.43	70		360	0.00013	230000
	254	0.5	60		365	0.00011	270000
	255	0.52	58		370	0.000093	320000
	260	0.65	46		375	0.000077	390000
	265	0.81	37		380	0.000064	470000
	270	1	30		385	0.000053	570000
	275	0.96	31		390	0.000044	680000
	280	0.88	34		395	0.000036	830000
UVB	285	0.77	39		400	0.00003	1000000
	290	0.64	47				
	295	0.54	56				
	297	0.46	65				
	300	0.3	100				
	303	0.19	250				
	305	0.06	500				
	308	0.026	1200				
	310	0.015	2000				
	313	0.006	5000				
	315	0.003	10000				
	316	0.0024	13000				
	317	0.002	15000				
	318	0.0016	19000				
	319	0.0012	25000				
	320	0.001	29000				

Note: This Table adopts the current convention in which UVB is defined as 280–320 nm and UVA is defined as 320–400 nm. Many sources still define UVB as 280–315 nm and UVA as 315–400 nm. The following subdivisions of UVA have also recently been defined: UVA2: 320–340 nm UVA1: 340–400 nm

sources are broadband, although single emission spectral lines can be produced by low-pressure gas discharges. These limits can be used for continuous UV sources and for pulsed UV sources provided the duration (or pulse width) is not less than 1 μs. These ELs do not apply to UV exposures that are part of medical treatment or cosmetic purposes.

The values in Table 12.1 should be considered absolute limits for direct exposure of the eye and advisory for skin exposure because of the wide range of susceptibility to skin injury depending on skin type. It should be cautioned that although this single set of exposure limits can apply for exposure of the eyes, there are different skin phototypes to which these limits may not apply equally and additional guidance may be required to apply these guidelines for skin protection in some cases (ICNIRP 2004). However, until such time as new research amends or revises these exposure limits, they will be considered to apply to all skin types in general, and should be adequate to protect lightly pigmented individuals. Juchem et al. (1998) provides a breakdown of UVB exposure limits by skin phototypes (I = White, VI = Black), including a recommended SPF for each, but the limits in Table 12.1 are conservative for all skin types. See Juchem et al. (1998) for subdivisions of skin phototypes.

The ELs in Table 12.1 should be applied to exposure directed perpendicular to those surfaces of the body facing the radiation source, measured with any device having a cosine angular response (see Chap. 13). When the irradiance field is highly non-uniform, the UV irradiance and UV dose can be averaged over a circular aperture no smaller than 3.5 mm in diameter for continuous UV sources, and no smaller than 1 mm in diameter for pulsed UV sources.

For selected wavelengths, the permissible dose levels may be read directly from Table 12.1. If the ultraviolet radiation is from a narrow-band or monochromatic UV source, the permissible dose levels for a daily 8 h period can be read directly from Fig. 12.3. Note that for narrow-band monochromatic UV at 254 nm (which is the most common type of UV lamp) the relative spectral effectiveness from Table 12.1 is 0.5, or exactly one-half that at 270 nm.

Ultraviolet radiant exposure in the spectral region between 180 and 400 nm incident upon the unprotected eye(s) should not exceed 30 J/m^2 (3000 μJ/cm^2), spectrally weighted using the weighting factors in Table 12.1, and the total (unweighted) ultraviolet radiant exposure in the region 315–400 nm should not exceed 10,000 J/m^2. For monochromatic UVC (254 nm) this limit is 60 J/m^2.

Ultraviolet radiant exposure in the spectral region between 180 and 400 nm incident upon the unprotected skin should not exceed 30 J/m^2 (3000 μJ/cm^2), spectrally weighted using the weighting factors in Table 12.1. This limit applies to the most sensitive, non-pathologic, skin phototypes (known as melano-compromised). This limit has a substantial safety factor for dark skin phototypes (melano-competent) and for individuals who have been previously conditioned by repeated exposures (melano-adapted). For monochromatic UVC (254 nm) this limit is 60 J/m^2.

In addition, the unweighted UVA exposure dose should not exceed a daily exposure limit value of 10,000 J/m^2 (Weiringa and Vermeulen 2005).

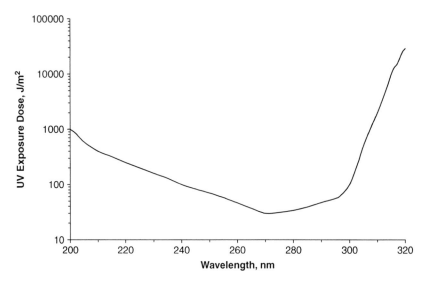

Fig. 12.3 Recommended ultraviolet radiation exposure standard. Figure is based on exposure limit data from ACGIH (2004)

The effective irradiance of a broadband UV source weighted in terms of the peak (270 nm) of the spectral effectiveness curve (see Fig. 12.1) can be determined using the following weighting formula:

$$E_{\text{eff}} = \sum E_\lambda \cdot S(\lambda) \cdot \Delta\lambda \tag{12.1}$$

where

E_{eff} = effective irradiance normalized to a monochromatic source at 270 nm, W/m^2

E_λ = spectral irradiance from measurements, W/m^2nm

$S(\lambda)$ = relative spectral effectiveness, dimensionless

$\Delta\lambda$ = bandwidth of the calculation or measurement intervals, nm

If we used normalized values of the spectral effectiveness and a normalized lamp spectrum in Eq. (12.1), we can obtain a relative indicator of the lamp safety on a per watt basis. Using the example spectrums from Fig. 2.1, we find that the LP lamp has a relative spectral effectiveness of 0.53 while for the MP lamp it is 0.73 – it would seem LP lamps are less hazardous to skin and eyes on a per watt basis, ignoring the factor of germicidal effectiveness.

The permissible exposure time in seconds, for both eyes and skin, may be computed by dividing 30 J/m^2 by the value of the effective irradiance, E_{eff}, in W/m^2. In equation form, the maximum exposure time, t_{max}, in seconds is:

$$t_{max} = \frac{30}{E_{eff}} \qquad (12.2)$$

The limiting exposure may also be found from Table 12.2, which provides representative exposure durations corresponding to the effective irradiances in W/m^2, for both total (broadband) UV and the single-wavelength case of 254 nm. If the irradiance from a point source is known at some specific distance from an individual, the attenuation of the UV radiation from that point to the person can be estimated from the inverse square law from illumination engineering (IESNA 2000) that states the radiation intensity decreases with the square of the distance. Except for near-field distances and far-field distances, the inverse square law provides and adequate estimate of irradiance.

Figure 12.4 plots the values of Duration (in seconds) vs. the Effective Irradiance on a log-log scale (natural log) for total UV. Wester (2000) provides interpolation functions that can be applied between 210 and 400 nm to determine the relative spectral effectiveness as follows:

$$S(\lambda) = 0.959^{(270-\lambda)} \qquad \text{for } 210 - 270 \text{ nm} \qquad (12.3)$$

Table 12.2 NIOSH/ACGIH maximum UV exposure times

UV (broadband)			Duration of exposure	Monochromatic UVC (254 nm)		
Measured irradiance				Measured irradiance		
$\mu W/cm^2$	W/m^2	mW/cm^2		$\mu W/cm^2$	W/m^2	mW/cm^2
0.1	0.001	0.0001	8 h	0.2	0.002	0.0002
0.13	0.0013	0.00013	6 h	0.26	0.003	0.00026
0.2	0.002	0.0002	4 h	0.4	0.004	0.0004
0.4	0.004	0.0004	2 h	0.8	0.008	0.0008
0.8	0.008	0.0008	1 h	1.6	0.016	0.0016
1.7	0.017	0.0017	30 min	3.4	0.034	0.0034
3.3	0.033	0.0033	15 min	6.6	0.066	0.0066
5	0.05	0.005	10 min	10	0.1	0.01
10	0.1	0.01	5 min	20	0.2	0.02
30	0.3	0.03	100 s	60	0.6	0.06
50	0.5	0.05	60 s	100	1	0.1
100	1	0.1	30 s	200	2	0.2
300	3	0.3	10 s	600	6	0.6
1000	10	1	3 s	2000	20	2
3000	30	3	1 s	6000	60	6
6000	60	6	0.5 s	12000	120	12
10000	100	10	0.3 s	20000	200	20
30000	300	30	0.1 s	60000	600	60

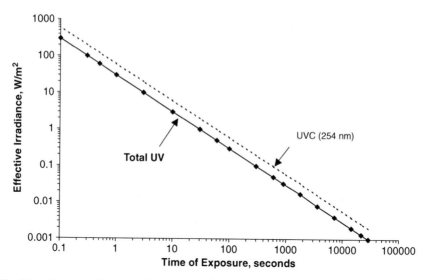

Fig. 12.4 Maximum Exposure Times vs. Effective Irradiance, log-log plot based on values in Table 12.3

$$S(\lambda) = 1 - 0.36 \cdot \left(\frac{\lambda - 270}{20}\right) \qquad \text{for } 270 - 300 \text{ nm} \qquad (12.4)$$

$$S(\lambda) = 0.3 \cdot 0.736^{(\lambda - 300)} + 10^{(2 - 0.0163\lambda)} \qquad \text{for } 300 - 400 \text{ nm} \qquad (12.5)$$

In general, the UV exposure is typically measured as broadband actinic UV (180–320 nm) without subdividing into individual wavelengths. This is a simple and conservative approach, provided the limit of 30 J/m^2 is not exceeded. If the UV dose limit is exceeded by this approach, then the measurements may have to be divided into the individual spectral bands of Table 12.1. In the event exposure to UV occurs piecemeal, or for multiple periods throughout the day, or at different locations and at different irradiance levels, the total exposure dose is computed as follows:

$$E_D = I_{R1}T_1 + I_{R2}T_2 + I_{R3}T_3 + \dots \qquad (12.6)$$

where

E_D = UV exposure dose, J/m^2
I_{Rn} = Irradiance at location or time n, W/m^2
T_n = time of exposure at location or time n

The total computed exposure dose from Eq. (12.6) would then be compared with the dose limit (30 J/m^2) for broadband exposure. If necessary, the exposure dose may have to be divided up by wavelengths per Table 12.1.

12.5 UV Measurement

Spectroradiometric measurements of indoor UV irradiance may be necessary to establish whether UV levels are within exposure limits. Various UV radiometers are available, although field safety survey meters that respond directly to UVB/UVC radiation following the $S(\lambda)$ function are reportedly not very effective. Measurements using standard UV radiometers may have to be gradated along spectral increments and then adjusted using the $S(\lambda)$ functions to calculate the effective irradiance E_{eff}. UV film badges are also available that integrate exposure of body sites that may move relative to the UV source but the spectral response of such film badges has not yet been shown to follow the $S(\lambda)$ function (Saunders and Diffey 1995).

The primary instruments used to measure UV irradiance are broadband radiometers and spectroradiometers. Measurements can also be made with photoelectric detectors, and with thermal detectors such as thermopiles and pyroelectric detectors. Radiometers are used to measure radiant power and have calibrated output in irradiance. These types of meters often have an integration function that permits radiant exposure to be measured over time. Detectors used for broadband measurements should display a wavelength-independent (flat) response in the spectral region of interest (UVB/UVC) and have a rapid response decrease outside the region of interest. Radiometers are generally smaller and easier to use than spectroradiometers but their accuracy depends on the proper selection of the spectral filters used. Radiometer optical lenses are susceptible to contamination by dust and fingerprints, and must be calibrated on a regular basis (NIOSH 2009).

Spectroradiometers measure UV irradiance in narrow wavelength bands. They normally have wavelength-specific calibration factors to determine the actual spectral irradiance. Spectroradiometers are often larger and more complicated to operate than radiometers but they may also provide a better estimate of UV irradiance.

The geometry of measurement is important. The UV exposure limits of Table 12.1 apply to exposures that are measured with an instrument having a cosine-response detector oriented perpendicularly to the most directly exposed surfaces of the body when assessing skin exposure (ICNIRP 2004). The detector should be oriented along, or parallel to, the line-of-sight of each exposed individual when assessing UV exposure of the eyes. Care should be taken to control any reflective surfaces that may interfere with UV measurements. See Chap. 13 for more detailed information on UV radiometers and measurement techniques.

The UVGI spectra of interest for measurement purposes correspond to the previously discussed actinic UVC/UVB regions. The UVB spectral band from 280 to 320 nm is referred to as the erythemal region with the range 295–298 having a peak effect. The main germicidal band, 220–280 nm, overlaps the erythemal region and has a maximum germicidal wavelength at 265 nm. There is also some erythemal effect in the region 250–260 nm. The remaining band, 170–220 nm is called the ozone region, or ozone-producing region, but since oxygen has a large absorption coefficient in this region it is of limited biological significance, being effectively absorbed over very short distances in air. Therefore, any spectral filter that removes

UV above 320 nm and below 220 nm is perfectly acceptable for use in measurements related to UV safety.

12.6 UV Protection and Control Measures

There are basically three types of protective measures that can be used to protect building occupants against the hazards associated with UV equipment, engineering controls, operational protocols, and clothing or personal protective equipment (PPE). Since it is assumed that the UV hazard comes from UVGI equipment for disinfecting air or surfaces, the topic of UV exposure from other sources (i.e. sunlight, welder's arcs etc.) is not addressed here and readers can refer to the various safety guidelines (in Chap. 11) for more information on these matters. The hazards posed by indoor equipment for air and surface disinfection depend greatly on the type of equipment used. A well-designed UVGI air disinfection system installed in a duct should produce no undue exposure hazards, although maintenance workers should take the necessary precautions. An Upper Room UVGI system, on the other hand, may pose a continuous low-level hazard that needs to be subject to periodic monitoring and control measures. These subjects and others are addressed separately in the following sections.

12.6.1 In-duct UVGI System Safety

In general, UVGI systems installed inside enclosed ducts should pose no hazard to building occupants provided they are designed in accordance with good practice and common sense. UVGI systems installed in air handling units (AHUs) have little or no chance of producing stray UV that may escape the AHUs provided there are no holes or gaps by which UV may escape. It is a simple test to turn out the lights while the AHU is operating and observe if any blue (violet) light is visibly leaking from any point around the AHU (other than the windows). Any holes found should be sealed. Upstream of any in-duct UVGI there will virtually always be an air filter, and most filters (minimum MERV 6-8 is recommended by various sources) will block UV light from penetrating upstream.

Downstream of the UVGI system there will typically be a fan, a 90° turn, and a supply duct extending for some distance, all of which contribute to reducing UV levels. UV irradiance tends to drop off roughly exponentially with distance from the source and each bend in the duct will tend to reduce the levels by at least one-half (assuming galvanized steel surfaces) and so the UV levels at the nearest supply air register are often negligible. In any event, the nearest supply and exhaust registers should be surveyed to verify that no hazardous levels of UV are exiting. Typically this means an 8 h exposure limit, or 0.001 W/m^2. If measured levels should exceed this value some modifications may be necessary, such as painting the ductwork black (4% UV reflectivity or 96% absorbance of UV) or installing UV light baffles.

Guidelines for in-duct UVGI air disinfection generally recommend that cutoff switches be installed to shut down any UV lamps if the nearest duct access door is opened. Alternate controls, such as motion detectors will also provide protection for maintenance workers. Procedural controls should also be in place that require systems to be powered down prior to performing any maintenance or inspection activities. A related concern is UV photodegradation, since any damage to organic or plastic components may result in damage to the air handling systems, and any UV damage to wiring or electrical components may result in a fire hazard. For more on photodegradation, see Chap. 15.

12.6.2 Unitary UVGI System Safety

When portable air recirculation units, or unitary UVGI systems, are placed inside rooms for additional air cleaning, it is generally assumed that they do not leak any stray UV radiation, by design. The design principles which will prevent stray UV rays from exiting such portable units are the same as those discussed above for in-duct systems. That is, there should be no line-of-sight exit points for UV from either the supply or exhaust registers. Often, such equipment has short travel distance from the UV lamps to the registers and this necessitates the use of light blocks or baffles to prevent UV from escaping at hazardous levels.

There are currently no standards that prevent equipment from being sold on the market that produce hazardous levels of stray UV, and so it is up to the end user to assure that levels around the portable UVGI units are within acceptable exposure limits. Surveys should be conducted around the unitary UVGI system at distances that occupants might be expected to be present. The risk of eye exposure should be considered paramount when such units are used, and, if necessary, such units could be positioned such that any UV leakage is directed to a surface (i.e. a wall) to provide room occupants with minimum exposure.

12.6.3 Upper Room UVGI Safety

Upper Room UVGI systems, or Upper Air UVGI systems, invariably produce low levels of stray UV radiation. These levels are typically kept at a minimum by design and surveys will be conducted upon installation to verify the levels are below exposure limits. A number of guidelines exist (see Chap. 11) that specifically address Upper Room UVGI (CDC 2005, NIOSH 2009, USACE 2000, CIE 2003, First et al. 1999, 2005). Irradiance measurements will typically be made to ensure that sufficient UV levels exist in the upper room to inactivate microbes, and separate readings will be taken in the occupied areas of the room to ensure that the system complies with NIOSH/ACGIH guidelines. Various measuring instruments may be used to conduct such surveys, including broadband radiometers, broadband spectroradiometers, and chemical actinometers. It is appropriate to take measurements at occupants' eye level and compare results with the NIOSH REL. Measurements should be taken again any time UV lamps are changed out.

The recommended TLV for an 8-hour exposure at the wavelength of 254 nm is 60 J/m^2 (or 30 J/m^2 at 270 nm or for broadband UV) per ACGIH. In current practice, designers attempt to limit the irradiance in the lower 6.5 ft of a room to no more than 20 W/m^2 (0.2 μW/cm^2), which ensures that anyone staring at the nearest emitting source for 8 h will not exceed the TLV dose to the eyes. In a study of various Upper Room installations by First et al. (2005) it was shown that the actual measured 8-hour dose at eye level was far lower than the TLV, typically less than a few percent in many areas. They suggest that a maximum eye-level irradiance of 20–40 W/m^2 (0.2–0.4 μW/cm^2) will allow safe use of higher levels of irradiance in the Upper Room zone without causing undue hazards to occupants.

Upper Room UVGI units are typically equipped with parallel blades (louvers) to direct the UV rays horizontally and away from the lower room areas. These louvers should be of sufficient width and have a sufficiently narrow exit height that no direct UV rays can be visible at normal occupant head height (standing height), or else they should be angled upwards such that only rays diminished by ceiling reflectivity could reach the lower room. In large rooms, it is possible for the UV lamp to be visible to occupants at a distance even through louvers. There are no formal mechanical guidelines yet that address such design issues, but if the UV lamp is visible to occupants, even at a distance, then there is a possibility that uninformed occupants may stare at the blue light without realizing the hazard. Certainly the UV levels should be established as being within NIOSH exposure limits, but if direct UV rays, however minimal the irradiance, cannot be blocked from direct line-of-sight, then warning signs (i.e. 'Do Not Stare At Light') may be placed on the units as a precautionary measure. For additional information on safety experience with Upper Room UVGI see Nardell et al. (2008) and First et al. (2005).

The UV reflectance hazards for Upper Room (and Lower Room) systems should not be ignored and care should be taken to ensure there are no highly UV-reflective materials in locations that may direct reflected UV rays towards the eyes or skin. Turner and Parisi (2009) show the results of studies of reflected UV from vertical surfaces on the human face – such reflections can bypass the natural defenses the eyes have against overhead light sources. Appendix F lists the UV reflectivities for a number of materials, including paints and construction materials that might be found in indoor environments. The effects of the UV reflectivity of metal materials on the human face has been studied in some detail (in relation to solar UV) by Turner and Parisi (2009), who found significant differences between specular and diffusely reflective materials, and whose methods may be useful for indoor UV studies.

12.6.4 Lower Room UVGI Safety

Lower Room UVGI systems have many of the same concerns as Upper Room units, although the hazards are somewhat lessened by the fact that they sit nearer the floor.

As with Upper Room units, Lower Room UVGI units should have louvers to prevent UV rays from reaching head height. A prudent design approach is to angle the louvers such that UV rays are directed about $10°–30°$ downwards to some target point on the floor. Whether the rays exit horizontally or towards the floor, there is minimal potential for UV to strike the eyes unless someone is sitting or laying on the floor. Appropriate warning signs may be placed on Lower Room units to guard against anyone looking into the UV light source. Even with downward-angled louvers, it is possible for Lower Room units to cause exposure to legs and feet and it may be necessary to restrict access to those with exposed legs or require leg coverings. This might be the case, for example, in hospital hallways or operating rooms, where full leg coverage might be required if irradiance levels were too high by design.

A alternate design approach for Lower Room units, however, is to limit the irradiance level at the floor or near the unit to the 8 h NIOSH REL (or ACGIH TLV) of 0.001 W/m^2 (broadband UV) or 0.002 W/m^2 (254 nm). In this situation there would be no specific requirement for leg coverage provided the occupants remained in these areas for no more than 8 h a day.

Appropriate procedures must be in place for maintenance workers who may stoop down to check or replace lamps in Lower Room units. Such procedures would normally require goggles of eye protection, and suitable full-body coverage if the unit irradiance levels were being measured.

12.6.5 Cooling Coil UVGI Safety

UVGI systems installed to disinfect cooling coils present essentially the same safety concerns as in-duct air disinfection systems, and the health hazards are minimal. Since cooling coil UVGI systems are essential identical in form to in-duct systems, all the recommendations of Sect. 12.6.1 apply, including control interlocks, cutoff switches, placement of warning signs, and the performance of surveys to ensure that no UV radiation is leaks from the duct housing.

12.6.6 Equipment Disinfection UVGI Safety

Many types of UVGI systems are used for equipment disinfection, including medical equipment disinfection, pharmaceutical disinfection, packaging disinfection, mail disinfection, etc. Such systems must be safe by design and although there are currently no standards or regulations for such systems, the same safety guidelines apply for these as for other UVGI systems (i.e. ACGIH 2004, NIOSH 1972, ICNIRP 2004). Cabinets employing internal UV lamps should be light-tight and make use of UV absorbing glass or plastic windows. At the least, such equipment should contain shutoff switches such that the UV lamps de-energize if the access doors are opened. In general, the manufacturers provide operation manuals with such equipment that specify safe operating practices and users should abide by these procedures.

12.6.7 Overhead Surgical Site System Safety

Overhead UVGI systems for use during surgical procedures in operating rooms pose
safety problems that need to be addressed through a combination of engineering
design, procedures, and personal protective equipment (PPE). There are clearly certain hazards from using naked UV lamps, or 'open air UV', but in the case of surgical site disinfection, the benefits can greatly outweigh the risks (see Chap. 16). The
current design approach is to create a horizontal irradiance field of 0.25–0.3 W/m^2
(25–30 μW/cm^2) at operating table height (about 1 m or 3 ft above the floor). This
level of irradiance implies an exposure limit of about 100 s (see Table 12.2) but doctors and nurses will typical have full body coverage. There may be gaps however, in
the face visors and clothing, and UV may reflect upwards from horizontal surfaces,
and so some consideration should be given to the level of reflected UV in operating
rooms. Often such reflected levels will be minimal due to a combination of low UV
reflectivity and the distances from the source.

Operating room personnel who employ overhead UVGI systems wear goggles,
eye shades, or visors, and use appropriate protective clothing to completely cover
their skin, including hoods to protect the neck, ears, and forehead. Eye protection
equipment should meet NIOSH recommendation HSM-73-11009 for protection
against UV (NIOSH 1972). Personnel also may apply a benzophenone-containing
lotion or other sunscreen to foreheads, cheeks, under surfaces of the chin and back
of the neck if these areas are exposed, or may become exposed due to activities. A
number 15 sunblock is suggested, but personnel with ultrasensitive skin should use
a sunblock with a rating of at least 25. Not every piece of clothing will necessarily
provide adequate protection from extended UV exposure (see the forthcoming
section on Personal Protective Equipment). Patients must be completely covered,
except around the surgical site, and should have eye protection or have eye ointment
applied by the anesthetist (Young 1991, Lowell and Kundsin 1980).

Proper training must be given to all personnel working in any operating room
(OR) equipped with an overhead UVGI systems. They must have some understanding of the eye and skin hazards associated with the levels of UV in the OR, must
be familiar with the NIOSH or ACGIH exposure limits, and should understand the
degree of protection afforded by the clothing and visors they use. Technicians and
maintenance workers must also be aware of the hazards so that they do not enter
an OR during surgery without PPEs and/or do not engage the UVGI system accidentally. It may be prudent to use UV lamps of the shatterproof variety in overhead
applications, although no problems have been reported in the past at the various
hospitals where such systems are in use. For more detailed information on safe
operation of Overhead UVGI systems, see Chap. 17.

12.6.8 Area Decontamination UVGI Safety

UVGI systems used for area or room decontamination are of two types – portable
and permanent. Portable UVGI systems that are placed in areas to disinfect them

will typically generate high levels of UV and these areas must be unoccupied during the disinfection process, except when appropriately attired personnel must enter during the procedure. Permanently installed UVGI systems for area disinfection are designed to continuously irradiate a room and will typically be used only when the room is unoccupied (i.e. After hours UVGI systems). Such equipment may contain interlocks or motion detectors to shut down when anyone enters the room, but personnel must still be trained to understand the hazards. Talbot et al. (2002) report on a case in which employees were unknowingly sent to work in a room with a UV area disinfection unit that had a wall switch but no labeling, resulting in eye and skin injuries to several workers. Some types of area disinfection equipment may be designed to produce levels of UV irradiance below ACGIH TLVs (i.e. the 8 h limit of 0.001 W/m^2), and such systems do not require shutoff switches or PPEs, but they do require personnel to be trained to understand the hazards.

12.7 Personal Protective Equipment (PPE)

When personnel are subject to UV exposure, personal protective equipment (PPE) should be used, and these include protective clothing, goggles, glasses, or other eye protection, gloves, hoods, and UV-proof sunblock lotions. Protective eyewear is typically designed to greatly reduce or completely prevent UV rays from reaching the eyes (NEHC 1992). Glass or plastic of at least 1/8 in. thickness provides adequate attenuation of UV rays. Optical density is a variable that determines the attenuation of eyewear, and it is a logarithmic function of the incident radiation vs. the transmitted radiation as follows:

$$OD = \log_{10}\left(\frac{E_0}{E}\right) \qquad (12.7)$$

where

E_0 = incident irradiance, W/m^2
E = transmitted irradiance, W/m^2

UV-blocking eyewear should be fitted with UV-blocking side shields to minimize the possibility of reflections hitting the eyes. Many types of eye protection, such as goggles, contain perforations for ventilation but may not provide complete protection against UV. Eyewear for UV protection should meet minimum standards such as CSA Standard Z94.3-07 (CSA 2007), which addresses transmittance requirements and test methods, ANSI Z87.1-2003 (ANSI 2003), or Sec. 4 of ANSI Z49.1 (ANSI 2005). Depending on the UV hazard, these documents recommend either Class 2C goggles or a Class 3 Welding Helmet. UV-blocking contact lenses are currently being developed, but there may be limits to their effectiveness (Cullen 2005).

The optical density in Eq. 12.8 can also be used to quantify the degree of protection afford by clothing and topical screening. Normal work clothing provides

Table 12.3 UV transmittance of fabrics

Material	Transmittance, %
White Muslin (Batiste)	50
Cotton voile	37–43
Kapron	31
Crepe de China (light grey)	32.5
Kapron and Nylon	26.6
Nylon	25–27
Silk stockings	25
Cotton stockings	18
Stockinet	14–16.5
Linen, coarse white	12
Rayon stockings	10.5
Satin, beige	10
Linen cambric	8–9.5
Rayon (linen type)	3.8–5.3
Wool stockinet	1.4–2.8
Flannelette	0.3
Poplin	0

adequate attenuation of UV radiation but some man-made materials (i.e. nylon) may transmit significant amounts of UV. Clothing of densely woven flannelette, poplin, or synthetic fabric will provide sufficient protection. Aluminized fabrics should be avoided as these may create unpredictable reflective hazards. Table 12.3 summarizes the UV transmittance of various fabrics based on NIOSH (1972).

Topical screening lotions can provide partial or full protection of skin against UV exposure. Most of these sunscreens concentrate on filtering out actinic wavelengths below 320 nm. Standard commercial sunscreens permit some UVA to be transmitted. Physical sunscreens are opaque and absorb UV rays. Absorbing skin creams typically contain benzophenones or p-aminobenzoic acid. Chemical sunscreens are non-opaque and reflect UV rays. These barrier creams typically contain titanium dioxide or zinc oxide. Sunscreens are classified according to the Sun Protection Factor (SPF), an index of protection against skin erythema. SPF ranges from 1 to 45 or above and quantifies UVB protection (Patel et al. 1992). A number 15 sunblock, the minimum recommended for UVGI hazards, filters 92% of UVB.

12.8 Labeling

Labels and warning signs should be used wherever UVGI systems are used, whether they are accessible or not. NIOSH (1972) provides the following recommendation for labeling: All sources, work areas, equipment, and housings that contain UV lamps or other UV sources should carry labels or signs with a warning as shown in Fig. 12.5. Figure 12.6 shows two warning signs are suggested by NEHC (1992).

Fig. 12.5 NIOSH recommended warning sign for UV hazards

> **CAUTION**
>
> **HIGH INTENSITY ULTRAVIOLET ENERGY**
>
> **PROTECT EYES AND SKIN**

Fig. 12.6 Examples of UV hazard warning signs suggested by NEHC

> **CAUTION**
>
> **EYE HAZARD**
>
> **Do Not Stare Into Light**

> **CAUTION**
>
> **ULTRAVIOLET SOURCE**
>
> **Eye and Skin Hazard**
>
> **Authorized Operators Only**

Fig. 12.7 Examples of UV warning signs used for areas with UVGI systems or with UV hazards

Various other warning signs and labels that carry similar warnings are used throughout the ultraviolet industry. Figure 12.7 shows some examples of warning signs used in other facilities. The Johns Hopkins Safety Manual (JHU 2003) provides the warning sign shown in Fig. 12.7 (second from left) and recommends its use for all rooms or areas accessible to faculty or students in which there is a potential for exposure to ultraviolet light above the NIOSH RELS for occupational exposure. Figure 12.8 shows one example of a UV warning sign used on the access door to an air handling unit with an installed in-duct UVGI system.

12.9 Ozone Generation Hazards

UVC radiation below wavelengths of 240 nm interacts with oxygen in the air and may form both ozone or oxides of nitrogen. In general the levels produced are low, the concentrations dilute, and these byproducts of the UVGI process tend to break down or dissipate rapidly. Ozone has a half-life of about 15 min in air. It is possible, however, for unitary UVGI systems to produce local levels of ozone (i.e. in a room) that pose a long term hazard to sensitive or susceptible individuals. The WHO sets

Fig. 12.8 Example of hazard warning signage for UVGI system installed in an air handling unit. Photo courtesy Immune Building Systems, Inc., New York

a maximum limit of 0.1 ppm for 1 h of exposure while the OSHA TLV for ozone is 0.1 ppm for 8 h exposure (ACGIH 2004).

Ozone produced locally around UV lamps may also be corrosive to certain materials (i.e. rubber) and cause them to degrade over time, possibly creating fire hazards or other dangers. It is a prudent measure to check local ozone levels after placing a UVGI system into service to ensure no hazards are created. Ozone sensors can be used for this purpose, but many people are capable of smelling ozone at or below the 0.1 ppm level, so a 'sniff test' may be used as a precursor for ozone measurements with a sensor. The presence of pollutants indoors may also result in certain hazardous by-products of ozone. See Kowalski (2006) for a list of indoor VOCs, precursors, and ozone byproducts.

12.10 Lamp Breakage and Mercury Risks

Mercury is a heavy metal and poses potential toxic hazards when ingested, including cellular necrosis. This is primarily a hazard in water systems but the breakage of UV lamps in air and surface disinfection can lead to both ingestion hazards if it should enter a water supply and inhalation hazards if it should evaporate. Mercury is rapidly absorbed through the lungs and approximately 75–80% of inhaled mercury is absorbed into the body (Borchers et al. 2008). The rate of dermal absorption, through the skin, is low and is estimated to be about 3%. The rate of absorption through the gastrointestinal tract is even lower, being about 0.01%. Clearly, the inhalation route poses the greatest risk. When a lamp breaks while it is not in use, the mercury is in liquid phase. If a lamp breaks during use, it will largely be in the vapor phase due to the high temperatures.

The USEPA has established a estimated limit of mercury intake called the Reference Dose (RfD) of '0.0003 mg/kg/day' (USEPA 1997, 2002). A typical medium pressure mercury vapor lamp will contain 400 mg or less of elemental mercury (Borchers et al. 2008). Although the hazard from ingestion and inhalation is relatively remote, operators and maintenance personnel should be educated to be aware of the problem and trained to clean up the mercury and dispose of it safely.

References

ACGIH. 2004. Threshold Limit Values and Biological Exposure Indices. Cincinnati, OH: American Conference of Governmental Industrial Hygienists.

AIHA. 2001. Nonionizing Radiation Guide Series, Ultraviolet Radiation. Akron, OH: American Industrial Hygiene Association. Report nr ISBN 0-932627-08-8.

ANSI. 2003. American National Standard Practice for Occupational and Educational Personal Eye and Face Protective Devices. New York: American National Standards Institute. Report nr ANSI Z87.1-2003.

ANSI. 2005. Safety in Welding, Cutting, and Allied Processes. New York: American National Standards Institute. Report nr ANSI Z49.1-2005.

Borchers H, Fuller A, Malley J. 2008. Assessing the risk of mercury from on-line lamp breaks. IUVA News 10(1):9–13.

CDC. 2005. Guidelines for preventing the transmission of *Mycobacterium tuberculosis* in health-care facilities. In: CDC, editor. Federal Register. Washington: US Govt. Printing Office.

CIE. 2003. Ultraviolet Air Disinfection. Vienna, Austria: International Commission on Illumination. Report nr CIE 155:2003.

Coblentz WW, Stair R, Hogue JM. 1932. Spectral erythemic reaction of the untanned human skin to ultraviolet radiation. J Res Nat Bur Stand 8:541–547.

CSA. 2007. Eye and Face Protectors. Mississauga, Ontario: Canadian Standards Association. Report nr CSA Standard Z94.3–07.

Cullen AP. 2005. UV radiation: Contact lenses and the opthalmohelioses. OT June 17:30–34.

Duchene AS, Lakey JRA, Repacholi MH. 1991. IRPA Guidelines on Protection Against Non-ionizing Radiation. New York: Pergamon Press.

EC. 1996. Non-ionizing radiation. Sources, exposure and health effects. Brussels: European Commission. Report nr CE-96-96-934-EN-C.

First MW, Nardell EA, Chaisson W, Riley R. 1999. Guidelines for the application of upper-room ultraviolet germicidal irradiation for preventing transmission of airborne contagion – Part II: Design and operational guidance. ASHRAE J 105:869–876.

First MW, Weker RA, Yasui S, Nardell EA. 2005. Monitoring human exposures to upper-room germicidal ultraviolet irradiation. J Occup Environ Hyg 2(5):285–292.

Griffiths HR, Mistry P, Herbert KE, Lunec J. 1998. Molecular and cellular effects of ultraviolet light-induced genotoxicity. Crit Rev Clin Lab Sci 35:189–237.

IARC. 1992. Solar and Ultraviolet Radiation, Volume 55. Cancer IAfRo, editor. Lyon, France: IARC.

ICNIRP. 2004. Guidelines on Limits of Exposure to Ultraviolet Radiation of Wavelengths Between 180 nm and 400 nm (Incoherent Optical Radiation). Health Phys 87(2):171–186.

IESNA. 2000. IESNA Lighting Handbook HB-9-2000, Reference and Application Chapter 6, Light Sources. New York: Illumination Engineering Society of North America.

IRPA. 1985. IRPA Guidelines on Limits of Exposure to Ultraviolet Radiation of Wavelengths Between 180 nm and 400 nm (Incoherent Optical Radiation). Health Phys 49:331–340.

IRPA. 1991. IRPA Guidelines on Protection Against Non-ionizing Radiation. Madrid: International Radiation Protection Association. Report nr Chapter 3.

JHU. 2003. Johns Hopkins Safety Manual. Baltimore, MD: Johns Hopkins University. Report nr HSE 803.

Juchem PP, Hochberg J, Winogron A, Ardenghy M, English R. 1998. Health risks of ultraviolet radiation. SBCP 13(2):31–60.

Kaji Y, Nagai R, Amano S, Takazawa Y, Fukayama M, Oshika T. 2007. Advanaced glycation end product deposits in climatic droplet keratopathy. BJO 91:85–88.

Kowalski WJ. 2006. Aerobiological Engineering Handbook: A Guide to Airborne Disease Control Technologies. New York: McGraw-Hill.

Lowell J, Kundsin R. 1980. Ultraviolet Radiation: Its Beneficial Effect on the Operating Room Environment and the Incidence of Deep Wound Infection Following Total Hip and Total Knee Arthroplasty. Murray Hill, NJ: American Ultraviolet Company. Report nr A-810.

Nardell EA, Bucher SJ, Brickner PW, Wang C, Vincent RL, Began-McBride K, James MA, Michael M, Wright JD. 2008. Safety of Upper Room Ultraviolet Germicidal Air Disinfection for Room Occupants: Results from the Tuberculosis Ultraviolet Shelter Study. Pub Health Rep 123:5260 .

NEHC. 1992. Ultraviolet Radiation Guide. Norfolk, VA: Navy Environmental Health Center, Bureau of Medicine and Surgery. Report nr Technical Manual NEHC-TM92-5.

NIOSH. 1972. Occupational Exposure to Ultraviolet Radiation. Cincinnati, OH: National Institute for Occupational Safety and Health. Report nr HSM 73-110009.

NIOSH. 2009. Environmental Control for Tuberculosis: Basic Upper-Room Ultraviolet Germicidal Irradiation Guidelines for Healthcare Settings. Washington, DC: Department of Health and Human Services, Centers for Disease Control and Prevention, National Institute for Occupational Safety and Health. Report nr NIOSH 2009-105.

NRL. 1999. Ultraviolet Radiation. Christchurch, New Zealand: National Radiation Laboratory. Report nr Information Sheet 10.

Parrish JA, Jaenicke KF, Anderson RR. 1982. Erythema and melanogenesis action spectra of normal human skin. Photochem Photobiol 36(2):187–191.

Patel NP, Highton A, Moy RL. 1992. Properties of topical sunscreen formulations. J Dermatol Surg Oncol 18:316–320.

Pathak MA. 1969. Basic aspects of cutaneous photosensitization. In: Urbach F, editor. The Biologic Effects of Ultraviolet Radiation with Emphasis on the Skin. New York: Pergamon Press, pp. 489–511.

Pitts DG, Tredici TJ. 1971. The effects of ultraviolet on the eye. Am Ind Hyg Assoc J 32:235–246.

Saunders PJ, Diffey BL. 1995. Ambulatory monitoring of ultraviolet erythema in photosensitive subjects. Photodermatol Photoimmunol Photomed 11:22–24.

Talbot E, Jensen P, Moffat HJ, Wells CD. 2002. Occupational risk from ultraviolet germicidal irradiation (UVGI) lamps. Int J Tuberc Lung Dis 6(8):738–741.

Taylor HR. 1981. Climatic droplet keratopathy and pterygium. Aust & NZ J Opthal 9(3): 199–206.

Turner J, Parisi A. 2009. Measuring the influence of UV reflection from vertical metal surfaces on humans. Photochem Photobiol Sci 8:62–69.

USACE. 2000. Guidelines on the Design and Operation of HVAC Systems in Disease Isolation Areas. Aberdeen Proving Ground, MD: U.S. Army Center for Health Promotion and Preventive Medicine. Report nr TG252.

USDHHS. 2002. Ultraviolet Radiation Related Exposures: Broad-spectrum ultraviolet (UV) radiation, UVA, UVB, UVC, Solar Radiation, and Exposure to Sunlamps and Sunbeds. Washington, DC: U.S. Department of Health and Human Services. Report nr 10th Report on Carcinogens.

USEPA. 1997. Volume V: Health Effects of Mercury and Mercury Compounds. Mercury Study Report to Congress. Cincinnati, OH: Office of Health and Environmental Criteria, and Assessment Office, Office of Air Quality Planning and Standards, Research Triangle Park, NC. Report nr EPA 452/R-97-007.

USEPA. 2002. National Primary Drinking Water Regulations: Announcement of the Result of EPA's Review of Existing Drinking Water Standards and Request for Public Comment. Fed Regis 67(74).

Weiringa F, Vermeulen A. 2005. An update on optical radiation safety in Europe. IUVA News 7(2):25–29.

Wester U. 2000. Analytic expressions to represent the hazard ultraviolet spectrum of ICNIRP and ACGIH. Radiat Protect Dosim 91:231–232.

Wikonkal NM, Brash DE. 1999. Ultraviolet radiation induced signature mutations in photocarcinogenesis. J Invest Dermatol Symp Proc 4:6–10.

Young DP. 1991. Ultraviolet Lights for Surgery Suites. Mooresville: St. Francis Hospital.

Chapter 13

Ultraviolet Radiometry

13.1 Introduction

Testing of Ultraviolet Germicidal Irradiation (UVGI) systems encompasses both irradiance measurement of UV lamps and UV dose measurements, which is a field known as radiometry. Measurement of individual UV lamp irradiance is often performed in the factory before, during, or after burn-in of the lamps. Traditionally the minimum testing required has been measurement of the UV Lamp Rating, or the irradiance at 1 m from the lamp axis. New approaches to UV lamp testing are currently under consideration by various professional organizations since the UV Lamp rating has limited practical value. This chapter addresses the types of equipment used for measuring irradiance, the traditional measurement methods, and describes various new measurement methods. The measurement of UV dose, both by UV sensors and via microbiological methods is addressed, as are the various applications in which measuring irradiance or UV dose provides performance information or safety related information. This chapter only covers irradiance and dose measurement by direct physical means. The measurement of UV dose via microbiological testing is covered in Chap. 14.

13.2 Lamp Electrical Characteristics

The measurement of lamp electrical characteristics, including voltage, current, and wattage, are performed according to long-established practices used for the fluorescent lamps. Many modern electronic ballasts have high frequencies and non-sinusoidal waveforms and accurate power measurements of these systems often require sophisticated and expensive power analyzers, or oscilloscopes. However, since the electrical characteristics of typical Medium Pressure or Low Pressure mercury UV lamps are virtually identical to those for fluorescent lamps, the same guidelines, standards, and measurement methods may often be applied. The standards and guidelines that are most applicable have been covered in Chap. 11 and

W. Kowalski, *Ultraviolet Germicidal Irradiation Handbook*,
DOI 10.1007/978-3-642-01999-9_13, © Springer-Verlag Berlin Heidelberg 2009

these include ANSI_IEC C78.81-2005 for double-capped lamps (ANSI 2005), and the various secondary documents referenced in this standard. The measurement of electrical characteristics of ballasts for UV lamps are also covered by existing standards and these too are covered in Chap. 11. Additional references of use include IESNA (2000) and UL (2004).

Some types of UV lamps, such as microwave UV lamps and UV LEDs, may not conform exactly to existing electrical measurement methods, but such testing methods can be adapted without difficulty. It should be noted that the measured UV lamp wattage is often used as the basis for analytical determinations of UV output, and the accuracy of such reported wattages has, in the past, left much to be desired. Accuracies of a single decimal place may be sufficient for lighting measurements (i.e. ±30% error) but greater accuracy is needed for UV lamps, and measurement to perhaps two or more decimal places accuracy would be more appropriate for UV lamps.

13.3   UV Irradiance and UV Dose

The primary quantities of interest in measuring UV lamp performance are irradiance and dose. *Irradiance* is defined as the total radiant power from all upward (hemispherical) directions incident upon an infinitesimal element of flat surface area and has the units W/m^2 (Bolton and Cotton 2001). Irradiance can also be defined as the total flux divided by the projected area upon which it is incident. Irradiance is the quantity most often measured and employed for use in estimating the UV dose. Whenever lamp irradiance is measured by a sensor, it will invariably be a hemispherical sensor. UV irradiance is often referred to as intensity, which is generally understood to mean irradiance but is not a technically precise term and is generally avoided in this text. For a precise definition of the term intensity and the proper units see Driscoll and Vaughan (2001). *Fluence rate* is defined as the total radiant power incident from all directions onto an infinitesimally small sphere but can also be defined as the total flux incident upon the projected area of a sphere, as shown in Fig. 13.1.

For a collimated beam, with parallel and perpendicularly incident rays, irradiance and fluence are identical. Technically, fluence rate is the appropriate term for defining the UV exposure of a three-dimensional microbe in air, but this convention is not yet widely used and irradiance is the term used most often for both air and surface applications.

The *UV dose*, more correctly termed *fluence*, represents the total radiant energy incident upon a surface or a microorganism and has the units J/m^2 ($W\text{-}s/m^2$). The UV dose is understood to refer to the *UV exposure dose*, it being the dose to which the surface or microbe is exposed, and not the absorbed dose, which is unknown. That is, it is not currently known how much UV energy is actually absorbed by a microbe, it is only known how much UV it was exposed to in any situation. UV rate constants for microbes are based on the exposure dose, and the amount of UV

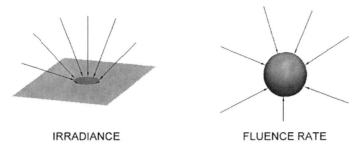

IRRADIANCE   FLUENCE RATE

Fig. 13.1 Illustration of the distinction between irradiance on a flat plane and fluence rate through a 3-dimensional sphere

absorbed is considerably less than that to which the microbes are exposed. In many cases, when UV dose is measured, the irradiance is the actual quantity being measured, over some measured time period. In other cases the absorbed dose is measured, usually as heat. When a test microbe is used to measure dose only the inferred exposure dose (not the absorbed dose) is measured.

13.4 Measurement Equipment

Ultraviolet light is a form of electromagnetic energy and it must be absorbed in some fashion for it to be detected. The measurement of light is called radiometry and the radiometer is the instrument normally used to measure UV light. Photometry is the measurement of visible light and the primary instrument for measuring visible light is known as a photometer (IESNA 2000). A photometer has a weighting function to mimic the response of the human eye, whereas a radiometer does not. There are two types of light detectors, thermal and photonic (Bolton 2001). In thermal light detectors, the light is converted to heat. In photonic light detectors, the absorption of photons drives an electronic circuit via the photoelectric effect. Measurement equipment include radiometers, spectroradiometers, UV photosensors, spherical actinometers, spectral filters, and other devices and methods.

A radiometer (aka radiometric photometer) is a device that senses the total irradiance incident upon a sensor element (see Fig. 13.2). They may employ wavelength-selective photoelectric detectors including photoconductors, photoemissive tubes, photovoltaic cells, and photodiodes, phototransistors, or other junction devices. Radiometers measure radiant power and have a calibrated output in irradiance (W/m^2).

A thermal radiometer detector consists of a black surface where all the incident light is converted into heat. A thermistor is placed in thermal contact behind the exposed black element and forms a Wheatstone bridge in conjunction with another black element kept in the dark. The current produced is proportional to the incident irradiance, and the output is calibrated to show the true irradiance. The accuracy and sensitivity of thermal radiometers is somewhat limited. Radiometers often include

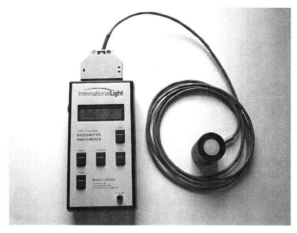

integration functions that allow the irradiance to be measured over time and thereby yield the UV dose.

A photonic radiometer typically uses a photocell with a UV-sensitive cathode that converts the incident energy into an electrical current. These detectors are very sensitive but their sensitivity may vary with wavelength. For monochromatic light sources the conversion to irradiance is straightforward. For broadband sources the spectral distribution must be evaluated along with the spectral sensitivity of the detector to obtain the actual value of irradiance.

Spectroradiometry refers to measurements that are made in discrete spectral bands. Spectroradiometers, or spectrophotometers, typically use wavelength-specific calibration factors to determine the actual spectral irradiance. Other radiometers need spectral filters to isolate the region of interest, which for UVGI is the UVC/UVB region (the actinic region). A detector's overall sensitivity is equal to the product of the responsivity of the sensor and the transmissivity of the filter (Ryer 1997). Filters operate by absorption or interference. Absorption filters are doped with materials that selectively absorb light by wavelength. Sharp-cut filters are used to block light above or below a particular wavelength. Interference filters rely on thin layers of dielectric materials to cause interference between wavefronts, providing for narrow bandwidths. The selection of which filter to use depends on the application. The measurement of germicidal irradiance for the purpose of assessing UV dose (i.e. on a cooling coil or to rate a lamp) will not be exactly the same as measuring the actinic UV irradiance for the purpose of assessing health hazards. Often, the difference is small enough that the same sensor/filter combination is used for both purposes, but it should be remembered that the germicidal effectiveness curve for microbes (see Fig. 2.1) is not quite the same as the actinic effectiveness curve (see Fig. 12.2).

UV photosensors typically employ semiconductors that absorb only in the UV region and are insensitive to visible light. Some of the materials employed include diamond (absorption range 190–230 nm), SiC (range 210–380 nm), and GaN (range 250–370 nm). These semiconductors can be incorporated into electronic circuits but still require periodic calibration (Bolton 2001).

The phototube is a light sensor that consists of a photoemissive cathode that emits electrons in proportion to incident light, and an anode that collects the emitted electrons (Ryer 1997). Solar-blind vacuum photodiodes use Cs-Te cathodes to limit sensitivity to ultraviolet light only. Vacuum photodiodes have an absorption range of about 180–360 nm.

Actinometers employ photochemical reactions for which the quantum yield is well defined. The quantum yield is a dimensionless measure of the efficiency of a photochemical reaction and is defined as the number of moles of product (or reactant formation) per Einstein of photons absorbed (see Sect. 2.8 in Chap. 2). Spherical actinometers are capable of absorbing photons from all directions, not just hemispherically, and so can be used to measure the true fluence incident upon microbes, which are often spherical or ovoid. There are three common types of actinometers that are useful for ultraviolet applications, ferrioxalate actinometers, persulfate actinometers, and iodide/iodate actinometers. The quantum yield of ferrioxalate (for Fe^{2+} generation) is about 1.25 in the 200–300 nm region. The quantum yield of persulfate (for H^+ production) is 1.8 below 300 nm. The quantum yield of iodide/iodate (for I_3^- generation) is 0.6 at 254 nm, and buffered solutions can be prepared that absorb only in the 200–300 nm region.

Fiber optics are available for use when direct or close access to the UV source is difficult or hazardous. Fiber optics consist of a core fiber and jacket with an index of refraction selected to maximize total internal reflection. Fiber optics can be used for remote sensing in operating systems where it is not practical to open access doors. There is typically some small percentage loss in fiber optics that can be accounted for by correcting the irradiance readings.

Two of the most common radiometers used in UV sensing are the model IL1400A Radiometer/Photometer (shown in Fig. 13.2) and the IL1700 Research Radiometer, both from International Light. These models display irradiance levels in units of W/cm^2. For purposes of measuring the ACGIH/NIOSH Actinic UV hazard, these units are provided with an actinic filter to match the actinic hazard function curve (see Fig. 12.2). For purposes of measuring the germicidal wavelengths per the IESNA/DIN germicidal effectiveness curve (see Fig. 2.1), the germicidal detector (SEL240/T2G) weights the effect of all light in this band to the actual germicidal effect (IL 1998). For narrow band sources a detector (SEL240/T2NS254) measuring only at 254 nm is also available. Other suppliers of commonly used UV radiometers include GigaHertz Optik and IL Metronic.

Spectrometers must be maintained and calibrated periodically. Maintenance typically consists of cleaning the optics with methyl alcohol and changing batteries. Some radiometers are capable of self-calibration (to internal programming). Calibration can be performed using working standard sources but these sources may themselves need to be calibrated per reference standards (IESNA 1996). Calibration is usually performed by a laboratory since it requires an absolute reference standard (usually a monochromatic light source). The responsivity is typically measured every 5 nm in the spectrum of interest. An overall uncertainty of 10% is considered good for radiometry equipment, while the limiting uncertainty seems to be about 1–6% in the ultraviolet (Ryer 1997, IL 1998). The measurement accuracy of radiometers in the UV region is limited by the reference source, which is usually from NIST, PTB, or some other national laboratory. Their uncertainty at 254 nm is approximately 6%, so any calibration that is "NIST Traceable" will have an uncertainty *greater* than this value. NIST-sourced deuterium lamps, however, may have an uncertainty as low as 3%. For an excellent review of ultraviolet detector accuracy see Reed et al. (2009).

13.5 UV Lamp Ratings

UV lamp ratings are defined by the irradiance measured in still air at 1 m perpendicular from the lamp axis at the lamp midpoint (see Fig. 13.3), per IESNA (2000). Lamp ratings are measured with lamps that have been burned in for at least 100 h. Ratings are published for all UV lamps and are commonly self-reported by manufacturers. They can be used to compare output of different lamps (i.e. of the same physical size) but otherwise serve little practical purpose. The lamp rating represents only one point in the lamp irradiance field, and it is not possible to accurately determine the lamp total UV output based on this single rating point. Furthermore, UV lamps are seldom used in still air and the cooling effects of airstreams can impact performance.

An alternative lamp rating method used in some sectors is to specify the UV output in terms of the watts per inch (W/in.) of lamp length (Driscoll and Vaughan 2001). The total UV output can then be estimated by multiplying this value by the lamp arclength. However, the watts per inch can vary along the length of any lamp axis, and so this method is primarily useful only for selecting between one lamp length and another. A more exact method of rating lamps is needed, and research towards this end is currently underway, such as the ASHRAE TG2.UVAS program, and standard methods will likely be accepted by the UV industry in the near future.

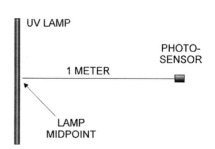

Fig. 13.3 IESNA method for UV Lamp Rating. Irradiance is measured at 1 m from lamp midpoint in still air at 70° ambient temperature

13.6 Lamp Irradiance Measurements

The UV output of a lamp, in terms of watts, is a function of the power input to the lamp, the lamp efficiency, and the operating conditions. If the power input to a UV lamp and the efficiency are known with any degree of accuracy, then the UV power output can easily be determined as the product of these two values. Often, the manufacturers provide this information for each lamp model but the accuracy of these numbers is not always certain. If the UV output power in watts is known, analytical methods can be used to determine the irradiance field (Kowalski et al. 2000). In any event, the only way to be absolutely certain of the UV output and the irradiance field is to take measurements in a laboratory setting under controlled conditions.

There are currently no formal standards available that define the test protocols, procedures, or methods for conducting such tests, although there are accepted methods for fluorescent lamps that can be adapted (IES 1988). but the principles are straightforward. In theory, if all the UV radiation that exits the glass of a cylindrical lamp were uniform, then a single measurement of the irradiance could be taken and multiplied by the surface area of the glass to obtain the total UV output in watts. Since the irradiance is measured in W/m^2, and the surface area of the lamp is in square meters, the product would yield watts of UV power output. There are two problems with this ideal approach. First, the irradiance field is not uniform around the glass surface, and second, the glass is curved and the detector or sensor will not sit flat against the glass. The approach can be improved by removing the sensor some specific distance from the glass surface, and taking multiple readings both around the glass and along the length, as shown in Fig. 13.4. In this approach an imaginary cylinder envelopes the UV lamp and represents a large surface area through which the UV radiations passes. Four angular test positions around the lamp axis are shown, at some constant radial distance (i.e. 10–20 cm), in which the photosensor is pointed directly at the lamp axis. Only three positions are shown along the lamp axis, the ends and the midpoint, but in order to obtain an accurate plot of the irradiance field several more test positions would be needed.

The additional test points along the axis length are needed since the irradiance field peaks at the lamp midpoint and tapers off towards the ends. One new problem introduced by this test plan is that the irradiance beyond the ends of the lamp are not accounted for. An additional set of readings would need to be taken in radial positions at successive diameters, pointing parallel to the lamp axis, to account for the radiation exiting the imaginary cylinder. Once all the test values have been read, each value can be multiplied by the increment of surface area (of the large imaginary cylinder) that the reading represents, and the values can be summed to determine the total UV output of the lamp.

There may be little variation in the irradiance around the lamp axis, and so perhaps two readings (0° and 180°) may suffice. Also, the irradiance field might be symmetric from one end to the center, and so half the readings might be omitted without seriously impacting the accuracy of the test. At a minimum, measurements

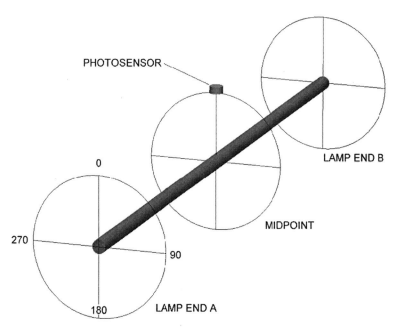

PHOTOSENSOR

LAMP END B

0

MIDPOINT

270

90

180   LAMP END A

Fig. 13.4 UV lamp and test photosensor positions for lamp ends and lamp midpoint, with four angular positions around the lamp axis

should be taken at a minimum of four positions along the axis from one end to the midpoint, and preferably about seven positions minimum from one end to the other.

All irradiance measurements should be made facing the source since the irradiance will vary with respect to the cosine of the angle between the optical axis and the normal to the detector. A bare silicon photocell has a near perfect cosine response, but once a filter is placed in front of the detector off-angle light becomes restricted. It is important for this reason to maintain the sensor perpendicular to the lamp axis when taking readings around the axis. For the measurements beyond the lamp ends, the sensor should, ideally, face the lamp axis midpoint, but since the angle of incidence will be small, the cosine may not contribute much error. Furthermore, the cosine angular response of the sensor may be sufficient to account for the difference even if the sensor is facing parallel to the lamp axis. The objective of such irradiance measurements is often to assess the lamp output power. Keitz (1971) presents the following equation for computing total UV output from a lamp with monochromatic output at 254 nm:

$$P = \frac{2\pi^2 EDL}{2\alpha + \sin 2\alpha} \tag{13.1}$$

where

P = UV output power
E = measured irradiance, W/m^2

D = distance from lamp axis to UV sensor, m
L = lamp arclength (from electrode tip to electrode tip), m
α = half angle (radians) subtended by lamp at sensor position

$$\alpha = \frac{L}{2D} \qquad (13.2.)$$

Testing has indicated that Eq. (13.1) is accurate to within about 5% (Laval et al. 2008).

Another approach to measuring UV lamp output is to use an integrating sphere such as is used in illumination engineering (IESNA 1985, Murdoch 1985, IESNA 2000). An integrating sphere is a large enclosed spherical surface coated inside with materials of extremely high reflectivity, such as magnesium oxide paint.

Sasges and Robinson (2005) have proposed a method using principles of goniometry to accurately measure the output of UV lamps using a virtual integrating sphere. This method is based on the divergence theorem, which states that the integral of irradiance over any closed surface is equal to the power output enclosed within the surface. The integrated irradiance must be known over the entire surface of the enclosing volume and therefore cannot be a single point measurement, but for axisymmetric lamps, the actual number of measurements is not great. The method involves rotating a horizontal lamp about a vertical axis through its center. The irradiance will be approximately constant for each circular strip of the integration volume as illustrated in Fig. 13.5. The sensor used for the goniometric measurements must have a Lambertian or cosine response versus the angle of incident light beams. This means that the detector will weight the incident irradiation by the cosine of the angle of incidence, which is a necessary condition for applying the divergence

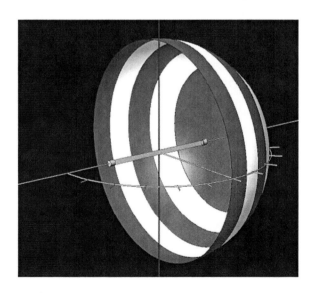

Fig. 13.5 Illustration of the spherical volume for goniometric measurement of UV lamp output. Image reprinted from Sasges and Robinson (2005) with permission from the International Ultraviolet Association (IUVA)

theorem. The Lambertian characteristic must be met for angles up to that subtended by the lamp at the detector. Measurements taken close to the lamp surface require an extremely good diffuser since the lamp subtends a very large angle at the detector. Manufacturer's claims regarding cosine or Lambertian response should be verified since they do not all achieve a cosine response in the ultraviolet range (Sasges and Robinson 2005).

A protocol has been provided for measuring UVGI lamp output by Sasges and Robinson (2005). It is summarized here with only minor edits for air applications:

1. Burn-in: Operate the lamp for 100–130 h at full rated electrical input power and with the amalgam spot (if any) oriented downwards.
2. Mount the lamp horizontally at least 75 cm from any surface. Support the lamp by the endcaps to minimize light blockage.
3. Attached a thermocouple to the bottom surface. Thermocouple must be at the amalgam spot, if any, or 35 cm away from the filament for a low pressure lamp, or in the middle of any lamp of less than 70 cm arclength.
4. Mount a detector at least 2.5 m from the center of the lamp at the same elevation as the lamp.
5. Ignite the lamp and monitor the current and voltage into the lamp.
6. Operate the lamp at full power until the temperature and electrical power have stabilized and uniform UV output has been obtained from the entire arc, or at least 15 min.
7. Measure the UV irradiance at the detector position.
8. Rotate the lamp about a vertical axis through the lamp center in increments of no more than 10° and repeat the measurements up to 90°. (Alternatively, the detector can be revolved around the lamp.)
9. Repeat the measurements from the normal position up to −90°.
10. Calculate the total lamp output.
11. Correct the lamp output to standard conditions for the local ambient temperature.

The total lamp output is calculated assuming that the UV irradiance is symmetric about the lamp axis. The irradiance may be integrated over an enclosing spherical surface of radius r by moving the sensor in an arc about the lamp (or by rotating the lamp). At each angle θ from the lamp normal, the differential area of the surrounding surface is given by:

$$dA = 2\pi r^2 \cos\theta d\theta \qquad (13.3)$$

The total lamp UV output is then given by the integral of the irradiance multiplied by the differential area element. The measurements at each angle can be tabulated and summed to approximate the integral. For additional detailed information on temperature corrections, equipment specifications, minimizing ambient reflectivity, and required documentation see Sasges and Robinson (2005) and also Laval et al. (2008).

13.7 Lamp Cooling and Heating Effects

The irradiance field around UV lamps can be affected by both air temperature and air velocity. UV lamps are typically designed to operate at an air temperature of approximately 21.5°C (70°F) and an air velocity of 2–2.54 m/s (400–500 fpm). Simple low pressure mercury lamps reach peak output at roughly 40°C (104°F), as a result of achieving the optimal mercury vapor pressure, a condition that may coincide with cooling conditions under design air flow and temperature. Air temperatures and air velocities beyond the normal design limits can cause UV output to decrease. The measurement of lamp cooling effects depends on measurements of the decrease in the lamp irradiance and the corresponding decrease in lamp UV output (in watts). The testing of lamp cooling effects is essentially a matter of measuring the irradiance field at various air temperatures and air velocities and plotting the results (see Figs. 5.11 and 5.12 in Chap. 5). The total UV output would be determined using some test plan such as detailed in the previous section, and the results compared to obtain the percentage decrease in output from the peak (presumably at design operating conditions of 70°F and 400 fpm air velocity). Lamp heating effects are measured in the same manner.

13.8 UV Reflectivity

There are two types of reflectivity, specular and diffusive. Specular reflectivity is mirror-like, in which the angle of reflection from the normal is equal to the incident angle of the light from the normal, and the reflected light remains largely cohesive. Diffuse, or cloud-like, reflectivity causes complete scattering in all hemispherical directions regardless of the angle of incidence of the incoming light, and the light becomes incoherent. Polished aluminum is almost completely specular, while white paper is almost completely diffuse. In real-world applications, reflectivity is often a combination of both specular and diffuse reflections.

Reflectometers are photometers specifically designed to measure reflectance (IESNA 2000). They can measure diffuse, specular, and total reflectance. The measurement of specular UV reflectivity requires special fixtures that can be swept in an arc of 180° about a rotational point that has a holder for the flat reflective sample (IL 1998). A steady light source is directed across the top of the rotational center where it is reflected at some standard distance to the detector. The reference condition is established by directing the light through the rotational point at the detector with the mirror removed. The reflector is then placed directly over the rotational point at an angle of 45° from the source. The detector is then rotated 90° about the same point until peak output is observed on the radiometer. Variations of this method are possible but the sample must be flat, not curved. For more information on reflectometry see NIST (1997), IESNA (1985), and also (IESNA 1994).

The measurement of diffuse reflectivity requires the use of either a goniometer or an integrating sphere to fully capture the components of diffuse reflectivity and integrate the reflections over diverse angles. A goniometer uses a computer driven, rotating mirror arm to measure total flux or total reflectance. The goniometer scans a reflector in at least a hemispherical solid angle to obtain the most accurate account of reflectance possible. For recommended practices for goniophotometry see ASTM E167 (ASTM 1996a). For more information on measuring reflectance in the UV spectrum see Zhuang and Yang (1989).

In an integrating sphere, a collimated light beam passes through the sphere and out a sample port on the opposite side. The photosensor is placed in the sphere surface orthogonal to this beam so as to be blind to either of the ports. The photosensor is zeroed out with the sample port open. A white reflectance standard (i.e. magnesium oxide) is placed in the sample port and provides the 100% reference value. The standard is removed and replaced by the reflectance sample, and then a reading is taken which represents the diffuse value of reflectance for the sample. This reading is divided by the reflectance of the standard (i.e. 98% or 0.98) to obtain the reflectivity of the sample (IL 1998). For additional information see the test method in ASTM E903 & E259 (ASTM 1996b).

For the purposes of assessing wall and surface reflectivities in rooms or other internal areas like ducts, UV reflectivity can be estimated by a simple procedure using a portable radiometer (Lindsey 1997). Appropriate PPE must be used if UV levels are expected to exceed NIOSH RELs. First, read the irradiance incident on the wall by placing the detector on the wall facing outwards. Then, read the reflected irradiance from the wall by pointing the sensor directly at the wall at a distance of about 12–18 in. The reflectivity (more properly termed reflectance) of a surface can then be computed as follows:

$$\rho = \frac{E_r}{E_i} \tag{13.4}$$

where

ρ = Reflectance or reflectivity, fractional
E_r = Reflected irradiance, W/m^2
E_i = Incident irradiance, W/m^2

Reflectance measured in this manner will often consist of both specular and diffusive components. The total reflectivity of any surface is defined as the sum of the diffuse and specular components (Modest 1993):

$$\rho = \rho_d + \rho_s \tag{13.5}$$

where

ρ_d = diffuse reflectivity
ρ_s = specular reflectivity

13.9 UV Transmittance

The UV transmittance (or transmissivity) of any material is determined as the ratio of the outlet (transmitted) irradiance to the inlet (incident) irradiance (Driscoll and Vaughan 2001). It may be expressed as a percent or as a fraction. When measuring the inlet or outlet irradiance, an aperture is often used to restrict and define the central optical area of the sample. Baffles may serve the same purpose. The incident irradiance is measured before the sample is placed behind the aperture. Then the transmitted irradiance is measured with the sample in place. The transmittance is the outlet irradiance divided by the inlet irradiance as follows:

$$\tau = \frac{E_t}{E_i} \tag{13.6}$$

where

τ = Transmittance, fractional
E_t = Transmitted irradiance, W/m^2

Transmittance is dimensionless. If the attenuation in transmittance is due primarily to absorption (internal transmittance, as in a gas), then the attenuation can be described by the following equation:

$$\tau = 10^{-Ad} \tag{13.7}$$

where

A = absorption coefficient, cm^{-1}
d = path length, cm

Optical density, which is used to define the UV attenuation of eyewear, is simply the base 10 log of the inverse of the transmittance (see Sect. 12.7). For more information on measuring UV transmittance see ASTM (1996a, 1996b). For information on measuring the UV transmittance of textiles see ASTM (2000).

13.10 UV Absorptance

The absorptance (or absorptivity) is defined as the base 10 logarithm of the ratio of the incident spectral irradiance to the transmitted spectral irradiance as follows:

$$\alpha = \log\left(\frac{E_i}{E_t}\right) \tag{13.8}$$

Absorptance is dimensionless. Absorptance and transmittance are related by the following equation:

$$T = 10^{-\alpha} \tag{13.9}$$

Finally, there is a relationship between absorptance, transmittance, and reflectance, of a partially transmittant surface, defined by the following relation:

$$\rho + \alpha + \tau = 1 \tag{13.10}$$

Equation (13.7) can be used to compute any one term if the other two are known. For example, Pyrex glass has a UV reflectivity of about 4% at an incident angle of 90° (Koller 1965). If the UV transmittance (which can vary with thickness) were 0.01%, the absorptance would be $100 - 4 - 0.01 = 95.99\%$. For opaque materials, $\tau = 0$, and either absorptance or reflectance can be determined from Eq. (13.10) if the other is known.

In all the above Eqs., (13.2) through (13.7), the relationships apply whether the radiation is broadband or whether it is broken down by spectral bands and specific wavelengths. It should be noted that the absorptivity of air is negligible (above VUV wavelengths) and that although the absorptivity of water is not negligible, the amount of water in humid air (even at 100% RH) is not sufficient to significantly attenuate UV radiation over the kinds of distances normally seen in indoor applications of UVGI.

13.11   Reflective Lamp Fixtures and Enclosures

UV lamp fixtures often include reflective surfaces for directing UV irradiance towards surfaces or for limiting the irradiance field. When an ordinary fluorescent lamp fixture with a white enamel surface coating is used with a UV lamp the reflectivity is so low (about 4–5%) that the extra projected irradiance can usually be neglected. If a UV lamp fixture with polished aluminum surfaces (about 75% reflectivity) is used, the projected irradiance can be boosted by up to the amount of reflectivity. Measurement of the irradiance produced from lamp fixtures is simply a matter of taking sensor readings in and around the area on which the irradiance field is projected. In the case of overhead UVGI systems used in operating rooms, for example, it may be desired to maximize the irradiance over the operating table and minimize it elsewhere. Parabolic reflectors and louvers in the lamp fixture can be used to achieve this end, and test measurements need to encompass both the area of maximum irradiance and the surrounding areas.

Certain types of lamp fixtures are used inside ducts and air handling units but these are intended merely to hold the lamps and not necessarily to project irradiance onto surfaces. These fixtures are often made from polished aluminum and cause limited interference with the distribution of irradiance. If a fixture consists of multiple lamps packed together, it can often be treated as a single large UV lamp for measurement purposes. That is, the irradiance field from the entire lamp array can be measured at some appropriate distance large enough to minimize any error caused by the presence of the fixture.

UV lamps installed inside ducts often have clearance distances less than 1 m around the lamps and so it becomes impossible to measure the traditional lamp rating inside ducts. In some case the lamps may be aligned perpendicular to the airflow through the duct and it may be possible to measure the irradiance at 1 m distance, but this reading is impacted by duct reflectivity, even if only a single lamp is present. A method like this was attempted in the series of test by the EPA on in-duct UV systems (see for example EPA 2006) but the usefulness of this approach is limited, even for comparative purposes. There currently exists no specific standardized method for verifying or testing lamp irradiance inside ducts, although computational methods do exist for measurement of lamp output that could be verified by limited sensor readings (Kowalski et al. 2000, Sasges et al. 2007).

Since the quantity of most interest is typically the average irradiance or fluence rate inside the duct, actinometric methods may be employed (Blatchley et al. 2007, Deguchi et al. 2005). Actinometry has not yet been applied to in-duct airstreams, although it has been used in room models (Rahn et al. 1999, Rahn 2004).

13.12 UV Dose Measurement

Direct measurement of the UV exposure dose produced inside a duct via photosensor readings is a matter of measuring the average irradiance (as discussed previously) and multiplying by the mean exposure time. In any duct in which the air is moving, the degree of turbulence ensures that the degree of air mixing is high and typically approaches complete mixing, and therefore it is appropriate to use the average irradiance to obtain the UV exposure dose (Kowalski et al. 2000). The design velocity of a typical UVGI system is about 2.54 m/s (500 fpm), corresponding to a Reynolds number of approximately 150,000, and turbulent mixing is the norm. Even if flow were laminar, mixing by diffusion would dominate and conditions will, at worst, lie somewhere between complete mixing and the idealized condition of completely unmixed flow, as shown by Severin et al. (1984) for water-based systems. The difficulty of this approach is that a large number of sensor readings must be taken not only at the duct inlet but at increments along the direction of airflow. The approach could be simplified by taking measurements only in a single quadrant, if the system is symmetrical, but still, this approach is probably only practical in a laboratory setting.

It may be possible to use computer analysis to determine a single point within any system (based on its geometry, lamp orientation, etc.) that represents the mean UV irradiance. The measurement of UV exposure dose could then be simplified greatly. Using several such symmetrical test points might provide a good indication of the mean UV dose.

13.13 Spherical Actinometry

The equipment for using spherical actinometry was described in Sect. 13.4. In air disinfection applications microbes typically receive a fluence (irradiation from all directions) rather than an irradiance dose as they may on flat surfaces. That is, a spherical microbe will be exposed to an omnidirectional radiant flux through its outer surface (Rahn et al. 1999, Kowalski 2001). In contrast, radiometric sensors are subject to differences in response when light arrives at sharp angles as opposed to coming from overhead. Spherical actinometry may be the most accurate means to quantify UV dose response in airborne microorganisms. The materials and methods for spherical actinometry have not, however, been simplified or standardized to the point that they are easily used or can yield fully reproducible results yet, but the approach is worth discussing for future application. Spherical quartz cells of about 1.35 cm diameter containing triiodide solution are placed within the exposure chamber at a stationary point and irradiated for some measured time period. For details of preparing the triiodide solution see Rahn et al. (1999). Following irradiation, the liquid contents of the spherical cells, about 0.2–1 ml, are removed and placed in optical absorbance cells in which the formation of triiodide is measured with a spectrophotometer. The number of moles of triiodide formed bear a direct relationship to the absorbed UV dose, and the dose can be calculated accordingly. By taking a large number of readings over the 3 dimensional space of the exposure chamber, an average can be calculated for the fluence rate.

Spherical actinometry could also be used to measure UV dose in a duct more directly by traversing a large number of sample points through the duct, in real time, to obtain an average value of the UV dose. If a matrix of sample points such as shown in Fig. 13.5 were used as starting points, the actinometers would have to traverse linear paths from the inlet to the outlet. The sum total exposure of each path would represent the dose for that path. The mean dose for all paths would represent the UV dose for the system. Again, this type of test may only be appropriate for laboratory settings.

13.14 Testing Applications

The measurement of irradiance and UV dose has a number of practical applications in both research and in applications, including laboratory testing of lamps and systems, and performance and safety testing of various types of systems, including UVGI air disinfection systems, cooling coil disinfection systems, Overhead UVGI systems, Upper Room systems, Lower Room systems, and area disinfection systems. This section only addresses measurement of irradiance and dose by physical means. For general information on UV photometric techniques see IESNA (1991).

13.14.1 Laboratory Testing of Equipment

Laboratory testing of UVGI equipment is generally conducted to determine the performance characteristics of UV lamps and unitary equipment. The methods for testing UV lamps or lamp fixture assemblies have been discussed in previous sections of this chapter. The testing of stand-alone or unitary UV equipment, primarily air disinfection systems, will involve an assessment of the UV dose produced based on the irradiance field and the airflow. Since it can be difficult to accurately assess the irradiance field inside an enclosed UVGI system, often only airflow measurements can be taken. These can be combined with computer analysis of the irradiance field to estimate the UV dose produced. One method for quantifying the UV dose is the UVGI Rating Value (URV), which categorizes UVGI systems in finite increments or ranges of UV dose (IUVA 2005, Kowalski 2006). See Appendix G for a tabulation of URV categories and the corresponding doses. The most common type of laboratory testing for UVGI systems is the airborne microbiological test, which is covered in the following chapter.

13.14.2 In-duct UVGI System Testing

There are no standards or recommended practices for measuring the irradiance of installed in-duct air disinfection systems although attempts have been made to formulate a uniform method of measuring upstream irradiance at a single point (see the example in EPA 2006). In systems where there is very limited internal interference and equipment (i.e. coils, fans, etc), it may be possible to take a series of irradiance measurements in a 3-dimensional matrix but this would be the exception in the real world. One approach that has been developed for water applications involves using dyed microspheres and Lagrangian actinometry to assess the dose inside a dynamically mixed vessel, but no such application is proposed for air systems (Blatchley et al. 2007). Similarly, there is no common method for measuring UV dose inside ducts other than through microbiological testing, except for the theoretical method discussed previously in which spherical actinometers would traverse a matrix of airflow paths from the inlet to the outlet. The in-place testing of in-duct UVGI systems with photosensors may be limited to simple verification of operation and safety testing.

Verification of UVGI system operation can be performed by taking one or more photosensor measurements inside the duct and comparing results with computational models to determine whether the UV output is in the appropriate design range. For example, if the free space around the UV lamp midpoint is less than 1 m, then a measurement should be taken and the distance measured. This could then be compared with a simple UV lamp model to determine if the irradiance levels were in accordance with expected levels.

13.14.3 Cooling Coil Systems

Cooling coil UVGI systems are simpler to test in place since the main objective of the system is to create an irradiance field on the cooling coil surface. There are currently no agreed-upon levels of cooling coil irradiance but current designs use anywhere from 0.5 W/m^2 to about 8 W/m^2. Whatever the target level of irradiance, the testing will consist of using a photosensor to measure the irradiance levels at several points across the cooling coil surface. Not all points may be accessible, but only a limited number of sample points (i.e. 5–10) are needed to verify the system is operating per specifications. Preferably, the test points will include points on the corners and sides where the irradiance may be at a minimum (see Fig. 13.6), and these minimum points should be identified or predicted prior to the test being performed. However, the entire coil surface could also be surveyed to find the minimum irradiance points. If none of the test points measure below the specified minimum irradiance the system can be considered verified.

If the design irradiance levels are specified as an average, which is an acceptable practice, then a series of equidistant test points must be taken and their values averaged to verify system performance. In this case the minimum irradiance points should still be identified and measured, but they will form part of a surface average. That is, each measurement point represents some surface area of the coil, depending on the number of measurements, and a weighted average is computed to find the mean irradiance across the coil surface. For example, if the face of the cooling coil spans an area of 2 m^2, and ten photosensor points are used, then each photosensor point represents 0.2 m^2 of surface area. If points do not represent equal areas then they must be weighted accordingly.

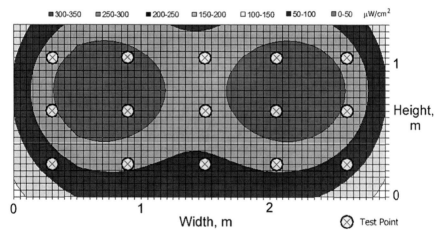

Fig. 13.6 Irradiance contour (with two UV lamps) across the face of a cooling coil showing equidistant test points. Minimum irradiance values in lower corners are 100–150 μW/cm^2 (1–1.5 W/m^2)

Most cooling coil systems are irradiated on only one side, which means the opposite side of the cooling coil will receive no direct irradiation. Some irradiance will find its way through the coils and therefore some low level of irradiance can be measured on the opposite coil face. There are no current standards for what levels are appropriate but this level should be recorded. In fact, the levels of irradiance on the opposite face of the coil could be expected to increase over time as the coil becomes cleaner, and the opposite face irradiance provides an indicator of when maximum cleaning has been achieved (when irradiance reaches a plateau).

13.14.4 Overhead Surgical Site Systems

Testing of Overhead UVGI systems is necessary to ensure adequate irradiance at the surgical site, which in typical systems means $0.25–0.3$ W/m^2 at a table height of 3 ft. Perhaps three or four measurements at the operating table will be sufficient to verify the design parameter, and several additional measurements should be taken around the room in order to assess the overall irradiance contour. If a single overhead UV lamp fixture is used, the highest irradiance will occur directly below the lamp, as shown in Fig. 13.7. If an array of UV lamps is positioned across the ceiling to obtain the design irradiance (i.e. $0.25–0.3$ W/m^2) across the entire room at table height, then the contour will be fairly even and constant across the three-foot height plane in the room, and the photosensor readings should assure that this design criteria is met. Many variations in Overhead UVGI system lamp placement are possible and they will each have distinctive irradiance

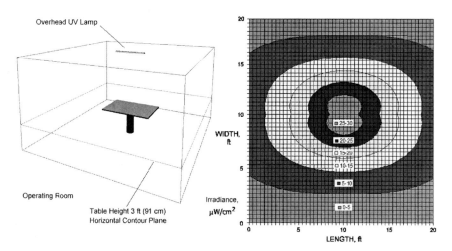

Fig. 13.7 Single lamp Overhead UVGI system. Arrangement (*left*) in a 20 ft^2 OR and irradiance contour (*right*) at 3 ft height (table height). Design irradiance at surgical site is approximately $25–30$ μW/cm^2 ($0.25–0.3$ W/m^2)

contours. Some systems have dual settings, one for operations and one for room disinfection during unoccupied periods. The irradiance levels under both settings should be tested.

Since the physicians, nurses, and technicians will presumably be using PPEs and full-body coverage, the photosensor measurements do not serve any direct safety purposes, however, the levels should be checked periodically to ensure the system is working according to design, and to maintain a record of irradiance levels for material evaluation purposes (see Chap. 15 on Photodegradation). If there are any gaps around the doors or in any pass-through doors, levels should be checked on the opposite side to ensure ACGIH TLVs are not being exceeded outside the OR.

Some sources recommend taking a UV photosensor reading on the table just prior to a surgical procedure, since lamps tend to degrade over time and lose irradiance. However, it is probably sufficient to check the irradiance levels once a week or even once a month, since minor decreases in output are unlikely to have any noticeable effect. Whether measuring irradiance weekly or before a surgical procedure, the bulbs should be wiped clean with alcohol or other cleaning agents and a non-abrasive lint-free cloth.

13.14.5 Upper Room Systems

Upper Room UVGI systems need to be tested to ensure that irradiance levels below the upper room are below ACGIH TLVs. For 8 h of permissible exposure the limit is 0.001 W/m^2 (0.1 μW/cm^2) for broadband UV or 0.002 W/m^2 (0.2 μW/cm^2) for UV at a wavelength of 254 nm (although other limits have been proposed – see Chap. 9). Figure 13.8 illustrates a typical irradiance profile of a room with an Upper Room UVGI unit. Above the designated height A (typically about 2.5–3 m) irradiance levels may be as high as 0.3–0.5 W/m^2 (30–50 μW/cm^2). Below the designated height irradiance levels should not exceed the ACGIH limit. Levels may exceed this limit by design if the occupancy times are limited according to the maximum permissible exposures (see Table 12.3).

Photosensor measurements should be taken both above and below the cutoff height (A) with several measurements taken at eye height, as suggested in Fig. 13.8, which shows eight test points in a vertical profile from the UVGI unit midpoint. Since the unit will produce an irradiance contour in the horizontal plane also, it would be prudent to repeat these eight test point measurements two or more times at other locations on either side of the UVGI unit. When multiple Upper Room UVGI units are installed, two or more additional vertical profile sets of test points should be taken for each unit. For centrally located (ceiling-mounted) circular UVGI units, additional test points will be needed. For additional examples of test plans for taking Upper Room UVGI measurements see Dumyahn and First (1999). Also see Rahn (2004) for information on actinometric measurements for Upper Room systems.

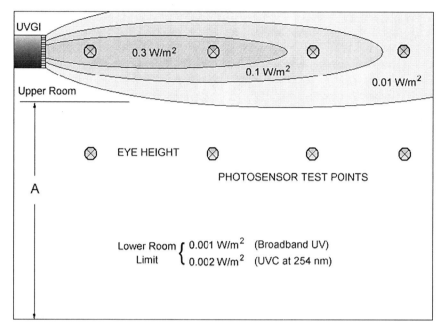

Fig. 13.8 Profile of Upper Room UVGI unit located at height A with typical irradiance field contours shown. A series of eight photosensor test points are shown in both the upper room and at eye height

13.14.6 Lower Room Systems

Lower Room UVGI systems are similar in form to Upper Room systems except that there is a maximum height (typically about 0.5 m or 18 in.) above which the irradiance levels should not exceed ACGIH TLVs. Test points should be taken below the Lower Room cutoff height to ensure design irradiance levels are met, and measurement should also be taken just above the cutoff (i.e. 0.6 m or 24 in.) and also at eye height to ensure levels are below the ACGIH limits (0.001 W/m^2 (0.1 μW/cm^2) for broadband UV or 0.002 W/m^2 (0.2 μW/cm^2) for UV at a wavelength of 254 nm). Higher limits proposed for Upper Room systems may also be applicable to Lower Room systems. In other respects test plans for photosensor measurements can duplicate those for Upper Room systems.

13.14.7 Area Disinfection Systems

Most area disinfection units use naked bulbs and are intended to produce moderate to high levels of UV irradiance in unoccupied rooms which expose all room surfaces continuously. For such equipment it is only necessary to take a few sample readings to verify that the expected levels of irradiance are being achieved on walls, floors, or room equipment surfaces. It will be necessary in such cases for the technicians to wear PPEs, and possibly to disable the UV system shutoff switches while performing measurements. It is prudent to also take measurements outside the irradiated area (i.e. around door gaps) to ensure no hazardous levels of UV are leaking out of the room.

Some area disinfection units may be intended for use in occupied rooms (i.e. in ORs) and produce very low levels of UV continuously. These levels will typically fall below ACGIH limits for 8 h of exposure (0.001 W/m^2 (0.1 μW/cm^2) for broadband UV or 0.002 W/m^2 (0.2 μW/cm^2) for UV at a wavelength of 254 nm). In some cases occupancy might be limited to less than 8 h and the UV irradiance limits will be as prorated from ACGIH limits (see Table 12.3). In such cases a series of photosensor measurements should be taken at diverse points around the room, including a few points at eye level, both near and far from the unit. Measured irradiance levels could be averaged, or the maximum level could be used as a conservative value.

References

ANSI. 2005. Double-Capped Fluorescent Lamps – Dimensional and Electrical Characteristics. New York: American National Standards Institute. Report nr ANSI_IEC C78.81-2005.
ASTM. 1996a. Recommended Practice for Gonidophotometry of Objects and Materials. West Conshohocken, PA: American Society for Testing and Materials. Report nr ASTM E259.
ASTM. 1996b. Standard Test method for Solar Absorptance, Reflectance, and Transmittance of Materials Using Integrating Spheres. West Conshohocken, PA: American Society for Testing and Materials. Report nr ASTM E903.
ASTM. 2000. Standard Practice for Preparation of Textiles Prior to Ultraviolet (UV) Transmission Testing. West Conshohocken, PA: American Society for Testing and Materials. Report nr ASTM D6544.
Blatchley E, Shen C, Scheible O. 2007. Lagrangian actinometry used dyed microspheres – A new validation method. IUVA News 9(1):17–21.
Bolton JR. 2001. Ultraviolet Applications Handbook. Ayr, Ontario, Canada: Bolton Photosciences, Inc.
Bolton JR, Cotton CA. 2001. The Ultraviolet Disinfection Handbook. Ayr, Ontario, Canada: American Water Works Association.
Deguchi K, Yamaguchi S, Ishida H. 2005. UV-disinfection reactor validation by computational fluid dynamics and relation to biodosimetry and actinometry. J Wat Technol Environ Technol 3(1):77–84.
Driscoll WG, Vaughan W. 2001. Handbook of Optics. New York: McGraw-Hill.
Dumyahn T, First M. 1999. Characterization of ultraviolet upper room air disinfection devices. Am Ind Hyg Assoc J 60(2):219–227.
EPA. 2006. Biological Inactivation Efficiency by HVAC In-Duct Ultraviolet Light Systems: Steril-Aire, Inc. Washington, DC: U.S. Environmental Protection Agency. Report nr EPA 600/R-06/052.
IES. 1988. IES Approved Method for the Electrical and Photometric Measurements of Fluorescent Lamps. New York: Illumination Engineering Society. Report nr LM-9-88.
IESNA. 1985. Method for Total and Diffuse Reflectometry. New York: Illumination Engineering Society of North America. Report nr IESNA LM-44.
IESNA. 1991. IESNA Guide to Spectroradiometric Measurements. New York: Illumination Engineering Society of North America. Report nr LM-58-1991.

IESNA. 1994. IESNA Approved Method for Photometric Testing of Reflector-Type Lamps. New York: Illumination Engineering Society of North America. Report nr LM-20-1994.

IESNA. 1996. Measurement of Ultraviolet Radiation from Light Sources. New York: Illumination Engineering Society of North America. Report nr IESNA LM-55-96.

IESNA. 2000. IESNA Lighting Handbook HB-9-2000, Reference and Application Chapter 6, Light Sources. New York: Illumination Engineering Society of North America.

IL. 1998. Instruction Manual: IL1400A Radiometer / Photometer. Newburyport, MA: International Light.

IUVA. 2005. General Guideline for UVGI Air and Surface Disinfection Systems. Ayr, Ontario, Canada: International Ultraviolet Association. Report nr IUVA-G01A-2005.

Keitz H. 1971. Light Calculations and Measurements. London: Macmillan and Co Ltd.

Koller LR. 1965. Ultraviolet Radiation. New York: John Wiley & Sons.

Kowalski WJ, Bahnfleth WP, Witham DL, Severin BF, Whittam TS. 2000. Mathematical modeling of UVGI for air disinfection. Quantitative Microbiology 2(3):249–270.

Kowalski WJ. 2001. Design and optimization of UVGI air disinfection systems [PhD]. State College: The Pennsylvania State University.

Kowalski WJ. 2006. Aerobiological Engineering Handbook: A Guide to Airborne Disease Control Technologies. New York: McGraw-Hill.

Laval O, Dussert B, Howarth C, Platzer K, Sasges M, Muller J, Whitby E, Stowe R, Adam V, Witham D and others. 2008. Proposed method for measurement of the output of monochromatic (254 nm) low pressure UV lamps. IUVA News 10(1):14–17.

Lindsey JL. 1997. Applied Illumination Engineering. Lilburn: The Fairmont Press, Inc.

Modest MF. 1993. Radiative Heat Transfer. New York: McGraw-Hill.

Murdoch JB. 1985. Illumination Engineering. New York: Macmillan.

NIST. 1997. Spectral Reflectance. Gaithersburg, MD: National Institute of Standards and Technology. Report nr Publication 250–8.

Rahn RO, Xu P, Miller SL. 1999. Dosimetry of room-air germicidal (254 nm) radiation using spherical actinometry. Photochem Photobiol 70(3):314–318.

Rahn RO. 2004. Spatial distribution of upper-room germicidal UV radiation as measured with tubular actinometry as compared with spherical actinometry. Photochem Photobiol 80(2): 346–350.

Reed NG, Wengraitis S, Sliney DH. 2009. Intercomparison of instruments used for safety and performance measurements of ultraviolet germicidal irradiation lamps. J Occup Environ Hyg 6:289–297.

Ryer AD. 1997. Light Measurement Handbook. Newburyport, MA: International Light, Inc.

Sasges M, Robinson J. 2005. Accurate measurement of UV lamp output. IUVA News 7(3):21–25.

Sasges M, vanderPool A, Voronov A, Robinson J. A standard method of quantifying the output of UV lamps; 2007. Ayr, ON: International Ultraviolet Association.

Severin BF, Suidan MT, Rittmann BE, Englebrecht RS. 1984. Inactivation kinetics in a flow-through UV reactor. J Water Pollution Control 56(2):164–169.

UL. 2004. Luminaires. Research Triangle Park, NC: Underwriters Laboratories. Report nr UL 1598.

Zhuang D-K, Yang T-L. 1989. Spectral reflectance measurements using a precision multiple reflectometer in the UV and VUV range. Appl Optics 28(23):5024–5028.

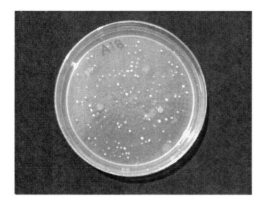

Chapter 14

Microbiological Testing

14.1 Introduction

Microbiological testing methods are the best means of verifying the effectiveness of ultraviolet germicidal irradiation systems. Testing may involve the irradiation of bacteria or fungi on surfaces or petri dishes, or it may involve airborne testing of bacteria and fungi inside UVGI units, ducts, or in buildings. Sampling methods for air and surface disinfection systems are fairly straightforward and include both surface sampling and airborne sampling. The equipment needed for such testing is relatively inexpensive and the user does not necessarily need a microbiology degree, just some training in the methods and some familiarity with the field of aerobiology. Often, the actual interpretation of air and surface sampling results is performed by a professional laboratory and the technician need only follow sampling procedures and deliver the samples to a lab. The procedures necessary for air and surface sampling are provided in this chapter and they are discussed in relation to the most common applications: Unitary UVGI, In-duct UVGI, Cooling Coil UVGI, Overhead UVGI, Upper Room UVGI, and Area Disinfection UVGI. This chapter also addresses UV rate constant measurement, in particular by the collimated beam method.

An additional form of testing, epidemiological testing, is also addressed in this chapter and involves the recording of illness or disease statistics as a measure of the performance of air disinfection systems installed in buildings.

14.2 Microbiological Testing

Microbiological testing may be the most practical, most feasible, and least expensive type of performance verification test for UVGI systems. In most cases the disinfection rate (or percentage reduction of microbial counts) is the parameter to be measured. In some cases, a particular microbe is of special interest (to the client) and the UV rate constant may not be known in advance, but the desired result of the test is still the disinfection rate. The procedures discussed in this chapter are generalized

W. Kowalski, *Ultraviolet Germicidal Irradiation Handbook*,   337
DOI 10.1007/978-3-642-01999-9_14, © Springer-Verlag Berlin Heidelberg 2009

so that any and all of these types of test results can be achieved, depending on the user (or client) preferences.

Sometimes it is desired to test a dangerous pathogenic microbe by using a physiologically similar or identical microbe known as a surrogate. For example, *Bacillus cereus* is virtually identical to *Bacillus anthracis* except that it is non-pathogenic as an airborne microbe (although it is an enteric pathogen), and so *B. cereus* spores are an acceptable surrogate for anthrax spores in air and surface testing

Most professional microbiology laboratories can perform air and surface sampling using the methods presented here or else using their own proprietary test protocols. Often, an engineer, microbiologist, or Certified Industrial Hygienist (CIH) will perform the air sampling and deliver the samples to a microbiology laboratory for culturing, counting, and speciation. Culturing and counting do not necessarily require a laboratory, provided either an incubator is available or when incubation is performed at room temperature. Counting, or the estimation of the number of cells in a sample, can be performed either by cultivation or microscopy. Microscopic methods provide the total number of single microorganisms while culturing determines only colony-forming units. Speciation is best performed by experts in laboratory settings, although many common bacteria and fungi can be identified either by culture or by microscopy

14.3 Biodosimetry

The measurement of the UV exposure dose through the use of microorganisms is known as biodosimetry (Bolton and Cotton 2001). Some microbes, such as *Serratia marcescens* and MS2, have been tested so often that their UV rate constants are fairly well established under indoor conditions. Such microorganisms serve as indicator microbes for establishing the UV exposure dose of any air or surface disinfection system. Microbial testing to establish the UV dose in air disinfection systems is conducted in the same manner as normal airborne testing for UV rate constants (see Chap. 4), only the analysis of results differs. In-room air sampling alone may not be able to produce a UV dose without extensive modeling. UV dose measurement testing is best suited for laboratories and generally requires microbes to be aerosolized and injected upstream of the UVGI system. For an example of a biodosimetric test see EPA (2006).

14.4 Test Microbes

Various microorganisms have been used in UVGI research, including bacteria, bacterial spores, fungal spores, fungi in yeast form, viruses, algae, and protozoa. In laboratory testing, any microbes may be used, but in field applications microbes are not normally injected or aerosolized because of health hazards, and so only ambient microbes, hailing naturally from the environment or from healthy humans, are used for field testing. In general, only bacteria and fungi are useful for field testing of UVGI systems, since they exist everywhere and are mostly innocuous to healthy individuals. Viruses are pathogenic and not suitable for use in field testing and, in any event, viruses are virtually impossible to sample from the air in indoor environments. In terms of UV susceptibility, viruses are comparable to bacteria and so it can often be assumed that if a UVGI system successfully eliminates bacteria then it will also be effective at eliminating viruses. This may not always be true, and the client's concerns with particular viruses may have to be specifically addressed through analysis or by further testing (i.e. in a laboratory).

The primary microbes useful for assessing performance of UVGI systems will include *viruses, bacteria, bacterial spores*, and *fungal spores*. Fungi in yeast form are rarely encountered, and are so much more vulnerable to UV than fungal spores, that they can be ignored for testing purposes (unless the client has some specific reason to evaluate them). Therefore, whenever test requirements refer to fungi, it can be assumed these are fungal spores. However, for UVGI in-duct air disinfection systems, which should always contain filters rated between MERV 8-15, the low rates of inactivation of spores is of little significance compared to the high rates of removal that could be expected through the filters, and so it should be understood that the performance of UVGI systems against airborne spores is of little consequence unless the filter/UVGI combination is tested as a whole (Kowalski 1997). The use of spores as test microbes is further complicated by the fact that at normal in-duct UVGI system doses, spore inactivation rates will be so low as to be marginal, which complicates interpretation (see the Chap. 8 review of the EPA test results). Though spores may be a poor choice for testing airstream disinfection, the opposite is true for UVGI cooling coil disinfection, in which fungal and bacterial spores form the bulk of contaminants that are to be removed from coil surfaces. In this case, the UV exposure is continuous, and very high UV doses are achieved in short periods of time, which ensures that significant disinfection rates can be assessed.

In most field applications only the total quantity of bacteria or fungi (i.e. airborne bacteria and fungi) is of interest, and it is not necessary for the samples to be speciated, to specifically identify each microbial species. In such tests the total bacteria or fungi are counted, either for surface or airborne samples. The total counts of bacteria provide an indicator of air or surface cleanliness, as do the total counts of fungi. In some cases the bacteria and fungi may be combined, but this is not a common practice and the two groups are usually assessed separately.

There are approximately 130 bacteria, fungi, and viruses of interest in UVGI air and surface applications (see Appendices A, B, and C). All of these microbes are included in the UV rate constant tables in Chap. 4 if they have been previously tested, and the reader should avoid water-based microbes in any test plans for UVGI systems. In fact, most of the time there are probably less than two dozen microbes that will appear in any air or surface sample in indoor environments (see Table 14.1), and so the identification of these microbes is greatly simplified. This does not apply to hospital environments, where a much greater variety of microbes may appear, or to specialized food or industrial environments.

Table 14.1   Most Common Indoor Microbes

Type	Family	Source	Type	Family	Source
Fungi	Penicillium	Environment	Bacteria	Micrococcus	Environment
	Cladosporium	Environment		Staphylococcus	Humans
	Aspergillus	Environment		Streptococcus	Humans
	Ulocladium	Environment		Pseudomonas	Humans
	Aureobasidium	Environment		Corynebacterium	Humans
	Acremonium	Environment		Bacillus	Environment

14.5 Test Materials and Equipment

For surface sampling, the necessary materials include sterile swabs, gloves, plastic bags, and growth media (typically petri dishes). Some types of sterile swabs contain integral sterile solution and this liquid is plated on media in a laboratory. Types of adhesive tape are also available that can lift fungal contamination off of surfaces (Aerotech 2001).

For air sampling it is necessary to use an air sampler and growth media (usually petri dishes) as shown in Fig. 14.1. Some types of air samplers use other materials, like blotter paper, to collect samples and these are transferred in laboratories to petri dishes. Filter type samplers usually involve the placement of a sterile coupon or filter cassette inside an air sampler. In general, samples will be sent to a laboratory for incubation, counting, and identification. Alternatively, they can be grown in an incubator and counted without identification if only the total counts are required, and this may not require a laboratory. If colonies are to be identified (speciated), it generally requires an experienced professional microbiologist using a microscope and other means to identify each microbial species.

Table 14.2 provides some examples of the types of media that may be used for air and surface sampling. The type of media depends on whether bacteria or fungi

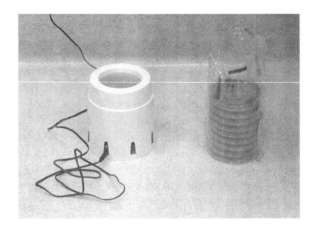

Fig. 14.1   Air sampler and prepackaged petri dishes ready for sampling. Image used with permission of Pathogenus, Inc. Etobicoke, Ontario

Table 14.2 Typical growth media

Medium	Microbe	Application notes	Incubation temperature
DG18 (Dichloran 18%) agar	Fungi	Slows fungal growth, low water activity (Aw=0.95)	20–25°C/35–37°C
Malt Extract Agar (MEA)	Fungi	Broad spectrum, saprophytic, allergenic and pathogenic fungi	20–25°C/35–37°C
Inhibitory mold agar	Fungi	Same as MEA but suppresses bacterial colonies	20–25°C/35–37°C
Malt extract agar with NaCl, sucrose, or dichloranglycerol	Fungi	Xerophilic fungi	20–25°C/35–37°C
Rose bengal agar	Fungi	Broad spectrum like MEA but suppresses bacterial colonies	20–25°C/35–37°C
Buffered charcoal yeast extract agar	Bacteria	Legionella	35–37°C
MacConkey agar	Bacteria	Gram negative bacteria	35–37°C
Heart infusion blood agar	Bacteria	Human commensal bacteria	35–37°C
R2A with cycloheximide	Bacteria	Environmental bacteria with fungal suppression	20–30°C
Soybean-casein digest agar	Bacteria	Environmental bacteria with fungal suppression	20–30°C
Soybean-casein digest agar	Bacteria	Thermophilic bacteria (*Actinomyces*)	50–55°C
Tripticase soy agar (TSA)	Bacteria	Bacteria	35–37°C

are to be grown and sometimes on which microbe species is to be grown. This list is not exclusive and many other media types may be acceptable.

14.6 Surface Sampling

Surface sampling will normally be performed to verify that exposed surfaces of equipment or buildings are disinfected by UV exposure. These surfaces can include the surfaces of medical equipment, pharmaceutical supplies, mail handling equipment, sterile containers or packaging, room floors and walls, HVAC duct internal surfaces, cooling coils, drain pans, etc. These surfaces are often subject to extended UV exposure, as in the case of ducts or cooling coils, and they are effectively sterilized. Sterilization, however, is difficult to prove absolutely, as even the act of sampling a surface can add slight amounts of contamination. Attempting to verify sterilization through sampling may require large numbers of samples, high sensitivity, or large-area sampling. Sensitivity of detection of microbes depends on the type of surface, microbial handling procedures and material used in quantification of particular microbe. For small-area sampling the number of microbes recovered may be low, while for large area-sampling (i.e. 1 m^2) low levels of pathogens may be detected with greater sensitivity (Buttner et al. 2004). For example the lower detection limits reported for BiSKit used on metal surfaces are approximately 40–100 cfu/m^2 for *B. atrophaeus* spores. Only in highly controlled environments like the pharmaceutical industry is it possible to demonstrate sterilization through testing, and even so there will be some level of error in the results.

Surface sampling is generally performed by wiping swabs across the surface or by pressing materials, including petri dishes of contact plates, against the surface. Adhesive tape and adhesive sheets are also available for use in surface sampling (Yamaguchi et al. 2003). The swabs will then either be drawn across a petri dish or will be inserted into sterile solutions which will later (in the lab) be plated on dishes. The actual area of the surface being sampled may not be critical in most cases since it is used for comparative purposes only. That is, when a Before condition (UV Off) is compared with an After condition (UV On), only the relative decrease in microbial counts is of interest, and not necessarily the absolute plate counts. Since there are no standards or guidelines for what levels of surface contamination are 'acceptable' or hazardous to health, such data have only a relative meaning.

Often, the sampling of surfaces such as walls, floors, HVAC duct, or cooling coils is intended to verify that high levels of disinfection have been achieved through prolonged UV exposure. In such cases a relative level of sterilization, or virtual sterilization may be defined as a matter of convenience. The traditional definition of sterilization in mathematical terms is six logs of reduction, or a 99.9999% reduction of surface microbial counts. This is a reasonable and practical approach and forms a workable target for disinfection processes, however, it can depend to some degree on the area that is sampled and the effectiveness of the sampling process. The surface area sampled may be about 2 in.^{2} (6.5 cm^2) or any other designated surface area, provided it is kept roughly constant for any series of samples to be compared. Sampling devices capable of sampling larger areas may have the advantage of requiring collection of fewer samples per site and greater detection sensitivity.

For any system that is expected to sterilize surfaces, the After sample is likely to have approximately zero survivors, regardless of the counts for the Before sample. This will be true for both bacteria and for fungal spores. It will also tend to be true regardless of the actual surface area sampled, and so it is not absolutely necessary to measure the sample area precisely. In such cases a visual estimate of the area (i.e. 2 in.^{2}) will be sufficient. If a template is used to mask of the area to be sampled, as shown in Fig. 14.2, it must be sterile. Prepackaged sterile paper templates are not reusable, but must be discarded after each use. Reusable templates (i.e. metal or plastic) must be disinfected before each use, *including before the first use*, or else they may contaminate all samples from the first to the last. Sterilization of reusable

Fig. 14.2 Surface swabbing with a reusable template. Image provided courtesy of SKC Inc., Eighty Four, PA

templates involves spraying or wiping a disinfectant such as alcohol on them and then drying them, a practice which is often not practical for many microbiological sampling situations. *Failure to sterilize a reusable template may result in all samples becoming contaminated with the same bacterium.* Such types of contamination will be obvious from the fact that the same bacteria (or fungi) will occur at high levels even when all other microbes are effectively sterilized. In one case witnessed by the author, a novice working for a chemical sampling company used an unsterilized sheet metal template pulled from a bag to sample for microbes on a UV-sterilized cooling coil surface – very high and disproportionate levels of *Bacillus subtilis* (a common indoor contaminant that likely came from the bag) occurred on all samples from first to last even though all other species of fungi and bacteria were nonexistent or found in only trace quantities. Reusable templates may be common for chemical sampling but they are poor practice for microbiological sampling, and one should preferably use either disposable sterile templates or none whatsoever.

Gloves should be worn during the sampling process, and it should be recognized that the person doing the sampling likely carries various environmental or commensal microbes on their person. They should be careful not to touch the surface with any part of their clothing. Gloves should preferably be sterile. Gloves can be sterilized beforehand by spraying with alcohol if necessary, but care should still be taken not to touch the swab heads with any part of the gloves.

Swabbing may be done with dry sterile swabs, typically made of cotton. These come prepackaged and the paper or plastic package can be torn open before each use. Both wet and dry sampling are possible, but one source reports that dry sampling is slightly more sensitive (Buttner et al. 2004). The area to be sampled should be swabbed with a back and forth motion, gently and with no more than a dozen or so strokes. Strokes can all be parallel, or they can be perpendicular (see Fig. 14.3). Once a swab is taken from a surface it is then brushed across a petri dish, and the

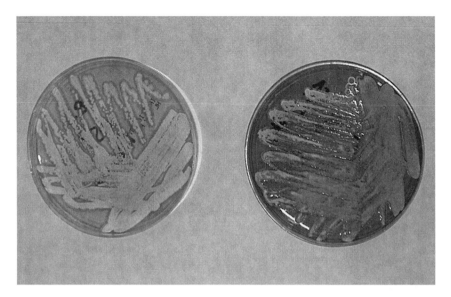

Fig. 14.3 Examples of petri dishes that have been swabbed with surface samples

dish is closed up and put away. Petri dishes should always be stored and carried upside down, whenever possible, so that no leakage will enter the plate and cause contamination. They can be swabbed in the vertical position so as to minimize contamination during the sampling process.

Some types of swabs come prepackaged with a sterile solution, and these may have to be crushed to break them open before each use. Once the surface sample is taken, the swab is merely inserted back into the sterile solution.

A Surface Sampling Protocol has been outlined on the following page that may be adapted to any one of a number of surface sampling applications. For any surface to be sampled, the Before samples should correspond roughly to the location for the After samples, but not exactly since the Before sampling process may actually clean the surface. The UVGI system must be turned off whenever surface sampling is being performed. For cooling coils and ducts the ventilation system should also be turned off, since blowing nonsterile air may create contamination problems.

In general, only the total counts of bacteria and fungi are of interest, as indicators of contamination or sterility. In certain settings, such as in health care facilities, it may be of interest to identify the actual species of microbes that were counted and the laboratory should be directed to speciate such results. Hospitals often are interested in sampling for Gram-positive versus Gram-negative bacteria (such as MRSA and VRE). One source reports that moistened swabs registered a sensitivity of 54% for Gram-positive cocci and 74.2% for Gram-negative bacteria while Rodac plates had 69.5% sensitivity for Gram-positive cocci and 42.7% for Gram-negative bacteria (Lemmena et al. 2001).

Test Protocol for Surface Sampling

The following procedure applies generally to surface sampling, including cooling coils. See the text for special information if cooling coil surfaces are to be sampled.

A.1: Materials and Preliminary Conditions

A.1.1: Procure surface sampling materials as necessary. These should include sterile swabs (wet or dry), sterile gloves, and petri dishes (if necessary). At least 3–9 plates (or swabs) will be needed for each condition, Before and After.

A.1.2: Gloves should be sterilized through the use of a disinfectant such as alcohol. If a sterile template is used to mask off a surface area, then the template must be discarded after each use. Keep at least 1 plate unused as a control.

B.1: Test Protocol

B.1.1: Disengage any UVGI systems and any operating equipment in the vicinity that may be a hazard to test personnel. If entry into an air handling unit is required, the fan should be shut down.

B.1.2: Identify and record the location of a suitable surface sampling point. The sample location should be approximately the same for subsequent samples. Either visually estimate an area of approximately 2 in.^{2} or use a sterile template for doing so.

B.1.3: Using the wet or dry swab, draw it gently across the sample area with a back and forth motion, and either insert the swab back into the sterile container or draw the swab across the petri dish. Cover the plate, seal as necessary, and label the swabs and/or the plates with a code or description of the location sampled.

B.1.4: Repeat the above process for all the pre-selected sample locations.

B.1.5: Start up the UVGI system, operate it for the specified time period (i.e. 24 hours–2 weeks) and repeat the test for the After condition.

C.1.1: Deliver the samples (either swabs or plates) to a laboratory, or place the plates in an incubator as soon as possible, and incubate for 24–48 hours at the required temperature.

C.1.2: After incubating for 24–48 hours, and before the plates become overgrown, remove them and count each plate. Alternatively, digital images of the plates can be made and used for counting.

C.1.3: Tabulate and summarize the results as necessary on appropriate forms.

C.1.4: If the client requires identification of the bacterial or fungal species, the laboratory should be directed to perform this function.

14.7 Air Sampling

The testing of UVGI systems in terms of the reduction of airborne counts of bacteria or fungi can be performed comparatively with settle plate sampling or quantitatively using air samplers. In each case it is desired to know the total reduction in airborne microbes resulting from the use of a UVGI air disinfection system. It is assumed in the following sections that the testing is performed in-place in an indoor environment. For laboratory testing, the methods described in Chap. 4 for UV rate constants are sufficient, and examples of lab tests of UVGI systems are available in the literature (VanOsdell and Foarde 2002, EPA 2006, Foarde et al. 1999).

Settle plates are simply petri dishes placed in rooms for specified time periods on which airborne microbes will settle. It is a common method for assaying the airborne microflora of a room but provides data that is only for comparative purposes (i.e. comparing a room with or without UVGI) and does not provide data that is absolute or quantitative (i.e. in terms of airborne cfu/m^3). Settle plates can be placed around a room in multiple locations, such as on the floor or at breathing height (sitting or standing), and in the corners, sides or center of a room. One set of plates would be used for bacteria and another set for fungi. In general, at least three plates are needed at each location to obtain an average value for consistency. Plates are best placed on the floor where airborne concentrations are highest. Testing three different locations in any room, with three plates at each test point, will mean at least nine plates total for each test. With one test run without UVGI (Before) and one run with UVGI (After) there will be at least 18 plates total required. See the suggested Test Protocol for Settle Plate Sampling on the following page.

Air Sampling is performed using any one of the available air samplers on the market today. These samplers can vary considerably in performance and although these differences can be an order of magnitude it should be remembered that they are often used for comparative purposes only, and therefore repeatability may be more important than absolute accuracy. Variations between air samplers of as much as an order of magnitude are not uncommon (Griffiths et al. 1993, Bradley et al. 1992, Straja and Leonard 1996, Ambroise et al. 1999, Jensen et al. 1992, Jensen and Schafer 1998, Li et al. 1999, Lin et al. 1999). Buttner and Stetzenbach (1993) report that Burkard and Andersen air samplers were the most accurate samplers for spores in the size range 1.8–3.5 μm but the Surface Air System (SAS) model had the highest mean level of retrieval. Andersen samplers had the highest level of repeatability, differing by about 7% between samples. If the data collected is used for absolute comparisons (i.e. to meet some indoor guideline or limit) then the choice of sampler, or test protocols, may be more critical.

Air samplers with petri dishes are operated for a specified period of time (i.e. 20 min), after which the plates are removed and incubated. Figure 14.4 shows some examples of locations where air samplers and settle plates can be placed. For additional information on sampling human pathogens, see Artenstein et al. (1967). For examples of sampling fungi see Flannigan (1997) and Flannigan et al. (1999). For more general sampling information see Boss and Day (2001), and Cox and Wathes (1995).

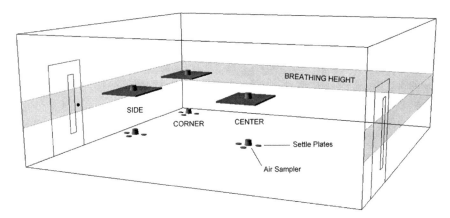

Fig. 14.4 Diagram of sample locations for air sampler and settle plates. Settle plates are placed at floor level. Air samplers can be placed at floor or at breathing height (3–5 feet)

Test Protocol for Settle Plate Sampling

The following test protocol is suggested for use when the settle plate method is used for sampling indoor air for Bacteria and/or Fungi in order to compare the performance of a UVGI air disinfection system Before (UVGI Off) vs. After (UVGI On).

A.1: Preliminary Conditions and Materials

A.1.1: Select and procure appropriate growth media in plates suitable for the intended application: Bacteria of Fungi. A list of typical growth media has been provided in Table 14.2. At least 18 plates will be needed: 9 for Before (UVGI Off) samples and 9 for After (UVGI On) samples, with three plates to be placed at each location. Keep at least 1 plate unused as a control.

A.1.2: Room conditions should be within normal indoor operating temperature and humidity ranges. Test locations should be selected and will preferably include room floor locations at the center, the side, and the corner.

A.1.3: Number the individual plates (either before they are placed at test locations, or after they are removed form the test locations). A minimum of three plates (Bacteria and/or Fungi) are recommended for each test location.

B.1: Test Protocol

B.1.1: Disengage the UVGI system for the Before (UVGI Off) condition.

B.1.2: Place the settle plates at the designated locations in the room. Each three-plate set of plates may be placed side-by-side, at each test location. Record each test location and the time of placement. Record room occupancy, including brief entrances or exits.

B.1.2: After a period of 1–2 hours, cover and remove the settle plates. Make sure the plates are labeled or numbered before or at the time they are removed, and record the removal time.

B.1.3: Incubate the plates at the required temperature or deliver them to a laboratory for incubation and counting.

B.2.1: Repeat the above procedure for the After condition, after engaging the UVGI system and operating for at least 4–8 hours or longer (i.e. 1–14 days).

B.2.2: Repeat the above procedure for Fungi (or Bacteria) if necessary. Fungi and Bacteria may be tested coincidentally, with six plates placed side-by-side at each test location, if desired.

C.1: Evaluation of Results

C.1.1: Incubate the plates for 24–48 hours, or deliver to a laboratory for culturing.

C.1.2: Count the colonies on the plates and record them on appropriate forms. Evaluate the counts and determine the net reduction, Before vs. After.

C.1.3: Speciate the samples if required by Client.

14.8 Sampling Applications

The general test protocols provided in the previous sections may be used for a variety of sampling applications for specific types of UVGI systems. The most common of these UVGI applications are discussed in the following sections and special considerations for each of these applications are discussed. For specific information on the type of building or facility in which sampling is to be performed, and the microbes of interest in each case, see the associated applications section in Chap. 16, and also Cox and Wathes (1995) and Vincent (1995).

14.8.1 UV Rate Constants

The factors affecting UV rate constant testing have been addressed in Chap. 4. In the event it is desired to measure UV rate constants in an application, either the surface sampling method or the air sampling method could be used. The surface sampling method will, of course, produce surface rate constants, and the air sampling method will yield airborne rate constants. The methods in each case are the same as described in the test protocols, only the analysis of results differs. The relative humidity should be recorded for such testing, and the number of plates used should be doubled or tripled when increased accuracy is desired.

Two main types of airborne UV rate constant tests can be performed – collimated beam testing and model system testing. The collimated beam test is considered the most accurate method and standard equipment is currently available for such testing. Model system testing, in which an actual full scale UV air disinfection system or a scaled down model of a real system is used for testing, is considered to be system-specific since the results may depend on the test apparatus or type of system used. Collimated beam testing is considered microbe-specific and if it is performed correctly it should provide an accurate and reproducible account of microbial UV susceptibility.

In order to establish the UV exposure dose in a model system, either an extensive series of measurements is needed throughout the interior of the duct to establish the irradiance field, or else computational methods can be used to estimate the irradiance field, from which the UV dose can be computed based upon the exposure time. UV rate constant testing must be performed by comparing the inlet concentrations vs. the outlet concentrations of airborne microbes. This requires drawing air samples from the duct upstream and downstream, rather than placing air samplers inside the duct (which may distort the sampler results). The normal method of sampling air inside a room served by a UVGI system (to determine performance) may not yield results accurate enough to establish a UV rate constant due to the complexities of recirculated airflow and the effects of deposition. Even if these factors are known or controlled, the variation of UV irradiance and airflow through the irradiance field impose a limit on the accuracy of UV rate constants measured by these methods.

The collimated beam method allows for extremely precise measurement of irradiance and exposure times, and avoids other problems associated with in-duct or room-sized exposure chambers (Ke et al. 2009). The UV rate constant determined in a collimated beam apparatus designed to provide uniform and invariable fluence rate (irradiance). In a collimated beam system, UV light is used to irradiate either a flat channel or a batch reactor through which a controlled airflow passes. Previous attempts to use collimated beams to study air disinfection often directed the UV beam onto a thin rectangular channel with a quartz glass top surface in much the same way as is performed for water-based UV studies (Sharp 1938, 1939, 1940, Luckiesh et al. 1947, Riley et al. 1976, Riley and Kaufman 1972, Lai et al. 2004, Fletcher et al. 2003). If the channel is thin, the lamp relatively high above the channel, and reflectivity is controlled, then the irradiance in the channel is effectively constant and can be measured with great precision. One of the problems with this approach is that the exposure time may be so short (i.e. less than 0.25 s) that shoulder effects can produce distortions in UV rate constant measurements. Attempts to correct for short exposure times by increasing the irradiance can be problematic since the assumption of reciprocity does not always hold at short exposure times.

A protocol for determining the UV susceptibility of microorganisms in air using collimated beams in a batch reactor has been developed by Ke et al. (2009). The protocol is partly based on successful test protocols for waterborne microbe studies (Bolton and Linden 2003, Kuo et al. 2005). The batch reactor in this protocol is a 10 cm diameter cylinder with a length of 20 cm which includes a mixing fan. Air is supplied to the batch reactor at a controlled rate and aerosolized microbes are fed into the airstream with a nebulizer. The exhaust air from the reactor is fed to an air sampler. Relative humidity is controlled in this apparatus by supplying dry air that is mixed with the saturated air from the nebulizer.

According to Bolton and Linden (2003), there are four critical factors necessary for accurately measuring the irradiance (or fluence rate) inside the exposure chamber: (1) the Petri Factor (PF), (2) the Divergence Factor (DF), (3) the Reflection factor (RF), and the Medium Factor (MF). In water-based studies, the PF is defined as the ratio of the average irradiance measured at different radial positions at the top quartz plane of the reactor to the irradiance measured at the central point of the top plane. The DF is derived by applying the inverse square law to the lamp irradiance between the distance from the top quartz plane to the bottom surface of the reactor. As long as the lamp is a sufficient distance from the top surface, the inverse square law can provide an accurate estimate of the variation in irradiance between the top of the reactor and the bottom. The Divergence Factor is defined as:

$$DF = \frac{L}{L+l} \qquad (14.1)$$

where

$L =$ distance from the UV lamp to the top (quartz) surface of the batch reactor
$l =$ distance from the top surface to the bottom of the reactor

The RF is determined by measuring the irradiance (using a radiometer) or fluence rate (using actinometry) with and without the top quartz plate in place. The ratio of these values is the RF of the quartz cap as follows:

$$RF = \frac{E_r}{E_n} \qquad (14.2)$$

where

$E_r =$ fluence rate at top of reactor with quartz plate
$E_n =$ fluence rate at top of reactor with no quartz plate

The MF applies to the medium in water-based studies, but in air the attenuation of UV is negligible and the MF $= 1$. The average irradiance or fluence rate can then be written as:

$$E_{avg} = E_0 \cdot PF \cdot DF \cdot RF \qquad (14.3)$$

where

$E_0 =$ the incident fluence rate (W/m^2) as read by a radiometer at the center of the top quartz plate of the reactor

The UV dose (fluence) is then calculated as the product of E_{avg} and the exposure time (seconds). The aerosolized microbes are fed into the reactor and then the airflow is cut off and the inlet and outlet valves are closed. The shutter on the collimated beam is then opened for a specified period of time (the exposure time) and then closed. Immediately afterwards the air mixture in the reactor is flushed with clean air to drive the reactor air volume into the air sampler. Ke et al. (2009) used this protocol to measure the UV rate constant of *Bacillus subtilis* spores in air at low relative humidity (50–60%), arriving at a value of k = 0.0155 m^2/J, which is fairly close to the average measured rate constant in water, or k = 0.0176 m^2/J (averaged from Appendix A), but lower than that measured in air at low humidity by other studies.

14.8.2 Cooling Coil Disinfection Systems

Cooling coil sampling presents a special case. The coil surfaces are actually fins and not smooth surfaces, but one should not change the surface area to be sampled. The typical fin spacing is between 8 and 12 fins per inch, meaning the area sampled is not a perfectly smooth surface. Although the actual physical surface area sampled will be less than about 2 in.^{2} (6.5 cm^2), due to the small spaces between the fins, this may be counterbalanced by the fact that spores and other debris tend to concentrate on the leading edges of the fins. In any event the surface area to be sampled is not critical or absolute, as discussed previously, and what matters is that the area sampled is consistently the same for all Before vs. After tests.

Samples should be taken on the side of the coil (upstream or downstream) subject to direct irradiation. Samples may be taken on the unirradiated side of the cooling coil if desired. If testing the unirradiated side of the coil it should be recognized that it may take much longer for the UVGI system to disinfect the opposite side of the coil, and the time between the Before and After test could require weeks or months of operation. Suggested sample points include the center, the side, and the corner most distant from the UV lamps, as shown in Fig. 14.5.

Samples may be taken in-between the fins, but this approach is not essential. Over time, the surfaces in-between the fins will become disinfected and follow-up tests may be conducted as required.

It is no problem if the Before samples prove to be so highly contaminated that they are uncountable, and these can merely be recorded as 'too high to count.' It is sufficient in such cases that the After samples show a significant reduction. Extremely high contamination levels may suggest, however, that manual coil cleaning might be warranted but testing may proceed prior to any coil cleaning. Coils should not be cleaned before testing since the surface sampling results in the Before condition might be invalidated. In such cases it is sufficient to demonstrate that the surface contamination levels in the After condition are either absolutely low (i.e. single digits of cfu/m^2) or that they are significantly lower than the condition before

Fig. 14.5 Suggested surface
sample locations for a typical
cooling coil, upstream side,
showing center, side, and
corner test points

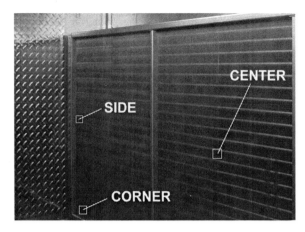

cleaning. Testing may be repeated after coil cleaning is complete to validate the
After condition.

Cooling coils tend to become contaminated with fungal spores and certain environmental bacteria like *Bacillus subtilis*. Since these microbes are generally no
health threat, to healthy individuals, it is not necessary to speciate them and they
may be dealt with in terms of total counts (either total fungi, bacteria, or total fungi
and bacteria). Hence, it is reasonable to test only for total fungi using fungal media
and total bacteria using bacterial media.

14.8.3 In-duct Air Disinfection Systems

Testing of an in-duct UVGI system could be conducted with any of the three aforementioned test protocols, but only air sampling will yield a quantitative accounting
of the effectiveness or performance of the system – one that may be used to assess
indoor aerobiology quantitatively.

Surface sampling may used to verify the operation of an air disinfection system by demonstrating that all internal exposed surfaces of the HVAC system (i.e.
duct, air handling unit components, etc.) have been sterilized (or highly disinfected). This approach does not demonstrate the level of air disinfection but does
demonstrate the UV lamps are working and may substitute for irradiance testing,
which can be more costly. Surface sampling cannot, however, substitute for air disinfection, although it may provide a certain confidence in the system's ability to
inactivate microbes. Sampling inside ducts will generally involve taking samples
of dust from the lowest flat surfaces, since fungal spores tend to settle by gravity
over time.

Settle plate sampling may be used to verify the effectiveness of the air disinfection system but only in terms of the Before vs. After condition. It could be
expected that the counts of the plates for the After condition would be much

lower than the Before condition (if the UVGI system works). The settle plate counts cannot be converted to airborne concentrations, by any currently known method.

Air sampling can be used for any in-duct application provided the test points are inside a room or enclosed area where air currents are minimal. Placing an air sampler inside a duct is not recommended since the air velocity may greatly distort test results. Air sampling can be used in any room equipped with a unitary UVGI system, or recirculation unit. Care must be taken not to place the air sampler in the exhaust flow path of the air disinfection unit since the air velocity may affect the air sampler accuracy. If the room has separate ventilation (supply and/or exhaust) in addition to the recirculation unit, the normal ventilation should be kept operating. Normal room occupancy should also be maintained.

14.8.4 Upper Room Systems

Only settle plate and air sampling are likely to be useful for Upper Room systems, unless they simultaneously provide surface disinfection by design (although surface disinfection is likely to be inevitable). Air sampling is preferred since the results will be quantitative. No special considerations are required for using air samplers to measure the effectiveness of Upper Room systems, except that care should be taken that no direct UV rays, or any significant levels of UV, reach the petri dishes or other sampling media (plates and samplers can be shielded from direct and reflected UV). Irradiance measurements may have to be taken to ensure this is true.

Samples should be taken at floor level, where microbial levels are highest, but samples taken at breathing height are also acceptable provided they are taken at the same location in both the Before and After conditions (see the example locations in Fig. 14.4). For examples of Upper Room sampling see Xu et al. (2003). Some additional useful information may be found in Beggs and Sleigh (2002) and Miller and Macher (2000).

14.8.5 Lower Room Systems

Lower Room systems may have as a design objective the sterilization of floors, and therefore surface sampling of the floor around the UVGI system is the appropriate test. Samples should be taken in Before and After conditions with a period of operation of at least 1–2 days, or weeks. Both bacteria and fungi should be sampled.

Air sampling of Lower Room UVGI systems present a slight problem in that the air samplers cannot be placed on the floor without risk of irradiating the sampling media. Therefore, the testing location may have to be above the irradiating plane (i.e. 18–24 in.), and irradiance measurements may have to be taken to ensure levels will not affect the sampling process. When testing air cleanliness in Before and After conditions, the Lower Room system should be allowed to operate for about

1–2 weeks, and indoor levels should be compared with outdoor levels. In all other respects, Lower Room UVGI system testing can be conducted the same as for Upper Room systems.

14.8.6 Overhead Surgical Site Systems

Surface sampling is an appropriate test method for Overhead UVGI systems since they will tend to sterilize all exposed surfaces in the OR. Samples should be taken at table height (approximately 3 feet or 1 m) or directly on the operating table. Samples may also be taken at locations around the operating room at floor height, as shown in Fig. 14.6, and on the walls. A more appropriate test, however, is to inoculate petri dishes with test bacteria such as *Staphylococcus* or *Streptococcus* and place them on the operating table. Plates should be inoculated with high titers of bacteria suitable for registering six logs of reduction and placed on the operating table for specified periods of time. If the design irradiance is 25–30 μW/cm^2, the plates could be expected to be sterilized within the normal time period for surgical procedures, or about 1–2 h. The exposed plates would then be compared with unexposed inoculated plates to determine the net reduction. Since plates are likely to be sterilized with 1–2 h exposure, it may be desired in some cases to test for shorter periods, so that some survivors remain to allow for biodosimetric computations, if such is required.

Air sampling is also appropriate for Overhead UVGI systems since the entire operating room is typically bathed in UV and the air will be disinfected. In this case the Before condition will have the UVGI system Off and the After condition will have the UVGI system On during sampling. Care should be taken not to expose the sampler to UV, and this may require shielding the sampler from direct and reflected

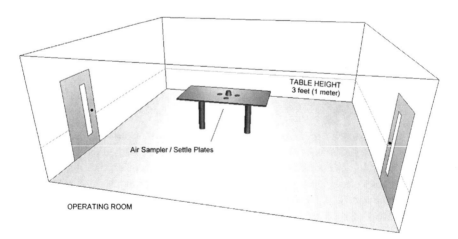

Fig. 14.6 Example of placement of air sampler or petri dishes in an Operating Room with an Overhead UVGI system. Samples could also be taken at floor level, where microbial levels are likely to be highest

UV. UV irradiance measurements might be needed to ensure that UV levels are negligible around the sampler. Air samplers can be placed on the floor and in the corners, as in Fig. 14.5, where the maximum concentrations are likely to occur, or else placed at table height. Air sampling can be performed during procedures, and the number of personnel should be recorded (including entry and exit times for transients). An unoccupied OR can be sampled for use as a baseline for comparison.

14.8.7 Area Disinfection Systems

The testing of an area disinfection system can be accomplished simply through the use of surface sampling of the floors and walls. The surface sampling test protocol given previously is adequate for this purpose. The Before condition represents the unirradiated room, prior to irradiation, and the After condition represents the room after some specified period of irradiation (i.e. 24 h). The UVGI system should be Off during surface sampling. Air sampling is also appropriate since air cleanliness is likely to occur when surfaces are disinfected on contamination.

14.9 Evaluating Sampling Results

The primary objective of surface or air sampling is to demonstrate system performance by measuring the reduction of microbes. In most cases the reduction of microbes is computed as percentage reduction, based on the ratio of the After plate counts to the Before plate counts. The Before plate counts can be averaged and compared with the average After plates. Alternatively, it may be of interest to compare each location in the Before condition with each of the after locations. In any event if three or more plates were placed at each location they should be averaged to represent that location. The counts from each individual plate may vary greatly and so averaging three plates for each location is an appropriate means of improving accuracy.

Bacterial samples should be incubated for 24–48 h, but they can typically be counted after 24 h. Fungal samples are also incubated for 24–48 h but may take longer to grow to the point that they can be counted, or about 48 h, depending on culture medium. Plates should be digitally photographed to provide a record and then disposed of, usually via incineration or steam sterilization. If the plates need to be kept, for whatever reason, they should be refrigerated, which will slow or stop further growth. Figure 14.7 shows examples of two air sampling plates taken by the author, each of which has multiple species growing.

If plates grow too fast, are too crowded, or overgrow other species colonies, then the plate is simply recorded as uncountable. Uncountable plates are not necessarily a problem if they are from the Before condition, although they may not provide a quantitative result or a value for the percent reduction.

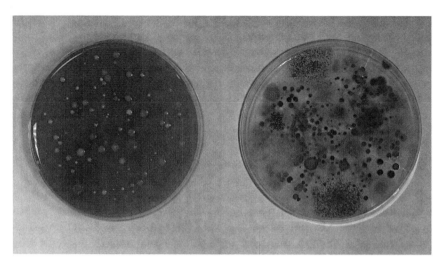

Fig. 14.7 Air sampled plates of bacteria (*left*) and fungi (*right*)

In most cases, bacteria and fungi will see similar net reductions. Furthermore, these reductions will apply to all individual species of bacteria and fungi encountered. In the event one species stands out for not having the same net average reduction as all the others, there may be at least two explanations. First, it may be an unusually resistant species (i.e. *Moraxella* or *Acinetobacter)*, in which case it is no fault or failure of the system. Second, it may be an environmental microbe (i.e. *Bacillus subtilis*) or a human commensal bacteria (i.e. *Streptococcus*) that contaminated the samples during the testing procedures. In the latter case the test procedures should be reviewed and the test repeated with extra care taken not to contaminate the samples.

Air sampling results will convert to airborne counts of bacteria and fungal spores, depending on the sampler used, and can be used as an indicator of air hygiene (Cox and Wathes 1995, Moschandreas et al. 1996, Reid et al. 1956, Yocom and McCarthy 1991). One criteria that has often been used to assess indoor aerobiology is the ratio of indoor fungi to outdoor fungi, which requires sampling of outdoor air (Rao and Burge 1996, Li and Kuo 1992). In general, the total counts of bacteria and fungi are the values that are most useful for verifying the effectiveness or performance of a UVGI system, and identifying or speciating the sampling results is not necessary. In some applications, such as hospitals, speciating bacteria may be of significant interest, and this is often best left to a laboratory. Health care facilities often have their own laboratories and these are the best places to get positive identification of microbes, especially when they are rare or problematic nosocomial microbes.

Bacteria and fungi, especially the common ones found in home and work environments, can often be roughly identified on the basis of their colony appearance alone. Corroboration of species identity can be pursued through the use of microscopy. Absolute identification may require microscopy, or even genetic testing. When multiple species occur on a petri dish, it is possible to create focused sample plates by removing individual colonies from crowded plates and regrowing them on fresh plates. These new plates will have a single species on them which can facilitate identification. Once single-species colonies are grown or images from microscopes are obtained, these images can be compared with published photos or bacteria and fungi from various texts, including Jordan and Burrows (1946), Flannigan et al. (2001), Sutton et al. (1998), and Ryan (1994). This approach is not guaranteed since some species, especially among the fungi, can take multiple forms and appear considerably different from published images.

Criteria for acceptability of test results are not standardized, but in general a properly designed and operated UVGI system should produced significant reductions in both air and surface microorganisms. Most surface disinfection systems will approach sterility, or a 99.9999% disinfection rate, although testing to this level of accuracy might be difficult to achieve. Air disinfection system will often perform at a level of 90–99% reduction of most common microorganisms (and this may include integral filter removal capacity). The minimum acceptable reduction for any airborne UVGI system would probably be about 50% – any lower than this and the error in the sampling process might not be exceeded.

14.10 Virus Sampling

Virus sampling in real world environments is a difficult venture. In fact, there seem to be no cases in which a human pathogenic virus has ever been successfully sampled and cultured from the air. Surface sampling of viruses has been successful, as in the recurring problem of cruise ships contaminated with Norwalk virus (Gunn et al. 1980, Wilson et al. 1982). Air sampling of animal viruses has been successful in agricultural environments (Thorne and Burrows 1960). Virus sampling is performed in much the same manner as bacterial sampling. The main difference is in the incubation process. Viruses are either incubated in host cells, such as chicken embryos (eggs) or cells suspended in some medium. The process can take two or three days. For more detailed information and examples of virus sampling procedures see Hermann et al. (2006), Hogan et al. (2005), Wilson et al. (1982).

References

Aerotech. 2001. IAQ Sampling Guide. Phoenix: Aerotech Laboratories.

Ambroise D, Greff-Mirguet G, Gorner P, Fabries JF, Hartemann P. 1999. Measurement of indoor viable airborne bacteria with different bioaerosol samplers. J Aerosol Sci 30(1):S699–S700.

Artenstein MS, Miller WS, Rust JH, Lamson TH. 1967. Large-volume air sampling of human respiratory disease pathogens. Am J Epidemiol 85(3):479–485.

Beggs CB, Sleigh PA. 2002. A quantitative method for evaluating the germicidal effects of upper room UV lights. J Aerosol Sci 33:1681–1699.

Bolton JR, Cotton CA. 2001. The Ultraviolet Disinfection Handbook. Ayr, Ontario, Canada: American Water Works Association.

Bolton JR, Linden KG. 2003. Standardization of methods for fluence (UV Dose) determination in bench-scale UV experiments. J Environ Eng 129(3):209–215.

Boss MJ, Day DW. 2001. Air Sampling and Industrial Hygiene Engineering. Boca Raton: Lewis Publishers.

Bradley D, Burdett GJ, Griffiths WD, Lyons CP. 1992. Design and performance of size selective microbiological samplers. J Aerosol Sci 23(S1):s659–s662.

Buttner M, Stetzenbach L. 1993. Monitoring of fungal spores in an experimental indoor environment to evaluate sampling methods and the effects of human activity on air sampling. Appl Environ Microbiol 59(1):219–226.

Buttner M, Cruz P, Stetzenbach L, Klima-Comba A, Stevens V, Emmanuel P. 2004. Evaluation of the biological sampling kit (BiSKit) for large-area surface sampling. Appl Environ Microbiol 70(12):7040–7045.

Cox CS, Wathes CM. 1995. Bioaerosols Handbook. Boca Raton: CRC Lewis Publishers.

EPA. 2006. Biological Inactivation Efficiency by HVAC In-Duct Ultraviolet Light Systems: Ultra-vIolet Devices, Inc.: U.S. Environmental Protection Agency. Report nr EPA 600/R-06/049.

Flannigan B. 1997. Air sampling for fungi in indoor environments. J Aerosol Sci 28(3):381–392.

Flannigan B, McEvoy EM, McGarry F. 1999. Investigation of Airborne and Surface Bacteria in Homes. Edinburgh, Scotland: IAIAS, pp. 884–889.

Flannigan B, Samson RA, Miller JD, editors. 2001. Microorganisms in Home and Indoor Work Environments. Andover, Hants, UK: Taylor and Francis.

Fletcher LA, Noakes CJ, Beggs CB, Sleigh PA, Kerr KG. 2003. The Ultraviolet Susceptibility of Aerosolised Microorganisms and the Role of Photoreactivation. Vienna: IUVA.

Foarde KK, Myers EA, Hanley JT, Ensor DS, Roessler PF. 1999. Methodology to perform clean air delivery rate type determinations with microbiological aerosols. Aerosol Sci Technol 30:235–245.

Griffiths WD, Upton SL, Mark D. 1993. An investigation into the collection efficiency & bioefficiencies of a number of aerosol samplers. J Aerosol Sci 24(S1):s541–s542.

Gunn RA, Terranova WA, Greenberg HB, Yashuk J, Gary GW, Wells JG, Taylor PR, Feldman RA. 1980. Norwalk virus gastroenteritis aboard a cruise ship: an outbreak on five consecutive cruises. Am J Epidemiol 112(6):820–827.

Hermann JR, Hoff S, Yoon KJ, Burkhardt A, Evans R, Zimmerman J. 2006. Optimization of a sampling system for recovery and detection of airborne porcine reproductive and respiratory syndrome virus and swine influenza virus. Appl Environ Microbiol 72(7):4811–4818.

Hogan CJ, Kettleson E, Lee M, Ramaswami B, Angenent L, Biswas P. 2005. Sampling methodologies and dosage assessment techniques for submicrometre and ultrafine virus aerosol particles. J Appl Microbiol 99(6):1422–1434.

Jensen PA, Todd WF, Davis GN, Scarpino PY. 1992. Evaluation of eight bioaerosol samplers challenged with aerosols of free bacteria. Am Ind Hyg Assoc J 53(10):660–667.

Jensen PA, Schafer MP. 1998. Chapter J: Sampling and Characterization of Bioaerosols. In: Cassinelli ME, O'Connor PF, editors. NIOSH Manual of Analytical Methods, NIOSH Publication, pp. 94–113. Atlanta: National Institute for Occupational Safety and Health, pp. 82–112.

Jordan EO, Burrows W. 1946. Textbook of Bacteriology. Philadelphia: W.B. Saunders Company.

Ke Q, Craik S, El-Din M, Bolton J. 2009. Development of a protocol for the determination of the ultraviolet sensitivity of microorganisms suspended in air. Aerosol Sci Technol 43(4):284–289.

Kowalski WJ. 1997. Technologies for controlling respiratory disease transmission in indoor environments: Theoretical performance and economics. [M.S.]: The Pennsylvania State University.

Kuo J, Chen C-L, Nellor M. 2005. Standardized collimated beam testing protocol for water/wastewater ultraviolet disinfection. J Environ Eng 131(5):828–829.

Lai KM, Burge H, First MW. 2004. Size and UV germicidal irradiation susceptibility of Serratia marcescens when aerosolized from different suspending media. Appl Environ Microbiol 70(4):2021–2027.

Lemmena S, Hafnera H, Zolldanna D, Amedicka G, Luttickenb R. 2001. Comparison of two sampling methods for the detection of Gram-positive and Gram-negative bacteria in the environment. Int J Hyg Environ Health 203(3):245–248.

Li C, and Kuo Y. 1992. Airborne characterization of fungi indoors and outdoors. J Aerosol Sci 23(S1):s667–s670.

Li C-S, Hao ML, Lin WH, Chang CW, Wang CS. 1999. Evaluation of microbial samplers for bacterial microorganisms. Aerosol Sci Technol 30:100–108.

Lin X, Reponen TA, Willeke K, Grinshpun SA, Foarde KK, Ensor DS. 1999. Long-term sampling of airborne bacteria and fungi into a non-evaporating liquid. Atmos Environ 33(26):4291–4298.

Luckiesh M, Taylor AH, Knowles T. 1947. Killing air-borne microorganisms with germicidal energy. J Franklin Instit Oct.:267–290.

Miller SL, Macher JM. 2000. Evaluation of a methodology for quantifying the effect of room air ultraviolet germicidal irradiation on airborne bacteria.Aerosol Sci Technol 33:274–295.

Moschandreas DJ, Cha DK, Qian J. 1996. Measurement of indoor bioaerosol levels by a direct counting method. J Environ Eng 122(5):374–378.

Rao CY, Burge HA. 1996. Review of quantitative standards and guidelines for fungi in indoor air. J Air Waste Mgt Assoc 46(Sep):899–908.

Reid DD, Lidwell OM, Williams REO. 1956. Counts of air-borne bacteria as indices of air hygiene. J Hygiene 54:524–532.

Riley RL, Kaufman JE. 1972. Effect of relative humidity on the inactivation of airborne *Serratia marcescens* by ultraviolet radiation. Appl Microbiol 23(6):1113–1120.

Riley RL, Knight M, Middlebrook G. 1976. Ultraviolet susceptibility of BCG and virulent tubercle bacilli. Am Rev Resp Dis 113:413–418.

Ryan KJ. 1994. Sherris Medical Microbiology. Norwalk: Appleton & Lange.

Sharp DG. 1938. A quantitative method of determining the lethal effect of ultraviolet light on bacteria suspended in air. J Bact 35:589–599.

Sharp G. 1939. The lethal action of short ultraviolet rays on several common pathogenic bacteria. J Bact 37:447–459.

Sharp G. 1940. The effects of ultraviolet light on bacteria suspended in air. J Bact 38:535–547.

Straja S, and Leonard RT. 1996. Statistical analysis of indoor bacterial air concentration and comparison of four RCS biotest samplers. Environ Internat 22(4):389.

Sutton DA, Fothergill AW, Rinaldi MG. 1998. Guide to Clinically Significant Fungi. Baltimore: Williams & Wilkins.

Thorne HV, Burrows TM. 1960. Aerosol sampling methods for the virus of foot-and-mouth disease and the measurement of virus penetration through aerosol filters. J Hygiene 58:409–417.

VanOsdell D, Foarde K. 2002. Defining the Effectiveness of UV Lamps Installed in Circulating Air Ductwork. Arlington, VA: Air-Conditioning and Refrigeration Technology Institute. Report nr ARTI-21CR/610-40030-01.

Vincent JH. 1995. Aerosol Science for Industrial Hygienists. New York: Pergamon.

Wilson R, Anderson LJ, Holman RC, Gary GW, Greeberg HB. 1982. Waterborne gastroenteritis due to the Norwalk agent: Clinical and epidemiologic investigation. AJPH 72(1):72–74.

Xu P, Peccia J, Fabian P, Martyny JW, Fennelly KP, Hernandez M, Miller SL. 2003. Efficacy of ultraviolet germicidal irradiation of upper-room air in inactivating airborne bacterial spores and mycobacteria in full-scale studies. Atmos Environ 37:405–419.

Yamaguchi N, Yoshida A, Saika T, Senda S, Nasu M. 2003. Development of an adhesive sheet for direct counting of bacteria on solid surfaces. J Microbiol Meth 53:405–410.

Yocom JE, McCarthy SM. 1991. Measuring Indoor Air Quality: A Practical Guide. New York: Wiley & Sons.

Chapter 15

UV Effects on Materials

15.1 Introduction

Ultraviolet exposure can cause a number of secondary effects on materials as a result of photochemical reactions and heat. These effects include solarization, photodiscoloration, photodegradation, damage to plants, and generation of ozone, which can itself degrade materials. These secondary effects are of concern to UVGI applications such as in-duct UVGI, cooling coil disinfection, Upper Room UVGI, hospital operating room systems, etc. This chapter addresses effects on materials not previously covered in other chapters, excluding health effects (see Chap. 12).

UV exposure, especially for prolonged periods of time, can cause photodegradation of organic and synthetic materials through photolysis, photo-oxidation, and other processes. Damage to materials can cause equipment problems, fire hazards, and unexpected costs. Photodegradation can impact system design and cause safety hazards. Much of what is known about UV photodegradation comes from studies of sunlight, which has UV components, and research is currently in progress to determine which materials can safely be used around UVGI systems. This chapter addresses these potential problems insofar as the available knowledgebase permits, providing a theoretical background and some basic mathematical modeling to facilitate ongoing research.

15.2 UV Effects on Materials

Much of the information of photodegradation and photodiscoloration of materials comes from studies of solar exposure. Much of the effect of sunlight on materials has been attributed to the UV component (IESNA 2000, Ellis and Wells 1941). UV can fade some wall paints, wallpapers, and drapery fabrics (GE 1950). Most common materials tend to be good absorbers of ultraviolet radiation and so may be subject to photodamage. Some materials may have high UV reflectivities, like aluminum, or have high transmissivities, like quartz glass and may absorb very little UV. Since reflectivity, transmissivity, and absorptivity are related quantities that sum to unity

W. Kowalski, *Ultraviolet Germicidal Irradiation Handbook*,
DOI 10.1007/978-3-642-01999-9_15, © Springer-Verlag Berlin Heidelberg 2009

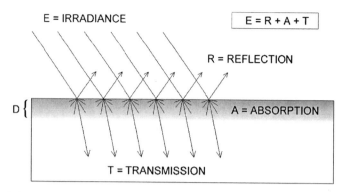

E = IRRADIANCE

E = R + A + T

R = REFLECTION

D {

A = ABSORPTION

T = TRANSMISSION

Fig. 15.1 Relationship of transmission, reflection, and absorption in a material surface exposed to UV irradiation. D is the depth to which photodegradation occurs

(see Fig. 15.1), materials that have low reflectivities and low transmissivities are likely to absorb UV at high rates. The absorption of UV by itself, is not necessarily an indicator that UV damage may occur, since it is the photochemistry which determines material effects. The absorption of UV may also result in heat, which can itself cause deterioration. The total absorption is, therefore, an indicator of the potential for photodegradation in materials, while reflectivity can indicate protective effects.

The refractive index of a material can affect the degree of reflection at acute angles. In the ultraviolet curing of pigments, the index of refraction at each wavelength is determined by the crystalline structure of the material (McGinniss 1976). The UV curing effect depends on internal light scattering, external reflection, and UV penetration, as well as the refractive index, the photochemical properties of the pigment, the density of the pigment, and film thickness. Titanium dioxide pigments, for example, differ from other crystalline pigments like silicon dioxide in the proportion of radiant energy that is transmitted, reflected, or absorbed by the film in which it is contained.

When a photon encounters a particle such as a molecule, one of three mechanisms will operate: (1) the photon may diffract and have its path altered, (2) a photon may reflect or refract off the particle, (3) the photon may penetrate and be refracted (Modest 1993). If any electromagnetic radiation is absorbed by the particle, it will occur in a quantum equal to the photon energy. As radiative energy penetrates a translucent material it gradually becomes attenuated over distance by absorption. The rate of attenuation is described adequately by exponential decay, and the transmissivity can be written as follows:

$$\tau_\eta = e^{-\kappa_\eta s} \tag{15.1}$$

where

τ_η = transmissivity, fractional
κ_η = absorption coefficient, 1/m
s = thickness, m

Both the transmissivity and the absorption coefficient are a function of the wavelength. In Eq. (15.1), 'η' is called the wavenumber and refers to a specific spectra or spectral range. Beyond the immediate surface of a material where most of the reflection will occur, the remaining incident radiation is either transmitted or absorbed, and the absorptivity, α_η, can be defined as:

$$\alpha_\eta = 1 - \tau_\eta = 1 - e^{-\kappa_\eta s} \tag{15.2}$$

When reflection, refraction, and diffraction, collectively known as scattering, all occur within a material, the transmissivity of the material layer must be rewritten as follows:

$$\tau_\eta = e^{-(\kappa_\eta s + \sigma_{s\eta})s} = e^{-\beta_\eta s} \tag{15.3}$$

where

$\sigma_{s\eta}$ = scattering coefficient
β_η = extinction coefficient

Absorbed photons of ultraviolet wavelengths can induce photochemical reactions that can modify existing polymers, induce cross-linking of polymers, and degrade polymers (Labana 1976). When polymers are exposed to ultraviolet radiation in the 100–400 nm spectrum, the wavelength exceeds the bond energy of the carbon bonds in the polymer or else exceeds the activation energy of chemical reaction (Moreau and Viswanathan 1976). The depth to which ultraviolet light penetrates the polymer creates a region of absorption where photochemical reaction may take place, and where photodegradation may occur. Since UV transmissivity tends to be very low for most materials, even at millimeter thicknesses, most of the photodegradation will occur on the immediate surface of a material, to a depth of a few millimeters. For polymers the depth of UV penetration is about 0.025 mm (25 μm). For glass the penetration depth may approach 0.5 mm (500 μm).

The absorption of a photon by a chemical substance may cause a direct breaking of a bond, a process called photolysis or scission. Wavelengths in the UV region have the greatest potential for inducing photolytic scission but photolysis is relatively rare (Feller 1994). More frequently, the absorption of a UV photon leads to excitation of electrons in a chemical bond, raising them to a higher level of energy. The excited molecule may then lose this energy by heat dissipation, by a chemical change in the molecule, by photolysis, or by transfer of energy to another molecule. A principle known as the Stark-Einstein Law, or the law of photochemical equivalence, states that one molecule is activated for every photon absorbed. In such a case the quantum yield (or quantum efficiency) is equal to unity. This ideal case is shown schematically in Fig. 15.2.

In the photodeterioration of paints, varnishes, and textiles, the quantum yield is often less than unity (Feller 1994). For the bleaching of certain dyes the quantum yield has been reported to be about 0.002, meaning a thousand photons must be absorbed before two molecules are bleached. Certain textiles have quantum yields

Fig. 15.2 Illustration of the
two stage process in which a
photon is absorbed and
excites an initiator molecule,
which then reacts with a
carbon bond in a polymer.
The result is photolysis or
photolytic scission of the
carbon backbone of the
polymer

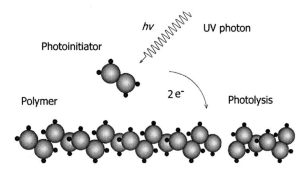

of about 0.03 (or 33 photons per molecule). Quantum yields as low as 0.0001 (10,000 photons per molecule) have been reported for some plastics.

Photosensitizers or photoiniators are light-sensitive catalysts that absorb energy and result in free radical species that can initiate polymerization (McGinniss 1976). In order for a light-sensitive reaction to take place, light energy, at the appropriate wavelength, must be absorbed by the reacting molecule, the photoinitiator. The degree to which the photoinitiator is consumed during the process is known as the molar extinction coefficient. If the molar extinction coefficient of the photoinitiator is large (or concentration is high), light energy will be absorbed in the same spectral region as the pigment. If the molar extinction coefficient of the photoinitiator is small (or concentration is low), light may not be absorbed or else it may be absorbed in a different region of the spectrum. In either case, the thickness of the films is an important determinant of the amount of energy absorbed, and for inks the UV-curable inks the thickness may be as low as 0.002–0.01 mm (2–10μ). Materials that will cure under prolonged UV exposure may be useful as protective coatings.

Many organic materials are susceptible to UV. UV irradiation can degrade wood surfaces to a greater extent than visible light, although the depth of UV penetration is limited (Kataoka et al. 2007, Hon and Ifju 1978). The depth of 1% transmittance of UV light in the 246–278 nm range was found to be 66–67μ (0.066–0.067 mm) while UV light in the 310–341 nm range transmitted to a depth of 99–131μ (0.099–0.131 mm). Hon and Ifju (1978) found that the penetration of UV into wood was negligible beyond 75μ (0.075 mm).

Polyurethane foams can degrade under UV exposure (Newman and Forciniti 2001). The thin cell membranes near the surface degrade such that only a thin network of polymer struts is left. This can lead to the escape of CFCs, which is an environmental problem.

Plastics are large molecules called polymers that are composed of repeated segments called monomers with carbon backbones. Often, the plastic is relatively resistant to UV but impurities and residual solvents in the plastic are responsible for photodegradation. UV energy absorbed by plastics can excite the creation of free radicals, which then cause secondary reactions and crosslinking.

Polymers tend to lose their properties under exposure to light, especially to ultraviolet wavelengths. A common polymer known as PADC (polyallyl diglycol carbonate) experiences surface hardening due to cross-linking and photo-oxidation under UV exposure (Tse et al. 2006). PADC has absorption spectra peaks at about 205 nm and 310 nm. Both the UVC band and the higher UVA/UVB spectral regions caused scission of the chemical bonds. Oxygen contributes to the reaction and the damage decreased with depth due to the lower rate of oxygen diffusion into the surface.

The photostability of polycarbonate (a common clear plastic with high tensile strength) was studied by Geretovszky et al. (2002). They found that polycarbonate absorbed strongly at 279 nm and 317 nm, and that morphological changes included chemical degradation and increasing roughness that was a linear function of irradiation time (i.e. UV dose).

The use of UV in health care facilities raises concerns about UV effects on hospital materials, including plastics and drugs. A variety of plastics are in use in hospitals and a number of these have been discussed above. Drugs are typically stored in glass or plastic containers that likely have low UV transmissivity, but some research in this area is needed. Commonly used drugs in hospitals like atropine, cefazolin, droperidol, epinephrine, and fentanyl must be protected from light exposure (Young 1991). The degree to which such drugs might be damaged, or to which their containers might be penetrated by UV, need to be considered in any application where such exposure is possible.

Materials that are used in conjunction with UVGI or that may be exposed to UV irradiation should preferably be resistant to UV damage, or *UV-proof*. UV-proof materials are those that will suffer no damage under extended UV exposure (IUVA 2005). Virtually all metals are UV-proof. *UV-resistant* materials are those that will suffer minimal damage under extended UV exposure. *UV-susceptible* materials are those likely to degrade significantly under extended UV exposure. No quantified measurements are currently available to classify the UV susceptibility of most materials. Values of UV transmittance and UV reflectivity may provide some indication of the UV absorptance, and therefore of potential photodamage. See Table 12.3 in Chap. 12 for the UV transmittance of fabrics, and see Appendix F for a list of UV reflectivities of materials, including paints, pigments, and coatings.

15.3 Activation Spectra

The spectra of light, including UV light, to which materials are exposed determines the photochemical mechanism and the type and degree of photodegradation (Rabek 1995). Just as with the erythemal spectra for human skin and the germicidal spectra for DNA molecules (see previous chapters), materials have an action spectra that determines their response. Often these action spectra can be described in terms of two or three peaks between 200 and 400 nm. Since the photon energies above 400 nm drop off rapidly, most of the photodegradation occurs with wavelengths below the UV cutoff of 400 nm (Feller 1994). For lignin paper, peak discoloration was observed at about 285 nm, followed by a decrease at lower wavelengths and a second peak below 265 nm (Nolan et al. 1945). Isoprene (rubber) shows peaks at about 355 nm and below 330 nm. The greatest bleaching effect on wool occurs around 450 nm but yellowing of wool is generated primarily by wavelengths below 360 nm (Lennox and King 1968). For nylon, the change in density and the elasticity as a function of UV wavelength has an apparent peak at 250 nm (Yano and Murayama 1980). Polymers may have absorption spectra with various peaks depending on the type and nature of the chromophores they contain (Gijsman et al. 1999). The reason for the variations in absorption spectra between materials relates primarily to the energies of dissociation of molecular bonds within the material or within the impurities in the material. Figure 15.3 shows an example of the action spectra for nylon 6.

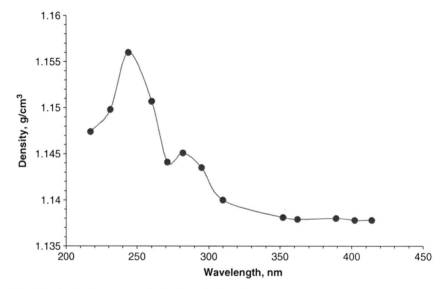

Fig. 15.3 Activation spectrum for density of nylon 6 irradiated in air for 6 h. Based on data from Yano and Murayama (1980)

High temperature alone can cause degradation of material properties for many materials, and the activation spectra can be increased at higher temperatures (Feller 1994). That is, higher temperatures can increase the photodegradation effects on materials. UV absorption can cause heating in materials, and this combination of effects can complicate the activation spectrum.

15.4 Solarization of Glass

Solarization is a process that can occur when certain types of transparent glass develop a pink or violet color under prolonged exposure to sunlight or UV. Solarization is the reduction in transmittance resulting from exposure to short wavelength UV radiation (Martin et al. 1999). Silica is the primary ingredient in most types of glass. Sand, a primary source of silica, contains iron as an impurity. Iron can produce a greenish color in glass, and manganese is often added to offset this tint and produce clear glass. Solarization can occur when colorless glass containing manganese oxide is exposed to ultraviolet light for extended periods of time and becomes photo-oxidized. The ultraviolet rays of the sun can induce this effect after a few years or decades.

Most types of glass have near-zero UV transmittance and very low UV reflectivity (i.e. 4%), and therefore they readily absorb UV radiation (Koller 1965). Although glass is a relatively stable material it can undergo a photochemical reaction under ultraviolet exposure that can result in a noticeable change in color. The extent of solarization depends primarily on the material composition of the glass, the UV wavelength, the intensity of the UV, and the time of exposure. In general, direct UV exposure from artificial sources causes greater solarization than natural sunlight. There is little effect on quartz glass, the kind used in UV lamps. In other types of glass, a visible color change can occur due to oxidation of the manganese ion from the weakly colored Mn^{2+} ion to the pink Mn^{3+} ion. The decrease in ultraviolet transmission is due to the corresponding change in iron ions from Fe^{2+} to Fe^{3+} (Koller 1965). The process of solarization can be reversed by heat treatment.

Solarization can cause a decrease in the UV transmission of glass at the wavelength of 254 nm. Decreases in transmittance are usually rapid at first, followed by a more gradual decrease in transmission. The greatest transmission loss occurs at wavelengths below 365 nm. Fused silica can experience a decrease in transmission of about 1% after a couple hundred hours of UV exposure, while Pyrex can decrease over 31% (Koller 1965). The depth of the solarization in glass may be less than 0.02 in (5 mm).

15.5 Photodiscoloration

Various surfaces and materials may experience surface discoloration due to long term exposure to low levels of UV, or to high levels of UV in the short term (Luckiesh and Taylor 1925). The rate at which dyes may fade under UV exposure depends on the photochemical properties of the dye, the physical structure of the material, and the UV spectrum (IESNA 2000). Ultraviolet energy in wavelengths below 300 nm may cause very rapid fading and even deterioration of materials. Bleaching tends to occur in regions of maximum spectral absorption and darkening can occur in regions of minimum spectral absorption. Fading can be exacerbated by the presence of moisture, especially in cellulose, and is inhibited in the absence of oxygen. The lignin component of paper may darken under UV exposure, but the degraded remnants of cellulose may be bleached by ultraviolet light, such that discoloration and bleaching may occur coincidentally (Feller 1994).

The exposure of wool, one of the most important fibers in the textile industry, to ultraviolet light can produce an initial stage of bleaching, followed by yellowing over longer periods of time (Lennox and King 1968). The photoyellowing of wool is tied to chromophores, or primary light absorbers (Millington 2006a). When a chromophore absorbs UV energy one of its electrons moves to a higher energy state, an excited singlet state, after which secondary reactions may take place. Wool contains a number of chromophores and the absorption spectrum contains peaks near 240 nm and 280 nm. Exposure of wool to UVC wavelengths can result in a green tint. UV radiation in the range 280–380 nm in the presence of oxygen results in rapid photoyellowing, but this can be offset by photobleaching at other wavelengths. Wool carpets are photobleached by sunlight because window glass attenuates UV below 350 nm. UV absorbers can be used to improve the lightfastness of wool exposed to sunlight (Holt and Milligan 1984).

Paints may often turn brownish near a UV source and cloths like draperies may lose their tints. Care should be taken to locate upper room and surface disinfection systems in locations where the least damage may be done to furnishings. Furnishings that are susceptible to UV damage should be removed and replaced with UV resistant materials. Some UV coatings are available that may protect indoor furnishings and the inclusion of UV stabilizers, like antioxidants, may prevent photodiscoloration, although they are primarily intended for sunlight protection (Millington 2006b). Plastic furniture coverings can offer good UV protection, as can other materials.

Yellowing of polymers from ultraviolet exposure is chiefly concentrated on the surface. Surface yellowing tends to block UV and protects the inner plastic. The fading of pigments and dyes can be evaluated in terms of the loss in concentration over time (Feller 1994). The depth of discoloration is reduced by the presence of pigments. As the concentration of pigments increases, the depth of discoloration or fading also decreases. A line of demarcation develops between the affected and unaffected portions, and the line may become more sharply developed at higher concentrations of pigment. This occurs because the penetration of light through translucent materials is reduced more sharply with distance than when the tint is less concentrated. A similar effect occurs for textiles, for which it has been noted that finer fibers deteriorate more rapidly under exposure than coarser fibers (Little 1964).

Dye molecules are excited by absorbed radiation and then take part in oxidation-reduction reactions. The dye may be oxidized or reduced, and the substrate may become reduced or oxidized as a result (Giles and McKay 1963). The oxidation of dye molecules takes place chiefly on cellulose substrates while reduction of dye molecules takes place chiefly on protein substrates like wool and gelatin. The fading of dyes on non-protein substrates is normally a photochemical oxidation process involving water and oxygen. Fading of dyes on protein substrates is normally a photochemical reduction process. The lightfastness of dyes can vary with the dye particle size. In fact, the rate of fading of dyes has been found to vary with $1/a^2$, where a=radius of dye particle (Giles et al. 1977).

15.6 Photodegradation

Photodegradation of materials can impact the service life of materials. If such materials are critical to equipment operation they can cause problems and, in the case of electrical insulation, may cause fire hazards. Various materials can be degraded by long-term exposure to UV without necessarily exhibiting any visible evidence of degradation. Visible light can induce photochemical changes but as the wavelength gets shorter towards the ultraviolet, the higher energy inherent in the photons is capable of inducing significant photochemical changes (Feller 1994). The rate of photodegradation of materials depends on the wavelength, the spectral absorptivity of the material, and the potential for photochemical reaction in the material components.

Figure 15.4 shows an example of how properties of materials may photodegrade under UV exposure. In this case the tensile strength and other properties of nylon 66 were studied under UV exposure and found to decrease over time relative to unexposed samples (Fornes et al. 1973). No data were available on the UV spectra or irradiance.

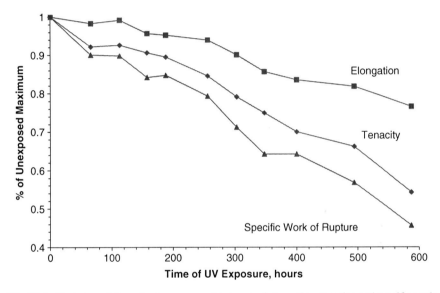

Fig. 15.4 Photodegradation of nylon measured in terms of elongation, tenacity, and specific work of rupture. Based on data from Fornes et al. (1973)

Many plastic materials deteriorate under extended ultraviolet exposure. Only limited data are available on the degree of photodegradation to which various plastics are subject. Plastic media filters, filter binding compounds and materials, wiring, duct insulation, and other materials that might be degraded by prolonged UV exposure to the point of failure should be avoided. Polymers vary in their response to UV absorption. Heat tends to increase UV photodegradation. Air filters composed of plastic fibers or bonded with certain glues are susceptible to UV damage. The plastic fibers may degrade and crumble to dust after extended UV exposure, and glues may lose their adhesion properties. Glass fiber filters are considered impervious to UV damage but sometimes the binding compounds used to hold the filter together may degrade and fail over time. Failed filters will leak and impact air disinfection system performance. Filters used with UVGI applications should be carefully selected so as to avoid such problems. Filters with plastic fibers can only be safely used when they are located outside the irradiance field of UVGI lamps.

The damage to paints and other organic coatings caused by UV exposure has been well known for a long time. The energy in ultraviolet radiation may initiate a chemical reaction leading to degradation of plastics. In addition to discoloration, other materials properties such as adhesion, flexibility, hardness, and toughness may be affected (Koller 1965). The effects are increased in the presence of oxygen. Ultraviolet radiation usually indices scission in carbon to carbon polymers, but in the phenyl methyl silicones cross-linking is more likely to occur when oxygen is present. The clear polymers in Table 15.1 are arranged in order of susceptibility to ultraviolet radiation in air based on their infrared absorptivity (adapted from Koller 1965).

Table 15.1 Relative UV Susceptiblity

UV Susceptibility Rank	Clear Polymer
1 (High)	Polyvinyl chloride
2	Soya oil alkyd resin
3	Polystyrene
4	Silicone resins (phenylmethyl type)
5	Styrenated alkyd resin (butylated)
6	melamin formaldehyde
7 (Low)	Urea formaldehyde (butylated)

Polyvinyl chloride (PVC) is a polymer that is widely used because of its high chemical resistance and relatively low cost. It is limited in applications (i.e. outdoors) because of its low thermal stability and its photochemical instability. The thermal and photochemical decomposition processes of PVC both take the form of a dehydrochlorination reaction that leads to discoloration and extensive changes in internal structure of the polymer, which adversely impacts the mechanical and electrical properties (Owen 1976).

Polypropylene is a polymer with wide commercial applications that is susceptible to degradation by light, heat, and oxygen (Grassie and Leeming 1976). It is widely used in carpeting, upholstery, and cordage. The rate of photodegradation of polypropylene can be measured in terms of weight loss. Polypropylene can degrade under sunlight exposure in as little as one year (Carlsson and Wiles 1970). The photodegradation of polypropylene is an oxygen-dependent process in which the primary photoinitiation steps involve backbone scission (Gijsman et al. 1999). The photodegradation of polypropyelene and other polymers is highly dependent on the type and concentration of chromophores, and the absorption of light is related to the extinction coefficient of the chromophores. Photo-oxidative degradation of polypropylene occurs primarily at the surface. The surface oxidation and associated backbone scission in polypropylene results in a reduction in surface toughness less resistance to stress, and loss in elongation properties. Resistance can be greatly improved by the use of UV stabilizers, including antioxidants (Hawkins 1972).

Kumar et al. (2002) studied the effects of UV exposure and moisture on carbon fiber-reinforced composites. Using 0.68 W/m^2 of UV at 340 nm, they alternated UV and moisture exposure for 1000 h, after which the transverse tensile strength was found to decrease by 29%. The specimen was also subject to weight loss, discoloration, and a decrease in thickness of about 7.5%. The combination of moisture and UV exposure operated synergistically to cause matrix erosion, microcracking, fiber debonding, and void formation.

Photodegradable plastics are those in which sunlight reduces the molecular weight of polymers so that they become brittle and disintegrate (Booma et al. 1994). Under UV exposure, chromophores such as carbonyl, ethylenic, or aromatic groups in the polymer absorb UV and undergo photochemical reactions resulting in photolysis and photo-oxidation. Due to antioxidants and UV stabilizers added during processing, conventional polymers do not degrade. UV sensitizers must be added to enhance their photodegradability. Photodegradable plastics include ethylene-carbon monoxide polymers, copolymers of ethylene, propylene, and styrene with a vinyl ketone, vinyl ketone copolymers, and ferric thiolates. Most commodity plastics undergo slow photodegradation. Polyethylene, polypropylene, polyacrylonitrile, and polyvinyl chloride absorb UV and photodegrade to some extent due to impurities. Polyethylene is more UV-resistant than polypropylene. Polyvinyl acetate and polyvinyl alcohol photodegrade.

There are as many as thirteen different properties of plastics which can be used as indicators of photodegradation, including coloration, tensile strength, elongation, hardness, degree of polymerization, infrared absorbance, etc. Experimental data indicates the response of most of these properties to extended ultraviolet exposure results in data that can be effectively modeled with exponential decay curves of one or more orders.

In air and surface disinfection, UV rarely, if ever, produces any by-products and those that do result from destruction of organic microbial materials or toxins are generally harmless. In water irradiation, especially contaminated water, some disinfection by-products (known as DPBs) are possible. UV irradiation at doses over 28,000 J/m^2 failed to demonstrate the production of any significant DPBs even in wastewater rich in organics (Trojan 2001). Kruithof and van der Leer (1990) confirmed that UV disinfection of water does not produce any mutagenic or carcinogenic by-products. Photodegradation of certain chemical contaminants can be a useful effect in some situations, such as in water treatment. Certain sensitive compounds such as endocrine-disrupting compounds (EDCs) can be photodegraded in water under UV exposure. UV irradiation, especially when combined with hydrogen peroxide, can degrade the known EDCs bisphenol-A, estradiol, ethinyl estradiol, and nonlyphenol (Linden and Kullman 2008). Degradation of chemicals by UV may also produce various by-products.

15.7 Damage to Houseplants

Prolonged UV exposure at levels that are below ACGIH limits for human exposure may cause plants to wilt and die (CIE 2003). Many houseplants are suited for shade and as a result they may not be able to handle even moderate levels of UV. Stray UVGI or low levels of UVGI from upper room systems has been reported, at least anecdotally, to kill houseplants.

Plants and the human visual system have evolved very different spectral sensitivities (IESNA 2000). Levels of UV that can be tolerated by human skin may eventually kill sensitive plants like ivy (Buttolph and Haynes 1950). Under solar exposure plants see negligible amounts of UVC band radiation, but in indoor environments, especially where local UVGI systems are present, plants may be exposed to levels of damaging UV irradiation that they cannot adequately deal with. The result is a gradual cessation of growth, browning of leaves, and eventually death, sometimes within only weeks or months.

Plants can be protected by removing them from the area, placing them in a UV shadow zone, or otherwise blocking them from direct and reflected UV rays with curtains. Designers and owners need to be aware of such possibilities and should take care to remove plants from the vicinity when installing systems that may shine UV on plants, even at levels that are below NIOSH TLVs. It may be possible, for example, to replace plants that may be subject to low level UV exposure with plants that can handle direct sunlight, as these may have a better chance of surviving prolonged low levels of UV exposure.

15.8 Ozone Production

Ozone is a hazardous gas that can be produced by UV lamps. If ozone is produced by UV lamps when it is not wanted, it is an indoor pollutant. If ozone is produced in the breathing zone of occupants, even at levels below OSHA limits for eight hours of exposure (0.1 ppm), it can still become a health hazard due to chronic inhalation. Ozone is corrosive and can oxidize organic materials and cause secondary hazards, including fire hazards if the ozone degrades wiring and insulation.

Ozone is generated by UV wavelengths of about 185 nm. Low pressure mercury lamps have a narrow spectral range of UV output around 254 nm and produce little or no ozone. Medium pressure mercury lamps that produce broad-range UV may produce ozone at sufficient concentrations that it can be smelled in the indoor air. Maximum ozone production rates of medium pressure UV lamps are on the order of 0.2% by weight of oxygen in the airstream. This level of ozone typically dissipates rapidly in the airstream of ductwork but local levels around lamps may persist through constant replenishment.

Materials in the vicinity of UV lamps in ducts, and in rooms equipped with Upper Room systems, may be subject to prolonged exposure to low levels of ozone, although this has no yet been demonstrated. Polymers can be modified by ozone, and rubber can be completely degraded from extended exposure to even low concentrations. Studies of ozonation on polyethylene, polypropylene, and polybutylene show chain scission without crosslinking and degradation of physical properties, including cracking (Murphy and Orr 1975). Ozone can cause the deterioration of paint films, and can fade fabrics and degrade textiles.

Although there are no known cases in which ozone produced by UV lamps has caused any major health problem, the possibility exists and any undue levels of indoor ozone measured or noted after a UV installation should be investigated. The measurement of ozone is a simple matter that requires a relatively inexpensive ozone detector or ozone sensor. Some people can smell ozone at levels below the detection limit of most sensors, or about 0.01 ppm, and therefore a 'sniff' test may be one means of verifying that no significant amount of ozone is entering the indoor breathing space.

15.9 Protective Coatings

The useful life of organic materials may be extended by protecting them with ultraviolet absorbers. Protective ultraviolet absorbers are typically effective in amounts of less than one percent (Koller 1965). One of the simplest UV-proof coatings is aluminum paint. With its high reflectivity and low transmissivity, very thin coatings will protect surfaces beneath. The addition of reflected irradiation in ductwork may enhance system efficiency, but it may also increase local UV levels. If the local UV exposure cannot be increased, then flat black paint (4% reflectivity) will also provide a UV-proof coating. The disadvantage of coatings that absorb, instead of reflect, UV, is that they may heat up and induce local radiant heating effects, which may not be desirable around electronics or wiring. Magnesium oxide coatings of at least 8 mm thickness have UV reflectivities of about 98%, which ensures minimal absorption (Tellex and Waldron 1955).

The use of photosensitized pigments or coatings for protection of equipment against prolonged UV exposure is a developing field of research. The properties of paint depend on the nature and concentration of pigment, and the addition of small amounts of colored pigment to white paint may result in a large decrease in ultraviolet reflection (Koller 1965). Oil-based paints usually have low reflectivity because of absorption by the oil. Zinc oxide paint reflects about 3% of UV, while white lead paint reflects about 60%. The choice of ceiling and wall paints, especially for Upper Room systems, can have a significant effect of scattering of UV rays, and the use of paints with very low UV reflectivity (i.e. about 5%) can enhance both the safety of such systems and minimize photodamage to furnishings.

Materials that would darken to UV after exposure could create a thin UV-proof film on the surfaces of polymers like PVC. This would enable them to develop resistance to further UV exposure, whether they were located outdoors in sunlight or inside an air handling unit equipped with UV.

Various stabilizers can be used to protect PVC and other polymers against photodegradation, including absorbers, quenchers, antioxidants, and HCL reactive species (Owen 1976). Strong ultraviolet absorbers dissipate light as thermal energy, and some of the best use hydrogen bond formation in the dissipative mechanism. The main classes of UV absorber compounds are the 2-hydroxybenzophenones and the 2-hydroxybenzotriazoles, which prevent or reduce the rate of formation of free radicals (Scott 1976).

Quenchers accept the energy of excited molecules and dissipate them as heat, and include oxygen (which can also oxidize) and a class of molecules known as Ni(II) chelates, which quench sensitizers as well as singlet oxygen. Also, the transition metal acetyl acetonates are among the most efficient quenchers of photo-excited states.

Antioxidants are effective because they intercept radicals or decompose radical initiators. They are among the most effective UV stabilizers. One class of antioxidants, nickel oximes, act both as screening agents and as radical trapping agents (Scott 1976). Some of the most powerful UV stabilizers belong to the class of peroxide decomposing preventive antioxidants, including metal dithiocarbamates, dithiophosphates, and cyclic phosphate esters.

HCL reactive species are a broad assortment of compounds and include inorganic lead salts, heavy metal soaps, organo-tin compounds, and epoxides. These compounds react with and deactivate potential sites of photoinitiation.

Table 15.2 summarizes the results of testing on UV absorbers for the protection of wool against photoyellowing by UVC radiation, based on data from Rose et al. (1961). Solvents were used to treat wool and the degree of protection was measured in terms of reflectance of the samples, and compared against water.

One of the most difficult problems with applying UV coatings is obtaining good adherence (Magny et al. 1996). Adherence to metals is difficult but not likely to be much of a problem since most metals tend to be UV-resistant. Polyolefines are the most difficult plastic substrates for UV coatings to adhere to, as they are inert and binding agents, like acrylates, cannot form a strong bond. Polystyrene is another

Table 15.2 Protection of Wool by UV Absorbers

UV Absorber Solvent	Proportion by Weight for 1 g wool		Protection
	Reagent	Solvent	%
Water	0.71	71	-10
Ethanol	0.71	56	20
Ethanol	0.17	23	30
Ethanol	0.17	23	40
5.8% maleic acid	-	24	40
2% maleic acid	0.071	36	60
2% aqueous H_2SO_4	0.13	19	65
0.25% aqueous H_2SO_4	0.5	50	80

polymer to which adherence of UV coatings is difficult to achieve, and can require applications of multiple acrylates for adhesion purposes.

15.10 UV Absorbers

Ultraviolet absorbers are commonly used to protect plastics and as coatings for materials that are used outdoors and exposed to ultraviolet radiation (Hamid 2000, Crews and Clark 1990). They function by preventing UV radiation from penetrating the coating and reaching the UV-sensitive substrate. They can be used on textiles like cotton and other dyed fabrics but the protective effects can vary greatly (Coleman and Peacock 1958). Protective absorbers can photodecompose, but usually at a much slower rate than the materials they are protecting (Hirt et al. 1961). Their photodecomposition is dependent on intrinsic properties of the materials and the wavelength of irradiation. To be useful, UV absorbers must have three properties: they must strongly absorb UV, they must dissipate the energy they absorb in a harmless manner, and they must persist for the lifetime of the substrate.

The most common types of UV absorbers include benzophenones, benzotriazoles, triazines, oxanilides, and cyanoacrylates (Hamid 2000). Table 15.3 shows these five types of UV absorbers, their approximate absorption spectrum peaks, and their extinction coefficients.

Table 15.3 Spectral Peaks for Classes of UV Absorbers

UV Absorber Class	Spectrum Peak	Exctinction Coefficient
	nm	
Benzophenone	285	15000
	325	10000
Benzotriazole	295	14000
	345	16000
Triazine	290	43000
	340	23500
Oxanilide	280	14000
	300	15000
Cyanoacrylate	305	13500

UV absorbers are most effective in relatively thin coatings on the surface layers of a polymer but, as a result, they are subject to surface damage, evaporation, and leaching. UV absorbers will lose absorbance over extended UV exposure and the rate of loss can be modeled through exponential decay models. Figure 15.5 shows an example of UV absorber loss vs. UV exposure dose.

In Fig. 15.5 a classic single-stage, multihit model, as used for virus inactivation modeling, was fit to the absorbance data. The model is given by:

$$A(t) = 1 - (1 - e^{-Dt})^n \qquad (15.4)$$

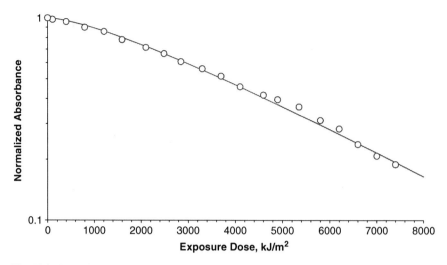

Fig. 15.5 Loss of absorber from a PMMA film during UV exposure. Line shows a multihit model curve fit. Based on data from Pickett and Moore (1995)

where

A = Absorbance, normalized to peak
D = UV exposure dose, J/m^2
t = time of exposure, s
n = multihit exponent, a constant for the process

In Fig. 15.4 the rate constant is 0.00028 m^2/kJ and the exponent is 1.6. These values were determined through curve-fitting the data to the shouldered, single-stage equation (15.4). A more extended data set might reveal a second stage at even higher doses, in which case a two-stage multihit model (see Eq. (3.6) in Chap.3) would be appropriate to mathematically define the entire process. In addition to modeling absorbance, a two-stage multihit model should be sufficient to model the photodegradation of almost any material property, including tensile strength, elasticity, hardness, etc., although some of these models might require a third stage, or more, due to secondary photochemical processes that can occur.

The initial absorbance due to a UV absorber in a coating can be estimated through the use of the following equation in which the extinction coefficient relates to the thickness, density, and molecular weight of the material (Hamid 2000).

$$A = \frac{\varepsilon C dr}{1000 M_W} \tag{15.5}$$

where

ε = extinction coefficient, m^2/kg

C = concentration of absorber in substrate, weight fraction
d = density of the polymer, kg/m^3
r = thickness, m
M_w = molecular weight

A variety of topical treatments for upholstery fabrics are available, including UV absorbers and antioxidants. Crews and Clark (1990) studied a number of commercially available UV absorbers and antioxidants and their effect on the lightfastness of fourteen different fabrics. They found that none of the fluorocarbon or silicone based soil repellent finishes containing UV absorbers significantly reduced fading, and that none of the UV absorbers or antioxidants improved lightfastness substantially.

15.11 Photodegradation Testing

Testing of materials is needed to determine what levels of UV exposure will cause photodegradation in common construction materials. This is especially true for in-duct UV air disinfection systems or cooling coil systems where interior HVAC materials and equipment are exposed to a degree that may result in structural failure or damage to electrical systems. Information culled from testing can provide designers with lists of materials that can be safely used for UV applications, what measures may be used to shield or protect components, and what UV coatings may be successfully applied. Examples of HVAC equipment that may be subject to UV photodegradation include synthetic filter media, filter bindings, plastic filter components, plastic or rubber gaskets, motor windings, electrical insulation, drive belts, internal duct insulation, and plastic piping. For other applications such as Upper Room systems and area disinfection units, it is of interest to know what indoor materials may be subject to discoloration.

Ultraviolet radiation exposure chambers have been used to generate weathering data for many commercial products including coatings, elastomers, plastics, and other construction materials (Martin et al. 1999). Traditional UV chambers consist of a UV light source and a specimen rack, and may include mirrors to direct irradiance onto specimens. Temperature must be controlled in such test chambers and often narrow-band spectral ultraviolet sources are used. The spectral UV dose must be known in order to compare performance of construction materials (Martin et al. 1996). Some guidelines and test protocols are available for exposing materials to light sources (ASTM 1996a, ASTM 2006a, b, ISO 1999).

Integrating spheres are commonly used for making reflectance or transmittance measurements and can provide a highly uniform irradiance field (IESNA 2000, ASTM 1996a, b, 2000). Integrating spheres have an inner surface of highly reflective material. The most reflective materials used include pressed polytetrafluoroethylene powder and bulk fluoropolymers which have reflectivities greater than 98% above wavelengths of 280 nm (Martin et al. 1999). When radiation is admitted to the interior of an integrating sphere it is diffusely reflected multiple times until it is spatially integrated and highly uniform. Approximately 5% of the interior surface area can be opened for specimen insertion without significantly affecting the internal irradiance field uniformity.

The effects of UV on materials are dose-dependent, as they are for microorganisms, and in testing it is generally desired to expose samples to much higher UV irradiance for much shorter periods of time. This approach carries with it the assumption of reciprocity, that higher UV dosing over short periods will approximate lower UV levels for extended periods of time. Some sources have validated the reciprocity law for photodegradation testing (Chin et al. 2005). Other sources have found that reciprocity does not apply at higher levels of irradiance (Delany and Makulec 1963). Based on the reciprocity law, the response of a material is dependent only on the total UV dose to which the specimen is subject, and is independent of the time of exposure or the UV irradiance. As with microorganism inactivation, it is likely that the response is nonlinear for very low doses and for very high doses, but this does not prevent accurate testing from being conducted in the mid-ranges of UV dose. Chin et al. (2005) found that an acrylicmalamine coating exposed to UV irradiance levels of 36–322 W/m^2 between 290 and 400 nm caused chemical changes in the coating, including chain scission, oxidation, and mass loss.

Photodegradability describes the ability of a material to degrade under UV exposure (Owen 1976). Presumably it would represent the fraction or percent of degradation that occurs from the peak strength. Associated with this there would be some limiting distance, a film thickness or penetration depth, to which UV would penetrate.

The rate at which photodegradation occurs is analogous to the rate at which inactivation of microorganism occurs, and an exponential decay model, for any given physical property, would suit quantification. Similar models, based on photochemical reaction rates, have been used in the past (Feller 1994). The photochemical degradation of materials is a dose-dependent function that depends only on the quantum yield and the molar absorption coefficient at the irradiation wavelength (Bolton and Stefan 2002). Rate constants for photodegradation applied to materials generally follow first-order kinetics (see Fig. 15.4) and, like bacteria, a single-stage rate constant associated with a D value can be used as an absolute indicator of material susceptibility. The presence of a shoulder (as in Fig. 15.5) can be accounted for with a multihit model. The presence of a second stage in photodegradation data has not yet been noted in material degradation studies.

References

ASTM. 1996a. Practice for operating light- and water-exposure apparatus (fluorescent UV condensation type) for exposure of nonmetallic materials. West Conshohocken, PA: American Society for Testing and Materials. Report nr ASTM G53-96.

ASTM. 1996b. Standard Test method for Solar Absorptance, Reflectance, and Transmittance of Materials Using Integrating Spheres. West Conshohocken, PA: American Society for Testing and Materials. Report nr ASTM E903.

ASTM. 2000. Standard Practice for Preparation of Textiles Prior to Ultraviolet (UV) Transmission Testing. West Conshohocken, PA: American Society for Testing and Materials. Report nr ASTM D6544.

ASTM. 2006a. Standard Practice for Operating Light Apparatus for UV Exposure of Nonmetallic Materials. West Conshohocken, PA: American Society for Testing and Materials. Report nr ASTM G-154-06.

ASTM. 2006b. Standard Practice for Exposing Nonmetallic Materials in Accelerated Test Devices that Use Laboratory Light Sources. West Conshohocken, PA: American Society for Testing and Materials. Report nr ASTM G-151-06.

Bolton J, Stefan M. 2002. Fundamental photochemical approach to the concepts of fluence (UV Dose) and electrical energy efficiency in photochemical degradation reactions. Res Chem Intermed 28(7–8):857–870.

Booma M, Selke SE, Giacin JR. 1994. Degradable plastics. J Elastomers Plastics 26:104–142.

Buttolph LJ, Haynes H. 1950. Ultraviolet Air Sanitation. Cleveland, OH: General Electric. Report nr LD-11.

Carlsson DJ, Wiles DM. 1970. Surface changes during the photo-oxidation of polypropylene. J Polymer Sci B8:419–424.

Chin J, Nguyen T, Byrd E, Martin J. 2005. Validation of the Reciprocity Law for Coating Photodegradation. JCT Res 01-JUL-05.

CIE. 2003. Ultraviolet Air Disinfection. Vienna, Austria: International Commission on Illumination. Report nr CIE 155:2003.

Coleman RA, Peacock WH. 1958. Ultraviolet absorbers. Textile Res J 28:784–791.

Crews PC, Clark DJ. 1990. Evaluating UV absorbers and antioxidants for topical treatment of upholstery fabrics. Textile Res J 60:172–179.

Delany WB, Makulec A. 1963. A review of the fading effects of modern light sources on modern fabrics. Illum Eng 58:76.

Ellis C, Wells AA. 1941. The Chemical Action of Ultraviolet Rays. New York: Rheinhold Pub Co.

Feller RL. 1994. Accelerated Aging: Photochemical and Thermal Aspects. Institute TGC, editor. Ann Arbor, MI: Edwards Bros.

Fornes RE, Gilbert RD, Stowe BS, Cheek GP. 1973. Photodegradation of Nylon 66 exposed to near-UV radiation. Textile Res J 43:714–715.

GE. 1950. Germicidal Lamps and Applications. USA: General Electric. Report nr SMA TAB: VIII-B.

Geretovszky Z, Hoppb B, Bertotic I, Boyd IW. 2002. Photodegradation of polycarbonate under narrow band irradiation at 172 nm. Appl Surf Sci 186(1–4):85–90.

Gijsman P, Meijiers G, Vitarelli G. 1999. Comparison of the UV-degradation chemistry of propylene, polyethylene, polyamide 6 and polybutylene terephthalate. Polymer Degrad Stab 65(3):433–441.

Giles CH, McKay RB. 1963. The lightfastness of dyes: A review. Textile Res J 33:527–577.

Giles CH, Walsh DJ, Sinclair RS. 1977. The relation between light fastness of colorants and their particle size. J Soc Dyers Colourists 93:348–352.

Grassie N, Leeming WBH. 1976. Influence fo UV Irradiation on the Stability of Polypropylene and Blends of Polypropylene with Polymethyl methacrylate. In: Labana SS, editor. Ultraviolet Light Induced Reactions in Polymers. Washington, DC: American Chemical Society, pp. 367–390.

Hamid SH. 2000. Handbook of Polymer Degradation. New York: Marcel Dekker.

Hawkins WL. 1972. Polymer Stabilization. New York: Wiley-Interscience.

Hirt RC, Searle NZ, Schmidt RG. 1961. Ultraviolet degradation of plastics and the use of protective ultraviolet absorbers. SPE Trans 1:26–30.

Holt L, Milligan B. 1984. Evaluation of the effects of temperature and UV-absorber treatments on the photodegradation of wool. Textile Res J 54(8):521–526.

Hon DN-S, Ifju G. 1978. Measuring penetration of light into wood by detection pf photo-induced radicals. Wood Sci 11:118–127.

IESNA. 2000. Lighting Handbook: Reference & Application IESNA HB-9-2000. New York: Illumination Engineering Society of North America.

ISO. 1999. Plastics – Methods of Exposure to Laboratory Light Sources. Geneva, Switzerland: International Organization for Standardization. Report nr ISO 4892-1.

IUVA. 2005. General Guideline for UVGI Air and Surface Disinfection Systems. Ayr, Ontario, Canada: International Ultraviolet Association. Report nr IUVA-G01A-2005.

Jepson JD. 1973. Disinfection of water supplies by ultraviolet radiation. Wat Treat Exam 22:175–193.

Johnson T. 1982. Flashblast: the light that cleans. Popular Science July:82–84.

Kataoka Y, Kiguchi M, Williams RS, Evans PD. 2007. Violet light causes photodegradation of wood beyond the zone affected by ultraviolet light. Holzforschung 61:23–27.

Koller LR. 1965. Ultraviolet Radiation. New York: John Wiley & Sons.

Kruithof J, van der Leer R. 1990. Practical Experience with UV Disinfection in the Netherlands. Cincinnati, OH: American Water Works Association, pp. 170–190.

Kumar BG, Singh RP, Nakamura T. 2002. Degradation of carbon fiber-reinforced epoxy composites by ultraviolet radiation and condensation. J Compos Mater 36(24):2713–2733.

Labana SS. 1976. Ultraviolet Light Induced Reactions in Polymers. Washington, DC: American Chemical Society.

Lennox FG, King MG. 1968. Studies of Wool Yellowing, Part XXIII, U.V. Yellowing and blue-light bleaching of different wools. Textile Res J 38:754–761.

Linden K, Kullman S. 2008. Impact of UV and UV/H_2O_2 AOP on EDC Activity in Water. Denver, CO: American Water Works Association.

Little AH. 1964. The effect of light on textiles. J Soc Dyers Colourists 80:527–534.

Luckiesh M, Taylor AH. 1925. Fading of colored materials in daylight and artificial light. Trans Illum Eng Soc 20:1078.

Magny B, Askienazy A, Pezron E. 1996. How to tackle adherence problems with UV-formulations. Proceedings: Radtech North America, p. 203.

Martin JW, Saunders SC, Floyd FL, Wineburg JP. 1996. Methodologies for predicting the service lives of coating systems. Blue Bell: Federation of Societies for Coatings Technologies.

Martin JW, Chin JW, Byrd WE, Embree E, Kraft KM. 1999. An integrating sphere-based ultraviolet exposure chamber design for the photodegradation of polymeric materials. Polymer Degrad Stab 63:297–304.

McGinniss VD. 1976. Ultraviolet Curing of Pigmented Coatings. In: Labana SS, editor. Ultraviolet Light Induced Reactions in Polymers. Washington, DC: American Chemical Society, pp. 135–149.

Millington KR. 2006a. Photoyellowing of wool. Part 1: Factors affecting photoyellowing and experimental techniques. Coloration Technol 122:169–186.

Millington KR. 2006b. Photoyellowing of wool. Part 2: Photoyellowing mechanisms and methods of prevention. Coloration Technol 122:301–316.

Modest MF. 1993. Radiative Heat Transfer. New York: McGraw-Hill.

Moreau W, Viswanathan N. 1976. Applications of Radiation Sensitive Polymer Systems. In: Labana SS, editor. Ultraviolet Light Induced Reactions in Polymers. Washington, DC: American Chemical Society, pp. 107–134.

Murphy JS, Orr JR. 1975. Ozone Chemistry and Technology. Philadelphia, PA: Franklin Institute Press.

Newman CR, Forciniti D. 2001. Modeling the ultraviolet photodegradation of rigid polyurethane foams. Ind Eng Chem Res 40:3346.

Nolan P, van den Akker JA, Wink WA. 1945. The fading of groundwood by light. Paper Trade J 121:101–105.

Owen ED. 1976. Photodegradation of Polyvinyl chloride. In: Labana SS, editor. Ultraviolet Light Induced Reactions in Polymers. Washington, DC: American Chemical Society, pp. 208–219.

Pickett JE, Moore JE. 1995. Photostability of UV screeners in polymers and coatings. In: Clough PL, Billingham NC, Gillen KT, editors. Polymer Durability: Degradation, Stabilization, and Lifetime Prediction. Washington, DC: American Chemical Society.

Rabek JF. 1995. Polymer Photodegradation. New York: Chapman & Hall.

Rose WG, Walden MK, Moore JE. 1961. Comparison of ultraviolet light absorbers for protection of wool against yellowing. Textile Res J 31(6):495–503.

Scott G. 1976. Mechanisms of Photodegradation and Stabilization of Polyolefins. In: Labana SS, editor. Ultraviolet Light Induced Reactions in Polymers. Washington, DC: American Chemical Society, pp. 340–366.

Tellex PA, Waldron JR. 1955. Reflectance of magnesium oxide. J Opt Soc Am 45:19–22.

Trojan. 2001. Disinfection By-Products (DPB). London, Ontario: Trojan Technologies, Inc. Report nr Trojan Technical Bulletin #55.

Tse KCC, Ng FMF, Yu KN. 2006. Photo-degradation of PADC by UV radiation at various wavelengths. Polymer Degrad Stab 91(10):2380–2388.

Yano S, Murayama M. 1980. Effect of photodegradation on dynamic mechanical properties of nylon 6. J Appl Polymer Sci 25:433–437.

Young DP. 1991. Ultraviolet Lights for Surgery Suites. Mooresville: St. Francis Hospital.

Chapter 16

Pulsed UV Systems

16.1 Introduction

Pulsed light systems using ultraviolet wavelengths have been shown to rapidly produce high levels of disinfection. The rate of disinfection is high due to the extreme UV power levels produced by pulsed UV lamps. The disinfection effect of pulsed light is primarily due to the UV content and can be modeled using the same basic UV rate constants previously presented for UV exposure. One exception exists – pulsed light induces a secondary effect, rapid heating due to the UVA content that can rupture microbial cells. This new disinfection process will also be addressed here after the basics of pulsed UV disinfection are covered.

Pulsed White Light (PWL), also called Pulsed Light or Pulsed UV Light (PUV), involves the pulsing of a high-power xenon lamp to produce broad spectrum light (100–1000 nm) with a large UVC component (MacGregor et al. 1998). Pulsed lamps require a separate power module as shown in the image in the chapter heading above. All forms of pulsed light are actually a subset of pulsed electromagnetic radiation, as illustrated by the breakdown in Fig. 16.1.

Pulsed light technology is currently being used to sterilize medical devices and is applied in the pharmaceutical packaging industry where translucent aseptically manufactured bottles and containers can be sterilized in a once-through light treatment chamber (Bushnell et al. 1998, Wallen et al. 2001). The chamber generates a light irradiance at the surface of the exposed containers of about 1.7 J/cm^2, or 1.7 W-s/cm^2. Two or three pulses are usually sufficient, depending on the fluence, to completely eradicate bacteria and fungal spores. Two pulses at 0.75 J/cm^2 each (1 J = 1 W-s) were sufficient to sterilize plate cultures of *Staphylococcus aureus* from an initial population of more than 7 logs of cfu (Dunn et al. 1997). Spores of *Bacillus subtilis, Bacillus pumilus, Bacillus stearothermophilus*, and *Aspergillus niger* were inactivated completely from an initial 6–8 logs of cfu with 1–3 pulses (Bushnell et al. 1998).

W. Kowalski, *Ultraviolet Germicidal Irradiation Handbook*,
DOI 10.1007/978-3-642-01999-9_16, © Springer-Verlag Berlin Heidelberg 2009

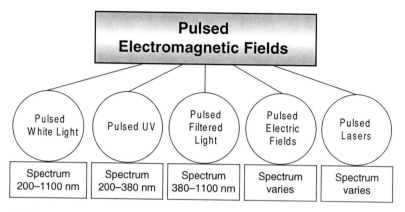

Fig. 16.1 Breakdown of the common types of pulsed electromagnetic radiation

16.2 Pulsed UV Disinfection

It has been suggested by various researchers that pulsed UV disinfection results from the same germicidal effects as UV lamps (Linden et al. 2000, Panico 2002). Others have claimed that the pulsed light disinfection process is more efficient than UV, but this is not borne out by detailed evaluation, and most direct comparisons indicate no significant differences exist between the disinfection processes (Marshall 1999, McDonald et al. 1999, Rice and Ewell 2001, Otaki et al. 2003). Some sources claim that pulsed light produces no tail, or second stage, but there is plenty of evidence that tailing exists in pulsed light studies when larger and more detailed data sets are developed and when similar UV doses are compared. It has been noted that pulsed UV has more penetrating ability, although this latter factor is primarily relevant to turbid liquids and certain foodstuffs (Krishnamurthy et al. 2004, Hillegas and Demirci 2003, Sharma and Demirci 2003). There are some differences between the spectrums of pulsed light and those of continuous wave UV lamps but these effects appear to be relatively minor. There are also some superheating effects from very high power pulsed light studies, but these effects are considered unique and independent from the germicidal effects, and are treated here separately.

Both the pulsed light and the continuous wave UV disinfection processes are, in fact, essentially identical and result from the germicidal effect of ultraviolet light. This germicidal effect, in turn, invariably produces a two-stage decay curve of survivability that is characteristic of each species. The observed differences between test results are primarily due to the fact that continuous wave UV systems operate at dosages that typically produce a single stage of decay, while pulsed light systems typically operate at dose levels that drive the survival curve into the second stage. A corollary of this hypothesis is that UV and pulsed light will produce approximately the same disinfection rate when they deliver the same total UV dose. The use of a complete two-stage mathematical disinfection model can bring both technologies into cohesion and make it evident that the use of a single stage rate constant to define pulsed light disinfection processes may be a misapplication.

16.3 The PUV Spectrum and Germicidal Effectiveness

The spectrum of pulsed light resembles the spectrum of sunlight but is momentarily up to 20,000 times as intense in the visible light spectrum (Bushnell et al. 1998). It is commonly believed that the ultraviolet light component of PWL and PUV is responsible for most of the biocidal effects since it is capable of causing thymine-thymine dimers, thymine-cytosine dimers, and '6-4 lesions' (Panico 2002, Rowan et al. 1999). Figure 16.2. shows the spectrum of pulsed light compared with the spectral distribution of germicidal effectiveness, which has a peak in the range from 263 to 266 nm (IESNA 2000). Pulsed light can also produce localized heating effects due to the infrared component (Krishnamurthy et al. 2008).

Another type of pulsed lamp, a surface discharge (SD) lamp, can operate at higher energy levels per pulse and greater UV efficiency (Schaefer et al. 2007). In an SD lamp, a high power electrical pulse discharges along the outer surface of a dielectric substrate, producing a plasma along the surface of the substrate. An outer envelope contains xenon gas but plays no part in the pulsed discharge. SD lamp life-times can be much longer than xenon flashlamps. Figure 16.3 shows a comparison of the spectrum of an SD lamp with that of a medium pressure UV lamp.

The spectrum of pulsed light differs markedly in form from those of low pressure and medium pressure UV lamps (see Fig. 2.1 in Chap. 2). Although there are significant differences in the spectra of MP and LP UV lamps, the differences in

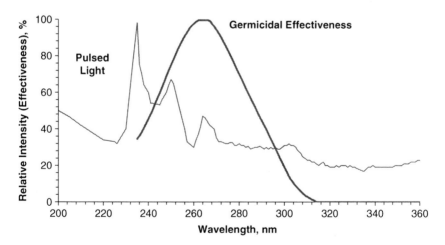

Fig. 16.2 Spectrum of a xenon pulsed light spectrum compared with the germicidal effectiveness of UV wavelengths. The pulsed lamp spectrum is representative only and may vary based on model or power input

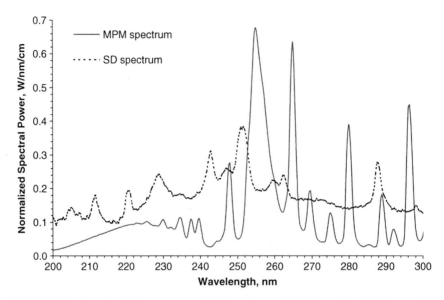

Fig. 16.3 Spectrum of an SD lamp compared with spectrum of a MP lamp. Spectral data provided by Phoenix Science and Technology, Chelmsford, MA

germicidal effects are relatively minor on a per watt basis, and differences that exist relate to secondary effects such as photoreactivation and ozone production (Masschelein 2002, Setlow 1966, Peccia and Hernandez 2001). Narrow-band UV lamps, in which the spectrum is confined to the UVC range, appear to allow more photoreactivation effects due to the fact that enzymes responsible for photorepair are subject to more damage by frequencies outside the UVC range. Broad-band UV lamps inhibit photorepair for this same reason. Since pulsed light spectrums are also broad-band, pulsed light also inhibits photoreactivation.

The spectrum of germicidal effectiveness varies between microbial species. Not all light emitted from lamps is equally germicidal for a given species, and to obtain a truly accurate account each wavelength should be weighted according to its germicidal effectiveness (Bolton and Linden 2003). Figure 16.4 compares the reported germicidal effectiveness obtained for UV and for pulsed light, scaled relative to 100% effectiveness at each respective peak. The differences between these curves may be due to experimental error or spectral differences in lamp output, but they are sufficiently similar to lend credence to the idea that pulsed light and continuous wave UV operate with the same disinfection processes. Otaki et al. (2003) found little significant difference between the pulsed light and UV inactivation of *E. coli,* as did Wang et al. (2005), and attributes the differences to the UV spectra, and also noted that PUV produces minor photoreactivation effects. Other researchers have also reported that the effects of PUV did not significantly differ from those of continuous-wave UV lamps (Hancock et al. 2004, Mofidi et al. 2001).

Bohrerova et al. (2008) compared the disinfection efficiency of pulsed light with that of continuous wave sources (LP and MP lamps) and, in contrast to previous

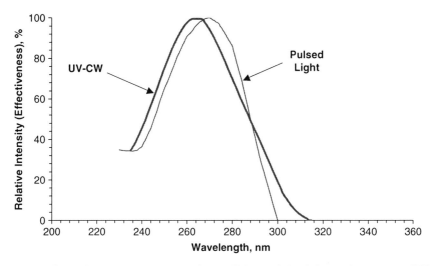

Fig. 16.4 Comparison of the reported germicidal efficiency of *E. coli* for continuous wave (CW) UV and pulsed light. Based on data from IESNA (2000) and Wang et al. (2005)

studies, found that inactivation was significantly faster with PUV at equivalent dose or fluence levels. A significant fraction of the enhanced PUV inactivation efficiency was found to be due to wavelengths above 295 nm. Overall PUV irradiation was approximately 2.4 times more effective at inactivating *E. coli* than MP irradiation, and 1.9 and 1.8 times more effective at inactivating phages T4 and T7 respectively. Because the phages are not able to reproduce or repair without a host, the improved disinfection rate under PUV can be explained almost exclusively by other DNA or protein damage, and possible rupture of the phage capsid from high-intensity irradiation at wavelengths greater than 400 nm. The visible portion of the PUV irradiation in these experiments was very intense and may be responsible for additional physical damage such as rupture or disintegration of the capsid. This hypothesis was supported by the fact that significant inactivation occurred in the phages when only wavelengths above 400 nm were used. *E. coli*, however, was not significantly affected by this portion of the spectrum, a fact that could be partly explained by damage to cell enzymes.

16.4 Modeling PUV Dose

A pulsed light disinfection model can be developed as a first order approximation (ignoring minor differences and photoreactivation effects) by accounting only for the wattage within the range of germicidal effectiveness, based on models used for standard UV lamps. This model excludes disintegration effects and only applies to UV damage. For UV lamps, the UV exposure dose (UV fluence) is computed as the irradiance multiplied by the time of exposure. PWL is not continuous like UV but can be approximated as a series of square waves. In some pulsed light systems the

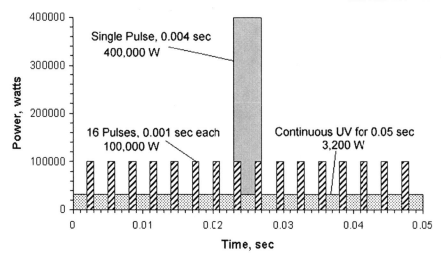

Fig. 16.5 Comparison of pulses at different power levels. In terms of total energy, the single pulse shown is equivalent to the sixteen pulses and both are equivalent to the continuous UV

duration of the pulse is about 0.1–3 ms (Johnson 1982). In other types of pulsed light systems the duration of the pulse is about 100 μs–10 ms (Wekhof 2000). Several pulses may be emitted per second. A figurative comparison of PWL and UVGI is shown in Fig. 16.5, including a single PWL pulse and a series of 16 PWL pulses. In terms of total dose all three exposures in Fig. 16.5 are equivalent.

The output power can be estimated by taking the total electrical input power and using conversion factors for system efficiency, % of bandwidth, and other factors as described by Capobianco (2000). A somewhat improved estimate of output power can be obtained by assuming that the light is pulsed in regular rectangular waves like those shown in Fig. 16.5. The energy in each pulse is the product of its amplitude in watts and duration in seconds:

$$E = A \cdot t_d \tag{16.1}$$

where

E = energy, W-s
A = amplitude of pulse, W
t_d = duration of pulse, s

As the frequency of pulses increases and their amplitude decreases proportionately, a limiting continuous output with the same total power is approached. Therefore, the power of a continuous source equivalent to a pulsed source can be computed by multiplying the energy per pulse by the pulse frequency:

$$P = E \cdot n \tag{16.2}$$

where

P = power, watts
n = frequency, number of pulses per second

In Fig. 16.4, the energy of the single pulse is:

$$E = 400{,}000(W) \cdot 0.004(s) = 1600\,(W \cdot s) \qquad (16.3)$$

The energy in the 16 smaller pulses is seen to be equivalent as follows:

$$E = 100{,}000(W) \cdot 0.001(s) \cdot 16 = 1600\,(W \cdot s) \qquad (16.4)$$

Both the single pulse and the 16 smaller pulses can be equated to the energy of continuous wave at 100 W for 16 s as follows:

$$E = 100(W) \cdot 16\,(s) = 1600\,(W \cdot s) \qquad (16.5)$$

The 16 smaller pulses in Fig. 16.5, which are distributed over 0.048 s, can also be equated to a continuous wave that produces the same power per unit of time as follows:

$$E = \frac{100{,}000(W) \cdot 0.001(s) \cdot 16}{0.048} = 33{,}333\,(W) \qquad (16.6)$$

The actual frequency of pulses in a pulsed light system is typically on the order of 1–10 per second, i.e., 1–10 Hz. Often, the power input to a pulsed light system will be specified in unit of W-s, which allows the equivalent or effective wattage to be computed. For example, if a pulsed power input of 500 W-s is used to generate 3 pulses per second for 1 s, the effective wattage is:

$$P_e = 500(W \cdot s) \cdot 3\,(Hz) = 1500\,(W) \qquad (16.7)$$

where

P_e = effective wattage, W

It should be noted here that P_e is the total wattage including white light and the UV components. These conversions facilitate determination of the UV rate constant under pulsed light exposure.

16.5 Modeling PUV Decay Curves

Most studies report that microbial cultures exposed to pulsed light have survival curves that exhibit no tailing or shoulder effects and therefore the tendency is to model pulsed light disinfection as a single stage decay curve with a simple log-linear model (Dunn et al. 1997). However, due to the extremely high UV dose produced, the disinfection rates for pulsed light systems are often on the order of 4–6 logs of reduction. For standard UV lamps, this level of disinfection has often been noted to produce a second stage decay curve, or a 'tail.' As a result, the use of a single stage model for pulsed light systems may not be accurate, although the degree of error introduced may or may not be significant. Table 16.2 summarizes test results from a variety of disinfection studies and Eq. (16.7) was used to estimate the wattage in each case. A number of other studies have been published besides those listed in Table 16.2, but in most of these studies the lamp output is either not reported or is not interpretable and no UV dose can be computed. In these PUV disinfection examples it was necessary to first determine the equivalent lamp wattage per the previously described methods, after which the irradiance field of the lamp was resolved using the computational methods previously developed for UV lamps as described in Chap. 7. The UV dose shown in Table 16.2 is determined based on the various authors' reported UV content. This approach has an inherent limitation – PWL has a considerable amount of UVA and UVB whereas low pressure and high pressure UV lamps have a much higher proportion of UVC. Studies in which pulsed light and UV are compared under controlled conditions indicate that these differences are minor or negligible (Rice and Ewell 2001, Otaki et al. 2003). In Table 16.1 the estimated percent of UV content is listed based on reported values, or is estimated where it was not specified.

PWL rate constants could easily be computed from the results in Table 16.2 but doing so would represent the modeling of a single stage of decay when, in all likelihood, the disinfection process is the result of two stages of decay – this is obvious from the high doses and low survival percentages. Most of the published data sets for pulsed light systems show no evidence of a shoulder, but the scale and dosages involved effectively preclude the possibility of elucidating the shoulder, which is likely insignificant. Microbes with high UV resistance, such as mold spores, are more likely to exhibit shouldering. Figure 16.6 shows a plot of a data set from Bushnell et al. (1998) in which *Aspergillus niger* spores were exposed to varying levels of pulsed light. The manifestation of the shoulder in this case is most easily explained by the relatively low doses used in the test.

The data set in Fig. 16.7, in which the dose was corrected for UV content, has been fitted to a shoulder curve of the form:

$$S = 1 - (1 - \exp^{-kD})^n \qquad (16.8)$$

The fitting of Eq. (16.9) to the data set in Fig. 16.7 was performed via trial and error, in which the exponent n and the rate constant k were adjusted until a reasonable fit was obtained. The resulting equation for the fitted curve in Fig. 16.7 is as follows:

$$S = 1 - (1 - \exp^{-0.0004D})^{5000} \qquad (16.9)$$

Table 16.1   Summary of representative PWL studies

Microbe	Medium	Dose J/m^2	Est. %UV	UV dose J/m^2	Survival frac	Reference
Staphylococcus aureus	plates	15,000	25	3,750	1.0E-07	Dunn (1996)
	solution	3,500	25	875	1.0E-08	Dunn (2000)
	plates	20,000	100	8,000	1.6E-07	Wekhof (2000)
	plates	168,000	100	168,000	1.3E-05	Krishnamurthy et al. (2003)
Staphylococcus epidermis	air	1,558	100	1,558	4.8E-04	UVDI (2002)
	plates	40,000	25	10,000	1.0E-07	Dunn (1996)
Bacillus subtilis spores	glass	11,000	30	3,300	1.0E-04	Wekhof et al. (2001)
	plates	80,000	40	32,000	1.00E-07	Wekhof (2000)
	solution	10,000	100	10,000	4.0E-07	Wekhof (1991)
	air	8,237	100	8,237	3.0E-03	UVDI (2002)
B. pumilis spores	plastic	45,000	56	25,200	1.0E-06	Bushnell et al. (1998)
	solution	53,000	25	13,250	1.0E-06	Dunn (2000)
Aspergillus niger spores	plastic	45,000	56	25,200	1.6E-01	Bushnell et al. (1998)
	plastic	45,000	65	29,250	1.0E-01	Bushnell et al. (1998)
	plastic	45,000	79	35,550	2.6E-03	Bushnell et al. (1998)
	glass	11,000	30	3,300	3.2E-06	Wekhof et al. (2001)
Aspergillus versicolor spores	air	1,781	100	1,781	2.2E-03	UVDI (2002)
Candida albicans	solution	9,500	25	2,375	1.0E-07	Dunn (2000)
E. coli	solution	800	100	800	1.0E-03	Wekhof (1991)
Salmonella enteritidis	eggshells	40,000	25	10,000	8.1E-07	Dunn (1996)
Cryptosporidium parvum	water	120,000	100	120,000	2.2E-02	Arrowood et al. (1996)
MS2 coliphage	plates	300	100	300	1.0E-02	Linden et al. (2000)

Some studies suggested that PWL eliminates the second stage of microbial decay curves, but this is a visual artifact resulting from the fact that the high doses involved quickly overwhelm the first stage, which disappears from the data plots. Sufficient data is available to show that second stages can be exhibited under PWL (Sonenshein 2001, Schaefer and Linden 2001, Rowan et al. 1999, MacGregor et al. 1998, Rice and Ewell 2001, Otaki et al. 2003). Figure 16.8 shows the results for pulsed light inactivation of spores of *Bacillus subtilis* in which the second stage is obvious.

The data in Fig. 16.7 was fitted to a multihit model two stage curve with a shoulder as defined by the following equation:

$$S = [1 - (1 - e^{-k_1 D})^n] + fe^{-k_2 D} \qquad (16.10)$$

Table 16.2 Combined PWL and UV results for *Aspergillus niger*

Dose J/m²	UV survival	Media	UV or PWL	Reference
0	1	–	both	(all)
315	0.1	surface	UV	Kowalski (2001)
750	0.1	–	UV	Gritz et al. (1990)
780	0.5	surface	UV	Luckiesh et al. (1949)
1000	0.1	water	UV	Jepson (1973)
1080	0.5	air	UV	Luckiesh et al. (1949)
1320	0.1	–	UV	Nagy (1964)
1600	0.125		UV	Zahl et al. (1939)
1781	0.0022	air	PWL	UVDI (2002)
1800	0.05	surface	UV	Luckiesh et al. (1949)
3384	0.00445	surface	UV	Fulton and Coblentz (1929)
4480	0.1	surface	UV	Chick et al. (1963)
5400	0.05	air	UV	Lucklesh et al. (1949)
9288	0.000003	water	UV	Begum et al. (2009)
11000	0.00000316	glass	PWL	Wekhoff et al. (2001)
45000	0.0001671	plastic	PWL	Bushnell et al. (1998)
45000	0.0000786	plastic	PWL	Bushnell et al. (1998)
45000	0.0000718	plastic	PWL	Bushnell et al. (1998)
53000	0.000001	solution	PWL	Dunn et al. (2000)

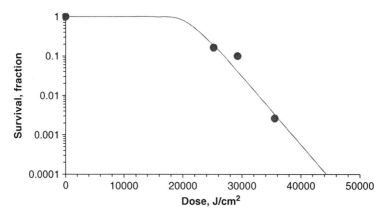

Fig. 16.6 Survival curve for *Aspergillus niger* fitted to a shoulder curve. Based on data from Bushnell et al. (1998)

The first and second stage rate constants in Fig. 16.7 were approximated by fitting to only the data within the well-defined first and second stages, respectively. The value of n was then found by trial and error with minor adjustments made to k_1 and k_2. The value of 'f' can be easily estimated by projecting the last two data points to the y axis. The resulting fitted curve is as follows:

$$S = [1 - (1 - e^{-0.05D})^{40}] + 0.000001e^{-0.00081D} \qquad (16.11)$$

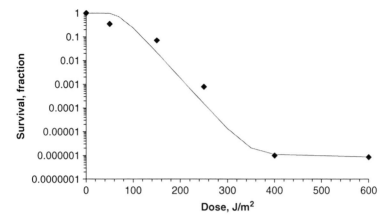

Fig. 16.7 Pulsed light inactivation curves for *Bacillus subtilis* spores. Based on data from Schaefer and Linden (2001). *Line* represents a multihit two stage model curve fit

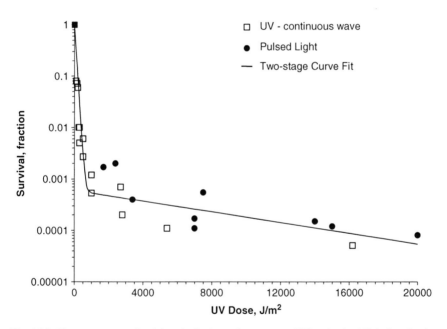

Fig. 16.8 Two-stage curve fit of data for both continuous wave UV and pulsed light inactivation of *Bacillus subtilis* spores. Based on data from Rice and Ewell (2001)

Equation (16.10) can be used to provide an accurate model of the two stage decay curve for pulsed light systems. It is usually necessary to have both the first and second stage rate constants but pulsed light test data typically lacks the detail necessary to define the first stage. However, UV data for the first stage should combine well with PWL data for the second stage provided the test methods from each are similar. An alternate single-stage model that addresses shoulder effects in thermal inactivation, called the Weibull model (VanBoekel 2002), has been adapted by Bialka et al. (2008) for use in evaluating pulsed light disinfection to account for penetration of partially opaque thicknesses of materials and liquids.

16.6 A General Model for PUV Disinfection

The complete model of Eq. (16.10) can be used for both pulsed light and UV disinfection applications. The shoulder can typically be neglected without introducing inaccuracy, and so the exponent 'n' can be assumed to be unity without impacting the overall prediction of disinfection rates. The resulting simplified two-stage curve is as follows:

$$S = (1 - f)e^{-k_1 D} + fe^{-k_2 D} \qquad (16.12)$$

Consider the data set provided by Rice and Ewell (2001), which includes both UV results (mostly first stage) and pulsed light results (mostly second stage) for *Bacillus subtilis*. Figure 16.8 shows the data points and the two-stage curve fit is as follows:

$$S = 0.9994e^{-0.012D} + 0.0006e^{-0.00012D} \qquad (16.13)$$

In Fig. 16.8 the two-stage model elegantly demonstrates the continuity between the pulsed light and UV inactivation process. In theory, this should be true for any given species, however, differences between testing methods and media, and species variants, can cause significant differences in results. Note that differences exist between these results and those of Schaefer and Linden (2001) although there is a general similarity in at least part of the data.

The problem of obtaining consistent results for the same species, whether with UV or with pulsed light, is a common one but it could be expected that as more data accumulates, and testing protocols become more standardized, the results should converge to similarity within normal ranges of error. Note, for example, that the first stage rate constant in Eq. (16.10), or $k_1 = 0.012$ m^2/J, is close to the average of reported values from Table 16.2, which is about 0.0123 m^2/J.

Consider another example of a species for which there is a variety of data for both UV and pulsed light systems. *Aspergillus niger* has been studied relatively extensively and the results of both UV and pulsed light studies are summarized in Table 16.2.

The results of Table 16.2 are shown plotted in Fig. 16.9, along with a fitted two-stage curve. Note that in spite of the varied test media and test conditions, the data produces a relatively intelligible two-stage curve.

The equation for the curve fit shown in Fig. 16.9 is as follows:

$$S = 0.99989e^{-0.0012D} + 0.00011e^{-0.00005D} \qquad (16.14)$$

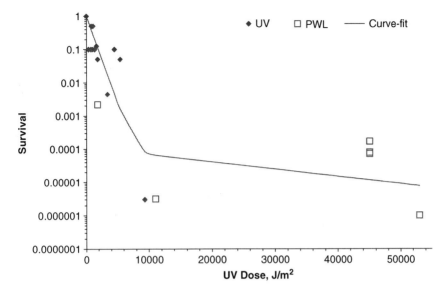

Fig. 16.9 Combined pulsed light and UV disinfection results for inactivation of *Aspergillus niger* spores. See Table 16.2 for data

It can be seen from the above analysis that there is no effective difference between pulsed light disinfection and LP or MP UV lamp disinfection processes. The differences in the rates of inactivation are due strictly to the higher power levels used with pulsed light and that fact that these high power levels drive the inactivation curve into the second stage. Previous comparisons of pulsed light inactivation with LP and MP lamps that claim higher inactivation rates are merely comparisons of first stage decay with second stage decay, nothing more, and the pulsed light inactivation process is seen to be due to the UV content alone, except when thermal disintegration effects occur.

16.7   Pulsed Light Disintegration

When PUV power levels are sufficiently high, two separate components operate in the disinfection process, the germicidal action of the UV band and bacterial cell wall rupture resulting from sudden heating. The heating results primarily from the UVA content and becomes manifest at higher overall power levels (Wekhof et al. 2001). At high dose rates the latter effect can be dominant. Per Krishnamurthy et al. (2008), absorption of pulsed light energy by bacterial cells may cause vaporization of the water leading to cell disruption, including cell wall damage, cytoplasmic membrane shrinkage, cell content leakage, mesosome rupture, and other effects that have been observed through microscopy. They categorize the inactivation mechanisms of pulsed light as either (1) photochemical effects from the UV,

(2) photothermal effects from heating during long-duration pulsed light treatment (>10 s), and (3) photophysical effects resulting from induced structural stresses that can occur with 5 s of pulsed light treatment or less. Theoretically, at moderate pulsed light doses only the effects of UV decay are manifest while at increasing dose a threshold will be passed at which the thermal disintegration effect becomes dominant.

If the pulsed light has the UVC/UVB spectrum filtered out, the UV decay effect and the disintegration effect will be due to the effects of UV-A alone. Filtering out the UV component of PWL and boosting total energy produces a form of pulsed light that, due to the absence of UV, is not considered hazardous to humans and yet maintains some degree of biocidal properties (Wekhof et al. 2001). Filtered pulsed light technology has potential applications in personnel decontamination since it can be used to expose contaminated skin surfaces without posing any known hazards. Applications of filtered pulsed light may also exist in the medical industry where the technology could be applied in operating rooms during procedures to control nosocomial surgical site infections. In theory, pulsed UVA light could disintegrate bacterial cells on tissue surfaces without doing damage to the tissue. It may be possible to manipulate this effect, by increasing the power and minimizing the pulse width, or by using a non-UV light source, to create more efficient surface sterilization systems.

The pulsed light disintegration effect should conform to the same disinfection models shown previously, but the associated rate constants will be different. Since the effects of pulsed light UV disinfection and pulsed light disintegration cannot be separated at present, no mathematical modeling is yet possible. Figure 16.10 shows a postulated graph of how the effects of pulsed light disintegration may be differentiated from UV effects based on power levels delivered per unit pulse. UV decay rates are purely a function of UV dose or fluence (above some minimum dose

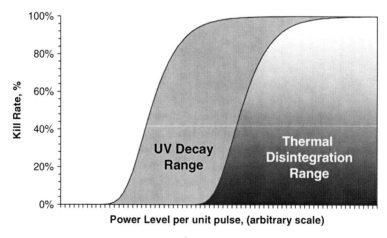

Fig. 16.10 Postulated graph of how the pulsed light disintegration effect may be differentiated from the UV effects based on power level per unit pulse

where effects become manifest), regardless of how fast they are delivered, but the disintegration effect is a function of how fast the dose is delivered (above some threshold where they become manifest) and the UVA content. Presumably, above this threshold there will be a point at which all of the decay will be due to thermal disintegration.

References

Arrowood MJ, Xie L-T, Rieger K, Dunn J. 1996. Disinfection of *Cryptosporidium parvum* oocysts by pulsed light treatment evaluated in an in vitro cultivation model. J Euk Microbiol 43(5):88S.

Begum M, Hocking A, Miskelly D. 2009. Inactivation of food spoilage fungi by ultraviolet (UVC) irradiation. Int J Food Microbiol 129:74–77.

Bialka K, Walker P, Puri V, Demirci A. 2008. Pulsed UV-light penetration of characterization and the inactivation of *Escherichia coli* K12 in solid model systems. Trans ASABE 51(1):195–204.

Bohrerova Z, Shemer H, Lantis R, Impellitteri C, Linden K. 2008. Comparative disinfection efficiency of pulsed and continuous-wave UV irradiation technologies. Wat Res 42:2975–2982.

Bolton JR, Linden KG. 2003. Standardization of methods for fluence (UV Dose) determination in bench-scale UV experiments. J Environ Eng 129(3):209–215.

Bushnell A, Cooper JR, Dunn J, Leo F, May R. 1998. Pulsed light sterilization tunnels and sterile-pass-throughs. Pharm Eng March/April:48–58.

Capobianco RA. 2000. Design Considerations for High-Stability Pulsed Light Systems. PerkinElmer Optoelectronics .

Chick EW, A.B. Hudnell J, Sharp DG. 1963. Ultraviolet sensitivity of fungi associated with mycotic keratitis and other mycoses. Sabouviad 2(4):195–200.

Dunn J. 1996. Pulsed light and pulsed electric field for food and eggs. Poult Sci 75(9):1133–1136.

Dunn J, Bushnell A, Ott T, Clark W. 1997. Pulsed white light food processing. Cereal Foods World 42(7):510–515.

Dunn J. 2000. Pulsed Light Disinfection of Water and Sterilization of Blow/Fill/Seal Manufactured Aseptic Pharmaceutical Products. Woodstock, IL: Automatic Liquid Packaging.

Fulton HR, Coblentz WW. 1929. The fungicidal action of ultraviolet radiation. J Agric Res 38:159.

Gritz DC, Lee TY, McDonnell PJ, Shih K, Baron N. 1990. Ultraviolet radiation for the sterilization of contact lenses. CLAO J 16(4):294–298.

Hancock P, Curry R, McDonald K, Altigilbers L. 2004. Megawatt, pulsed ultraviolet photon sources for microbial inactivation. IEEE Trans Plasma Sci 32:2026–2031.

Hillegas SL, Demirci A. 2003. Inactivation of *Clostridium sporogenes* in clover honey by pulsed UV-light treatment. Las Vegas, NV: ASAE.

IESNA. 2000. Lighting Handbook: Reference & Application IESNA HB-9-2000. New York: Illumination Engineering Society of North America.

Kowalski WJ. 2001. Design and optimization of UVGI air disinfection systems [PhD]. State College: The Pennsylvania State University.

Krishnamurthy K, Demirci A, Irudayaraj J. 2003. Paper # 03-037: Inactivation of *Staphylococcus aureus* using pulsed UV treatment. Storrs, CT: ASAE.

Krishnamurthy K, Demirci A, Irudayaraj J. 2004. Inactivation of *Staphylococcus aureus* by pulsed UV-light sterilization. J Food Prot 67(5):1027–1030.

Krishnamurthy K, Tewari J, Irudayaraj J, Demirci A. 2008. Microscopic and spectroscopic evaluation of inactivation of *Staphylococcus aureus* by pulsed UV light and infrared heating. Food Bioprocess Technol, published online 25 April 2008.

Linden KG, Shin G-A, Sobsey MD. 2000. Comparison of Monochromatic and Polychromatic UV Light for Disinfection Efficacy. American Water Works Association.

Luckiesh M, Taylor AH, Knowles T, Leppelmeier ET. 1949. Inactivation of molds by germicidal ultraviolet energy. J Franklin Instit 248(4):311–325.

MacGregor SJ, Rowan NJ, McIlvaney L, Anderson JG, Fouracre RA, Farish O. 1998. Light inactivation of food-related pathogenic bacteria using a pulsed power source. Lett Appl Microbiol 27:67–70.

Marshall T. 1999. Deadly pulses. Water Environ Technol 11(2):37–41.

Masschelein WJ. 2002. Ultraviolet Light in Water and Wastewater Sanitation. Boca Raton: Lewis Publishers.

McDonald K, Curry R, Clevenger T, Brazos B, Unklesbay K, Eisenstark A, Baker S, Golden J, Morgan R. 1999. Comparison of pulsed vs. continuous ultraviolet light sources for the decontamination of surfaces. Monterey, CA.

Mofidi A, Baribeau H, Rochelle P, DeLeon R, Coffey B. 2001. Disinfection of *Cryptosporidium parvum* with polychromatic UV light. Am Wat Works Assoc J 93:95–109.

Nagy R. 1964. Application and measurement of ultraviolet radiation. AIHA J 25:274–281.

Otaki M, Okuda A, Tajima K, Iwasaki T, Kinoshita S, Ohgaki S. 2003. Inactivation differences of microorganisms by low pressure UV and pulsed xenon lamps. Wat Sci Technol 47(3):185–190.

Panico LR. 2002. Instantaneous Surface Sanitization with Pulsed UV. July 8–9, Brussels, Belgium, p. 12.

Peccia J, Hernandez M. 2001. Photoreactivation in airborne *Mycobacterium parafortuitum*. Appl Environ Microbiol 67:4225–4232.

Rice JK, Ewell M. 2001. Examination of peak power dependence in the UV inactivation of bacterial spores. Appl Environ Microbiol 67(12):5830–5832.

Rowan NJ, MacGregor SJ, Anderson JG, Fouracre RA, McIlvaney L, Farish O. 1999. Pulsed-light inactivation of food-related microorganisms. Appl Environ Microbiol 65(3):1312–1315.

Schaefer RB, Linden K. 2001. Innovative Ultraviolet Light Source for Disinfection of Drinking Water. Washington, DC.

Schaefer R, Grapperhaus M, Schaefer I, Linden K. 2007. Pulsed UV lamp performance and comparison with UV mercury lamps. J Environ Eng Sci 6:103–310.

Setlow JK. 1966. Photoreactivation. Radiat Res Suppl 6:141–155.

Sharma RR, Demirci A. 2003. Inactivation of *Escherichia coli* 0157:H7 on inoculated alfalfa seeds with pulsed ultraviolet light and response surface modeling. J Food Sci 68(4):1448–1453.

Sonenshein AL. 2001. Killing of *Bacillus* spores by high-intensity ultraviolet light. Boston, MA: Tufts University School of Medicine/Xenon Corp.

UVDI. 2002. Report on pulsed light disinfection of microorganisms prepared by K. Foarde and Research Triangle Institute. Valencia: Ultraviolet Devices Incorporated.

VanBoekel M. 2002. On the use of the Weibull model to describe thermal inactivation of microbial vegetative cells. Int J Food Microbiol 74(1–2):139–159.

Wallen RD, May R, Reiger K, Holoway JM, Cover WH. 2001. Sterilization of a new medical device using broad-spectrum pulsed light. Biomed Instr Tech 35(5):323–330.

Wang T, MacGregor SJ, Anderson JG, Woolsey GA. 2005. Pulsed ultra-violet inactivation spectrum of E. coli. Wat Res 39:2921–2925.

Wekhof A. 1991. Treatment of contaminated water, air, and soil with UV flashlamps. Environ Prog 10(4):241–247.

Wekhof A. 2000. Disinfection with flashlamps. PDA J Pharm Sci Technol 54(3):264–267.

Wekhof A, Trompeter I-J, Franken O. 2001. Pulsed UV-Disintegration, a New Sterilization Mechanism for Broad Packaging and Medical-Hospital Applications. Washington, DC.

Zahl PA, Koller LR, Haskins CP. 1939. The effects of ultraviolet radiation on spores of the fungus *Aspergillus niger*. J Gen Physiol 16:221–235.

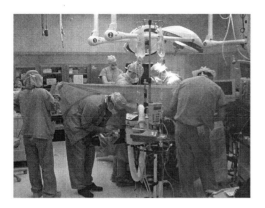

Chapter 17

Health Care Facilities

17.1 Introduction

This chapter addresses applications of ultraviolet germicidal irradiation (UVGI) in health care facilities, including laboratories and animal or veterinary facilities. This chapter does not review the applicable guidelines or standards for air quality in these facilities, which typically do not mention UVGI, but the references and Chapter 11 may be consulted for more detailed information. The types of UV systems covered in this chapter have been addressed in detail in previous chapters and these designs are not revisited here. Instead, this chapter discusses the various applications of UV to the indicated facilities, how they have been applied in the past, what effectiveness they have previously demonstrated, and how new UV systems may be applied to reduce the microbiological hazards associated with each type of facility. Also addressed here are the types of pathogens that are unique to certain facilities and, in particular, nosocomial pathogens and the problem of increasing drug resistance.

The use of ultraviolet germicidal irradiation in hospitals to control hospital acquired, or nosocomial, infections represents some of the earliest and most important applications of this technology. UVGI has been used for the disinfection of medical equipment, entire rooms, ventilation air, and surgical sites for well over half a century, often with definitive results. New applications are being developed even today and although UVGI is not a total solution to the problem of disease transmission in health care facilities, it can be an effective and economic component of any program designed to reduce hospital-acquired infections. This chapter discusses the various ways in which UVGI systems can be applied in hospitals and related health care facilities, including dental offices, laboratories, and veterinary facilities. Limited mention is made in most health care literature of UVGI, although some recent guidelines have acknowledged its potential effectiveness (CDC 2003, ASHRAE 2008). Chapter 11 can be consulted for more detailed information on the various standards and guidelines that address the use of UVGI in health care facilities. For a complete review of codes and standards for hospital ventilation and air quality see, for example, Kowalski (2006) or ASHRAE (2003).

17.2 Nosocomial Infections

Hospital acquired, or nosocomial, infections include any type of microbiological infections acquired in hospital environments, and since some of these have spread to communities, this category may be considered to include community-acquired infections also. Airborne and surface borne microbiological hazards in health care facilities can cause infections in both patients and health care workers. Nosocomial infections have proven to be a persistent problem in hospitals and some drug-resistant infections have transitioned from being hospital-acquired to community-acquired, including Methicillin-resistant *Staphylococcus aureus* (MRSA). Nosocomial infections can have complex, multifaceted etiologies that involve one or more routes of transmission and so the solution may involve more than one type of disinfection system (i.e. air disinfection, surface disinfection, water disinfection, equipment disinfection, personnel decontamination, etc.), procedural methods, and formal standards for air quality (Kowalski 2008a).

Drug resistance among nosocomial microbes is a growing problem. Bacteria that cause respiratory infections have developed increased drug resistance over the past ten years. Drug resistance is defined in terms of an IC_{50} value, or the concentration that causes 50% growth inhibition, and resistance is defined as a ten-fold increase in the IC_{50} value (Andrei et al. 2004, Andersson et al. 2004). The number of microbes that have demonstrated increased drug resistance in recent years is extensive and growing (Kowalski 2007b). The drug resistance of Streptococcal infections, which can cause Scarlet Fever, has increased from 0.8% to 28% in the past decade. MRSA has shown up repeatedly outside hospital settings and has become a contamination problem in athletic environments. Multidrug-resistant tuberculosis (XTB) has caused a resurgence in TB infections worldwide, and close to a million people die each year from this disease. Strains of multiple drug resistant (MDR) *Haemophilus influenzae* are being increasingly reported from around the world (Jain and Agarwal 1996). Multidrug resistant strains of *Acinetobacter baumannii, Pseudomonas aeruginosa*, and *Klebsiella pneumoniae*, have been recognized among casualties returning from battlefields (Davis et al. 2005). Various fungi can also cause nosocomial infections and unlike viruses and bacteria, most fungi hail from outdoors but can grow indoors (Kowalski and Bahnfleth 1998).

Evidence has been mounting over the years that the airborne transmission route plays a significant role in nosocomial infections (Fletcher et al. 2004). It has been estimated that the airborne transmission route may be responsible for as much as 10–20% of all endemic nosocomial infections (Brachman 1970). Airborne concentrations of bacteria in the OR bear a direct relationship to surgical site infections or sepsis. Figure 17.1 plots the rate of joint sepsis versus the airborne bacterial count in operating rooms from six hospitals. The data has been fit to a logarithmic equation as shown.

Mycobacterium tuberculosis, long known for its airborne transmission potential, is now a multi-drug resistant airborne pathogen that can cause outbreaks in hospitals (Breathnach et al. 1998). Evidence exists for airborne nosocomial transmission of *Acinetobacter, Pseudomonas*, and MRSA (Allen and Green 1987, Farrington et al.

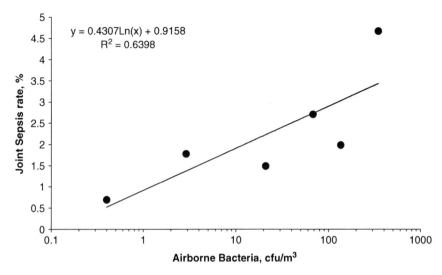

Fig. 17.1 The relationship between airborne bacteria and the incidence of joint sepsis. Adapted from Lidwell et al. (1983)

1990, Grieble et al. 1970). Table 17.1 summarizes the various nosocomial microbes that have airborne potential. Most nosocomial infections have been identified as having at least some potential for airborne transmission although most of them are primarily spread by other routes, such as direct contact. These microbes are ranked by estimated order of occurrence and are classified as Contagious, Noncontagious, and Endogenous (present as part of normal human flora). The last column in this table indicates whether or not the microbe has demonstrated any evolving drug resistance. It is clear from this tabulation that drug resistance is a growing and widespread problem and that all three microbial categories, viruses, bacteria, and fungi, have been developing such resistance. As drugs for treating these nosocomial infections become less effective and less available, increasing reliance on engineering methods such as UVGI may be one of the few remaining means of effectively dealing with the problem.

17.3 Operating Rooms and ICUs

Operating rooms (ORs), surgery suites, procedure rooms, treatment rooms, intensive care units (ICUs), and related facilities generally have very high levels of surface and air cleanliness, but these facilities are still far from being sterile. Many managers assume that if the design requirements for ventilation are met then the air is sterile, but this is rarely, if ever, the case. Levels of airborne contaminants in ORs are often lower than in the general wards, but not significantly so. For hospital air, WHO recommends the limits of 100 cfu/m^3 for bacteria and 50 cfu/m^3 for fungi (WHO 1988). There are currently no standards for OR aerobiology in the USA, but the

Table 17.1 Potentially airborne nosocomial microbes

Pathogen	Group	Type	Annual cases (USA)	Primary infections	Increasing resistance
Corynebacterium diphtheriae	Bacteria	Contagious	10	diphtheria	Yes
Acinetobacter	Bacteria	Endogenous	147	SSI, meningitis	Yes
Serratia marcescens	Bacteria	Endogenous	479	SSI, pneumonia, bacteremia	Yes
Aspergillus	Fungi	Noncontagious	666	Aspergillosis	Yes
Histoplasma capsulatum	Fungi	Noncontagious	1,000	Histoplasmosis	Yes
Haemophilus influenzae	Bacteria	Contagious	1,162	SSI, pneumonia, meningitis	Yes
Legionella pneumophila	Bacteria	Noncontagious	1,163	pneumonia	?
Klebsiella pneumoniae	Bacteria	Endogenous	1,488	SSI, pneumonia	Yes
Pseudomonas aeruginosa	Bacteria	Noncontagious	2,626	SSI, pneumonia	Yes
Staphylococcus aureus	Bacteria	Endogenous	2,750	SSI, pneumonia	Yes
Rubella virus	Virus	Contagious	3,000	rubella	?
Bordetella pertussis	Bacteria	Contagious	6,564	Whooping cough	Yes
Mycobacterium tuberculosis	Bacteria	Contagious	20,000	TB	Yes
Parainfluenza virus	Virus	Contagious	28,900	flu, pneumonia	?
Varicella-zoster virus	Virus	Contagious	46,016	VZV	Yes
Respiratory Syncytial Virus	Virus	Contagious	75,000	RSV	No
Streptococcus pyogenes	Bacteria	Contagious	213,962	Scarlet fever, SSI	Yes
Streptococcus pneumoniae	Bacteria	Contagious	500,000	pneumonia, meningitis	Yes
Measles virus	Virus	Contagious	500,000	measles	No
Influenza A virus	Virus	Contagious	2,000,000	flu	Yes
SARS virus	Virus	Contagious	10 (China)	SARS	?
Cryptococcus neoformans	Fungi	Noncontagious	high	cryptococcosis	Yes
Alcaligenes	Bacteria	Endogenous	rare	SSI	Yes
Bacteroides fragilis	Bacteria	Endogenous	rare	bacteremia, SSI	Yes
Blastomyces dermatitidis	Fungi	Noncontagious	rare	Blastomycosis	?

Table 17.1 (continued)

Pathogen	Group	Type	Annual cases (USA)	Primary infections	Increasing resistance
Burkholderia pseudomallei	Bacteria	Noncontagious	rare	melioidosis	Yes
Cardiobacterium	Bacteria	Endogenous	rare	endocarditis	Yes
Chlamydia pneumoniae	Bacteria	Contagious	rare	pneumonia	No
Coccidioides immitis	Fungi	Noncontagious	rare	coccidioido-mycosis	?
Haemophilus parainfluenzae	Bacteria	Endogenous	rare	pneumonia, meningitis	Yes
Moraxella	Bacteria	Endogenous	rare	otitis media	Yes
Mucor plumbeus	Fungi	Noncontagious	rare	mucormycosis	No
Nocardia asteroides	Bacteria	Noncontagious	rare	nocardiosis	Yes
Nocardia brasiliensis	Bacteria	Noncontagious	rare	nocardiosis	Yes
Pneumocystis carinii	Fungi	Noncontagious	rare	pneumocystosis	Yes
Rhizopus stolonifer	Fungi	Noncontagious	rare	zygomycosis	No

Note: *SSI = Surgical Site Infections*

current standard used in China is 200 cfu/m^3, while the EU has suggested a limit of 10 cfu/m^3, based on the ISO Class 7 cleanroom limit (EU Grade B) used in the pharmaceutical industry and as a target for ultra clean ventilation (UCV) systems (Durmaz et al. 2005, Kowalski 2007a). It is doubtful any ORs could achieve a limit of 10 cfu/m^3 unless the OR was unoccupied, but it is a target worth striving for.

Figure 17.2 is a figurative diagram showing the various sources of microbiological contamination in an operating room. If the supply air is assumed to be sterile (which is not always the case), then the main sources are the occupants, local internal surfaces, and infiltration. Although ORs are generally under positive pressure with respect to external areas, this may not always be the case and even if it is, the opening of doors can allow contaminants to enter.

The concentration of airborne bacteria in any OR is proportional to the number of personnel in the room (Mangram et al. 1999, Duvlis and Drescher 1980, Moggio et al. 1979, Kundsin 1976). The amount of surface contamination is also likely to be related to the level of airborne contamination since microbes aerosolize and settle continuously during occupation and activity. Air supplied to ORs, especially through HEPA filters, may be highly disinfected or even sterile, but most of the airborne bacteria hail from the room occupants, including the patient, and so increasing the rates of supply air above design guidelines is an approach that brings diminishing returns, often at high economic cost. Figure 17.3 plots the airborne concentrations of bacteria for six operating rooms representing measurements from 13

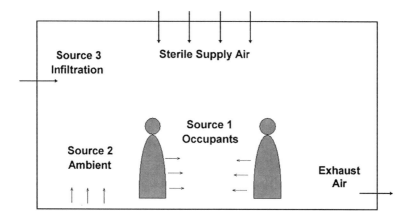

Fig. 17.2 Schematic of an OR with three major sources of contamination, the occupants, the ambient surfaces, and outside infiltration

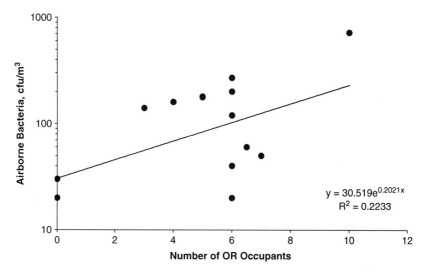

Fig. 17.3 Relationship between occupancy and airborne bacteria in ORs during surgery. Data taken by author (Kowalski 2008b)

operations and two empty ORs that had been cleaned and disinfected. A curve-fit of the data is shown.

Options for dealing with air and surface contamination coming from inside ORs include Upper and Lower Room UVGI systems, local recirculation units, continuous UV exposure systems, equipment disinfection systems, barrier UV systems, and Overhead Surgical UVGI systems. Overhead UVGI systems have been in use in some operating rooms since at least 1936 (Hart and Sanger 1939, Brown et al. 1996). Duke University has successfully used overhead UVGI systems since 1940 to maintain a low level of orthopedic infections (Lowell et al. 1980, Goldner and Allen 1973). Table 17.2 summarizes the various studies that have been performed

Table 17.2 Results of UV hospital field trials in operating rooms

System	Location	Infection/Operation	Infection Cases		Decrease		Reference
			Before (%)	After (%)	Net (%)	%	
Overhead	Duke University Hospital	SSI	5	1	4	80	Kraissl et al. (1940)
Overhead	NE Deaconess Hospital	SSI	13.8	2.7	11.1	80	Overholt and Betts (1940)
Barrier	Infant & Children's Hospital, Boston	SSI	12.5	2.7	9.8	78	Del Mundo and McKhann (1941)
Overhead	Montreal Neurological Inst	SSI	1.1	0.36	0.7	67	Woodhall et al. (1949)
Overhead	MA General Hospital	Craniotomies	5.3	0.70	4.6	87	Wright and Burke (1969)
Overhead	MA General Hospital	Laminectomies	4.1	0.30	3.8	93	Wright and Burke (1969)
Overhead	Duke University Hospital	Hip arthroplasty infection	5	0.5	5	90	Lowell et al. (1980)
Overhead	Brigham Hospitals	Hip & Knee	3.5	0.89	3	75	Young (1991)
Overhead	Watson Clinic, FL	Mediastinitis	1.4	0.23	1.2	84	Brown et al. (1996)
Overhead	St. Francis Hospital	SSI	1.77	0.57	1.2	68	Ritter et al. (2007)
Overhead	Average Reduction					80	

on operating rooms, including all those equipped with Overhead UV systems, and these show a net average reduction of 80% (barrier system results not included). One study showed that UVGI can reduce airborne microbial concentrations to below 10 cfu/m^3 in the operating room (Berg et al. 1991, Berg-Perier et al. 1992). Moggio et al. (1979) demonstrated a 49% decrease in airborne bacteria with an Overhead UV system. Lowell and Kundsin (1980) reports on an Overhead UV system that produced a 99–100% decrease in aerosolized *E. coli*, and that resulted in a 54% decrease in airborne bacteria during procedures.

The Overhead Surgical System implemented by Ritter et al. (2007) consists of a series of UV lamp fixtures suspended overhead in the OR in recessed lighting troffers. This system includes 8 UV lamps that produce a net average of about 25 μW/cm^2 at operating table height. According to Ritter et al. (2007), this system was able to reduce the surgical site infection rate from 1.77 to 0.5%. The study did not report airborne concentrations of bacteria in the ORs but it is possible that this system could reduce airborne levels significantly, even with personnel present. It is likely that this UV system inhibited both airborne transport and survival of bacteria on surfaces, including microbes that settle on equipment, on personnel and on floors. Since the irradiance of the UV in this system exceeds ACGIH limits, personnel are required to be completely covered, including eye protection, during operating procedures. Systems can be operated for short periods during surgery, but can also be operated for longer periods when the operating room is unoccupied to provide area decontamination between procedures. Figure 17.4 shows a diagram of how the UV lamp fixtures are located relative to the operating table. In older systems, UV lamp fixtures were often hung below the ceiling directly above the operating table.

Lower Room UV systems also have potential value in ORs, since most nosocomial bacteria will gravitate towards the floor, may settle, and may be stirred up again by activity. Figure 17.5 shows an example of how a Lower Room UV system may be applied to an operating or procedure room. Such systems will keep the floor and lower air (below about 18 in.) virtually sterile and turn the most contaminated portion of the room, the floor, into the cleanest area. Legwear would be required, depending on the irradiance produced, but UV levels above 18 in. would be below ACGIH/NIOSH 8-hour limits and upper body coverings and eyewear would not necessarily be required.

17.4 Isolation Rooms

Isolation rooms, like ORs, incorporate pressurization control to protect those inside or outside the room and often include supply air filtration. Isolation room systems are essentially 100% outside air purge air systems, and their performance characteristics are similar to ORs except that airflow rates and filtration levels may be different. Isolation rooms can be classified in three basic categories:

- Negative Pressure Isolation Rooms
- Positive Pressure Isolation Rooms
- Dual Purpose Isolation Rooms (Positive or Negative)

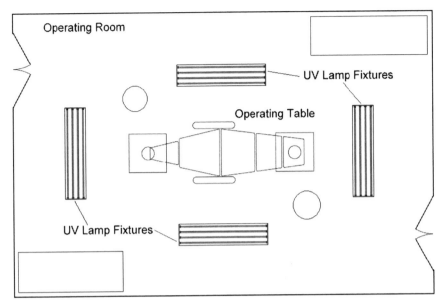

Fig. 17.4 Diagram of typical placement of recessed UV lamp fixtures in an operating room. Lamps are often located in recessed troffers in the ceiling

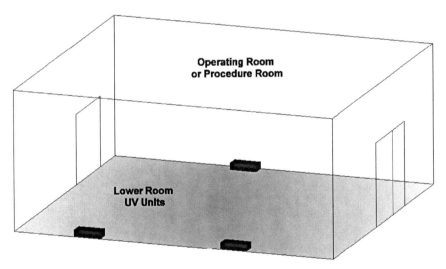

Fig. 17.5 Lower Room UV units can be used to disinfect floors and maintain high levels of disinfection in ORs without undue exposure hazards to occupants. UV levels above 18 in. would be below ACGIH TLVs and NIOSH RELs

The modern approach to designing isolation rooms is to include an anteroom that separates the isolation room from the corridor of the facility, thereby maintaining pressurization integrity during access (ASHRAE 1999). Air is supplied to the isolation room and typically exhausted from both the isolation room and the anteroom.

TB rooms are isolation rooms that maintain negative pressure so as to protect those outside the room. TB rooms often include internal recirculation units situated above the patient's bed which draw air across the bed. These recirculation units may include HEPA filters and UV lamps.

Schneider et al. (1969) applied in-duct UV for the supply air of an isolation ward and simultaneously irradiated the surrounding corridors, effectively controlling pathogens in the wards. UV recirculation units are routinely used in TB isolation rooms. Options for applying UV in isolation rooms are the same as those for ORs: Upper Room systems, Lower Room systems, Area Disinfection systems, After Hours systems, UV Barrier systems, and UV recirculation units can all be applied to improve conditions and reduce the risks to both patients and health care workers.

17.5 General Areas

Various types of UV systems have been applied successfully to hospitals to reduce infection rates in the General Areas, including In-duct airstream disinfection systems, Upper and Lower Room systems, and UV Barrier systems (Kowalski 2007a, Dumyahn and First 1999). Barrier systems in doorways between isolation wards were found to be effective in preventing the spread of chickenpox (Wells 1938). Barrier systems were found to reduce cross-infections across patient cubicles (Del Mundo and McKhann 1941, Sommer and Stokes 1942, Robertson et al. 1943). Upper room UVGI systems have been used successfully to control disease transmission in hospitals (see Fig. 17.6).

Upper Room systems were used at The New England Deaconess Hospital, The Infant and Children's Hospital in Boston, The Cradle in Evanston, and St. Luke's Hospital in New York, for the control of respiratory infections, which decreased by a net average of 50% (Overholt and Betts 1940, Del Mundo and McKhann 1941, Sauer et al. 1942, Higgons and Hyde 1947). Table 17.3 summarizes the results

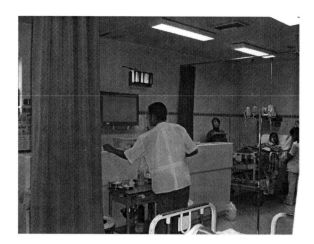

Fig. 17.6 Upper Room UV system (located on wall below ceiling) in a hospital ward. Image provided courtesy of Chuck Dunn, Lumalier, Memphis, TN

Table 17.3 Results of hospital field trials of Upper Room systems

System	Location	Infection	Infection cases		Decrease		Reference
			Before	After	Net	%	
	The Cradle, Evanston	Respiratory infection	14.5	4.6	9.9	68	Sauer et al. (1942)
	St. Luke's Hospital, NY	Respiratory infection	10.0	6.6	3.4	33	Higgons and Hyde (1947)
Upper room UVGI	Home for Hebrew Infants, NY	Varicella epidemic	97	0	97	100	Wells (1955)
	Livermore, CA Veteran's Hospital	Influenza epidemic	19.0	2.0	17.0	89	McLean (1961)
	North Central Bronx Hospital	TB conversions among staff	2.5	1	2	60	EPRI (1997)
	Average Reduction					70	

of field trials of Upper Room systems, and these show a net average reduction of infections of 70%. The Home for Hebrew Infants in New York successfully brought a halt to a Varicella epidemic using UVGI (Wells 1955). The Electric Power Research Institute (EPRI) Community Environmental Center funded the installation Upper Room fixtures in the VA Medical Center in Memphis in 1993 (EPRI 1996). The Memphis VA hospital found that the UV installations provided cost-effective protection against airborne pathogens. New York Central Bronx Hospital installed Upper Room UVGI systems in 1995 to successfully control TB and nosocomial infections, in a project that was supported by The New York Power Authority (NYPA), who provide electricity to all the NYC Health and Hospitals Corporation facilities (EPRI 1997).

Lower Room UV systems have not previously been used in hospitals (to this author's knowledge) but Wheeler et al. (1945) and Miller et al. (1948) used lower room UV systems to reduce respiratory infections at a Naval barracks, in conjunction with upper room systems. Since most bacteria and spores tend to settle downwards over time (and to get re-aerosolized by foot traffic) it is likely that Lower Room UV systems could have a major impact on airborne microbial contamination and should be able to help reduce infection rates.

In-duct air disinfection systems have been used in hospitals but there is no epidemiological data available on their effectiveness, although it is likely they would contribute to overall improvements in air quality of general areas (Luciano 1977). Another application in which hospitals may find UVGI beneficial is to reduce microbial contamination of cooling coils, an approach which pays energy dividends and saves costs (Keikavousi 2004). The latter approach can be combined with UV air disinfection to provide both cost savings and improved air quality. Also available for use in health care facilities are area decontamination units such as the one shown in Fig. 17.7, which is designed for rapidly disinfecting areas of MRSA and other microorganisms.

17.6 Hallways and Storage Areas

Hallways and storage areas surrounding operating rooms and ICUs can be a source of biocontamination that may be tracked into the ORs and isolation rooms by foot traffic. Hallways are also often used as storage areas (see Fig. 17.8) and both types of areas can accumulate microbiological contamination due to the greater surface area, which can act as both a microbial substrate and as protection (i.e. from sunlight or desiccation). The greater the total surface area in any given environment, the greater the potential for accumulated microbial contamination, which may include bacteria, fungal spores, and viruses.

Storage areas for supplies and equipment are often located adjacent to ORs and ICUs so that materials may be delivered expediently. Such storage areas can provide vast amounts of surface area on which microbial contamination may accumulate over time. Since such areas are only transiently occupied, UV area decontamination systems may be appropriately applied to provide high levels of cleanliness and

Fig. 17.7 Portable UV area disinfection system suitable for decontaminating entire hospital rooms. Photo of Tru-D disinfection unit provided courtesy of Lumalier, Memphis, TN

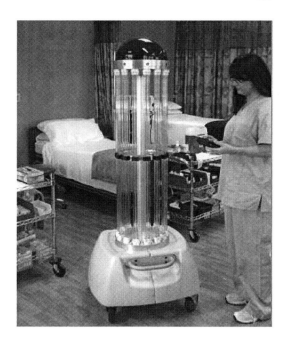

Fig. 17.8 Hospital hallways and other areas used for storage provide increased surface area for the accumulation of microbiological contamination

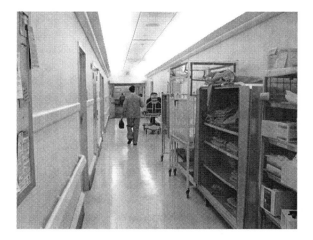

sterility, and to augment manual cleaning and disinfection procedures. In such areas it is also feasible to apply Upper Room systems which will not only disinfect the air but will tend to disinfect the lower room surfaces over time via the stray irradiance. Although the stray irradiance is below ACGIH/NIOSH limits for human safety, the accumulated dose to surfaces over time will provide fairly high levels of disinfection (see Chap. 9).

Bacteria and spores tend to settle downwards over time and accumulate near the floor, and are re-aerosolized by traffic or tracked into hallways by foot. The

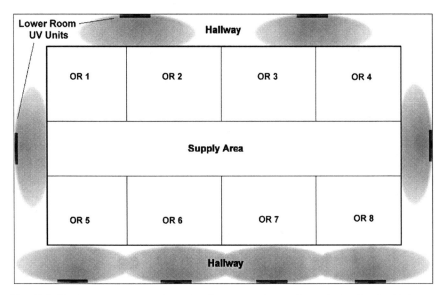

Fig. 17.9 Hallways surrounding Operating Rooms can be continuously irradiated with Lower Room UV Systems

placement of Lower Room UV units along the walls in several places will maintain the hallway floors under sterile conditions. NIOSH RELs for human exposure would not be exceeded above a height of about 18 in. if the system is designed appropriately. Lower Room units will produce no hazardous levels of UV above 18–24 in. and do not require any special protection other than normal leg attire and footwear. Figure 17.9 shows an example of how Lower Room UV systems may be applied in hallways surrounding a group of ORs, and which will help protect the ORs against contamination brought in from foot traffic.

17.7 Dental Offices

Dental offices have needs that are similar to operating suites except that the hazards are generally considered to be less severe. These hazards include airborne pathogens settling on open wounds during dental surgery or procedures, settling on equipment and being transferred to open wounds, and inhalation of microbes from patient to dentist. Face masks are in common use by dental workers and these are adequate for the most common threats. Dental workers are at risk for infection with various airborne pathogens such as *Mycobacterium tuberculosis*, influenza, and cold viruses (Araujo and Andreana 2002). The various types of UV systems that may be appropriately applied in dental offices are the same as those that are used in other health care applications: UV recirculation units, Upper and Lower Room systems, equipment disinfection systems, and area disinfection systems.

17.8 AIDS Clinics

Acquired Immune Deficiency Syndrome (AIDS) patients need heightened levels of protection against airborne microbes and other sources of contamination by bacteria, viruses, and fungal spores. Ideally, AIDS patients require sterile air, but this is difficult to achieve even in hospital settings. The use of air filtration combined with UV can go a long way towards providing a near-sterile environment.

Both ambient environmental microbes and normal commensal human microflora can present health threats to AIDS patients. A positive pressure isolation room, as described previously, can protect immunodeficient patients from possible contaminants and pathogens that might otherwise enter from the ambient environment (Linscomb 1994). Design criteria for HIV Rooms are similar to those for TB Rooms and isolation wards (see Fig. 17.10). Air supplied to, or recirculated in, HIV Rooms is normally filtered through HEPA filters, and UVGI systems are sometimes used in combination. The requirements for maintaining air pressure differential are the same as those for negative pressure rooms – airflow direction must be maintained from the positive pressure area to the negative pressure area.

Approximately 15% of AIDS patients also suffer from tuberculosis infection, and this presents a unique design problem (ASHRAE 1999). One possible solution is to nest a positive pressure (HIV) room within a negative pressure (TB) room or vice-versa (Gill 1994). Another approach is to modify a house such that the entire building is under positive pressure while the outdoor air (which is self-sterilizing) acts as a barrier to protect outsiders. The use of various UVGI systems can greatly enhance the ability of any building system to reduce airborne and surface microbial contamination. Recirculation units, Upper Room systems, Lower Room systems, and area decontamination units can all provide greater levels of protection to AIDS patients.

Fig. 17.10 Example of an Upper Room UV system (located at the ceiling) in a TB clinic. Photo provided courtesy of Pablo R. Antonio Designs and Consultancy, Inc. Makati City, Philippines

17.9 Hospital Laboratories

Any laboratories that deal with biological agents face potential inhalation hazards from handling mishaps and casual exposure (Kowalski 2006). Hospital laboratories have serious risks from pathogenic microorganisms brought in with infected patients and sometimes these risks are unknown until analyzed and identified. Therefore it is essential that the highest levels of air and surface cleanliness be maintained. Biological laboratories normally have a variety of systems and protocols to protect workers from such laboratory hazards, including laboratory hoods, air cleaning systems, pressurization zones, sterilization equipment, biohazard-rated facilities, personnel protective suits, and strict procedures for handling hazardous agents.

Existing procedures are considered adequate to protect workers and these are typically applied rigidly and diligently. All of the existing guidelines and standards offer similar guidance about the design and operation of the ventilation or air cleaning systems (for a review of these documents see, for example, Kowalski 2006 or ASHRAE 2003). Typically these guidelines recommend about 6–15 air changes per hour (ACH). The use of filtered 100% outside air is generally specified as an option and this is the most common approach taken today. Air is typically exhausted to outside, and certain codes may require HEPA filtration of the exhaust air, although the necessity for this is open to question. For systems that recirculate air, a minimum of 50% outside air (or maximum 50% return air) is suggested by some of the guidelines. HEPA filtration is also recommended for recirculated or exhaust air from biosafety cabinets (ASHRAE 1999).

There are four levels for categorizing containment laboratories, Biosafety Level 1, 2, 3, and 4 (DHHS 1993, CDC 2003). The basic characteristics of these laboratories are summarized in Table 17.4. There are no specific requirements for the use of UVGI in any biosafety laboratories, but UV systems are often used in them, especially in biosafety cabinets and for equipment disinfection, and all BSL containment laboratories may benefit from UV in various applications.

Since the use of 100% outside air systems consumes considerable energy in warm and cold climates the question may be raised as to whether it is not more economical to recirculate disinfected air, something that may be accomplished using UVGI combined with filtration. Filtration in combination with UVGI can also offer performance comparable to HEPA filtration without increasing risks (Kowalski 2006). Other applications for UVGI in laboratories include biosafety cabinets, equipment disinfection, surface disinfection systems, area decontamination systems, Upper and Lower Room systems, in-duct air disinfection, and unitary or local recirculation systems.

17.10 Animal Laboratories and Veterinary Facilities

Animal laboratories and veterinary facilities have unique hazards from zoonotic diseases that may be transmitted not from animals to humans but also between animals (Besch 1980, Tuffery 1995). Laboratories that handle animals are subject

Table 17.4   Basic characteristics of BSL containment laboratories

BSL	Requirements	Recommendations (ACH = Air Changes per Hour)	Application
1	No specific HVAC requirements	3–4 ACH, slight negative pressure	Microbial agents of no known hazard or minimal hazard
2	No specific HVAC requirements	100% OA, 6–15 ACH, slight negative pressure, use of safety cabinets	Microbial agents of moderate potential hazard
3	Physical barrier, double doors, no recirculation, maintain negative pressure	Exhaust may require HEPA filtration	Microbial agents that pose a serious hazard via inhalation
4	Physical barrier, double doors, no recirculation, maintain negative pressure, etc.	Requirements determined by biological safety officer	Microbial agents that pose a high risk of lethality via inhalation

to occupational hazards from a wide variety of infections, especially respiratory infections (Kowalski et al. 2002, Benirschke et al. 1978). In addition to pathogenic disease hazards, laboratory workers can develop allergies from prolonged or chronic exposure to animals (Hunskaar and Fosse 1993).Many species of animal diseases have the ability to transmit to humans or vice versa since the biological and physiological similarities between humans and animals are sufficient to permit such exchanges. Whenever such interspecies transmission occurs, secondary transmissions are rare. Often, such interspecies transmissions occur by direct contact or very close proximity, but airborne transmission is an ever-present possibility. Contact transmissions may be controlled procedurally, but the control of airborne transmission requires engineered systems. Table 17.5 lists the wide variety of zoonoses that can potentially transmit by the airborne route in animal laboratory or other animal facilities (adapted from Kowalski et al. 2002). Most of these pathogens and allergens can also transmit by direct contact and other means, such as ingestion or vie blood contamination. Few of the zoonotic microbes in the list have been evaluated for their UV susceptibility, but most of them are sufficiently physiologically similar to various human pathogens that the UV rate constants for the human pathogenic species may be used as an approximation. Genomic analysis of UV susceptibility is also possible (Kowalski et al. 2009).

Applications for UVGI systems in animal laboratories and veterinary facilities are much the same as for biological laboratories and are likely to be just as effective. These applications include biosafety cabinets, equipment disinfection systems, surface disinfection systems, area decontamination systems, Upper and Lower Room systems, in-duct air disinfection, and unitary or local recirculation systems.

Table 17.5 Airborne pathogens and allergens in animal laboratories

Airborne pathogen	Source or infected animal	Airborne pathogen	Source or infected animal
Acinetobacter	Env., soil, sewage, Rats, swine bldgs	Mucor plumbeus	Env., sewage, Guinea Pigs
Actinomyces bovis	Hamsters	Mumps virus	Humans, primates, Rodents
Actinomyces israelii	Humans, cattle, Rabbits, hamsters	Mycobacterium africanum	Monkeys
Aerococcus viridans	Rodents, Rabbits	Mycobacterium avium	Env., water, Mice
Aeromonas spp.	Env., rodents, soil	Mycobacterium bovis	Monkeys
Alcaligenes	Humans, soil, water, swine bigs	Mycobacterium lepraemurium	Rodents
Animal dander	Rats, dogs, cats, horses, etc.	Mycobacterium microti	Rodents
Avian adenovirus (FAV)	Birds	Mycobacterium tuberculosis	Humans, sewage, Monkeys
Bacillus anthracis	Cattle, sheep, Mice, Horses	Mycoplasma pulmonis	Rats & mice

Table 17.5   (continued)

Airborne pathogen	Source or infected animal	Airborne pathogen	Source or infected animal
Bacteroides fragilis	Humans, Rodents, Rabbits	Newcastle Disease Virus (NDV)	Birds
Bordetella bronchiseptica	Rabbits, Cats	Nocardia asteroides	Env., sewage, Rodents, Rabbits
Bovine adenovirus	Bovines	*Paecilomyces variotii*	Env., Rats
Brucella	Goats, cattle, swine, dogs.	Parainfluenza virus	Humans, Monkeys, dogs, rats
Burkholderia cepacia	Env., Rabbits	Paravaccinia	Cattle, humans
Burkholderia mallei	Env., Horses, mules, nosocomial	*Pasteurella lepisceptica*	Rabbits
Burkholderia pseudomallei	Env., rodents, soil, nosocomial	*Pasteurella multocida*	Rabbits, Rodents
Canine distemper virus (CDV)	Dogs	*Pasteurella pneumotropica*	Rodents
Chlamydia psittaci	Birds, fowl	*Pasteurella* spp.	Monkeys
Clostridium perfringens	Env., Humans, Animals, soil	Pneumococcus Type II	Rats, Guinea Pigs
Coccidioides immitis	Env., soil, Guinea Pigs, Rabbits	Pneumocystis carinii	Env., monkeys, animals
Corynebacterium bovis	Mice	Pneumonia Virus of Mice (PVM)	Mice
Corynebacterium kutscheri	Mice	Poxviruses	Rabbits, Sheep, Swine, Mice, Horses, Fowl, Goats, Cows
Coxiella burnetii	Cattle, sheep	*Pseudomonas aeruginosa*	Env., sewage, swine bldgs
Coxsackievirus	Humans, Mice, Rabbits, Hamsters, swine, primates	*Pseudomonas diminuta*	Rats, Guinea Pigs
Diplococcus pneumoniae	Monkeys	Reovirus	Humans, birds, mice
Echovirus	Humans, Mice. primates	Respiratory Syncytial Virus	Humans, Chimpanzees
Enterobacter cloacae	Humans, Env., Rabbits	Reston Virus	Monkeys
Equine rhinop-neumonitis	Horses	Rubella virus	Humans, Monkeys
Feline picomavirus	Cats	Sendai virus	Rodents, Hamsters
Francisella tularensis	Animals, Hamsters	Sialodacryoadenitis virus (SDAV)	Rats

Table 17.5 (continued)

Airborne pathogen	Source or infected animal	Airborne pathogen	Source or infected animal
Guineapig adenovirus	Guinea Pigs	Simian adenovirus	Primates
Haemophilus spp.	Rodents, Guinea Pigs, Rabbits	*Staphylococcus aureus*	Humans, sewage, rodents
Hantaan virus	Rodents	*Staphylococcus cohnii*	Rats
Influenza A virus	Humans, birds, pigs, nosocomial	*Staphylococcus haemolyticus*	Rats
Junin virus	Rodents	*Staphylococcus sciuri*	Rats
Klebsiella orthinolytica	Rodents, Rabbits	*Staphylococcus xylosus*	Rats
Klebsiella oxytoca	Rodents, Rabbits	*Streptobacillus moniliformis*	Rats
Klebsiella planticola	Rodents, Rabbits	*Streptococcus pneumoniae*	Rodents, Guinea Pigs, Rabbits
Klebsiella pneumoniae	Env., soil, Humans, Monkeys, Mice, swine bigs	*Streptococcus pyogenes*	Humans, Guinea Pigs
Marburg virus	Humans, monkeys	Theiler's virus	Mice
Measles virus	Humans, Monkeys	Vaccinia virus	Agricultural
Micromonospora faeni	Agricultural, moldy Hay, indoor growth	*Yersinia pestis*	Rodents, fleas, Humans
Micropolyspora faeni	Agricultural, indoor growth	*Yersinia pseudotu-berculosis*	Rodents, Rabbits, Guinea Pigs

References

Allen K, Green H. 1987. Hospital outbreak of multi-resistant *Acinetobacter anitratus*: An airborne mode of spread? J Hosp Infect 9:110–119.

Andersson E, Horal P, Vahlne A, Svennerholm B. 2004. No cross-resistance or selection of HIV-1 resistant mutants in vitro to the antiretroviral tripeptide glycyl-prolyl-glycine-amide. Antiviral Res 61(2):119–124.

Andrei G, DeClercq E, Snoeck R. 2004. In vitro selection of drug-resistant varicella-zoster virus (VZV) mutants (OKA strain): Differences between acyclovir and penciclovir? Antiviral Res 61(3):181–187.

Araujo MW, Andreana S. 2002. Risk and prevention of transmission of infectious diseases in dentistry. Quintessence Int 33(5):376–382.

ASHRAE. 1999. Handbook of Applications. Atlanta: ASHRAE.

ASHRAE. 2003. HVAC Design Manual for Hospitals and Clinics. Atlanta: American Society of Heating, Ventilating, and Air Conditioning Engineers.

ASHRAE. 2008. Handbook of Applications: Chapter 16: Ultraviolet Lamp Systems. Atlanta, GA: American Society of Heating, Refrigerating, and Air-Conditioning Engineers.

Benirschke K, Garner FM, Jones TC, editors. 1978. Pathology of Laboratory Animals. New York: Springer-Verlag.

Berg M, Bergman BR, Hoborn J. 1991. Ultraviolet Radiation Compared to an Ultra-Clean Air Enclosure. Comparison of Air Bacteria Counts in Operating Rooms. JBJS 73(5):811–815.

Berg-Perier M, Cederblad A, Persson U. 1992. Ultraviolet radiation and ultra-clean air enclosures in operating rooms. J Arthroplasty 7(4):457–463.

Besch EL. 1980. Environmental quality within animal facilities. Lab Animal Sci 30(2):385–398.

Brachman P. 1970. Nosocomial infection – airborne or not?; American Hospital Association, pp. 189–192.

Breathnach A, deRuiter A, Holdworth G, Bateman N, O-Sullivan D, Rees P, Snashall D, Milburn H, Peters B, Watson J et al. 1998. An outbreak of multi-drug-resistant tuberculosis in a London teaching hospital. J Hosp Infect 39:11–17.

Brown IWJ, Moor GF, Hummel BW, Collins JP. 1996. Toward further reducing wound infections in cardiac operations. Ann Thorac Surg 62(6):1783–1789.

CDC. 2003. Guidelines for environmental infection control in health-care facilities. MMWR 52(RR-10).

Davis KA, Moran KA, McAllister CK, Gray PJ. 2005. Multi-drug resistant *Acinetobacter* extremity infections in soldiers. Emerg Infect Dis 11(8):1218–1224.

Del Mundo F, McKhann CF. 1941. Effect of ultra-violet irradiation of air on incidence of infections in an infant's hospital. Am J Dis Child 61:213–225.

DHHS. 1993. Biosafety in Microbiological and Biomedical Laboratories. Cincinnati, OH: U.S. Department of Health and Human Services.

Dumyahn T, First M. 1999. Characterization of ultraviolet upper room air disinfection devices. Am Ind Hyg Assoc J 60(2):219–227.

Durmaz G, Kiremitci A, Akgun Y, Oz Y, Kasifoglu N, Aybey A, Kiraz N. 2005. The relationship between airborne colonization and nosocomial infections in intensive care units. Mikrobiyol Bul 39(4):465–471.

Duvlis Z, Drescher J. 1980. Investigations on the concentration of air-borne germs in conventionally air-conditioned operating theaters. *Zentralbl Bakteriol [B]* 170(1–2):185–198.

EPRI. 1996. UVGI for Infection Control in Hospitals. Palo Alto, CA: Electric Power Research Institute. Report nr TA-106887.

EPRI. 1997. UVGI for TB Infection Control in a Hospital. Palo Alto, CA: Electric Power Research Institute. Report nr TA-107885.

Farrington M, Ling T, French G. 1990. Outbreaks of infection with methicillin-resistant *Staphylococcus aureus* on neonatal and burns units of a new hospital. Epidem Infect 105:215–228.

Fletcher LA, Noakes CJ, Beggs CB, Sleigh PA. 2004. The Importance of Bioaerosols in Hospital Infections and the Potential for Control Using Germicidal Ultraviolet Radiation. Murcia, Spain.

Gill KE. 1994. HVAC design for isolation rooms. HPAC July:45–52.

Goldner JL, Allen BL. 1973. Ultraviolet light in orthopedic operating rooms at Duke University. Clin Ortho 96:195–205.

Grieble H, Bird T, Nidea H, Miller C. 1970. Chute-hydropulping waste disposal system: A reservoir of enteric bacilli and *Pseudomonas* in a modern hospital. J Infect Dis 130:602.

Hart D, Sanger PW. 1939. Effect on wound healing of bactericidal ultraviolet radiation from a special unit: Experimental study. Arch Surg 38(5):797–815.

Higgons RA, Hyde GM. 1947. Effect of ultra-violet air sterilization upon incidence of respiratory infections in a children's institution. New York State J Med 47(7).

Hunskaar S, Fosse RT. 1993. Allergy to laboratory mice and rats: A review of its prevention, management, and treatment. Lab Anim 27:206–221.

Jain DL, Agarwal V. 1996. Multi-drug resistant invasive isolates of *Haemophilus influenzae*: A multi-center study in India: indiaclen ibis study group. J Clin Epidem 50(Suppl.1):16S.

Keikavousi F. 2004. UVC: Florida hospital puts HVAC maintenance under a new light. Engin Sys March:60–66.

Kowalski W, Bahnfleth WP. 1998. Airborne respiratory diseases and technologies for control of microbes. HPAC 70(6):34–48.

Kowalski WJ, Bahnfleth WP, Carey DD. 2002. Engineering control of airborne disease transmission in animal research laboratories. Contemporary Topics in Lab Animal Sci 41(3):9–17.

Kowalski WJ. 2006. Aerobiological Engineering Handbook: A Guide to Airborne Disease Control Technologies. New York: McGraw-Hill.

Kowalski WJ. 2007a. Air-treatment systems for controlling hospital-acquired infections. HPAC Eng 79(1):28–48.

Kowalski WJ. 2007b. Airborne superbugs: Can hospital-acquired infections cause community epidemics? Consult Specif Eng 41(3):28–36, 69.

Kowalski W. 2008a. UVGI for hospital applications. IUVA News 10(4):30–34.

Kowalski W. 2008b. Operating Room Aerobiology in the Alaska Native Medical Center: Air and Surface Sampling Results and Recommendations for Reducing Surgical Site Infections. Anchorage, AK: Alaska Native Medical Center.

Kowalski W, Bahnfleth W, Hernandez M. A Genomic Model for the Prediction of Ultraviolet Inactivation Rate Constants for RNA and DNA Viruses; 2009 May 4–5; Boston, MA. International Ultraviolet Association.

Kraissl CJ, Cimiotti JG, Meleney FL. 1940. Considerations in the use of ultra-violet radiation in operating rooms. Ann Surg 111:161–185.

Kundsin R. 1976. Operating Room as a Source of Wound Contamination and Infection. National Research Council, National Academy of Sciences, pp. 167–172.

Lidwell OM, Lowbury EJL, Whyte W, Blowers R, Stanley SJ, Lowe D. 1983. Airborne contamination of wounds in joint replacement operations: the relationship to sepsis rates. J Hosp Infect 4:111–131.

Linscomb M. 1994. AIDS clinic HVAC system limits spread of TB. HPAC February.

Lowell J, Kundsin R. 1980. Ultraviolet Radiation: Its Beneficial Effect on the Operating Room Environment and the Incidence of Deep Wound Infcetion Following Total Hip and Total Knee Arthroplasty. Murray Hill, NJ: American Ultraviolet Company. Report nr A-810.

Lowell JD, Kundsin RB, Schwartz CM, Pozin D. 1980. Ultraviolet radiation and reduction of deep wound infection following hip and knee arthroplasty. In: Kundsin RB, editor. Airborne Contagion, Annals of the New York Academy of Sciences. New York: NYAS, pp. 285–293.

Luciano JR. 1977. Air Contamination Control in Hospitals. New York: Plenum Press.

Mangram AJ, Horan TC, Pearson ML, Silver LC, Jarvis WR, HICPAC. 1999. Guideline for prevention of surgical site infection. Am J. Infect Control 27(2): 97–132.

McLean R. 1961. The effect of ultraviolet radiation upon the transmission of epidemic influenza in long-term hospital patients. Am Rev Resp Dis 83:36–38.

Miller WR, Jarrett ET, Willmon TL, Hollaender A, Brown EW, Lewandowski T, Stone RS. 1948. Evaluation of ultra-violet radiation and dust control measures in control of respiratory disease at a naval training center. J Infect Dis 82:86–100.

Moggio M, Goldner JL, McCollum DE, Beissinger SF. 1979. Wound Infections in Patients Undergoing Total Hip Arthroplasty. Ultraviolet Light for the Control of Airborne Bacteria. Arch Surg 114(7):815–823.

Overholt RH, Betts RH. 1940. A comparative report on infection of thoracoplasty wounds. J Thoracic Surg 9:520–529.

Ritter M, Olberding E, Malinzak R. 2007. Ultraviolet Lighting During Orthopaedic Surgery and the Rate of Infection. J Bone Joint Surg 89:1935–1940.

Robertson EC, Doyle ME, Tisdall FF, Koller LR, Ward FS. 1943. Use of ultra-violet radiation in reduction of respiratory cross-infections in a children's hospital. JAMA 121:908–914.

Sauer LW, Minsk LD, Rosenstern I. 1942. Control of cross infections of respiratory tract in nursery for young infants. JAMA 118:1271–1274.

Schneider M, Schwartenberg L, Amiel JL, Cattan A, Schlumberger JR, Hayat M, deVassal F, Jasmin CL, Rosenfeld CL, Mathe G. 1969. Pathogen-free isolation unit – three years' experience. Brit Med J 29 March:836–839.

Sommer HE, Stokes J. 1942. Studies on air-borne infection in a hospital ward. J Pediat 21:569–576.

Tuffery AA, editor. 1995. Laboratory Animals: An Introduction for Experimenters. Chichester: John Wiley & Sons.

Wells WF. 1938. Air-borne infections. Mod Hosp 51:66–69.

Wells WF. 1955. Airborne Contagion and Air Hygiene. Cambridge, MA: Harvard University Press.

Wheeler SM, Ingraham HS, Hollaender A, Lill ND, Gershon-Cohen J, Brown EW. 1945. Ultraviolet light control of airborne infections in a naval training center. Am J Pub Health 35: 457–468.

WHO. 1988. Indoor air quality: Biological contaminants. Copenhagen, Denmark: World Health Organization. Report nr European Series 31.

Woodhall B, Neill R, Dratz H. 1949. Ultraviolet radiation as an adjunct in the control of post-operative neurosurgical infection. Clinical experience 1938–1948. Ann Surg 129:820–825.

Wright R, Burke J. 1969. Effect of ultraviolet radiation on post-operative neurosurgical sepsis. J Neurosurg 31:533–537.

Young DP. 1991. Ultraviolet Lights for Surgery Suites. Mooresville: St. Francis Hospital.

Chapter 18

Commercial Buildings

18.1 Introduction

Applications for UVGI systems in commercial buildings vary with the type of building, but virtually every type of building can benefit from the use of in-duct air disinfection and many buildings can benefit from the use of other types of UV systems. The health hazards and microbiological problems associated with various types of commercial buildings are often unique to the type of facility. The problem of air quality is paramount in commercial office buildings while the problem of biocontamination is of the highest concern in the food industry. Other types of buildings have their own microbial concerns and even their own standards. The pharmaceutical industry has the highest aerobiological air quality standards (and lowest airborne microbial levels) while the other extreme, the agricultural industry, has the highest airborne microbial levels and unique aerobiological concerns. The individual problems of these facilities are addressed in the following sections, and industry experience relating to UV applications are discussed, along with recommendations for how UV systems can be applied. The specific types of UVGI systems are described in previous chapters and these should be referred to for detailed information on such applications.

18.2 Office Buildings

Commercial buildings are the single largest and most common type of building in the United States today, in terms of total floor area. Many common respiratory infections are regularly transmitted inside these structures due to daily occupancy and the extensive interaction of people within office buildings. Proximity and duration of exposure are major factors in the transmission of respiratory infections between office workers (Lidwell and Williams 1961). The risk of catching the common cold is increased by shared office space (Jaakkola and Heinonen 1995). Tuberculosis has also been show to transmit in office buildings between co-workers (Kenyon et al.

W. Kowalski, *Ultraviolet Germicidal Irradiation Handbook*,
DOI 10.1007/978-3-642-01999-9_18, © Springer-Verlag Berlin Heidelberg 2009

2000). An association between respiratory tract symptoms in office workers and exposure to fungal and house dust mite aeroallergens was established by Menzies et al. (1998). The recent appearance of SARS virus highlights the susceptibility of office workers to the spread of airborne viruses (Yu et al. 2004). Sick Building Syndrome (SBS), often referred to as Building Related Illness (BRI) is a general category for a number of ailments, allergies, and complaints that are due to low levels of pollutants, synthetic irritants, fungi, bacteria or other factors that cause reactions in a certain fraction of building occupants (Lundin 1991).

Office buildings become contaminated with microbes brought in with the outside air as well as microbes that hail from indoor sources, including the occupants themselves. Accumulation of microbes on the cooling coils and ductwork can contribute to indoor air quality problems (Fink 1970). The aerobiology of office buildings can depend on both indoor sources (i.e. occupancy and cleanliness) as well as the composition and concentration of microbes in the outdoor air. Environmental microbes can enter via the ventilation system, via occupants (i.e. on clothes and shoes), and infiltration, especially from high-traffic lobbies that are under high negative pressure. Office buildings provide an environment that sustains microorganisms long enough for them to transmit infections to new hosts, or sometimes allows them to persist indefinitely, as in the case of fungal spores (Kowalski 2006). The indoor airborne concentrations of bacteria and fungi increase with the presence of occupants, due to the fact that people release bacteria continuously and also because human activity stirs up dust that may contain fungi and bacteria. Sessa et al. (2002) measured average office airborne concentrations of 493 cfu/m^3 for bacteria and 858 cfu/m^3 for fungi while levels were about four times lower when there were no occupants.

Most office buildings use dust filters in the MERV 6–8 range to filter outdoor or recirculated air but these have a limited effect on reducing fungal spores, bacteria, and viruses. Increasing filtration efficiency is one possible solution but retrofitting MERV filters in the 9–13 range, which can be highly effective against most airborne pathogens and allergens, may be difficult in most buildings because of increased pressure losses and possible reduction in airflow (Kowalski and Bahnfleth 2002). A suitable alternative to replacing filters is to add UV systems in the ductwork, which will not necessarily have any impact on airflow or pressure drop (see Fig. 18.1). In-duct UVGI is the most economic approach to disinfecting the air of commercial office buildings provided there is space in the ventilation system.

The economics of in-duct air disinfection can be greatly improved by locating the UV lamps at or around the cooling coils, thereby disinfecting the coils and improving their heat transfer capabilities. Cooling coils can first be steam cleaned to remove biological contamination and then a UVGI system can be added to continuously irradiate the coils. This not only prevents existing microbial growth but will tend to restore cooling coils to their original design operating level of performance, thereby saving energy costs (Kelly 1999).

Another alternative for improving air quality in office buildings and reducing the incidence of disease transmission between office workers is to locate recirculating UV units or Upper Room systems around the building to deal with local problems.

Fig. 18.1 In-duct UVGI
system retrofit into an
existing ventilation duct.
System consists of four
modular axial UV lamp
fixtures. Photo courtesy of
Virobuster Electronic Air
Sterilization, Marthalaan, The
Netherlands

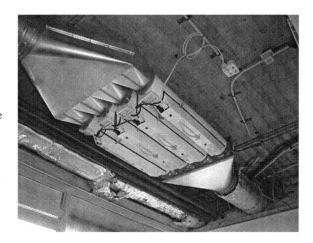

Lower Room systems may also be used to control microbial contamination at floor level, especially in hallways or lobbies that see heavy foot traffic. Other applications of UVGI in office buildings may include overnight decontamination of kitchens, bathrooms, and storage areas with UV area disinfection units.

UVGI may not be a complete solution to aerobiological problems if the cause is due to some other factor like moisture or water damage. For more detailed information on dealing with air quality problems in office buildings see Kowalski (2006).

18.3 Industrial Facilities

Industrial facilities cannot be generalized in terms of the types of respiratory diseases to which workers may be subject because of their wide variety. Industrial facilities that handle organic materials have greater microbiological hazards while those that process inorganic products usually have less microbial hazards and more pollution hazards. Pollution, however, is a contributing factor in infectious diseases, including respiratory diseases. There are many types of occupational diseases, including asthma, and allergic reactions or hypersensitivity pneumonitis, which are not necessarily due to microbiological causes. Air disinfection systems that use UV have little or no effect on non-microbiological contaminants but if pathogens or allergens are the cause of the problem then UV systems may be applied to reduce the hazard.

Airborne particulate dust in industrial environments may contain organic materials, chemicals, microbes, biological compounds, or inert materials, and form a substrate on which microbes can grow. Allergen-bearing airborne particulates are causative agents of lung inflammation via their immunotoxic properties and can induce inflammatory alveolitis (Salvaggio 1994). Occupational asthma occurs in

2–6% of the asthmatic population and respiratory infection can be a predisposing factor for occupational asthma (Bardana 2003). Workers who clean or enter rarely-used rodent infested structures may be at increased risk of exposure to rodent-borne viruses such as hantavirus (Armstrong et al. 1995). Industries that process wood or paper, or that make use of paper products, have airborne allergen hazards since cellulose provides a nutrient source for a variety of allergenic fungi. Microbiological sampling in furniture factories found the most common airborne microorganisms included *Corynebacterium, Arthrobacter, Aspergillus, Penicillium,* and *Absidia* (Krysinska-Traczyk et al. 2002). Microbiologial studies of the air in sawmills were conducted by Dutkiewicz et al. (1996), who found that the most common organisms were *Arthrobacter, Corynebacterium, Brevibacterium, Microbacterium, Bacillus* spp., Gram-negative bacteria (*Rahnella*) and filamentous fungi (*Aspergillus, Penicillium*). Inhalation anthrax and cutaneous anthrax still occur occasionally in the textile industry. In a mill in the USA where outbreaks had occurred, anthrax levels were measured at up to 300 cfu/m^3 (Crook and Swan 2001). Q fever, which can result from inhalation or exposure to *Coxiella burnetti,* has been reported among workers in a wool and hair processing plant (Sigel et al. 1950).

Tuberculosis continues to be an occupational hazard in some parts of the world and is a growing problem elsewhere because of multi-drug resistance. In a study at Russian plants by Khudushina et al. (1991) the incidence of tuberculosis was highest at a foundry plant at 39.9 per 100,000 workers and in an automatic-assembly plant at 64.4 per 100,000 workers. Health care workers, prison guards, and prison inmates may all be at higher risk for tuberculosis. A one-year prospective study of inmates in Geneva showed that the prevalence of active and residual tuberculosis is 5–10 times higher among prisoners than in the general population (Chevallay et al. 1983). Many prisons in the US incorporate filtration and UVGI systems to help control tuberculosis transmissions.

Occasional outbreaks of Legionnaire's disease have continued to be sporadically reported in the work environment, including a large outbreak among exhibitors at a floral trade show where a whirlpool spa was on display (Boshuizen et al. 2001). Water damage in any building increases the risk of workers to allergens if fungal mold growth has occurred. Trout et al. (2001) documented extensive fungal contamination, including *Penicillium, Aspergillus,* and *Stachybotrys,* in a water-damaged building.

Control measures for aerobiological contamination in the workplace are similar to those for any buildings in general, including improved air filtration, improved air distribution, and increased outside air combined with energy recovery systems. The use of UVGI for air disinfection can have positive benefits for almost any type of industrial building depending on the nature of the problem. In-duct UV air disinfection systems are likely to be the most economic recourse since they provide centralized control of the air quality, and can simultaneously be applied to the cooling coils (in most applications) to save energy while disinfecting the air. The use of local recirculation units can aid in resolving local microbial contamination problems, and both Upper and Lower Room UVGI systems can be used to control biocontamination.

18.4 Food Industries

The food industry comprises food and beverage processing facilities, food handling industries, food storage facilities other than agricultural, and kitchens in the restaurant and hospitality industry. There are at least four types of potential health hazards associated with food handling – foodborne human pathogens, spoilage microbes, microbial allergens, and food allergens. Foodborne pathogens are generally transmitted by the oral route and almost exclusively cause stomach or intestinal diseases but some foodborne pathogens may be airborne at various stages of processing, storage, cooking, or consumption. The processing and handling of foods may also create opportunities for airborne mold spores to germinate and grow, resulting in secondary inhalation hazards. Applications for UVGI in the food industry include both air and surface disinfection, including surface disinfection of packaging materials and food handling equipment. UV has been used for air and surface disinfection in cheese plants, bakeries, breweries, meat plants, for food preservation, and for the decontamination of conveyor surfaces, and packaging containers (Koutchma 2008). One of the earliest applications of UV in the food industry involved the use of bare UV lamps to irradiate surfaces in the brewing and cheese making industries to control mold (Philips 1985). Liquid food disinfection with UV has been approved by the FDA but although UV has been highly effective for water disinfection, successful applications in disinfecting other liquids depends on the transmissivity of the liquid.

   The direct disinfection of food by UV is generally only effective in cases where only the surface of foodstuffs requires disinfection since UV has limited penetrating ability (FDA 2000, Yaun and Summer 2002). Some foods can be effectively disinfected of certain food spoilage microbes, and UV is most effective on food products that have smooth and clean surfaces (Shama 1999, Seiler 1984). UV irradiation can significantly reduce the mold population on shells of eggs in only 15 min (Kuo et al. 1997). Studies have demonstrated that UV can reduce levels of *E. coli* and *Salmonella* on pork skin and muscle, *Listeria* on chicken meat, and *Salmonella* on poultry (Wong et al. 1998, Kim et al. 2001, Wallner-Pendleton et al. 1994). Begum et al. (2009) demonstrated that UV irradiation can inactivate food spoilage fungi like *Aspergillus, Penicillium*, and *Eurotium* but that the type of surface impacted the degree of disinfection. Marquenie et al. (2002) showed that *Botrytis cinerea* and *Monilinia fructigena*, two major post-harvest spoilage fungi of strawberries and cherries, could be reduced 3–4 logs by UV irradiation by doses of about 1000 J/m^2. Stevens et al. (1997) reported that UV could effectively reduce the incidence of storage rot disease on peaches due to *Monilinia fructicola*, reduce green mold (*Penicillium digitatum*) on tangerines, and reduce soft rot due to *Rhizopus* on tomatoes and sweet potatoes. Liu et al. (1993) showed that tomato diseases caused by *Alternaria alternata, Botrytis cinerea*, and *Rhizopus stolonifer* were effectively reduced by UV treatment. Hidaka and Kubota (2006) demonstrated a 90% reduction of *Aspergillus* and *Penicillium* species on the surfaces of wheat grain. Seiler (1984) reported increases in mold-free shelf life of clear-wrapped bakery products after moderate levels of UV exposure. Treatment of baked loaves in UV conveyor belt tunnels resulted in significantly increased shelf life (Shama 1999). Valero et al. (2007) found that UV irradiation of harvested grapes could prevent germination of fungi during storage or the dehydration process for raisins, using exposure times of up to 600 s.

One specialized application of UV in the food industry involves overhead tank disinfection, in which the airspace at the top of a liquid storage tank is disinfected by UV to control bacteria, yeast, and mold spores. Condensation of vapors in the head space of tanks, such as sugar syrup tanks, can produce dilute solutions on the liquid surface that provide ideal conditions for microbial growth. Two types of tank top systems are in use – systems which draw the air through a filter and UV system and return sterilized air to the tank space, and systems in which UV lamps are located directly in the head space of the tank and irradiate the liquid surface and internal tank surfaces. See Fig. 6.6 in Chap. 6 for an example of an overhead tank UV system.

Foodborne and waterborne pathogens represent the largest group of microorganisms that present health hazards in the food industry. In general, these do not present inhalation hazards, only ingestion hazards. However, some of these microbes can become airborne during processing and settle on foods, thereby becoming amenable to control by UV air and surface disinfection systems. Modern foodborne pathogens are often uniquely virulent and hazardous, like *Salmonella* and *Shigella*, which are contagious and depend on either on excretion in feces or vomiting to facilitate epidemic spread. Some agents of food poisoning, like *Staphylococcus* and *Clostridium*, are opportunistic or incidental contaminants of foods. Many molds like *Aspergillus* and *Penicillium* are common contaminants of the outdoor and indoor air that can grow on food and although they are not food pathogens they are potential inhalation hazards for food industry workers. Foodborne pathogens are predominantly bacteria but one virus has recently emerged and joined this class, Norwalk virus. Norwalk virus is a waterborne pathogen and has caused outbreaks on cruise ships (Marks et al. 2000). Food spoilage microbes are less of a health hazard than they are a nuisance in the food industry because of the damage they can cause to processed food (Samson et al. 2000). A wide variety of yeasts may also be causes of spoilage.

Table 18.1 lists the most common foodborne and waterborne pathogens in the food industry along with the type of hazard they present – pathogenic, toxic, allergenic, or spoilage. Many of the species listed in Table 18.1 have UV rate constants as given in Appendices A, B, and C. Virtually all of the fungal spores listed in Appendix C are potential contaminants in the food industry. Toxins produced by microbes may be endotoxins, exotoxins, or mycotoxins, and these will grow only if the conditions (moisture, temperature, etc.) are right. The key to controlling these toxins is to control microbial growth and to control the concentration of microbes in the surrounding areas. Ambient microbial levels can be controlled by food plant sanitation, adequate ventilation, and by various UV air disinfection technologies.

Table 18.1 Major microbial hazards in the food industry

Microbe	Type	Food class or source	Pathogen	Toxin	Allergen	Spoilage
Acinetobacter	Bacteria	various, humans	Yes	No	No	No
Aeromonas	Bacteria	various, humans	Yes	Yes	No	No
Alcaligenes	Bacteria	protein-rich	Yes	No	No	Yes
Aspergillus	Fungal spore	Grains, Vegetables, Meat	Yes	Yes	Yes	Yes
Bacillus cereus	Bacterial spore	various, environment	Yes	Yes	No	No
Botrytis	Fungal spore	Fruits, Vegetables, Meat	No	No	Yes	Yes
Brucella suis	Bacteria	various, humans	Yes	No	No	No
Campylobacter	Bacteria	various, humans	Yes	No	No	No
Cladosporium	Fungal spore	Dairy, Vegetables, Meat	Yes	Yes	Yes	Yes
Clostridium botulinum	Bacterial spore	various, humans	Yes	Yes	No	No
Clostridium perfingens	Bacterial spore	various, humans	Yes	Yes	No	No
Colletotrichum	Fungal spore	Fruits, Vegetables,	No	–	No	Yes
Corynebacterium	Bacteria	various, humans	Yes	No	No	Yes
Coxiella	Bacteria	various, humans	Yes	No	No	No
Cryptosporidium	Protozoa	various, humans	Yes	No	No	No
Cyclospora	Protozoa	various, environment	Yes	No	No	No
Diplodia	Fungal spore	Fruits, Vegetables,	No	–	No	Yes
Enterobacter	Bacteria	various, humans	Yes	Yes	No	No
Escherichia coli	Bacteria	various, humans	Yes	Yes	No	No
Fusarium	Fungal spore	Cereal, Fruits, Vegetables	No	Yes	Yes	Yes
Geotrichum	Fungal spore	Dairy,	No	No	No	Yes
Klebsiella pneumoniae	Bacteria	various, humans	Yes	Yes	No	No
Listeria	Bacteria	various, humans	Yes	No	No	No
Monila	Fungal spore	Grains, Dairy, Veg., Meat	No	–	No	Yes
Moraxella	Bacteria	various, humans	Yes	No	No	No
Mucor	Fungal spore	Grains, Fruits, Veg., Meat	Yes	No	Yes	Yes
Norwalk virus	Virus	various, humans	Yes	No	No	No
Oospora	Fungal spore	Dairy, Vegetables, Meat	No	–	No	Yes

Table 18.1 (continued)

Microbe	Type	Food class or source	Pathogen	Toxin	Allergen	Spoilage
Penicillium	Fungal spore	Grains, Dairy, Fruits, Veg.	No	Yes	Yes	Yes
Phomopsis	Fungal spore	Fruits, Vegetables	No	–	No	Yes
Phytophthora	Fungal spore	Fruits, Vegetables	No	–	No	Yes
Pseudomonas	Bacteria	various	Yes	No	No	Yes
Rhizopus	Fungal spore	Grains, Fruits, Veg., Meat	Yes	No	Yes	Yes
Salmonella	Bacteria	various, humans	Yes	Yes	No	No
Sclerotinia	Fungal spore	Fruits, Vegetables	No	–	No	Yes
Serratia	Bacteria	various, humans	Yes	No	No	Yes
Shigella	Bacteria	various, humans	Yes	Yes	No	No
Sporotrichum	Fungal spore	Bakery, Meat	No	–	Yes	Yes
Staphylococcus aureus	Bacteria	various, humans	Yes	No	No	No
Streptococcus	Bacteria	various, humans	Yes	No	No	No
Streptococcus suis	Bacteria	various, humans	Yes	No	No	No
Thamnidium	Fungal spore	Meat,	No	–	Yes	Yes
Trichoderma	Fungal spore	Fruits, Vegetables	No	Yes	Yes	Yes
Vibrio	Bacteria	various, humans	Yes	Yes	No	No
Yersinia enterolitica	Bacteria	various, humans	Yes	No	No	No

Most common food spoilage microbes have the capacity to transport in the air and therefore they are controllable to some degree by ventilation and by UV air disinfection systems. Evidence exists to suggest *Listeria monocytogenes*, a cause of many recent outbreaks, can settle on foodstuffs via the airborne route (1998). Studies indicate that *Salmonella* can survive in air for hours (Stersky et al. 1972). If air is recirculated through plant ventilation systems, organic material may accumulate inside ductwork and on air handling equipment, where fungi and bacteria may grow (1998). Air filtration can go along way towards controlling the amount of organic debris that accumulates, but UV is an ideal technology for use in controlling microbial and fungal growth inside air handling units, ductwork, and on cooling coils.

Concerns about the breakage of bare UV lamps and mercury hazards has led to new lamps that are sealed in an unbreakable plastic coating, such as shown in Fig. 18.2.

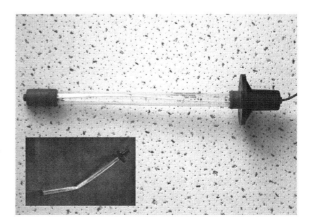

Fig. 18.2 Example of a waterproof, shatterproof UV lamp suitable for food industry applications. Insert shows a broken lamp in which the glass and mercury are fully contained by the plastic enclosure. Images courtesy of UVC Manufacturing and Consulting, Inc., Minden, NV

Mold and biofilms can develop on surfaces and equipment in the food and beverage industry, including tanks and vats, cooking equipment, walls and floors, and cooling coils (Carpentier and Cerf 1993). In general, standard cleaning and disinfection procedures are adequate to contain these problems but alternatives are available, including antimicrobial coatings like copper. UV irradiation of food processing equipment and surfaces, cooling coil disinfection systems, whole-area UV disinfection, and after-hours irradiation of rooms when personnel are not present are all viable options for maintaining high levels of disinfection in food industry facilities (Philips 1985, Kowalski and Dunn 2002). UV air disinfection systems may also be useful in controlling airborne hazards that result from hazards are created by industrial food processes that forcibly aerosolize contaminants. Pulsed UV light has seen increasing application in the food industry because of its rapid disinfection capabilities – see Chap. 16 for more information.

18.5 Educational Facilities

Educational facilities for children and students are focal points for disease transmission and as such they are ideally suited to UV applications that inhibit the transmission of contagious diseases. The concentration of young people in rooms causes cross-infections when even a single child comes to school with an infection such as a cold or a flu. Inevitably children who attend school bring home contagious infections which are then transmitted to family members, thereby continuing the process of epidemic spread in the community.

There are basically five types of educational facilities for children and students: Day Care centers, Preschools or Kindergartens, Elementary schools, Middle or High schools, and Colleges and Universities. Each type of facility and age group is subject to particular types on infections. Day care centers or nursery schools are commonly subject to high rates of Adenovirus, Echovirus, and Rhinovirus infections. Preschools or Kindergartens are commonly subject to upper respiratory infection (URI) outbreaks caused by Influenza, *Chlamydia*, and *Mycoplasma*. Elementary schools or Primary schools are commonly subject to outbreaks caused by *Bordetella*, Influenza, *Chlamydophila*, Mumps, Measles, *Mycoplasma*, Varicellazoster, and *Streptococcus*. High schools and Military Academies are commonly subject to respiratory infections caused by Adenovirus, *Chlamydia*, Coxsackievirus, Influenza, *Mycobacterium tuberculosis*, and *Streptococcus* species. Colleges and Universities are commonly subject to outbreaks caused by Adenovirus, *Bordetella*, *Chlamydia*, Influenza, *Mycoplasma*, Parainfluenza, *Neisseria*, and *Streptococcus*. Medical schools and colleges have also been subject to outbreaks of Mumps and Respiratory Syncytial Virus (RSV). Table 18.2 summarizes many of the various common infections that have been associated with school facilities, with reference sources as indicated. Many of the microbes in Table 18.2 are susceptible to UV disinfection in air and on surfaces (see the UV rate constants for these species given in Appendix A).

Fungal spores also accumulate in schools and are responsible for a variety of respiratory illnesses including allergies and asthma. The most frequently encountered fungal taxa identified in a study of dust samples from twelve schools in Spain were *Alternaria, Aspergillus*, and *Penicillium* (Austin 1991).

Guidelines for ventilation air in school buildings are provided in ASHRAE Standard 62-2001 (ASHRAE 2001) but many schools do not meet the minimum recommendations either because the buildings are older or because of attempts to save energy costs due to endless budget cutting. One way to make up for the lack of ventilation air is to provide UV air disinfection systems, including Upper Room systems and in-duct UV systems.

A number of studies have been performed in schools to evaluate the effectiveness of UVGI systems on respiratory infections in schools. All of these studies involved Upper Room systems (see Chap. 9 for a summary of the results). Most of the studies produced a reduction of disease transmission with a net average reduction in respiratory infections of approximately 30%, and were effective against measles, mumps, varicella, chickenpox, cold viruses and other respiratory infections. Upper Room systems can be especially effective in classrooms that have little of no ventilation air and that have high ceilings. There are no studies that have addressed In-duct UVGI systems, which would be likely to prove even more effective if they were adequately designed.

18.6 Museums and Libraries

Materials and books that are kept in museums, libraries, and related archives may be both a source of allergens and a nutrient source for the growth of microorganisms, especially fungi. The growth of fungi and certain bacteria on stored materials contributes to biodeterioration, which is a major concern, and the aerobiology in

Table 18.2 Common pathogenic microbes in schools

Agent	Type	Disease	Type of school	Reference
Adenovirus	Virus	colds	Nursery school	Blacklow (1968)
		colds	College	Jackson (2000)
		colds	Medical school	Gerth (1987)
		URI	Naval Academy	Gray (2001)
Bordetella pertussis	Bacteria	whooping cough	Elementary	Gonzalez (2002)
		whooping cough	College	Jackson (2000)
Chlamydia pneumoniae	Bacteria	URI	Preschool/Elementary	Schmidt (2002)
		pneumonia	College	Jackson (2000)
		URI	Naval Academy	Gray (2001)
Coxsackievirus	Virus	pleurodynia	High school	Ikeda (1993)
Echovirus	Virus	colds	Nursery school	Hartmann (1967)
Herpes simplex	Virus	herpes simplex	College	McMillan (1993)
Influenza	Virus	flu	College	Jackson (2000)
		flu	Naval Academy	Gray (2001)
		flu	Preschool/Elementary	Neuzil (2002)
		flu	Medical school	Gerth (1987)
Measles virus	Virus	measles	Elementary	Perkins (1947)
Mumps virus	Virus	mumps	Medical school	Gerth (1987)
Mycobacterium tuberculosis	Bacteria	TB	Elementary	Watson (2001)
		TB	High school	Kim (2001)

Table 18.2 (continued)

Agent	Type	Disease	Type of school	Reference
Mycoplasma pneumoniae	Bacteria	pneumonia	College	Jackson (2000)
		URI	Medical school	Gerth (1987)
		URI	Naval Academy	Gray (2001)
		URI	Preschool/Elementary	Bosnak (2002)
Neisseria menigitidis	Bacteria	meningitis	University dormitory	Round (2001)
Parainfluenza	Virus	URI	Medical school	Gerth (1987)
		URI	Nursery school	Zilisteanu (1966)
Respiratory Syncytial Virus	Virus	RSV	College	Jackson (2000)
		RSV	Medical school	Gerth (1987)
Rhinovirus	Virus	colds	Nursery school	Beem (1969)
Streptococcus pneumoniae	Bacteria	URI	Naval Academy	Gray (2001)
Streptococcus pyogenes	Bacteria	URI	College	McMillan (1993)
		Rheumatic fever	Elementary	Dierksen (2000)
		Pharyngitis	Elementary	Hoebe (2000)
		Rheumatic fever	Elementary/High school	Olivier (2000)
Varicella-zoster virus	Virus	Chickenpox	Elementary	Bahlke (1949)

museums and libraries may impact both the occupants as well as the stored materials. Most books are made from cellulose and can be degraded by a variety of microbes and insects which may include allergens and a few potential pathogens.

Over 84 microbial genera, representing 234 species, have been isolated from library materials like books, paper, parchment, feather, textiles, animal and vegetable glues, inks, wax seals, films, magnetic tapes, microfilms, photographs, papyrus, wood, and synthetic materials in books (Zyska 1997). Many of the fungi identified in library materials can produce mycotoxins and cause respiratory and other diseases. Table 18.3 summarizes the variety of microbes that have been found in libraries and museums, and indicates the type of materials that they can grow on. The term 'deteriogen' in the table refers to microbes that can cause biodeterioration of materials.

Libraries often use carpeting to quiet the noise of foot traffic, but carpets tend to accumulate fungi and bacteria over time due to the fact that they settle in air or are brought in from the outdoors on shoes and clothes. In one study of carpeted buildings, carpet dust was found to contain 85,000 cfu/gm of fungi, 12,000,000 cfu/gm of mesophilic bacteria, and 4,500 cfu/gm of thermophilic bacteria (Cole et al. 1993). Foot traffic tends to aerosolize microbes and causes them to settle in other locations, including on books and materials. Aerosolized microbes can also be inhaled, leading to allergies and asthma. Carpets have a high equilibrium moisture content that favors microbial growth (IEA 1996).

Fabrics kept in museums, especially ancient cloth, are subject to biodeterioration from fungi and bacteria. Fungi can grow extensively on cotton fibers at the right humidity and growth rates are highest on 100% cotton (Goynes et al. 1995). Studies of biodeterioration caused by fungi that grow on paintings have isolated over one hundred species of fungi, with the most common species being *Alternaria, Cladosporium, Fusarium, Aspergillus, Trichoderma*, and *Penicillium* (Inoue and Koyano 1991).

Ultraviolet light is an option for disinfecting surfaces and materials, but it can cause damage and discoloration to pigments and paint and therefore is not necessarily suitable for decontaminating books that are already damaged by microbial growth. UVGI is best used for disinfecting the air and especially for removing mold from air conditioning systems. In-duct UVGI can greatly aid in reducing airborne levels of bacteria and spores and in removing accumulated mold spores from air conditioning cooling coils, which are used extensively to control humidity in libraries and museums. Keeping cooling coils disinfected will minimize the spread of spores through a building as well as reducing energy costs (see Chap. 10). UVGI systems can be used selectively in areas of museums and libraries that do not contain sensitive materials, such as cafeterias. Lower Room UV systems may also be ideal for hallways and entryways (where no materials are stored) where outdoor spores may be tracked into buildings. For more detailed information on technologies and techniques for dealing with biocontamination of sensitive materials in libraries and museums see Kowalski (2006).

Table 18.3 Air and surface microbes in museums and libraries

Microbe	Type	Location	Hazard
Acremonium	Fungi	feathers, leather	allergen
Actinomyces	Bacteria	cotton textiles	allergen
Alternaria	Fungi	paintings, paper	allergen
Arthrinium	Fungi	library air	allergen
Arthrobacter	Bacteria	museum air, paintings, frescoes, building materials	deteriogen
Aspergillus	Fungi	paintings, feathers, leather, books, documents, textiles	allergen
Aureobacterium	Bacteria	museum air, paintings	deteriogen
Aureobasidium	Fungi	paintings, stained glass, limestone, library air	allergen
Bacillus	Bacteria	paintings, frescoes, building materials, cotton textiles	deteriogen
Chaetomium	Fungi	ancient documents, textiles	allergen
Chrysosporium	Fungi	feathers, leather	allergen
Citrobacter	Bacteria	books	deteriogen
Cladosporium	Fungi	paintings, limestone, frescoes, cotton, paper, library air	allergen
Corynebacterium	Bacteria	cotton textiles, books	pathogen
Curvularia	Fungi	cotton textiles	allergen
Cyanobacteria	Bacteria	building materials	deteriogen
Enterobacter	Bacteria	books	pathogen
Epicoccum	Fungi	library air, paper	allergen
Eurotium	Fungi	books	allergen
Exophiala	Fungi	limestone	allergen
Fusarium	Fungi	paintings, textiles, library air, paper	allergen
Geomyces	Fungi	stained glass	allergen
M. tuberculosis	Bacteria	books	pathogen
Microbacterium	Bacteria	museum air, paintings	deteriogen
Micrococcus	Bacteria	museum air, paintings, cotton textiles	deteriogen
Micromonospora	Bacteria	cotton textiles	allergen
Mucor	Fungi	cotton textiles	allergen
Paecilomyces	Fungi	limestone, cotton textiles, library air	allergen
Paenibacillus	Bacteria	books	deteriogen
Penicillium	Fungi	paintings, feathers, leather, documents, limestone, textiles	allergen
Phialaphora	Fungi	frescoes, building materials	deteriogen
Phoma	Fungi	limestone, cotton textiles	allergen
Pithomyces	Fungi	library air, paper	allergen
Proteus vulgaris	Bacteria	paintings	deteriogen
Rhizopus	Fungi	paintings, textiles	allergen
Rhodotorula	Fungi	stained glass	allergen
Smallpox	Bacteria	books	pathogen
Stachybotrys	Fungi	textiles	allergen
Staphylococcus	Bacteria	books	pathogen
Streptococcus	Bacteria	books	pathogen
Streptomyces	Bacteria	frescoes, building materials	deteriogen

Table 18.3 (continued)

Microbe	Type	Location	Hazard
Trichoderma	Fungi	paintings, limestone, textiles, paper, library air	allergen
Tritirachium	Fungi	frescoes, building materials, library air	deteriogen
Ulocladium	Fungi	paintings	allergen
Ustilago	Fungi	stained glass	allergen
Verticillium	Fungi	stained glass, limestone, frescoes	deteriogen

18.7 Agricultural and Animal Facilities

Agricultural facilities pose a variety of occupational microbial hazards including infectious diseases from farm animals, allergies from animal dander and food-stuffs, and health threats from mold spores or actinomycetes. Animal facilities like barns, poultry houses, swine houses, and kennels can have the highest levels of bioaerosols seen in any indoor environments. Animal facilities may also include non-agricultural buildings used to house or service animals such as pet shelters and zoos. Many animal pathogens can transmit to humans by direct contact or by the airborne route. The microorganisms of greatest concern are those that can become airborne in animal facilities and these are often respiratory pathogens and allergens. Allergens can be produced as a byproduct of animal husbanding or from animal waste, animal feeds, or other farm produce. The actinomycetes are particularly common type of bacteria found in agriculture and can grow on moldy hay. Farmers may be routinely exposed to very high concentrations of actinomycetes and may inhale as many as 750,000 spores per minute (Lacey and Crook 1988). Farmer's Lung represents a group of respiratory problems that often afflict farm workers who receive chronic exposure to high concentrations of actinomycetes (Pepys et al. 1963).

Table 18.4 lists some of the most common microorganisms that occur in animal and agricultural environments, and that can transmit to man by various routes. The diseases caused and natural source are identified. For information on the UV susceptibility of these microbes see Appendices A, B, and C. Most of these microbes can transmit by direct contact or by the airborne. Table 18.5 provides a list of microbes found in sewage, many of which can become airborne.

Natural ventilation is common in agricultural facilities but mechanical ventilation will generally provide superior control of airborne microorganisms. However, filtration is often necessary to clean both outdoor air and indoor recirculated air. Recirculated air in animal facilities can be cleaned more effectively through the use of UVGI combined with filtration, and this approach works well for viruses, which may not filter out easily.

Other applications for UVGI in animal facilities include Upper Room systems, and area disinfection systems. After-hours UV disinfection systems are also appropriate for animal facilities provide the animals can be periodically removed from

Table 18.4 Common agricultural and animal pathogens and allergens

Microbe	Type	Disease or infection	Natural source
Acinetobacter	Bacteria	opportunistic/septic infections,	Environmental, soil, sewage
Actinomyces israelii	Bacteria	actinomycosis	Humans, cattle
Aeromonas	Bacteria	Non-respiratory opportunistic infections, gastroenteritis	Environmental, water, soil
Alcaligenes	Bacteria	opportunistic infections	Humans, soil, water,
Alternaria alternata	Fungi	allergic alveolitis, rhinitis, irritation, asthma, toxic	Environmental, indoor growth on paint, dust
Aspergillus	Fungi	aspergillosis, alveolitis, asthma, allergic fungal sinusitis, ODTS, toxic	Environmental, indoor growth on insulation and coils.
Bacillus anthracis	Bacteria	anthrax, woolsorter's disease	Cattle, sheep, other animals,
Brucella	Bacteria	Brucellosis, undulant fever	Goats, cattle, swine, dogs, sheep, caribou, elk, coyotes, camels.
Chlamydia psittaci	Bacteria	Psittacosis	Birds
Cladosporium	Fungi	chromoblastomycosis, allergic reactions, rhinitis, asthma	Environmental, indoor growth on dust,
Clostridium	Bacteria	tetanus, gas gangrene, toxic	Sheep, cattle
Corynebacteria	Bacteria	diphtheria	Rabbits, guinea pigs
Coxiella burnetii	Bacteria	Q fever	Cattle, sheep, goats.
Cryptostroma corticale	Fungi	alveolitis, asthma, maple bark pneumonitis, maple bark	Environmental, found on maple and sycamore bark.
Dander, hair	Allergen	cows, horses, farm animals	Agricultural
Foot and Mouth Disease	Virus	Foot and mouth disease (FMD)	Cattle
Hantavirus	Virus	Hanta virus	Rodents
Influenza A virus	Virus	flu, secondary pneumonia	Humans, birds, pigs,

Table 18.4 (continued)

Microbe	Type	Disease or infection	Natural source
Leptospira	Bacteria	jaundice	Dogs, rats
Lymphocytic choriomeningitis	Virus	LCM, lymphocytic meningitis	House mouse, swine, dogs, hamsters, guinea pigs.
Micromonospora faeni	Bacteria	Farmers Lung, pulmonary fibrosis, allergic reactions, UR irritation	Agricultural, moldy hay, indoor growth
Mucor	Fungi	allergen	dairy products
Mycobacterium kansasii	Bacteria	cavitary pulmonary disease	Water, cattle, swine
Pasteurella tularensis	Bacteria	Bubonic, pneumonic,	Rodents, rabbits, birds
Penicillium	Fungi	alveolitis, rhinitis, asthma, allergic reactions, irritation, ODTS, toxic reactions, VOCs	Environmental, indoor growth on paint, filters, coils, and humidifiers.
Rabies virus	Virus	Rabies	Dogs
Rhizopus	Fungi	allergen	Agricultural
Rickettsia	Bacteria	Epidemic typhus	Rodents, ticks
Saccharopolyspora rectivirgula	Bacteria	Farmers Lung, alveolitis,	Agricultural
Salmonella	Bacteria	Poultry, eggs	Agricultural
Sporothrix schenckii	Fungi	sporotrichosis, rose gardeners	Environmental, plant material.
Staphylococcus aureus	Bacteria	staphylococcal pneumonia	Humans, sewage
Thermoactinomyces sacchari	Bacteria	bagassosis, alveolitis, HP	Agricultural, bagasse
Thermoactinomyces vulgaris	Bacteria	Farmers Lung, pulmonary fibrosis, allergic reactions, asthma, HP	Agricultural, indoor growth in air conditioners
Thermomonospora viridis	Bacteria	Farmers Lung, HP	Agricultural
Vaccinia virus	Virus	cowpox	Agricultural
Vesicular Stomatitis Virus	Virus	VSV	Cattle, pigs, horses

Table 18.5 Microbes that may grow or occur in sewage

Microbe	Type	Group
Acinetobacter	Bacteria	Pathogen
Adenovirus	Virus	Pathogen
Aeromonas	Bacteria	Pathogen
Alcaligenes	Bacteria	Pathogen
Alternaria	Fungi	Allergen
Aspergillus	Fungi	Allergen
Bacillus anthracis	Bacteria	Pathogen
Brucella spp.	Bacteria	Pathogen
Calicivirus	Virus	Pathogen
Campylobacter spp.	Bacteria	Pathogen
Candida spp.	Bacteria	Pathogen
Cladosporium	Fungi	Allergen
Clostridium botulinum	Bacteria	Pathogen
Clostridium perfringens	Bacteria	Pathogen
Corynebacterium spp.	Bacteria	Pathogen
Coxsackievirus	Virus	Pathogen
Cryptococcus neoformans	Bacteria	Pathogen
Cryptosporidium	Protozoa	Pathogen
E. coli	Bacteria	Pathogen
Echovirus	Virus	Pathogen
Enterobacter	Bacteria	Pathogen
Exophiala	Fungi	Allergen
Fusarium	Fungi	Allergen
Giardia lamblia	Protozoa	Protozoa
Hepatitis A	Virus	Pathogen
Hepatitis B	Virus	Pathogen
Hepatitis E	Virus	Pathogen
Klebsiella spp.	Bacteria	Pathogen
Listeria monocytogenes	Bacteria	Pathogen
Mucor	Fungi	Allergen
Mycobacterium	Bacteria	Pathogen
Nocardia	Bacteria	Pathogen
Norwalk virus	Virus	Pathogen
Parvovirus	Virus	Pathogen
Penicillium	Fungi	Allergen
Phialaphora	Fungi	Allergen
Poliovirus	Virus	Pathogen
Pseudomonas aeruginosa	Bacteria	Pathogen
Rotavirus	Virus	Pathogen
Reovirus	Virus	Pathogen
Rhizopus	Fungi	Allergen
Salmonella	Bacteria	Pathogen
Serratia	Bacteria	Pathogen
Shigella spp.	Bacteria	Pathogen
Staphylococcus spp.	Bacteria	Pathogen
Streptococcus faecalis	Bacteria	Pathogen
Thermoactinomyces spp.	Bacteria	Pathogen

the indoor areas in order to disinfect them. Such an approach can be more economical than scrubbing with disinfectants, and can enhance the effectiveness of
manual scrubbing procedures. For more detailed information on dealing with the
complex bioaerosol and dust problems in agricultural and animal facilities see
Kowalski (2006).

18.8 Malls, Airports, and Places of Assembly

Facilities like malls and airports are characterized by large enclosed volumes where
large crowds may be concentrated and where heavy cyclical occupancy may occur.
Places of Assembly may include auditoriums, stadiums, theaters, gymnasiums, nata-
toriums, arenas, town halls, churches, cathedrals, temples, mosques, industrial halls,
convention centers, atriums, shopping centers, and other places where large public gatherings may occur indoors. In such buildings infectious diseases may be
exchanged by direct contact, indirect contact, or via inhalation, and large numbers
of people may be exposed simultaneously. The large volumes of air enclosed in such
facilities often ensure good mixing of air, even in naturally ventilated stadiums, and
as a result the air quality is often acceptable. However, it is difficult to provide ventilation to all corners and therefore the actual airflow distribution is often uncertain.

One of the most famous outbreaks of a respiratory disease in a heavily occupied building was the eponymous Legionnaire's Disease outbreak at a convention in
Philadelphia in 1976 that resulted in 29 deaths (Spengler et al. 2001). It was traced
to *Legionella* contamination in a cooling tower that apparently wafted from outdoors into a crowded hallway through open doors. An outbreak of measles occurred
inside a domed stadium in the Minneapolis-St. Paul metropolitan area during July
1991 that resulted in sixteen associated cases of measles in seven states (Ehresmann
et al. 1995). Several tuberculosis outbreaks have occurred in churches and mosques
due to a single infectious person. Dutt et al. (1995) reports one outbreak of TB in
a church where one man exposed 42% of his congregation. A norovirus outbreak
at a concert hall was reviewed by Evans et al. (2002) in which a concert attendee
who vomited in the auditorium and a toilet. Gastrointestinal illness occurred among
members of school children who attended the following day. Transmission was
most likely through direct contact with contaminated fomites that remained in the
toilet area.

Large facilities that use 100% outside air provide little or no opportunity for
the use of UV systems, since outside air is generally clean and free of airborne pathogens, but facilities that recirculate air may benefit from in-duct UV air
disinfection systems. Large facilities often have high ceilings and this provides an
opportunity to use Upper Room systems for air disinfection, especially in cases
where natural ventilation is employed and there is no other means for disinfecting
the air. High power UV systems can be safely used in such applications since the
UV lamps can be located far from the occupied floor areas. The use of UV area disinfection systems in public toilets is an appropriate means of dealing with potential
fomites, or infectious particles left on surfaces.

18.9 Aircraft and Transportation

Aircraft, trains, cars, and other compact enclosed environments can pose microbiological hazards from extended exposure due to the fact that risks due to proximity are increased regardless of whether the pathogen transmits through the air or by direct contact. Large cruise ships may resemble hotels and apartment buildings in terms of their ventilation systems and health risks but smaller craft like cars and planes create extended opportunities for infectious exchanges due to the close quarters, shared breathing air, potentially extended periods of occupancy, and the limited amount of outside air that may be brought in, especially in cold climates. Other microenvironments like elevators and city buses are unlikely to play a major role in the spread of contagious diseases due to their brief occupancy times.

Aircraft are one of the most crowded environments in which people remain for extended periods of time and the potential for airborne disease transmission is fairly obvious, except perhaps to airline owners who fly in private jets. Airplanes are potential vectors for the transmission of airborne diseases between continents and play a role in the global dissemination of epidemic diseases (Masterson and Green 1991). Airline crews and passengers have a higher risk of contracting infections on long flights (NRC 2002, Ungs and Sangal 1990). Respiratory pathogens that have been identified onboard airlines include Adenovirus, Chickenpox, Coronavirus, Influenza, Measles, Mumps, *Mycobacterium tuberculosis, Neisseria meningitidis,* and SARS virus (Kowalski 2006).

It has been reported that 85% of newer airplanes have HEPA filters (GAO 2004). However, there is a common practice in the industry of referring to MERV 14 filters (95% DSP filters) as 'HEPA-like' or 'HEPA type' filters, which is misleading. MERV 14 filters, while being excellent filters for controlling spores and most bacteria, cannot guarantee protection against viruses and smaller bacteria. Coupling a MERV 14 filter with a UVGI system can, however, can provide superior performance (Kowalski and Bahnfleth 2002). Furthermore, the energy costs associated with using HEPA filters may not be justified when a simple combination of a MERV 13–15 filter and an URV 13–15 UVGI system will provide comparable results at lower cost. UVGI systems can be installed in the recirculation ducts of airplanes as well as being located at individual seats (in the overhead 'gaspers').

The transmission of infectious disease aboard ships is a recurring phenomenon, with onboard transmission of some diseases such as Norwalk virus and Legionnaire's Disease being favored. The incidence of respiratory disease aboard military ships increases as ship size decreases (Blood and Griffith 1990). A study of a tuberculosis outbreak aboard a Navy ship found that although proximity and direct contact played a role in transmission, airborne transmission of droplet nuclei, including via the ventilation system, was responsible for most of the secondary infections (Kundsin 1980). It is common for outbreaks aboard ships to consist of multiple viral and bacterial infections, including diarrheal illness and influenza (Ruben and Ehreth 2002). Noroviruses are responsible for 23 million cases of illness each year and Norwalk viruses have recently caused numerous outbreaks of gastroenteritis on cruise ships. They are largely attributed to fomites on ship surfaces.

During 2002, cruise ships with foreign itineraries sailing into US ports reported 21 gastroenteritis outbreaks on 17 cruise ships, of which most were identified as noroviruses (Cramer et al. 2003). In an influenza outbreak aboard a cruise ship from Hong Kong in 1987, 38% of passengers came down with acute respiratory illnesses (Berlingberg et al. 1988).

UV applications for ships include in-duct UV systems to interdict recirculated pathogens and allergens, and surface irradiation systems to control fomites. Noroviruses may be particularly susceptible to After Hours UV systems placed in hallways, bathrooms, and other locations where occupancy is intermittent. Cruise ships can also be irradiated and decontaminated between voyages by portable UV systems.

18.10 Sewage and Waste Facilities

Most of the microorganisms associated with waste are either water borne or food borne pathogens and allergens and workers in sewage, wastewater, and waste processing industries are subject to occupational hazards from these microbes. Airborne hazards also exist since aerosolization of microbial pathogens, endotoxins, and allergens is an inevitable consequence of the generation and handling of waste material. Table 18.4 lists the most common microorganisms that have been found to grow or to occur in sewage and waste (Kowalski 2006). Many of these are human pathogens and they include bacteria, viruses, fungi, and protozoa. For information on the UV susceptibility of these pathogens refer to Appendix A, which addresses most of these species. Many of these species are potentially airborne and can cause respiratory infections, while most of the remainder are either waterborne or foodborne stomach pathogens.

Wastewater treatment workers are exposed to a variety of infectious agents and Khuder et al. (1998) examined the prevalence of infectious diseases and associated symptoms in wastewater treatment workers over a 12-month period. The wastewater workers exhibited a significantly higher prevalence of gastroenteritis, gastrointestinal symptoms, and headaches over those in a control group but no significant differences were found with regard to respiratory and other symptoms. Thorn et al. (2002), however, found that sewage workers had significantly increased risks for respiratory symptoms, including chronic bronchitis, and toxic pneumonitis, as well as central nervous system problems, over workers in non-sewage industries.

Reduction of microbial hazards in the sewage and waste industries is probably best approached using source control methods, but there are potential applications for UVGI. In-duct air disinfection can promote healthier breathing air for workers inside plants and both Upper Room and Lower Room UVGI systems can provide disinfection of both air and surfaces where mold and bacteria may accumulate. Area disinfection systems, like After-hours UV systems, may also provide a means of decontaminating areas during periods when they are not occupied (i.e. overnight).

References

Armstrong LR, Zaki SR, Goldoft MJ, Todd RL, Khan AS, Khabbaz RF, Ksiazek TG, Peters CJ. 1995. Hantavirus pulmonary syndrome associated with entering or cleaning rarely used, rodent-infested structures. J Infect Dis 172(4):1166.

ASHRAE. 2001. Standard 62: Ventilation for acceptable indoor air quality. Atlanta: ASHRAE.

Austin B. 1991. Pathogens in the Environment. Bacteriology TSfA, editor. Oxford: Blackwell Scientific Publications.

Bahlke AM, Silverman HF, Ingraham HS. 1949. Effect of ultra-violet irradiation of classrooms on spread of mumps and chickenpox in large rural central schools. Am J Pub Health 41: 1321–1330.

Bardana EJJ. 2003. 8. Occupational asthma and allergies. J Allergy Clin Immunol 111(2 Suppl):S530–S539.

Beem M. 1969. Rhinovirus infections in nursery school children. J Pediatr 74(5):818.

Begum M, Hocking A, Miskelly D. 2009. Inactivation of food spoilage fungi by ultraviolet (UVC) irradiation. Int J Food Microbiol 129:74–77.

Bell C, Kyriakides A. 1998. Listeria: A practical approach to the organism and its control in foods. London: Blackie Academic & Professional.

Berlingberg CD, Kahn FH, Chun LY, et al. 1988. Acute respiratory illness among cruise ship passengers – Asia. MMWR 259(9):13051306.

Blacklow NR, Hoggan MD, Kapikian AZ, Austin JB, Rowe WP. 1968. Epidemiology of adenovirus-associated virus infection in a nursery population. Am J Epidemiol 88(3):368–378.

Blood CG, Griffith DK. 1990. Ship size as a factor in illness incidence among U.S. Navy vessels. Mil Med 155(7):310–314.

Boshuizen HC, Neppelenbroek SE, vanVliet H, Schellekens JF, denBoer JW, Peeters MF, Spaendonck MAC-v. 2001. Subclinical Legionella infection in workers near the source of a large outbreak of Legionnaires disease. J Infect Dis 184(4):515–518.

Bosnak M, Dikici B, Bosnak V, Dogru O, Ozkan I, Ceylan A, Haspolat K. 2002. Prevalence of Mycoplasma pneumoniae in children in Diyarbakir, the south-east of Turkey. Pediatr Int 44(5):510–512.

Carpentier B, Cerf O. 1993. Biofilms and their consequences, with particular reference to hygiene in the food industry. J Appl Bact 75:499–511.

Chevallay B, Haller Rd, Bernheim J. 1983. Epidemiology of pulmonary tuberculosis in the prison environment. Schweiz Med Wochenschr 113(7):261–265.

Cole EC, Foarde KK, Leese KE, Franke DL, Berry MA. 1993. Biocontaminants in carpeted environments. Helsinki, Finland.

Cramer EH, Forney D, Dannenberg AL, et al. 2003. Outbreaks of Gastroenteritis associated with Noroviruses on cruise ships – United States, 2002. MMWR 289(2):167–169.

Crook B, and Swan, JRM. 2001. Bacteria and other bioaerosols in industrial workplaces. In: B Flannigan RAS, and JD Miller, editor. Microorganisms in Home and Indoor Work Environments. New York: Taylor and Francis.

Dierksen KP, Inglis M, Tagg JR. 2000. High pharyngeal carriage rates of Streptococcus pyogenes in Dunedin school children with a low incidence of rheumatic fever. N Z Med J 113(1122): 496–499.

Dutkiewicz J, Krysinska-Traczyk E, Skorska C, Milanowski J, Sitkowska J, Dutkiewicz E, Matuszyk A, Fafrowicz B. 1996. Microflora of the air in sawmills as a potential occupational hazard: concentration and composition of microflora and immunologic reactivity of workers to microbial aeroallergens. Pneumonol Alergol Pol 64(Suppl 1):25–31.

Dutt AK, Mehta JB, Whitaker BJ, Westmoreland H. 1995. Outbreak of tuberculosis in a church. Chest 107(2):447–452.

Ehresmann KR, Hedberg CW, Grimm MB, Norton CA, MacDonald KL, Osterholm MT. 1995. An outbreak of measles at an international sporting event with airborne transmission in a domed stadium. J Infect Dis 171(3):679–683.

Evans MR, Meldrum R, Lane W, Gardner D, Ribeiro CD, Gallimore CI, Westmoreland D. 2002. An outbreak of viral gastroenteritis following environmental contamination at a concert hall. Epidemiol Infect 129(2):355–360.

FDA. 2000. Ultraviolet radiation for the processing and treatment of food. Food and Drug Administration. Report nr 10CFR21, Section 179.39 & 179.41.

Fink JN. 1970. Mold in air conditioner causes pneumonitis in office workers. JAMA 211(10):1627.

GAO. 2004. More Research Needed on the Effects of Air Quality on Airliner Cabin Occupants. Washington, DC: General Accounting Office. Report nr GAO-05-54.

Gerth HJ, Gruner C, Muller R, Dietz K. 1987. Seroepidemiological studies on the occurrence of common respiratory infections in paediatric student nurses and medical technology students. Epidemiol Infect 98(1):47–63.

Gonzalez MF, Moreno CA, Amela HC, Pachon AI, Garcia BA, Herrero CC, Herrera GD, Martinez NF. 2002. A study of whooping cough epidemic outbreak in Castellon, Spain. Rev Esp Salud Publica 76(4):311–319.

Goynes WR, Moreau JP, DeLucca AJ, Ingber BF. 1995. Biodeterioration of nonwoven fabrics. Textile Res J 65(8):489–494.

Gray GC, Schultz RG, Gackstetter GD, McKeehan JA, Aldridge KV, Hudspeth MK, Malasig MD, Fuller JM, McBride WZ. 2001. Prospective study of respiratory infections at the U.S. Naval Academy. Mil Med 166(9):759–763.

Hartmann D. 1967. Silent passage of Echo virus type 20 together with dyspepsia coli strains in a day nursery school. Zentralbl Bakteriol 200(2):274–276.

Hidaka Y, Kubota K. 2006. Study on the sterilization of grain surface using UV radiation–development and evaluation of UV irradiation equipment. Japan Agricult Res Quarterly 40(2):157–161.

Hoebe CJ, Wagenvoort JH, Schellekens JF. 2000. An outbreak of scarlet fever, impetigo and pharyngitis caused by the same Streptococcus pyogenes type T4M4 in a primary school. Ned Tijdschr Geneeskd 144(45):2148–2152.

IEA. 1996. Heat and Moisture Transfer: International Energy Agency.

Ikeda RM, Kondracki SF, Drabkin PD, Birkhead GS, Morse DL. 1993. Pleurodynia among football players at a high school. An outbreak associated with coxsackievirus B1. JAMA 270(18):2205–2206.

Inoue M, Koyano M. 1991. Fungal contamination of oil paintings in Japan. Int Biodeter 28(1–4):23–35.

Jaakkola JJ, and Heinonen, O. P. 1995. Shared office space and the risk of the common cold. Eur J Epidemiol 11(2):213–216.

Jackson LA, Cherry JD, Wang SP, Grayston JT. 2000. Frequency of serological evidence of Bordetella infections and mixed infections with other respiratory pathogens in university students with cough illnesses. Clin Infect Dis 31(1):3–6.

Kelly TJ. 1999. Shedding Some Light on IAQ. RSES J November.

Kenyon TA, Copeland, J. E., Moeti, T., Oyewo, R., and Binkin, N. 2000. Transmission of Mycobacterium tuberculosis among employees in a US government office, Gaborone, Botswana. Int J Tuberc Lung Dis 4(10):962–967.

Khuder SA, Arthur T, Bisesi MS, Schaub EA. 1998. Prevalence of infectious diseases and associated symptoms in wastewater treatment workers. Am J Ind Med 33(6):571–577.

Khudushina TA, Altynova MP, Sagirova GM. 1991. The optimal system of ambulatory examination of groups with high risk of tuberculosis at a large industrial plant. Probl Tuberk 5:10–13.

Kim SJ, Bai GH, Lee H, Kim HJ, Lew WJ, Park YK, Kim Y. 2001. Transmission of Mycobacterium tuberculosis among high school students in Korea. Int J Tuberc Lung Dis 5(9):824–830.

Kim T, Silva JL, Chen TC. 2002. Effects of UV irradiation on selected pathogens in peptone water and on stainless steel and chicken meat. J Food Prot 65(7):1142–1145.

Kowalski WJ, Dunn CE. 2002. Current trends in UVGI air and surface disinfection. INvironment Professional 8(6):4–6.

Kowalski WJ, Bahnfleth WP. 2002. MERV filter models for aerobiological applications. Air Media Summer:13–17.

Kowalski WJ. 2006. Aerobiological Engineering Handbook: A Guide to Airborne Disease Control Technologies. New York: McGraw-Hill.

Krysinska-Traczyk E, Skorska C, Cholewa G, Sitkowska J, Milanowski J, Dutkiewicz J. 2002. Exposure to airborne microorganisms in furniture factories. Ann Agric Environ Med 9(1): 85–90.

Kundsin RB. 1980. Airborne Contagion. Boston. New York Academy of Sciences.

Kuo F, Carey J, Ricke S. 1997. UV irradiation of shell eggs: Effect on populations of aerobes, molds, and inoculated *Salmonella typhimurium*. J Food Prot 60:639–643.

Lacey J, Crook B. 1988. Fungal and actinomycete spores as pollutants of the workplace and occupational illness. Ann Occup Hyg 32:515–533.

Lidwell OM, Williams REO. 1961. The epidemiology of the common cold. J Hygiene 59:309–334.

Liu J, Stevens C, Khan V, Lu J, Wilson C, Adeyeye O, Kabwe M, Pusey P, Chalutz E, Sultana T et al. 1993. Application of ultraviolet-C light on storage rots and ripening of tomatoes. J Food Prot 56:868–872.

Lundin L. 1991. On Building-related Causes of the Sick Building Syndrome. Stockholmiensistx AU, editor. Stockholm: Almqvist & Wiksell Intl.

Marks PJ, Vipond IB, Carlisle D, Deakin D, Fey RE, Caul EO. 2000. Evidence for airborne transmission of Norwalk-like virus (NLV) in a hotel restaurant. Epidemiol Infect 124(3):481–487.

Marquenie D, Lammertyn J, Geeraerd A, Soontjens C, VanImpe J, Nicolai B, Michiels C. 2002. Inactivation of conidia of *Botrytis cinerea* and *Monilinia fructigena* using UV-C and heat treatment. Int J Food Microbiol 74:27–35.

Masterson RG, Green AD. 1991. Dissemination of human pathogens by airline travel. In: Austin B, editor. Pathogens in the Environment. Oxford: Blackwell Scientific Publications.

McMillan JA, Weiner LB, Higgins AM, Lamparella VJ. 1993. Pharyngitis associated with herpes simplex virus in college students. Pediatr Infect Dis J 12(4):280–284.

Menzies D, Comtois, P., Pasztor, J., Nunes, F., and Hanley, J. A. 1998. Aeroallergens and work-related respiratory symptoms among office workers. J Allergy Clin Immunol 101(1 Part 1): 38–44.

Neuzil KM, Hohlbein C, Zhu Y. 2002. Illness among schoolchildren during influenza season: effect on school absenteeism, parental absenteeism from work, and secondary illness in families. Arch Pediatr Adolesc Med 156(10):986–991.

NRC. 2002. The Airliner Cabin Environment and the Health of Passengers and Crew. Council NR, editor. Washington, DC: National Academy Press.

Olivier C. 2000. Rheumatic fever–is it still a problem? J Antimicrob Chemother 45(Suppl):13–21.

Pepys J, Jenkins P, Festenstein G, Gregory PH, Lacey ME, Skinner F. 1963. Farmer's Lung: Thermophilic Actinomycetes as a Source. Lancet 2.

Perkins JE, Bahlke AM, Silverman HF. 1947. Effect of ultra-violet irradiation of classrooms on the spread of measles in large rural central schools. Am J Pub Health 37:529–537.

Philips. 1985. UVGI Catalog and Design Guide. Netherlands: Catalog No. U.D.C. 628.9.

Round A, Evans MR, Salmon RL, Hosein IK, Mukerjee AK, Smith RW, Palmer SR. 2001. Public health management of an outbreak of group C meningococcal disease in university campus residents. Eur J Public Health 11(4):431–436.

Ruben FL, Ehreth, J. 2002. Maritime health: a case for preventing influenza on the high seas. Int Marit Health 53(1–4):36–42.

Salvaggio JE. 1994. Inhaled particles and respiratory disease. J Allergy Clin Immunol 94(2 Part 2):304–309.

Samson RA, Hoekstra ES, Frisvad JC, Filtenborg O. 2000. Introduction to food and airborne fungi. Waganingen, The Netherlands: Ponsen Looyen.

Schmidt SM, Muller CE, Mahner B, Wiersbitzky SK. 2002. Prevalence, rate of persistence and respiratory tract symptoms of Chlamydia pneumoniae infection in 1211 kindergarten and school age children. Pediatr Infect Dis J 21(8):758–762.

Seiler D. 1984. Preservation of bakery products. Inst Food Sci Technol Proc (UK) 17:31–39.

Sessa R, Di PM, Schiavoni G, Santino I, Altieri A, Pinelli S, Del PM. 2002. Microbiological indoor air quality in healthy buildings. New Microbiol 25(1):51–56.

Shama G. 1999. Ultraviolet Light. In: Robinson R, Batt C, Patel P, editors. Encyclopaedia of Food Microbiology. London: Academic Press.

Sigel MM, MvNair TF, Henle W. 1950. Q Fever in a wool and hair processing plant. Am J Pub Health 40:524–532.

Spengler JD, Samet JM, McCarthy JF. 2001. Indoor Air Quality Handbook. New York: McGraw-Hill.

Stersky AK, Heldman DR, Hedrick TI. 1972. Viability of airborne *Salmonella newbrunswick* under various conditions. J Dairy Sci 55(1):14–18.

Stevens C, Khan V, Lu J, Wilson C, Pusey P, Igwegbe E, Kabwe K, Mafolo Y, Liu J, Chalutz E et al. 1997. Integration of ultraviolet (UV-C) light with yeast treatment for control of post harvest storage rots of fruits and vegetables. Biological Control 10:98–103.

Thorn J, Beijer L, Rylander R. 2002. Work related symptoms among sewage workers: a nationwide survey in Sweden. Occup Environ Med 59(8):562–566.

Trout D, Bernstein J, Martinez K, Biagini R, Wallingford K. 2001. Bioaerosol lung damage in a worker with repeated exposure to fungi in a water-damaged building. Environ Health Perspect 190(6):641–644.

Ungs TJ, Sangal SP. 1990. Flight crews with upper respiratory tract infections: epidemiology and failure to seek aeromedical attention. Aviation, Space & Environ Med 61:938–941.

Valero A, Begum M, Leong S, Hocking A, Ramos A, Sanchis V, Marin S. 2007. Effect of germicidal UVC light on fungi isolated from grapes and raisins. Lett Appl Microbiol 45(3):238–243.

Wallin P, and Haycock, P. 1998. Foreign Body Prevention, Detection and Control. London: Blackie Academic & Professional.

Wallner-Pendleton E, Summer S, Froning G, Stetson L. 1994. The use of ultraviolet radiation to reduce Salmonella and psychrotrophic bacterial contamination on poultry carcasses. Poul Sci 73:1327–1333.

Watson JM. 2001. TB in Leicester: out of control, or just one of those things? BMJ 322:1133–1134.

Wong E, Linton R, Gerrard D. 1998. Reduction of *Escherichia coli* and *Salmonella senftenberg* on pork skin and pork muscle using ultraviolet light. Food Microbiol 15:415–423.

Yaun BR, Summer SS. 2002. Efficacy of Ultraviolet Treatments for the Inhibition of Pathogens on the Surface of Fresh Fruits and Vegetables. Blacksburg, VA: Virginia Polytechnic Institute and State University.

Yu ITS, Li Y, Wong TW, Tam W, Chan AT, Lee JHW, Leung DYC, Ho T. 2004. Evidence of airborne transmission of the Severe Acute Respiratory Syndrome virus. N Engl J Med 350(17):1731–1739.

Zilisteanu E, Nafta I, Cretesco L, Nicoulesco I, Focsaneanu M. 1966. Strains of parainfluenza virus type 4, isolated in a day nursery. Arch Roum Pathol Exp Microbiol 25(2):459–464.

Zyska B. 1997. Fungi isolated from library materials: A review of the literature. Int Biodeter Biodegrad 40(1):43–51.

Chapter 19

Residential Applications

19.1 Introduction

UVGI systems can be implemented in the various types of buildings in which people live to reduce the transmission of pathogens and to control the levels of allergens. Living accommodations vary from single unit houses to large condominiums and apartments where thousands of people may live. In houses and apartments with separate ventilation systems, the exposure risks are generally due to family members and visitors only, but in large apartments and dormitories with central ventilation, pathogens can be recirculated between residential units. Older homes and buildings often have natural ventilation and this has the risk of concentrating microorganisms and allergens, although natural ventilation in mild climates can also offer an abundance of clean, fresh air. This chapter focuses on the basic microbiological contamination problems in housing and the possible applications for UVGI in the control of these hazards.

19.2 Residential Homes

Most individuals spend most of their time in their residences, especially if they own their own home. For families the home may be the source of most of the respiratory infections that they will contract in their lives, especially if they have children. Most older homes are naturally ventilated with separate heating and air conditioning systems that are used only intermittently. Newer homes often have central air handling units with outside refrigeration coils. Home ventilation systems rarely employ filtration or other air cleaning devices, other than simple low-efficiency dust filters.

The microflora in occupied homes normally consists of endogenous bacteria and seasonal viruses that come family members, and environmental bacteria and fungal spores that come from the outdoors. Families with pets will also have dander in their homes, and dust mites are also common due to the extensive use of carpeting and bedding. The fungal aerobiology of indoor air in homes will often resemble that

W. Kowalski, *Ultraviolet Germicidal Irradiation Handbook*,
DOI 10.1007/978-3-642-01999-9_19, © Springer-Verlag Berlin Heidelberg 2009

of the outdoor air, except in the case of water-damaged or problem homes, where certain fungal species may become predominant. The most frequently isolated fungi in homes, in order of their occurrence, are *Penicillium, Cladosporium, Aspergillus, Ulocladium, Aureobasidium,* and *Acremonium,* based on data from Flannigan and Hunter (1988). Godish et al. (1993) also identified *Botrytis, Phoma, Alternaria, Fusarium, Rhizopus, Epicoccum,* and *Mucor* as additional common fungal species found in homes. The levels of bacteria in homes varies depending on occupancy, since most bacteria hail from human sources. Table 19.1 identifies the fungi that have been found on typical home furnishings (Kemp et al. 1999). Similar fungi are also sometimes found growing on other building materials like gypsum, mineral wool, base boarding, and inside moisture damaged concrete walls (Kujanpaa et al. 1999, Pessi et al. 1999). The most common bacteria found in homes include, in order of their occurrence, *Micrococcus, Pseudomonas, Staphylococcus, Corynebacterium, Flavobacterium, Acinetobacter, Bacillus,* and *Streptococcus,* based on data from Flannigan et al. (1999).

All of these aforementioned fungal and bacterial species are susceptible to UV exposure, as are the various viruses that may transmit between family members (see Appendices A, B, and C for specific UV rate constants).

Central air handling units in homes can become a source of microbiological hazards if the cooling coils become contaminated with mold spores. Spores may germinate and grow on the coils, on moist filters, or in ductwork, and the air handling unit may serve to disseminate these microbes throughout the house. Perhaps the best solution to this problem is to install a UV lamp in the air handling unit to irradiate the coils, filters, or internal duct surfaces. For the typical one-family house, a UVGI lamp or about 8–24 W of UV output should be sufficient if it can be located so as to irradiate most of the coil surface area. Smaller lamps of 4 UV watts are available and these can be installed at multiple locations (i.e. upstream and downstream of the coils) to provide good exposure. It is necessary for a filter of at least a MERV 6 rating (20% DSP) to be present to keep the lamp free of dust.

Table 19.1 Fungi found on typical home furnishings

Fungal genera	Carpet
Alternaria	Carpets, pillows, bathroom mats
Acremonium	Carpets
Aspergillus	Carpets, futons, latex mattress, pillows
Botrytis	Carpets, latex mattress, pillows
Cladosporium	Carpets, mats, futons, latex mattress, pillows, bath mat, sofas
Epicoccum	Mats, futon mattress, sofas
Fusarium	Carpets, latex mattress
Mucor	Carpets
Paecilomyces	Carpets, latex mattress
Penicillium	Carpets, mats, latex mattress, bath mat
Rhizopus	Carpets, mats, futons, latex mattress, pillows, bath mat, sofas
Trichoderma	Carpets
Other fungi	Carpets, futons, latex mattress, pillows

Fig. 19.1 UV lamp assembly for side-mounting on ductwork. Image courtesy of Lumalier, Memphis, TN

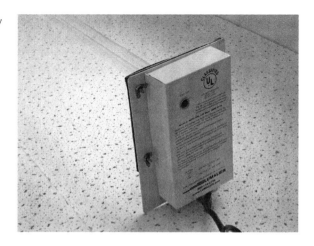

Any UV lamps installed on home central air handling units should be UL rated (see Fig. 19.1). Careful consideration be given to the possible presence of water or condensation inside the air handling unit as this could create an electrical fire hazard. Always consult the lamp manufacturer or an electrician for specific directions for such installations. Care should also be taken to ensure that no leakage of UV occurs and that no stray UV rays exit the ductwork (try turning out the lights and observing any leakage). UV lamp watts may be estimated from Fig. 19.2. The total airflow is

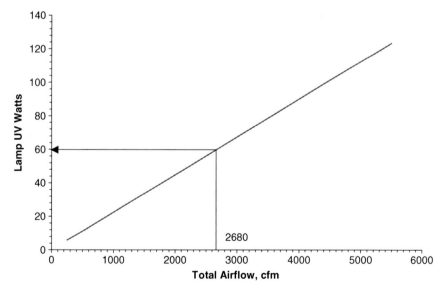

Fig. 19.2 Estimated lamp UV watts required for residential airflows. Sizing example at 2680 cfm and 60 UV watts is indicated. Adapted from Kowalski (2006)

the airflow through the fan, which may or may not be mixed with outside air. Total input power in watts will be the UV watts divided by the lamp efficiency, which is typically about 31%. For the indicated example at 2680 cfm, the UV watts is seen to be 60 W. The total input watts will then be 60/0.31 = 194 W. For estimating the size of recirculating units for a home, see the following section on apartments.

19.3 Apartments

Apartment buildings vary widely in their size and occupancy, the presence or absence of central ventilation systems, and other factors. Many apartment buildings often have individual air conditioning units and operable windows. The microflora of apartment buildings is generally similar to that of residential homes, with seasonal viruses and endogenous bacteria coming from the residents, and outdoor mold spores and environmental bacteria finding their way in by various routes from Spring through Fall.

The indoor concentration of airborne bacteria and fungi in apartments is higher in the presence and absence of furnishings and occupants, based on a study by Sessa et al. (2002). The average airborne concentration of bacteria in an apartment with people and furnishings was 92–182 cfu/m^3, while in their absence the levels were 66–80 cfu/m^3. The average fungal airborne concentrations was 147–297 cfu/m^3, while in their absence the levels were 102–132 cfu/m^3.

Many apartments are naturally ventilated and have few options for improving indoor air quality except for the use of local recirculating air treatment units. Two typical stand-alone recirculating units with UV lamps are shown in Fig. 19.3. As a general rule, at least 2 air changes per hour are necessary to obtain adequate air cleaner performance. The size of a recirculating unit can be estimated from the floor area of an apartment with a typical 8 foot ceiling height from the following relation:

$$\text{Airflow} = 1.1 \cdot \text{Area} \qquad (19.1)$$

where

   Airflow = total airflow, CFM
   Area = Floor Area in ft^2

The following relation applies for SI units:

$$Q = 0.335 \cdot FA \qquad (19.2)$$

where

   Q = Airflow in m^3/min
   FA = Floor Area in m^2

Fig. 19.3 Examples of unitary UVGI air cleaners suitable for use in homes, apartments, and other facilities. *Left image* courtesy of Virobuster Electronic Air Sterilization, Marthalaan, The Netherlands. *Right image* courtesy of Pathogen Solutions, Ltd. United Kingdom

In the event a UV lamp is being sized for a local ventilation unit in an apartment, the airflow estimated from Eqs. (19.1) or (19.2) can be used to estimate the lamp wattage using Fig. 19.2.

19.4   Hotels and Dormitories

Hotels and dormitories are often distinct from apartments and homes in that the living quarters are often small and usually do not have direct supply air. Air conditioners or unitary heaters are often installed and are under occupant control. Hotels typically have such separate heating and cooling units but they also often have one or more central air handling units providing air to the lobbies, hallways, restaurants, and sometimes to the rooms themselves. Unless air is supplied directly to the rooms, it is assumed to infiltrate into them via doorways or grilles. Sometimes the unitary heaters or air conditioning units have individual outside air dampers so that fresh air can be brought in locally.

Unoccupied hotel rooms have low levels of bacteria (usually environmental) and higher levels of fungal spores, usually due to either spores that have settled into the carpeting and furnishings, or that have accumulated on the air conditioner. In one air sampling survey of a hotel taken by the author, the fungal spores identified, in order of their prevalence, were *Cladosporium, Penicillium, Aureobasidium, Pithomyces, Aspergillus, Fusarium, Alternaria,* and a variety of yeast.

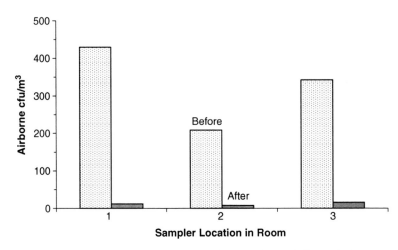

Fig. 19.4 Airborne concentrations of fungal spores in a hotel room before and after two weeks of operation of a recirculating UV air cleaner. Author's data, used courtesy of Immune Building Systems, Inc. New York

Central air handling units in hotels typically have medium-to-low efficiency filters, but the room air conditioners rarely have more than a simple fabric dust filter. As a result, the coils on these air conditioners tend to accumulate mold spores over time and have to be cleaned periodically, otherwise they may impart a moldy odor to room air when they are turned on. UV lamps can disinfect large cooling coils but small air conditioners do not typically have enough space to allow for installing lamps. Certain models of air conditioners are reportedly now available that have the option for installing UV lamps to keep the air conditioner coils clean. Hotel rooms that have separate wall-mounted or ceiling mounted air recirculating units can usually be retrofitted with UV lamps and such applications have been demonstrated to reduce room levels of bacteria and fungi significantly (see Figure 19.3). The continuous operation of UV air cleaners will tend to remove fungal spores as well as bacteria from the air since room activity will stir up spores from the carpeting and furnishings and these spores will be drawn into the air cleaner. Although UV has a limited effect on mold spores, the chronic dosing through repeated passes will eventually destroy the spores at very high rates. In the example shown in Figure 19.4, a UV air cleaner, with no filter, was operated for two weeks and dramatically reduced the airborne levels of mold spores.

References

Flannigan B, Hunter CA. 1988. Factors affecting airborne moulds in domestic dwellings. In: Perry R, Kirk PW, editors. Indoor Air and Ambient Air Quality, pp. 461–468.
Flannigan B, McEvoy EM, McGarry F. 1999. Investigation of Airborne and Surface Bacteria in Homes. Edinburgh, Scotland. IAIAS, pp. 884–889.

Godish D, Godish T, Hooper B, Panter C, Cole M, Hooper M. 1993. Airborne Mold and Bacteria Levels in Selected Houses in the Latrobe Valley. Victoria, Australia: Helsinki, pp. 171–175.

Kemp PC, Neumeister-Kemp HG, Nickelmann A, Murray F. 1999. Fungi in the Dust Extracted from Fabric Covered Furnishings: Preliminary Results During Method Standardization. Edinburgh, Scotland: IAIAS, pp. 890–891.

Kowalski WJ. 2006. Aerobiological Engineering Handbook: A Guide to Airborne Disease Control Technologies. New York: McGraw-Hill.

Kujanpaa L, Haatainen S, Kujanpaa R, Vilkki R, Reiman M. 1999. Microbes in Material Samples Taken from Base Boardings, Gypsum Boards and Mineral Wool Insulation. Edinburgh, Scotland: IAIAS, pp. 892–896.

Pessi AM, Helkio K, Suonketo J, Pentti M, Rantio-Lehtimaki A. 1999. Microbial Growth Inside Exterior Walls of Precast Concrete Buildings as a Possible Risk Factor for Indoor Air Quality. Edinburgh, Scotland: IAIAS, pp. 899–904.

Sessa R, Di PM, Schiavoni G, Santino I, Altieri A, Pinelli S, Del PM. 2002. Microbiological indoor air quality in healthy buildings. New Microbiol 25(1):51–56.

Chapter 20

Bioterrorism Defense

20.1 Introduction

Various applications for UVGI exist in the field of bioterrorism defense, or biodefense, including air disinfection, surface disinfection, material disinfection, and remediation of the aftereffects of a bioweapon attack. The specific types of microbes that may be used as biological weapons are finite in number and are addressed here insofar as they are known to exist or predicted to have bioweapon potential. Nor are toxins or poisons specifically addressed here since these are not controllable by UV technologies. As with other air and surface applications, UV technologies can be highly effective against microbial pathogens whether they are disseminated by natural means or intentionally. This chapter reviews the various ways in which UV systems may be used to protect buildings against bioweapons attacks or to remediate buildings that have been contaminated with bioweapon agents.

20.2 Bioweapons

There are several hundred pathogenic that may cause disease in humans but only a few dozen of these are feasible or adaptable for use as biological weapon (BW) agents. In addition to microbes, there are dozens of organically produced toxins that may also be used as biological weapons, but toxins are not easily destroyed by ultraviolet light and so the focus of this chapter is strictly on pathogenic bacteria, virus, and fungi that can be killed or inactivated by UVGI systems.

Bioweapons can be foodborne, waterborne, or airborne pathogens, and many of these may also be surface-borne (i.e. fomites). Contagious pathogens may also be considered human-borne, since an infected individual may deliberately infect others, becoming a human vector. Vector-borne pathogens, however, are not completely outside the current scope since vector-borne microbes may be weaponized for use as airborne agents (Kowalski 2003). Waterborne pathogens are not exactly

W. Kowalski, *Ultraviolet Germicidal Irradiation Handbook,*   457
DOI 10.1007/978-3-642-01999-9_20, © Springer-Verlag Berlin Heidelberg 2009

within the scope of this text either, but because there is so much overlap between waterborne microbes and the topic of air and surface disinfection, many waterborne pathogens are addressed in this chapter. Other types of biological weapons exist, such as those that have been used to destroy crops, cattle, vegetation, and agricultural products in North Korea, Cuba, Vietnam, South Africa, Zimbabwe, and Palestine (CNS 2002), but these are mostly not human pathogens and agroterrorism is not the subject of this text. The pathogens addressed here are specifically pathogens that might be disseminated in buildings, and that can be treated or remediated by ultraviolet light exposure. Airborne, surface-borne, and waterborne biocontamination can be remediated effectively with UV systems, but foodborne biocontamination is not necessarily remediable unless it is strictly on the exterior surface of the foodstuff.

The characteristics that typically define whether or not a pathogen is suitable for use as a bioweapon agent are lethality, pathogenicity, and ease of dissemination. Many bioweapon agents have been developed for the purpose of incapacitating troops on the battlefield and these types of agents are often fast-acting. Bioweapon agents that are thought to be desirable as terroristic weapons are those which cause a high rate of fatality and that can be easily disseminated. Airborne contamination is by far the most serious type of biological weapons threat since an airborne agent can be spread faster and more widely than would a foodborne or waterborne agent. Chief among such airborne agents are anthrax, influenza, hantavirus, and smallpox. Table 20.1 summarizes the various pathogens that have been cited, by the indicated source, as potential bioweapon agents. Many of these pathogens have known UV susceptibilities (see Appendices A, B, and C) but for those that are unknown, methods exist by which they can be predicted (Kowalski et al. 2009).

One of the factors that distinguishes naturally occurring pathogens from bioweapon agents is that the latter are typically weaponized. Weaponization is a process that may involve various types of pre-treatment and preparation. The weaponization of microbial pathogens usually involves culturing the most virulent strains and then selecting those variants that can withstand the delivery process. The delivery of airborne bioweapons may involve the use of aerosolization devices or slow-detonating explosives, and the pathogens must survive the process in sufficient numbers to make the weapon functional. The product of the weaponization process may be a liquid or a powder, and in either case it may have to be processed or ground to micron-sized particles in order for it to be effectively aerosolized. Biological agents may also tend to clump in liquid or powder form, which reduces their ability to remain aerosolized, and so compounds like silicone may be added to solutions or dry mixtures to prevent clumping.

The UV rate constant can be used to determine the removal rate of any system based on the UV dose. If a UV system has an URV rating then the URV can be used to determine the estimated minimum dose (see, for example, Kowalski and Bahnfleth 2004) The filtration rate of these microbes can be determined directly from their mean size per various references (see, for example, Kowalski and Bahnfleth 2002a, or Kowalski 2006). The combined removal rate of a UV system

Table 20.1 Potential biological weapon agents

Microbe	Group	Type	Reference
Bacillus anthracis spores	Airborne	Bacterial spore	Wald (1970)
Blastomyces dermatidis	Airborne	Fungal spore	Kowalski (2003)
Bordetella pertussis	Airborne	Bacteria	Thomas (1970)
Brucella	Airborne	Bacteria	Wald (1970)
Burkholderia mallei	Airborne	Bacteria	Clarke (1968)
Burkholderia pseudomallei	Airborne	Bacteria	Clarke (1968)
Chikungunya virus	Vector-borne	Virus	Wald (1970)
Chlamydia psittaci	Airborne	Bacteria	Clarke (1968)
Clostridium botulinum	Foodborne	Bacterial spore	Clarke (1968)
Clostridium perfringens	Foodborne	Bacterial spore	Ellis (1999)
Coccidioidis immitis	Airborne	Fungal spore	Wald (1970)
Corynebacterium diphtheria	Airborne	Bacteria	Paddle (1996)
Coxiella burnetti	Airborne	Bacteria	Wald (1970)
Crimean-Congo hemorrhagic fever	Airborne	Virus	Paddle (1996)
Dengue fever	Vector-borne	Virus	Clarke (1968)
Ebola (GE)	Airborne	Virus	Paddle (1996)
Francisella tularensis	Airborne	Bacteria	Wald (1970)
Hantaan virus	Airborne	Virus	Paddle (1996)
Hepatitis A	Int	Virus	Paddle (1996)
Histoplasma capsulatum	Airborne	Fungal spore	Paddle (1996)
Influenza	Airborne	Virus	McCarthy (1969)
Japanese encephalitis	Vector-borne	Virus	Paddle (1996)
Junin virus	Airborne	Virus	Paddle (1996)
Lassa fever virus	Airborne	Virus	Paddle (1996)
Legionella pneumophila	Airborne	Bacteria	Wright (1990)
Lymphocyte choriomeningitis	Vector-borne	Virus	Paddle (1996)
Machupo virus	Airborne	Virus	Paddle (1996)
Marburg virus	Airborne	Virus	Paddle (1996)
Mycobacterium tuberculosis	Airborne	Bacteria	Paddle (1996)
Mycoplasma pneumoniae (GE)	Airborne	Bacteria	Kowalski (2003)
Nocardia asteroides	Airborne	Bacteria	Paddle (1996)
Paracoccidioides brasiliensis	Airborne	Fungal spore	Kowalski (2003)
Rickettsia rickettsii	Vector-borne	Bacteria	Wald (1970)
Rickettsiae prowazeki	Airborne	Bacteria	Wald (1970)
Rift Valley fever	Vector-borne	Virus	Hersh (1968)
Russian spring-summer encephalitis	Vector-borne	Virus	Paddle (1996)
Salmonella typhi (Typhoid Fever)	Foodborne	Bacteria	Wald (1970)
Shigella	Foodborne	Bacteria	McCarthy (1969)
Stachybotrys chartarum	Airborne	Fungal spore	Kowalski (2003)
Streptococcus pneumoniae	Airborne	Bacteria	Ellis (1999)
Tick-borne encephalitis	Vector-borne	Virus	Wald (1970)
Variola (Smallpox & Camelpox)	Airborne	Virus	Clarke (1968)
VEE (EEE, WEE)	Vector-borne	Virus	Wald (1970)
Vibrio Cholerae (Cholera)	Foodborne	Bacteria	Wald (1970)
West Nile virus	Vector-borne	Virus	Wright (1990)
Yellow Fever virus	Vector-borne	Virus	Wald (1970)
Yersinia pestis	Airborne	Bacteria	Wald (1970)

can then be determined algebraically by combining the removal rate of the filter with the removal rate of the UV system as follows:

$$R_T = 1 - (1 - R_F)(1 - R_U) \qquad (20.1)$$

where

R_T = Total Removal Rate
R_F = Removal Rate of Filter
R_U = Removal Rate of UV system

Figure 20.1 illustrates the effect of the combined removal rate through UVGI and filtration for a range of bioweapon agents, including botulinum toxin. It can be observed that relatively high removal rates are obtained for the entire array of microbes, including those that are not easily filterable and those that are resistant to UV. Figure 20.1 represents a single pass through such a combined MERV/URV system and considerably higher removal rates are possible after several passes. The effect of dilution, when outside make-up air is mixed in, is not accounted for in this simple approach, but will further reduce the inhalation threat. Under dilution ventilation, all microbes are removed at virtually identical rates, and so the total outside

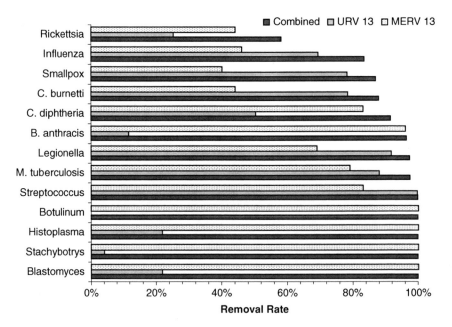

Fig. 20.1 Removal rates of various bioweapon agents by an URV 13 UVGI, a MERV 13 filter, and the combination of both. Adapted from Kowalski et al. (2003)

air (or make-up flow rate) will be the primary determinant of the final airborne concentration after some period of time (Kowalski et al. 2003).

20.3 Biodefense of Buildings

All modern buildings have forced ventilation systems that can be retrofitted to incorporate biodefense technologies such as high efficiency filtration and UVGI. The degree to which such retrofits are effective will depend on several factors, including the available space for UV installations, the total outside airflow, and the distribution of air throughout the building. It is essential that both filtration and UV be combined with filtration in order to achieve the best possible performance for any given ventilation system. Some sources recommend using HEPA filters (without UVGI) for protecting buildings, but it is not possible to simply add a HEPA filter to most ventilation systems without changing out the fan motor and reconstructing ductwork to allow for the air velocity of 250 fpm that is the design specification for HEPA filters. The expense and energy costs associated with a HEPA filter retrofit are prohibitive and ultimately unnecessary, considering that UV combined with filtration (at the MERV 11–15 level) is sufficient to offer considerable protection to buildings against bioweapon agents.

UV complements filtration performance by removing those microbes that may penetrate the filter. This is true for biological weapon agents as well as naturally occurring pathogens. The fact remains that bioweapon agents occur in various sizes and may be either spores, or bacteria and viruses. Although bioweapons may be ground to a specific size of about 2 µm, the resulting powder will invariably consist of a range of particle sizes that form a lognormal distribution. The lognormal size distribution of micron-sized particles, whether biological or inorganic, is a natural phenomenon (Koch 1969). The normal range of sizes necessary to produce aerosolizable particles is about 1–5 µm. The log mean size of a particle in this range may be about 1.5 µm.

Although the specific particle size has limited impact on the UV susceptibility of the pathogen, it will impact the performance of any filter that is associated with a UVGI system. The smaller microbes tend to be more susceptible to UV, while the larger microbes like spores tend to be susceptible to filtration. Figure 20.2 illustrates this point with an example that employs a MERV 13 filter combined with an URV 13 UVGI system to remove an array of bioweapon agents. In this example it is assumed there is a prefilter, which has a marginal effect in removing some of the larger pathogens. The MERV 13 filter (approximately 80% DSP) removed most of the larger microbes, like the spores, and some of the smaller viruses. The second stage of the air cleaner is an URV 13 system (UV dose 20 J/m^2 minimum) which removes most of the smaller bacteria and viruses. The combined effect of these components is an overall high rate of removal.

The overall removal rates of the air cleaner may be increased even further if desired by increasing the filtration and/or the UV removal (i.e. to MERV 15 and

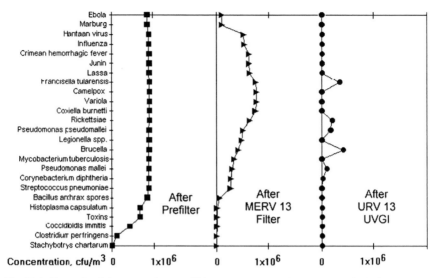

Fig. 20.2 Example of the removal rates of bioweapon agents when an URV 13 UVGI system is combined with MERV 13 filters

URV 15). The ultimate determinant of whether the removal rates are adequate is the degree of protection offered to building occupants. One way to assess the protection provided to building occupants is to analyze the airborne concentrations in the building that may result from a release and determine how many infections might be caused. One method for performing such analysis is the Building Protection Factor (BPF) for buildings or the Zonal Protection Factor (ZPF) for individual building zones (Kowalski 2003, 2006). The BPF represents the percentage of building occupants that are protected from infection under a design basis release. When different areas of a building have different levels of protection, the protection in each zone can be defined by the ZPF. For more detailed information on computing the ZPF and estimating the protection afforded by buildings, see Chap. 8.

Figure 20.3 illustrates the effect of the MERV filter rating and the URV rating against smallpox, for a model attack on typical 40 story office building in which smallpox is released in the outside air intakes. It can be observed that the use of a MERV 13 filter combined with an URV 13 UVGI system provides near complete protection (approximately 99% BPF), while the protection afforded by a MERV/URV 15 system is the same as that provided by a HEPA filter. HEPA filters are often suggested for use in biodefense applications but it is clear from this and many other examples that the same performance can be obtained by a combination of a MERV 15 filter and an URV 15 UV system, which are considerably less expensive and easier to implement.

Figure 20.4 shows the results of a number of simulations with various pathogens and toxins on a multistory building with a large atrium/plaza. The attack scenario is a release of the agent in the outside air intakes. It can be observed that although some

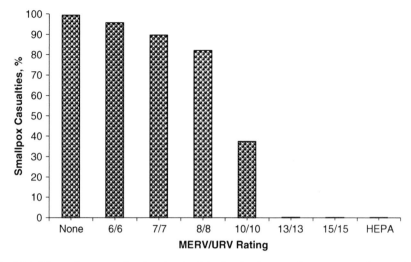

Fig. 20.3 The performance of various levels of air cleaning in protecting occupants of a model 40 story building subject to a smallpox release

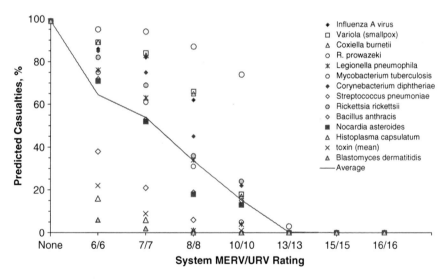

Fig. 20.4 Results of simulated attacks (outside air intake release) on a multistory office building with increasing levels of air treatment

agents penetrate at low levels of air treatment (specified by MERV/URV ratings), they all converge towards complete removal at the higher levels of air treatment. The protection offered to occupants levels off at approximately 99% with a MERV/URV 13/13 system and beyond a MERV/URV 15/15 system there are no added benefits (i.e. a HEPA filter would add no additional building protection). These results are typical for all types of large office buildings.

The complement of the casualties in the previous examples is known as the protection factor. At MERV/URV 15/15 the BPF for both of the examples is approximately 99%, meaning 99% of the occupants are protected from infection. For entire buildings it is called a Building Protection Factor (BPF) and for individual zones within a building it is called the Zonal Protection Factor (ZPF). Methods for computing the BPF have been detailed in Chap. 8, and the method is the same regardless of whether the infections produced occur naturally or by intentional release, only the scenarios and agents involved are different. A corollary of this, as noted previously, is that any system designed to protect against normal airborne infections will also protect against bioweapon agents, and air treatment systems that produce high disinfection rates will work equally well regardless of the microbial threat.

20.4 UV Systems for Biodefense

Ultraviolet air and surface disinfection systems can be used for biodefense as an integral component of a whole-building protection system, and operate in conjunction with the building envelope and the building ventilation system. The ventilation system of any building will normally perform the function of air cleaning when it brings in outside air and exhausts any contaminated air. If the ventilation system includes recirculation filters, it will clean and recirculate building air. When operating with 100% outside air and at a sufficient airflow rate, the ventilation system may be capable of cleaning the interior spaces without air disinfection technology, except when bioweapons are released in the outside air or in the outside air intake (Skistad 1994). Hundred percent of outside air is, however, is an exception, since most ventilation system recirculate indoor air and therefore the use of air cleaning devices will enhance building protection against bioweapons.

There is essentially no difference between an air disinfection system intended to protect against natural airborne pathogens and one that is intended to protect against intentional releases of pathogens. Any system that employs dilution ventilation (exchanging fresh outdoor air for indoor air) and uses some form of air cleaning (filtration, UV, or both) will defend against any kind of airborne microbiological contamination.

Any well-designed air treatment system employs both filtration and UVGI. Filtration represents the primary defense against a wide array of pathogens and allergens. Most buildings use only prefilters for dust control but it is increasingly common today to find buildings with MERV 11–13 (DSP 65–85%) filters installed.

UVGI systems have a demonstrated ability to disinfect air of viruses and most bacteria when they are properly designed and installed. UVGI can play an important part of any building protection system when used in conjunction with high efficiency filters. Although the installation of UV lamps is relatively straightforward, the sizing of a UVGI system for any given application can be a complex function of air velocity, geometry, lamp placement, reflectivity, and other factors. The location of UV lamps inside an air handling unit (AHU) or duct is often constrained by

Fig. 20.5 Example of a UV system placed upstream of the cooling coils (*left*) and downstream of the filters (*right*). Reflective diamond plate aluminum has been added to distribute the more irradiance evenly across all surfaces. Photo courtesy of Immune Building Systems, Inc., New York

space. The ability to simultaneously expose cooling coils and other internal surfaces while disinfecting air is a desirable design characteristic that can protect against the accumulation of bioweapon agents on internal surfaces. Figure 20.5 shows a system designed for air treatment and cooling coil disinfection in which the UV lamps are located overhead and reflective aluminum diamond plate is used to ensure coil coverage and decrease shadowing effects. Diamond plate tends to diffuse and scatter the specular reflections of polished aluminum, producing a more even distribution of irradiance. Appropriate placement of UV lamps can also ensure that the filters are sterilized by UV exposure so that they pose no hazard to maintenance workers if the filters become contaminated with dangerous pathogens or toxins.

The specific type of bioweapon attack scenario affects the way in which bioweapon agents are distributed throughout a building. Various attack scenarios are possible, including the release in the outside air intakes discussed previously, a release inside the building, a release inside the ductwork or air handling equipment, and others. An external release (a release in the outside air) is virtually identical to a release in the outside air intakes if the outside airborne concentrations are high enough. The outside air intake release is often the scenario of greatest concern to building owners, mainly because of ease of access. Each of these scenarios can be studied in detail for any particular building. Some scenarios produce greater hazards than other, but regardless of the scenario chosen for analysis, air cleaning systems can provide high levels of protection to building occupants (Kowalski et al. 2003).

Figure 20.5 illustrates two basic types of attack scenarios, the internal release and the outside air intake release. The internal release, if it occurs in the main first floor lobby, will heavily contaminate the immediate area, and if the lobby air is recirculated then the contaminants will be spread throughout the building. The second example in Fig. 20.6 shows a release in the outside air intakes. In this case the air handling unit is employed as a weapon and contaminants are spread evenly throughout the entire building, depending on air distribution. In both cases there will be some contaminant removal in the AHU that depends on the level of filtration or

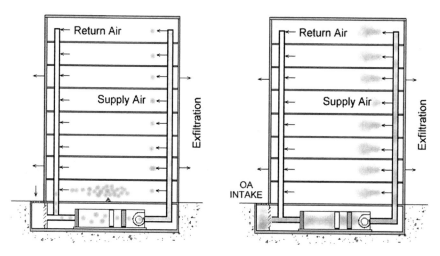

Fig. 20.6 Two basic attack scenarios: *Left* represents an internal release in the first floor lobby, *Right* represents a release inside the outside air intake

air cleaning, but in the absence of air cleaning the second scenario is by far the most dangerous to occupants.

Sheltering zones are areas that are isolated and may have independent air cleaning systems, and may be pressurized against surrounding areas or the outside air. Any area that can be physically isolated can be used as a temporary sheltering zone, especially if the airflow can be shut off completely or if it has no airflow. In many office buildings, lobbies are separately ventilated and the lobby is often pressurized. In such cases the lobby can serve as a buffer zone to protect upper floors against local internal releases and infiltration from the street, and this function can be enhanced through the use of filtration and UV air cleaning (Kowalski and Bahnfleth 2003). The buffer zone concept involves isolating the first floor or lobby and increasing air change rate so as to prevent migration of contaminants to other floors and to rapidly purge the lobby. In such implementations, the lobby will also operate as a sheltering zone if bioweapon agents are released outside or on upper floors. In such cases the retrofit may required the fan and motor to be replaced to boost the airflow and permit the addition of higher efficiency (i.e. MERV 13) filters. Many first floor lobbies have high ceilings (see Fig. 20.7) and enclose such volumes of air that simply isolating them during a threat will provide adequate sheltering even if the ventilation is shut down (Kowalski 2003).

Automatic isolation of a building envelope is a desirable characteristic that can protect occupants against an outside release, or that can protect most occupants from an indoor release (excluding those at the point of release). Some biosensors and biodetectors are available today that can identify a few specific bioweapon agents, but the response times are insufficient to provide automatic isolation of a building (Kowalski and Bahnfleth 2002b). Particle detectors can provide automatic isolation upon detection of airborne contaminants in the micron-size range, but the expense

Fig. 20.7 First floor lobbies can be converted into buffer zones by isolating and pressurizing them with an independent ventilation and air disinfection system

of such systems may not be justified in comparison with an air cleaning system that offers full-time protection against all bioweapon agents, which is what a combined filtration/UVGI system would provide.

20.5 UV Bioremediation

UV technology can be used for the remediation of buildings contaminated with pathogenic microorganisms just as it has been used in the past to remediate buildings contaminated with mold. Remediation of buildings with mold contamination usually involves scrubbing the surfaces with bleach or other disinfectants, or else tearing out the walls and replacing them (Kowalski and Burnett 2001). After manual scrubbing, UV systems, such as Upper Room or After Hours systems, can be permanently installed to provide continued eradication of spores and suppression of mold growth.

UV can also be used for the remediation of buildings contaminated with pathogens such as anthrax spores provided the contamination is primarily on surfaces and not embedded deep in materials such as carpeting. Portable UV disinfection units can be placed inside rooms and operated for a period of hours or more to disinfect all exposed surfaces. Given sufficient UV dose or exposure time, any exposed pathogen can be inactivated. Even the low levels produced by an Upper Room system are capable of inactivating the hardiest spores given sufficient exposure time. Area disinfection systems are designed to be used, however, in unoccupied rooms and the levels of UV irradiance produced are hazardous to humans. Figure 20.8 illustrates the sterilization times (assuming sterilization as a six log reduction) for a variety of bioweapon agents. It can be observed that at a level of irradiance of 100 J/m^2, a six log reduction can be achieved for all pathogens within a few hours. Typically, the irradiance levels will vary with the distance from the UV lamps and may be higher or lower, but the exposure times are often much longer, and

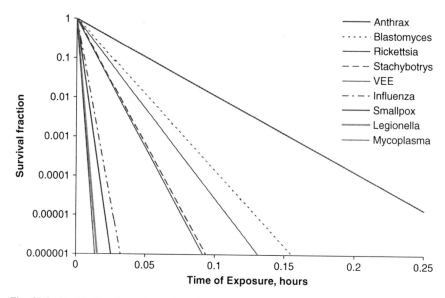

Fig. 20.8 Sterilization times for various bioweapon pathogens under continuous UV exposure at 1 J/m^2

may be as long as 24 h. For more detailed information on surface disinfection, see Chap. 10. For more detailed information on bioweapons, disinfection technologies, and analysis methods, see Kowalski (2003).

References

Clarke R. 1968. We All Fall Down. London: Allen Lane The Penguin Press.

CNS. 2002. Agro-Terrorism: Chronology of CBW Attacks Targeting Crops & Livestock 1915–2000. Center for Non-Proliferation Studies.

Ellis JW. 1999. Police Analysis and Planning for Chemical, Biological and Radiological Attacks: Prevention, Defense and Response. Springfield, IL: Charles C. Thomas.

Hersh SJ. 1968. Chemical & Biological Warfare: America's Hidden Arsenal. Garden City, New York: Doubleday & Co., Inc.

Koch AL. 1969. The logarithm in biology: Distributions simulating the log-normal. J Theoret Biol 23:251–268.

Kowalski WJ, Burnett E. 2001. Mold and Buildings. University Park, PA: The Pennsylvania Housing Research Center. Report nr Builder Brief BB0301.

Kowalski WJ, Bahnfleth WP. 2002a. Defending Buildings Against Bioterrorism. HPAC Engineering.

Kowalski WJ, Bahnfleth WP. 2002b. MERV filter models for aerobiological applications. Air Media Summer:13–17.

Kowalski WJ. 2003. Immune Building Systems Technology. New York: McGraw-Hill.

Kowalski WJ, Bahnfleth WP. 2003. Immune-Building Technology and Bioterrorism Defense. HPAC Engineering 75 (Jan)(1):57–62.

Kowalski WJ, Bahnfleth WP, Musser A. 2003. Modeling immune building systems for bioterrorism defense. J Arch Eng June:86–96.

Kowalski WJ, Bahnfleth WP. 2004. Proposed Standards and Guidelines for UVGI Air Disinfection. IUVA News 6(1):20–25.

Kowalski WJ. 2006. Aerobiological Engineering Handbook: A Guide to Airborne Disease Control Technologies. New York: McGraw-Hill.

Kowalski W, Bahnfleth W, Hernandez M. 2009. A Genomic Model for predicting the Ultraviolet Susceptibility of Viruses. IUVA News 11(2):15–28.

McCathy RD. 1969. The Ultimate Folly: War by Pestilence, Asphyxiation, and Defoliation. New York: Alfred A. Knopf.

Paddle BM. 1996. Biosensors for chemical and biological agents of defence interest. Biosens Bioelectron 11(11):1079–1113.

Skistad H. 1994. Displacement Ventilation. New York: John Wiley & Sons, Inc.

Thomas AVW. 1970. Legal Limits on the Use of Chemical and Biological Weapons. Dallas: Southern Methodist University Press.

Wald G. 1970. Chemical and Bacteriological (Biological) Weapons and the Effects of Their Possible Use, Report E.69.I.24. New York: Ballantine.

Wright S. 1990. Preventing a Biological Arms Race. Cambridge, MA: The MIT Press.

APPENDIX A: UV Rate Constants for Bacteria

Microbe	Type	D_{90} J/m²	UVGI k m²/J	Media	RH %	Sh	St	UL J/m²	Dia. μm	Base Pairs kb	Source (see Chapter 4 Refs)
Acinetobacter baumannii	Veg	18	0.12800	S	-	N	1	90	1.225	3598	Rastogi 2007
Acinetobacter baumannii	Veg	33	0.19200	W	-	Y	2	48	1.225	3598	Templeton 2009
Aeromonas	Veg	11	0.20310	W	Wat	-	-	-	2.098	4740	Sako 1985
Aeromonas hydrophila	Veg	16	0.14100	W	Wat	Y	1	590	2.098	4740	Liltved 1996
B. atrophaeus (B. globigii)	Sp	144	0.01600	Air	(Lo RH)	-	-	-	1.12	4140	EPA 2006
B. atrophaeus spores	Sp	1323	0.00174	W	Wat	N	1	4000	1.12	4140	Shafaat 2006
Bacillus anthracis spores	Sp	411	0.00560	W	Wat	Y	1	600	1.118	5220	Nicholson 2003
Bacillus anthracis spores	Sp	45	0.05094	S	-	N	1	52	1.118	5220	Sharp 1939
Bacillus anthracis spores	Sp	743	0.00310	S	-	Y	2	1890	1.118	5220	Knudson 1986
Bacillus cereus spores	Sp	267	0.00863	S	-	-	-	-	1.118	5700	Weinberger 1984
Bacillus cereus spores	Sp	210	0.01098	S	-	-	-	-	1.118	5700	Weisova 1966
Bacillus cereus spores	Sp	116	0.01979	S	-	-	-	-	1.118	5700	Germaine 1973
Bacillus cereus spores	Sp	408	0.00564	S	-	-	-	-	1.118	5700	Benoit 1990
Bacillus megatherium	Sp	273	0.00843	S	-	-	-	-	1.12	4600	Hercik 1937
Bacillus megatherium	Veg	113	0.02038	S	-	-	-	-	-	4600	Hercik 1937
Bacillus pumilis spores	Sp	50	0.04600	W	Wat	Y	2	100	-	-	New combe 2005
Bacillus subtilis	Veg	25	0.09210	W	Wat	-	-	-	-	4210	Lojo 1995
Bacillus subtilis	Veg	14	0.16858	Air	(Lo RH)	N	1	940	-	4210	Nakamura 1987
Bacillus subtilis spores	Sp	250	0.00920	W	Wat	Y	1	600	1.12	4140	Nicholson 2003
Bacillus subtilis spores	Sp	161	0.01430	W	Wat	N	1	600	1.12	4140	Hoyer 2000
Bacillus subtilis spores	Sp	116	0.01982	W	Wat				1.12	4140	Sommer 1989
Bacillus subtilis spores	Sp	220	0.01047	W	Wat	N	1	500	1.12	4140	Sommer 1998
Bacillus subtilis spores	Sp	199	0.01155	W	Wat	N	1	810	1.12	4140	Sommer 1999
Bacillus subtilis spores	Sp	77	0.03000	W	Wat	Y	1	400	1.12	4140	Qualls 1983
Bacillus subtilis spores	Sp	155	0.01490	W	Wat	Y	1	400	1.12	4140	Mamane-Gravetz 2005
Bacillus subtilis spores	Sp	89	0.02580	W	Wat	-	-	-	1.12	4140	Horneck 1985
Bacillus subtilis spores	Sp	200	0.01150	W	Wat	Y	2	800	1.12	4140	Chang 1985
Bacillus subtilis spores	Sp	80	0.02880	W	Wat	Y	2	400	1.12	4140	DeGuchi 2005
Bacillus subtilis spores	Sp	94	0.02460	S	-	-	-	220	1.12	4140	Rentschler 1941
Bacillus subtilis spores	Sp	68	0.03370	S	-	-	-	-	1.12	4140	Munakata 1975
Bacillus subtilis spores	Sp	113	0.02030	S	-	-	-	-	1.12	4140	Munakata 1972
Bacillus subtilis spores	Sp	89	0.02600	Air	Hi RH	N	1	45.0	1.12	4140	Peccia 2001a
Bacillus subtilis spores	Sp	149	0.01550	Air	Lo RH	Y	1	550	1.12	4140	Ke 2009
Bacillus subtilis spores	Sp	85	0.02700	Air	Lo RH	N	1	45.0	1.12	4140	Peccia 2001a
Bacillus thuringiensis	Sp	2303	0.00100	W	Wat	N	2	10000	-	-	Griego 1978
Burkholderia cenocepacia	Veg	58	0.03956	W	Wat	N	1	60	0.707	7270	Abshire 1981
Burkholderia cepacia	Veg	11	0.21150	Air	Lo RH	N	1	23	0.77	7700	Fletcher 2004
Burkholderia cepacia	Veg	22	0.10520	Air	Hi RH	N	1	23	0.77	7700	Fletcher 2004
Campylobacter jejuni	Veg	11	0.20933	W	Wat	-	-	-	2.12	1641	Wilson 1992
Campylobacter jejuni	Veg	29	0.07940	W	Wat	-	-	-	2.12	1641	Butler 1987
Citrobacter diversus	Veg	32	0.07140	W	Wat	N	1	130	1.2	-	Giese 2000
Citrobacter freundii	Veg	42	0.05482	W	Wat	-	-	-	1.2	-	Zemke 1990
Citrobacter freundii	Veg	46	0.05010	W	Wat	N	1	130	1.2	-	Giese 2000
Clostridium perfringens	Veg	38	0.06000	W	Wat	-	-	-	5	3031	Hijnen 2006
Clostridium perfringens	Veg	135	0.01700	-	-	-	-	-	5	3031	Jepson 1973
Clostridium tetani	Veg	49	0.04699	-	-	-	-	-	5	2790	Jepson 1973
Corynebacterium diphtheriae	Veg	33	0.07010	S	-	N	1	46	0.698	2480	Sharp 1939
Coxiella burnetii	Veg	15	0.15350	W	Wat	-	-	-	0.283	2030	Little 1980
Deinococcus radiodurans	Veg	365	0.00630	W	Wat	Y	1	1200	-	3280	Setlow 1964
Enterobacter cloacae	Veg	64	0.03598	W	Wat	-	-	-	1.414	-	Zemke 1990
Escherichia coli	Veg	21	0.10900	W	Wat	N	1	21	0.5	5490	Zelle 1955
Escherichia coli	Veg	53	0.04320	W	Wat	N	1	24	0.5	5490	Tyrrell 1972
Escherichia coli	Veg	20	0.11510	W	Wat	N	1	60	0.5	5490	Oguma 2001
Escherichia coli	Veg	47	0.04940	W	Wat	N	1	900	0.5	5490	Kim 2002
Escherichia coli	Veg	43	0.05300	W	Wat	Y	1	60	0.5	5490	Hofemeister 1975
Escherichia coli	Veg	13	0.18000	W	Wat	Y	2	-	0.5	5490	Harris 1987
Escherichia coli	Veg	20	0.11500	W	Wat	Y	1	120	0.5	5490	Harm 1968
Escherichia coli	Veg	24	0.09600	W	Wat	Y	1	200	0.5	5490	David 1973
Escherichia coli	Veg	81	0.02832	W	Wat	N	1	83	0.5	5490	Abshire 1981
Escherichia coli	Veg	25	0.09398	S	-	N	1	45	0.5	5490	Sharp 1939
Escherichia coli	Veg	19	0.12000	S	Hi RH	N	1	4.4	0.5	5490	Rentschler 1942
Escherichia coli	Veg	12	0.19300	S	Lo RH	N	1	4.4	0.5	5490	Rentschler 1942
Escherichia coli	Veg	25	0.09210	S	-				22	0.5	5490	Rentschler 1941

APPENDIX A: UV Rate Constants for Bacteria

Microbe	Type	D₉₀	UVGI k	Media	RH	Sh	St	UL	Dia.	Base Pairs	Source
		J/m^2	m^2/J		%			J/m^2	μm	kb	(see Chapter 4 Refs)
Escherichia coli	Veg	20	0.11670	S	-	Y	1	130	0.5	5490	Quek 2008
Escherichia coli	Veg	51	0.04540	S	-	N	1	66	0.5	5490	Luckiesh 1949
Escherichia coli	Veg	34	0.06720	S	-	N	1	900	0.5	5490	Kim 2002
Escherichia coli	Veg	55	0.04187	S	-	-	-	-	0.5	5490	Hollaender 1955
Escherichia coli	Veg	8	0.28300	S	-	N	1	48	0.5	5490	Collins 1971
Escherichia coli	Veg	3	0.72300	Air	Lo RH	-	-	6	0.5	5490	Webb 1970
Escherichia coli	Veg	11	0.21800	Air	Hi RH	-	-	6	0.5	5490	Webb 1970
Escherichia coli	Veg	11	0.21900	Air	Hi RH	N	1	4.4	0.5	5490	Rentschler 1942
Escherichia coli	Veg	13	0.18100	Air	Lo RH	N	1	4.4	0.5	5490	Rentschler 1942
Escherichia coli	Veg	15	0.15611	Air	Lo RH	N	2	4	0.5	5490	Luckiesh 1949
Escherichia coli	Veg	2	0.96500	Air	Lo RH	N	1	0.5	0.5	5490	Koller 1939
Escherichia coli	Veg	11	0.20500	Air	Hi RH	N	1	0.5	0.5	5490	Koller 1939
Francisella tularensis	Veg	256	0.00900	Air	Lo RH	N	1	1	0.2	1890	Beebe 1959
Francisella tularensis	Veg	288	0.00800	Air	Hi RH	N	1	1	0.2	1890	Beebe 1959
Haemophilus influenzae	Veg	38	0.05990	S	-	Y	2	16	0.285	1910	Mongold 1992
Haemophilus influenzae Rd	Veg	13	0.17700	W	Wat	N	2	55	0.285	1910	Barnhart 1970
Halobacterium sp. NRC-1	Veg	25	0.09210	S	-	N	1	150	-	2571	Crowley 2006
Halobacterium salinarum	Veg	68	0.03390	-	-	N	1	200	-	-	Martin 2000
Halomonas elongata	Veg	13	0.18090	-	-	N	1	10	-	-	Martin 2000
Helicobacter pylori	Veg	33	0.06900	W	Wat	N	1	80	2.1	1780	Hayes 2006
Klebsiella pneumoniae	Veg	42	0.05480	W	Wat	-	-	-	0.671	5315	Zemke 1990
Klebsiella pneumoniae	Veg	68	0.03390	W	Wat	N	1	200	0.671	5315	Giese 2000
Klebsiella terrigena	Veg	33	0.07000	W	Wat	N	1	110	-	-	Wilson 1992
Legionella dumoffi	Veg	24	0.09594	S	-	N	1	72	0.52	3400	Knudson 1985
Legionella bozemanii	Veg	19	0.17400	W	Wat	Y	1	97	0.52	3400	Yamamoto 1987
Legionella bozemanii	Veg	15	0.15351	S	-	N	2	72	0.52	3400	Knudson 1985
Legionella gormanii	Veg	26	0.08856	S	-	N	1	72	0.52	3400	Knudson 1985
Legionella jordanis	Veg	11	0.20933	S	-	N	1	72	0.52	3400	Knudson 1985
Legionella longbeach	Veg	11	0.20933	S	-	N	1	72	0.52	3400	Knudson 1985
Legionella micdadei	Veg	15	0.15351	S	-	N	1	72	0.52	3400	Knudson 1985
Legionella oakridgensis	Veg	22	0.10466	S	-	N	1	72	0.52	3400	Knudson 1985
Legionella pneumophila	Veg	13	0.17400	W	Wat	Y	1	97	0.52	3400	Yamamoto 1987
Legionella pneumophila	Veg	12	0.19298	W	Wat	Y	1	0.5	0.52	3400	Gilpin 1985
Legionella pneumophila	Veg	9	0.24849	W	Wat	N	1	30	0.52	3400	Antopol 1979
Legionella pneumophila	Veg	5	0.44613	S	-	N	2	72	0.52	3400	Knudson 1985
Legionella pneumophila	Veg	25	0.09110	W	Wat	N	1	72	0.52	3400	Wilson 1992
Legionella pneumophila	Veg	16	0.14390	W	Wat	N	1	80	0.52	3400	Oguma 2004 (LP)
Legionella pneumophila	Veg	19	0.12020	W	Wat	N	1	96	0.52	3400	Oguma 2004 (MP)
Legionella wadsworthii	Veg	4	0.57565	S	-	N	2	72	0.52	3400	Knudson 1985
Listeria monocytogenes	Veg	181	0.01270	W	Wat	N	1	900	0.707	3130	Kim 2002
Listeria monocytogenes	Veg	156	0.01480	S	-	N	1	900	0.707	3130	Kim 2002
Listeria monocytogenes	Veg	10	0.23030	S	-	Y	1	48	0.707	3130	Collins 1971
Micrococcus candidus	Veg	61	0.03806	S	-	-	-	-	1.2	4050	Hollaender 1955
Micrococcus piltonensis	Veg	81	0.02843	S	-	-	-	132	2.2	-	Rentschler 1941
Micrococcus sphaeroides	Veg	100	0.02303	S	-	-	-	154	1.2	4050	Rentschler 1941
Moraxella	Veg	10965	0.00022	W	Wat	N	1	5940	1.225	1940	Keller 1982
Mycobacterium avium-intra.	Veg	84	0.02740	W	Wat	Y	1	200	1.118	5470	David 1973
Mycobacterium avium	Veg	60	0.03840	W	Wat	Y	1	20	1.118	5470	Shin 2008
Mycobacterium avium	Veg	35	0.06580	W	Wat	-	-	-	1.118	5470	McCarthy 1974
Mycobacterium bovis BCG	Veg	22	0.10550	S	-	N	1	48	0.637	4340	Collins 1971
Mycobacterium bovis BCG	Veg	10	0.24200	Air	50	N	1	5.0	0.637	4340	Riley 1976
Mycobacterium bovis BCG	Veg	12	0.19000	Air	-	-	-	-	0.637	4340	Peccia 2002
Mycobacterium bovis BCG	Veg	19	0.12000	Air	Lo RH	N	1	8.3	0.637	4340	Ko 2000
Mycobacterium bovis BCG	Veg	33	0.07000	Air	Hi RH	N	1	8.3	0.637	4340	Ko 2000
Mycobacterium flaviscens	Veg	120	0.01919	W	Wat	Y	1	200	0.637	-	David 1973
Mycobacterium fortuitum	Veg	68	0.03390	W	Wat	Y	1	200	0.637	5000	David 1973
Mycobacterium fortuitum	Veg	96	0.02400	W	Wat	Y	1	891	0.637	5000	David 1971
Mycobacterium kansasii	Veg	80	0.02880	W	Wat	Y	1	200	0.637	4345	David 1973
Mycobacterium marinum	Veg	76	0.03030	W	Wat	Y	1	200	0.637	6485	David 1973
Mycobacterium marinum	Veg	743	0.00310	W	Wat	Y	1	1782	0.637	6485	David 1971
Mycobacterium parafortuitum	Veg	13	0.18000	Air	50	N	1	45.0	0.637	-	Peccia 2001
Mycobacterium parafortuitum	Veg	46	0.05000	Air	95	N	1	45.0	0.637	-	Peccia 2001
Mycobacterium parafortuitum	Veg	19	0.12000	Air	50	N	1	-	0.637	-	Xu 2003

APPENDIX A: UV Rate Constants for Bacteria

Microbe	Type	D90 J/m²	UVGI k m²/J	Media	RH %	Sh	St	UL J/m²	Dia. μm	Base Pairs kb	Source (see Chapter 4 Refs)
Mycobacterium phlei	Veg	76	0.03030	W	Wat	Y	1	200	0.637	6000	David 1973
Mycobacterium phlei	Veg	63	0.03650	Air	50	N	1	5.0	0.637	6000	Riley 1976
Mycobacterium phlei	Veg	23	0.10000	Air	50	-	-	-	0.637	6000	Kethley 1973
Mycobacterium phlei	Veg	16	0.14000	Air	50	-	-	-	0.637	6000	Gillis 1974
Mycobacterium smegmatis	Veg	108	0.02130	W	Wat	Y	1	200	0.637	6980	David 1973
Mycobacterium smegmatis	Veg	1047	0.00220	W	Wat	Y	1	2430	0.637	6980	David 1971
Mycobacterium smegmatis	Veg	68	0.03400	W	Wat	Y	2	500	0.637	6980	Boshoff 2003
Mycobacterium smegmatis	Veg	12	0.19000	Air	50	-	-	-	0.637	6980	Gillis 1974
Mycobacterium terrae	Veg	50	0.04610	W	Wat	N	1	100	0.637	-	Bohrerova 2006
Mycobacterium tuberculosis	Veg	28	0.08220	W	Wat	Y	1	200	0.637	4400	David 1973
Mycobacterium tuberculosis	Veg	77	0.03000	W	Wat	Y	1	567	0.637	4400	David 1971
Mycobacterium tuberculosis	Veg	74	0.03100	W	Wat	Y	2	500	0.637	4400	Boshoff 2003
Mycobacterium tuberculosis	Veg	11	0.21320	S	-	N	1	48	0.637	4400	Collins 1971
Mycobacterium tuberculosis	Veg	5	0.47210	Air	50	N	1	5.0	0.637	4400	Riley 1976
Mycoplasma arthritidis	Veg	7	0.31240	S	-	Y	1	22	0.177	816	Furness 1977
Mycoplasma fermentans	Veg	9	0.25220	S	-	Y	1	22	0.177	816	Furness 1977
Mycoplasma hominis	Veg	7	0.32710	S	-	Y	1	22	0.177	816	Furness 1977
Mycoplasma Orale type 1	Veg	11	0.21800	S	-	Y	1	22	0.177	816	Furness 1977
Mycoplasma Orale type 2	Veg	6	0.38760	S	-	Y	1	22	0.177	816	Furness 1977
Mycoplasma pneumoniae	Veg	8	0.27910	S	-	Y	1	22	0.177	816	Furness 1977
Mycoplasma salivarium	Veg	11	0.21140	S	-	Y	1	22	0.177	816	Furness 1977
Myxobolus cerebralis	Veg	10011	0.00023	W		Y	2	10000	-	-	Hedrick 2000
Neisseria catarrhalis	Veg	44	0.05233	S	-	-	-	121	0.177	816	Rentschler 1941
Nocardia asteroides	Veg	280	0.00822	S	-	-	-	280	1.118	6021	Chick 1963
Phytomonas tumefaciens	Veg	44	0.05233	S	-	-	-	110	-	-	Rentschler 1941
Proteus mirabilis	Veg	8	0.28900	W	Wat	N	1	60	0.494	4063	Hofemeister 1975
Proteus vulgaris	Veg	30	0.07675	S	-			70	0.291	3462	Rentschler 1941
Pseudomonas aeruginosa	Veg	10	0.22692	W	Wat	Y	1	0.4	0.494	5900	Gilpin 1985
Pseudomonas aeruginosa	Veg	172	0.01340	W	Wat	N	2	770	0.494	5900	Dolman 1989
Pseudomonas aeruginosa	Veg	36	0.06600	W	Wat	N	1	340	0.494	5900	Abshire 1981
Pseudomonas aeruginosa	Veg	55	0.04190	W	Wat	N	1	55	0.494	5900	Zelle 1955
Pseudomonas aeruginosa	Veg	55	0.04187	S	-	-	-	-	0.494	5900	Hollaender 1955
Pseudomonas aeruginosa	Veg	22	0.10470	S	-	N	1	20	0.494	5900	Elasri 1999
Pseudomonas aeruginosa	Veg	10	0.23750	S	-	N	2	48	0.494	5900	Collins 1971
Pseudomonas aeruginosa	Veg	4	0.57210	Air	(Lo RH)	N	1	248	0.494	5900	Sharp 1940
Pseudomonas diminuta	Veg	96	0.02391	W	Wat	N	1	118	0.5	-	Abshire 1981
Pseudomonas fluorescens	Veg	35	0.06579	S	-	N	1	70	0.5	6438	Rentschler 1941
Pseudomonas fluorescens	Veg	3	0.47730	Air	50	N	1	13	0.5	6438	vanOsdell 2002
Pseudomonas maltophilia	Veg	70	0.03294	W	Wat	N	1	71	0.5	-	Abshire 1981
Pseudomonas putrefaciens	Veg	87	0.02662	W	Wat	N	1	89	0.5	-	Abshire 1981
Rickettsia prowazekii	Veg	13	0.17600	W	Wat	N	2	6700	0.6	1110	Allen 1954
Salmonella spp.	Veg	11	0.21380	W	Wat	N	2	20	0.8	4746	Yaun 2003
Salmonella anatum	Veg	60	0.03840	W	Wat	N	1	150	0.8	-	Tosa 1998
Salmonella derby	Veg	36	0.06360	W	Wat	N	1	75	0.8	-	Tosa 1998
Salmonella enteritidis	Veg	10	0.22100	S	-	N	1	48	0.8	4746	Collins 1971
Salmonella enteritidis	Veg	33	0.07010	W	Wat	N	1	100	0.8	4746	Tosa 1998
Salmonella infantis	Veg	20	0.11510	W	Wat	N	1	60	0.8	-	Tosa 1998
Salmonella typhi	Veg	21	0.10760	W	Wat	N	1	21	0.806	4791	Zelle 1955
Salmonella typhi	Veg	30	0.07675	W	Wat	Y	1	100	0.806	4791	Chang 1985
Salmonella typhi	Veg	21	0.10760	S	-	N	1	40	0.806	4791	Sharp 1939
Salmonella typhi	Veg	9	0.25580	W	Wat	N	2	18	0.806	4791	Wilson 1992
Salmonella typhimurium	Veg	295	0.00780	W	Wat	N	1	900	0.8	4950	Kim 2002
Salmonella typhimurium	Veg	18	0.12830	W	Wat	N	2	50	0.8	-	Tosa 1998
Sarcina lutea	Veg	197	0.01169	S	-	-	-	264	1.48	-	Rentschler 1941
Serratia indica	Veg	209	0.01100	Air	42-51	N	1	370	0.632	-	Harstad 1954
Serratia marcescens	Veg	22	0.10490	W	Wat	N	1	22	0.632	5114	Zelle 1955
Serratia marcescens	Veg	105	0.02194	W	Wat	-	-	-	0.632	5114	Harris 1993
Serratia marcescens	Veg	22	0.10470	S	-	N	1	39	0.632	5114	Sharp 1939
Serratia marcescens	Veg	22	0.10466	S	-	-	-	70	0.632	5114	Rentschler 1941
Serratia marcescens	Veg	8	0.27742	S	-	-	-	-	0.632	5114	Hollaender 1955
Serratia marcescens	Veg	10	0.22080	S	-	N	1	48	0.632	5114	Collins 1971
Serratia marcescens	Veg	2	0.93900	Air	Lo RH	Y	2	40	0.632	5114	Fletcher 2003
Serratia marcescens	Veg	24	0.09500	Air	Hi RH	Y	2	40	0.632	5114	Fletcher 2003

APPENDIX A: UV Rate Constants for Bacteria

Microbe	Type	D₉₀ J/m²	UVGI k m²/J	Media	RH %	Sh	St	UL J/m²	Dia. μm	Base Pairs kb	Source (see Chapter 4 Refs)
Serratia marcescens	Veg	8	0.28670	Air	25-57	N	1	31	0.632	5114	UVDI 2001
Serratia marcescens	Veg	4	0.57500	Air	22-33	N	1	8.3	0.632	5114	Ko 2000
Serratia marcescens	Veg	115	0.02000	Air	Hi RH	N	1	8.3	0.632	5114	Ko 2000
Serratia marcescens	Veg	5	0.44490	Air	(Lo RH)	Y	1	248	0.632	5114	Sharp 1940
Serratia marcescens	Veg	20	0.11300	Air	(Lo RH)	Y	1	940	0.632	5114	Nakamura 1987
Serratia marcescens	Veg	33	0.07000	Air	95	N	1	45.0	0.632	5114	Peccia 2001
Serratia marcescens	Veg	3	0.92000	Air	68	N	1	2	0.632	5114	Lai 2004
Serratia marcescens	Veg	3	0.43050	Air	50	N	1	13	0.632	5114	vanOsdell 2002
Serratia marcescens	Veg	5	0.45000	Air	50	N	1	45.0	0.632	5114	Peccia 2001
Serratia marcescens	Veg	1	2.20000	Air	36	N	1	2	0.632	5114	Lai 2004
Shigella dysenteriae	Veg	18	0.13080	W	Wat	-	-	-	0.801	4369	Wilson 1992
Shigella paradysenteriae	Veg	17	0.13706	S	-	N	1	40	0.801	-	Sharp 1939
Shigella sonnei	Veg	18	0.12500	W	Wat	Y	1	100	0.801	-	Chang 1985
Spirillum rubrum	Veg	44	0.05233	S	-	-	-	88	-	-	Rentschler 1941
Staphylococcus albus	Veg	18	0.12514	S	-	Y	1	25	1.06	2900	Sharp 1939
Staphylococcus albus	Veg	33	0.06978	S	-	-	-	62	1.06	2900	Rentschler 1941
Staphylococcus albus (1)	Veg	23	0.09950	Air	(Lo RH)	N	1	4.4	1.06	2900	Rentschler 1942
Staphylococcus albus (2)	Veg	52	0.04400	Air	(Lo RH)	N	1	4.4	1.06	2900	Rentschler 1942
Staphylococcus aureus	Veg	52	0.04400	W	Wat	N	2	770	0.866	2800	Dolman 1989
Staphylococcus aureus	Veg	27	0.08531	W	Wat	Y	2	150	0.866	2800	Chang 1985
Staphylococcus aureus	Veg	56	0.04134	W	Wat	N	1	58	0.866	2800	Abshire 1981
Staphylococcus aureus	Veg	30	0.07700	S	-	N	1	4	0.866	2800	Sturm 1932
Staphylococcus aureus	Veg	50	0.04652	S	-	-	-	-	0.866	2800	Hollaender 1955
Staphylococcus aureus	Veg	66	0.03500	S	-	N	1	30	0.866	2800	Gates 1934
Staphylococcus aureus	Veg	26	0.08860	S	-	N	1	35	0.866	2800	Sharp 1939
Staphylococcus aureus	Veg	37	0.06240	S	-	N	1	48	0.866	2800	Luckiesh 1949
Staphylococcus aureus	Veg	19	0.11840	S	-	N	2	33	0.866	2800	Gates 1929
Staphylococcus aureus	Veg	20	0.11300	Air	(Lo RH)	N	1	940	0.866	2800	Nakamura 1987
Staphylococcus aureus	Veg	7	0.34760	Air	(Lo RH)	N	1	248	0.866	2800	Sharp 1940
Staphylococcus aureus	Veg	2	0.96020	Air	-	N	2	3	0.866	2800	Luckiesh 1949
Staphylococcus aureus	Veg	2	0.96200	Air	(Lo RH)	-	-	-	0.866	2800	Luckiesh 1946
Staphylococcus epidermis	Veg	161	0.01433	W	Wat	-	-	-	0.866	2640	Harris 1993
Staphylococcus epidermis	Veg	14	0.16210	Air	50	N	1	10	0.866	2640	vanOsdell 2002
Staphylococcus epidermis	Veg	29	0.00800	Air	85	N	1	10	0.866	2640	vanOsdell 2002
Staphylococcus epidermis	Veg	20	0.11300	Air	(Lo RH)	N	1	940	0.866	2640	Nakamura 1987
Staphylococcus epidermis	Veg	22	0.10500	Air	(Lo RH)	N	1	56	0.866	2640	Furuhashi 1989
Streptococcus agalactiae	Veg	5	0.43420	Air	-	N	2	7	0.707	2127	Luckiesh 1949
Streptococcus faecalis	Veg	55	0.09200	W	Wat	Y	2	150	0.707	-	Chang 1985
Streptococcus faecalis	Veg	195	0.01180	W	Wat	N	2	500	0.707	-	Sanz 2007
Streptococcus faecalis	Veg	31	0.07540	W	Wat	Y	2	150	0.707	-	Harris 1987
Streptococcus faecalis	Veg	120	0.01919	W	Wat	N	1	121	0.707	-	Abshire 1981
Streptococcus faecium	Veg	45	0.05100	W	Wat	N	1	350	0.632	5114	Martiny 1988
Streptococcus haemolyticus	Veg	22	0.10660	S	-	N	1	35	0.707	2680	Sharp 1939
Streptococcus lactis	Veg	62	0.03744	S	-	-	-	88	0.707	-	Rentschler 1941
Streptococcus pneumoniae	Veg	468	0.00492	S	-	-	1	3000	0.707	-	Gritz 1990
Streptococcus pyogenes	Veg	4	0.06161	S	-	N	2	94	0.894	1900	Lidwell 1950
Streptococcus pyogenes	Veg	1	1.56100	Air	-	N	2	2	0.894	1900	Luckiesh 1949
Streptococcus viridans	Veg	20	0.11513	S	-	N	1	32	0.707	-	Sharp 1939
Streptomyces coelicolor	Veg	60	0.03840	W	Wat	N	1	120	-	8667	Jagger 1970
Streptomyces griseus	Veg	129	0.01780	W	Wat	N	1	672	-	8545	Kelner 1949
Streptomyces griseus	Veg	60	0.03840	W	Wat	N	1	120	-	8545	Jagger 1970
Vibrio anguillarum (fish)	Veg	10	0.23820	W	Wat	-	-	-	2.12	-	Sako 1985
Vibrio anguillarum (fish)	Veg	5	0.42600	W	Wat	N	2	27	2.12	-	Liltved 1995
Vibrio cholerae	Veg	17	0.13400	W	Wat	-	-	-	2.12	4148	Wilson 1992
Vibrio ordalii	Veg	18	0.12560	W	Wat	-	-	-	2.12	-	Sako 1985
Vibrio parahaemolyticus	Veg	8	0.30700	W	Wat	N	1	30	2.12	5165	Nozu 1977
Vibrio salmonicida (fish)	Veg	5	0.42600	W	Wat	N	2	27	-	-	Liltved 1995
Yersinia enterocolitica	Veg	15	0.15351	W	Wat	-	-	-	0.707	4615	Butler 1987
Yersinia enterocolitica	Veg	28	0.08127	W	Wat	-	-	-	0.707	4615	Carlson 1975
Yersinia enterocolitica	Veg	11	0.20467	W	Wat	-	-	-	0.707	4615	Butler 1987
Yersinia enterocolitica	Veg	13	0.17170	W	Wat	N	1	46	0.707	4615	Wilson 1992
Yersinia ruckeri (fish)	Veg	5	0.42600	W	Wat	N	2	-	-	-	Liltved 1995
Yersinia ruckeri (fish)	Veg	10	0.23020	W	Wat	N	2	30	-	-	Liltved 1996

APPENDIX B: UV Rate Constants for Viruses

Microbe	Type	D90 J/m²	UVGI k m²/J	Media	RH %	Sh	St	UL J/m²	Dia. μm	Base Pairs kb	Source (see Chapter 4 Refs)
Adenovirus	dsDNA	34	0.06800	Air	Hi RH	N	1	26	0.079	35.937	Walker 2007
Adenovirus	dsDNA	59	0.03900	Air	Lo RH	N	1	26	0.079	35.937	Walker 2007
Adenovirus	dsDNA	42	0.05500	Air	50	N	1	68	0.079	36.001	Jensen 1964
Adenovirus	dsDNA	903	0.00255	W	Wat	N	1	900	0.079	36.001	Wasserman 1962
Adenovirus type 1	dsDNA	299	0.00770	W	Wat	-	-	300	0.079	36.001	Battiggelli 1993
Adenovirus type 1	dsDNA	350	0.00658	W	Wat	N	1	1200	0.079	36.001	Nwachuku 2005
Adenovirus type 2	dsDNA	400	0.00576	S	-	N	1	1200	0.079	35.937	Day 1974
Adenovirus type 2	dsDNA	640	0.00360	W	Wat	N	1	480	0.079	35.937	Rainbow 1970
Adenovirus type 2	dsDNA	490	0.00470	W	Wat	N	2	400	0.079	36.001	Rainbow 1973
Adenovirus type 2	dsDNA	533	0.00432	W	Wat	N	1	1200	0.079	35.937	Linden 2007 (LP lamp)
Adenovirus type 2	dsDNA	150	0.01540	W	Wat	N	1	1200	0.079	35.937	Linden 2007 (MP lamp)
Adenovirus type 2	dsDNA	300	0.00768	W	Wat	N	1	60	0.079	35.937	Shin 2005
Adenovirus type 2	dsDNA	400	0.00576	W	Wat	N	1	3000	0.079	35.937	Gerba 2002
Adenovirus type 2	dsDNA	276	0.00834	W	Wat	N	1	1000	0.079	35.937	Ballester 2004
Adenovirus type 4	dsDNA	921	0.00250	W	Wat	N	1	1200	0.079	35.937	Nwachuku 2005
Adenovirus type 15	dsDNA	396	0.00581	W	Wat	N	1	2100	0.079	35.937	Thompson 2003
Adenovirus type 40	dsDNA	300	0.00768	S	-	N	1	1240	0.069	36.001	Meng 1996
Adenovirus type 40	dsDNA	546	0.00422	W	Wat	N	1	200	0.069	36.001	Thurston-Enriquez 2003
Adenovirus type 41	dsDNA	240	0.00976	S	-	N	1	1118	0.069	36.001	Meng 1996
Adenovirus type 41	dsDNA	425	0.00542	W	Wat	-	-	-	0.069	36.001	Malley 2004
Adenovirus type 41	dsDNA	555	0.00415	W	Wat	N	1	300	0.069	36.001	Ko 2005
Adenovirus type 41	dsDNA	600	0.00384	W	Wat	N	1	12	0.069	36.001	Durance 2005
Adenovirus type 5	dsDNA	400	0.00576	W	Wat	N	1	12	0.084	35.938	Durance 2005
Adenovirus type 5	dsDNA	541	0.00426	W	Wat	N	1	2160	0.084	36.598	Wang 2004
Adenovirus type 5	dsDNA	720	0.00320	W	Wat	N	1	1200	0.084	35.598	Nwachuku 2005
Adenovirus type 6	dsDNA	390	0.00590	W	Wat	N	1	1200	0.079	35.937	Nwachuku 2005
Adenovirus type 6	dsDNA	400	0.00576	W	Wat	-	-	-	0.079	35.937	Battiggelli 1993
AHNV (fish virus)	ssRNA	349	0.00660	W	Wat	-	-	-	-	-	Liltved 2005
Avian Influenza virus	ssRNA	22	0.10600	W	Wat	N	2	97	0.09	-	Lucio-Forster 2006
Avian Influenza virus	ssRNA	30	0.07680	W	Wat	-	-	-	0.098	-	Deshmukh 1968
Avian Leukosis virus (RSA)	ssRNA	631	0.00365	W	Wat	N	1	1620	0.107	7.286	Levinson 1966
Avian Sarcoma virus	ssDNA	155	0.01490	W	Wat	N	1	372	0.098	7	Owada 1976
Avian Sarcoma virus	ssDNA	381	0.00604	W	Wat	N	1	768	0.098	7	Bister 1977
B. subtilis phage 029	dsDNA	70	0.03289	W	Wat	-	-	-	-	-	Freeman 1987
B. subtilis phage SP02c12	dsDNA	100	0.02303	W	Wat	-	-	-	0.087	44.01	Freeman 1987
B. subtilis phage SPP1	dsDNA	195	0.01181	W	Wat	-	-	-	0.087	44.01	Freeman 1987
Bacteriophage B40-8	dsDNA	137	0.01679	W	Wat	Y	1	400	-	-	Sommer 2001
Bacteriophage F-specific	dsRNA	292	0.00789	W	Wat	N	1	300	0.025	-	Havelaar 1987
Bacteriophage MS2	ssRNA	26	0.04800	Air	Hi RH	N	1	26	0.02	3.569	Walker 2007
Bacteriophage MS2	ssRNA	61	0.03800	Air	Lo RH	N	1	26	0.02	3.569	Walker 2007
Bacteriophage MS2	ssRNA	3	0.81000	Air	Lo RH	N	1	12	0.02	3.569	Tseng 2005
Bacteriophage MS2	ssRNA	4	0.64000	Air	Hi RH	N	1	12	0.02	3.569	Tseng 2005
Bacteriophage MS2	ssRNA	606	0.00380	W	Wat	N	1	110	0.02	3.569	Furuse 1971
Bacteriophage MS2	ssRNA	135	0.01710	W	Wat	N	1	301	0.02	3.569	Tree 1997
Bacteriophage MS2	ssRNA	427	0.00539	W	Wat	N	1	600	0.02	3.569	Sommer 2001
Bacteriophage MS2	ssRNA	193	0.01190	W	Wat	N	1	360	0.02	3.569	Sommer 1998
Bacteriophage MS2	ssRNA	419	0.00550	W	Wat	Y	1	600	0.02	3.569	Mamane-Gravetz 2005
Bacteriophage MS2	ssRNA	368	0.00625	W	Wat	N	1	600	0.02	3.569	Templeton 2006
Bacteriophage MS2	ssRNA	295	0.00780	W	Wat	N	1	201	0.02	3.569	Ko 2005
Bacteriophage MS2	ssRNA	40	0.05760	W	Wat	N	2	40	0.02	3.569	Weidenmann 1993
Bacteriophage MS2	ssRNA	173	0.01330	W	Wat	N	1	1090	0.02	3.569	Wilson 1992
Bacteriophage MS2	ssRNA	275	0.00837	W	Wat	N	1	200	0.02	3.569	Thurston-Enriquez 2003
Bacteriophage MS2	ssRNA	217	0.01060	W	Wat	N	1	920	0.02	3.569	Batch 2004
Bacteriophage MS2	ssRNA	250	0.00920	W	Wat	N	1	250	0.02	3.569	Battiggelli 1993
Bacteriophage MS2	ssRNA	217	0.01060	W	Wat	N	1	1500	0.02	3.569	Simonet 2006
Bacteriophage MS2	ssRNA	217	0.01063	W	Wat	N	1	800	0.02	3.569	deRodaHusman 2004
Bacteriophage MS2	ssRNA	213	0.01080	W	Wat	N	1	400	0.02	3.569	Butkus 2004
Bacteriophage MS2	ssRNA	187	0.01230	W	Wat	-	-	-	0.02	3.569	Oppenheimer 1997
Bacteriophage MS2	ssRNA	169	0.01360	W	Wat	N	1	800	0.02	3.569	Nuanualsuwan 2002
Bacteriophage MS2	ssRNA	164	0.01402	W	Wat	N	1	900	0.02	3.569	Rauth 1965
Bacteriophage MS2	ssRNA	150	0.01540	W	Wat	N	1	30	0.02	3.569	Shin 2005
Bacteriophage MS2	ssRNA	140	0.01640	W	Wat	N	1	-	0.02	3.569	Meng 1996
Bacteriophage MS2	ssRNA	198	0.01160	W	Wat	N	1	1520	0.02	3.569	Nieuwstad 1994

APPENDIX B: UV Rate Constants for Viruses

Microbe	Type	D_{90} J/m²	UVGI k m²/J	Media	RH %	Sh	St	UL J/m²	Dia. μm	Base Pairs kb	Source (see Chapter 4 Refs)
Bacteriophage MS2	ssRNA	228	0.01010	W	Wat	N	1	1550	0.02	3.569	Lazarova 2004
Bacteriophage MS2	ssRNA	245	0.00940	W	Wat	N	1	-	0.02	3.569	Thompson 2003
Bacteriophage Qβ	ssRNA	125	0.01840	W	Wat	N	1	1500	-	-	Simonet 2006
Bacteriophage Qβ	ssRNA	1919	0.00120	W	Wat	N	1	2500	-	-	O'Hara 1980
Berne virus	ssRNA	13	0.18420	W	Wat	-	-	-	0.13	20	Weiss 1986
BF-NNV (fish virus)	ssRNA	501	0.00460	W	Wat	-	-	-	-	-	Yoshimizu 2005
BLV	ssRNA	1799	0.00128	W	Wat	Y	2	400	0.1	8.419	Shimizu 2004
BLV	ssRNA	221	0.01040	W	Wat	N	1	1000	0.1	8.419	Guillemain 1981
Borna virus	ssRNA	79	0.02920	W	Wat	-	-	-	0.09	8.91	Danner 1979
Bovine Calicivirus	ssRNA	95	0.02420	W	Wat	-	-	-	0.02	7.45	Malley 2004
Bovine Parvovirus	ssDNA	35	0.06580	W	-	-	-	-	0.02	5.517	vonBrodorotti 1982
Canine Calicivirus	ssRNA	67	0.03450	W	Wat	N	1	800	0.037	8.513	deRodaHusman 2004
Canine hepatic Adenovirus	dsDNA	265	0.00869	W	Wat	-	-	-	0.08	36.5	vonBrodorotti 1982
CCHV (fish virus)	dsDNA	5	0.46050	W	Wat	-	-	-	-	130	Yoshimizu 2005
Cholera phage Kappa	dsDNA	634	0.00363	W	Wat	N	1	1919	-	-	Samad 1987
Coliphage f2	ssRNA	310	0.00743	W	Wat	-	-	-	-	-	Severin 1983
Coliphage fd	ssDNA	23	0.09940	W	Wat	N	1	900	-	-	Rauth 1965
Coliphage φX-174	ssDNA	3	0.71000	Air	Lo RH	N	1	12	0.025	5.386	Tseng 2005
Coliphage φX-174	ssDNA	4	0.53000	Air	Hi RH	N	1	12	0.025	5.386	Tseng 2005
Coliphage φX-174	ssDNA	18	0.12800	W	Wat	N	1	42	0.025	5.386	Yarus 1964
Coliphage φX-174	ssDNA	21	0.11140	W	Wat	N	1	90	0.025	5.386	Setlow 1960
Coliphage φX-174	ssDNA	21	0.11090	W	Wat	N	1	900	0.025	5.386	Rauth 1965
Coliphage φX-174	ssDNA	30	0.07650	W	Wat	-	-	-	0.025	5.386	Proctor 1972
Coliphage φX-174	ssDNA	25	0.09200	W	Wat	N	2	2000	0.025	5.386	Gurzadyan 1981
Coliphage φX-174	ssDNA	14	0.16060	W	Wat	-	-	-	0.025	5.386	David 1964
Coliphage φX-174	ssDNA	25	0.09350	W	Wat	N	1	105	0.025	5.386	Sommer 1998
Coliphage φX-174	ssDNA	57	0.04013	W	Wat	N	1	130	0.025	5.386	Sommer 2001
Coliphage φX-174	ssDNA	177	0.01300	W	Wat	N	1	800	0.025	5.386	Nuanualsuwan 2002
Coliphage φX-174	ssDNA	23	0.10230	W	Wat	N	1	150	0.025	5.386	Battigelli 1993
Coliphage φX-174	ssDNA	40	0.05760	W	Wat	N	1	120	0.025	5.386	Oppenheimer 1993
Coliphage φX-174	ssDNA	18	0.12910	W	Wat	N	1	70	0.025	5.386	Giese 2000
Coliphage lambda	dsDNA	57	0.04050	W	Wat	N	2	600	0.05	168.9	Gurzadyan 1981
Coliphage lambda	dsDNA	70	0.03310	W	Wat	-	-	-	0.05	168.9	Harm 1961
Coliphage lambda	dsDNA	72	0.03200	W	Wat	Y	1	-	0.05	168.9	Weigle 1953
Coliphage lambda	dsDNA	184	0.01250	W	Wat	N	2	1100	0.05	168.9	Davidovich 1991
Coliphage PRD1	dsDNA	87	0.02650	S	-	N	1	-	0.062	14.925	Meng 1996
Coliphage PRD1	dsDNA	20	0.11500	W	Wat	N	1	10	0.062	14.925	Shin 2005
Coliphage T1	dsDNA	6	0.36970	W	Wat	-	-	-	0.05	48.836	Hotz 1969
Coliphage T1	dsDNA	38	0.06000	W	Wat	N	1	200	0.05	48.836	Harm 1968
Coliphage T1	dsDNA	40	0.05800	W	Wat	N	1	60	0.05	48.836	Fluke 1949 (265 nm)
Coliphage T2	dsDNA	5	0.48400	W	Wat	N	1	900	0.065	-	Rauth 1965
Coliphage T2	dsDNA	9	0.25600	W	Wat	-	-	-	0.065	-	Jagger 1956
Coliphage T2	dsDNA	133	0.01730	W	Wat	Y	1	927	0.065	-	Dulbecco 1952
Coliphage T3	dsDNA	10	0.23100	W	Wat	Y	1	-	0.045	-	Winkler 1962
Coliphage T4	dsDNA	7	0.34500	W	Wat	N	1	60	0.089	168.9	Otaki 2003
Coliphage T4	dsDNA	14	0.16850	W	Wat	N	1	19	0.089	168.9	Ross 1971
Coliphage T4	dsDNA	15	0.15400	W	Wat	N	1	40	0.089	168.9	Harm 1968
Coliphage T4	dsDNA	29	0.08000	W	Wat	N	1	50	0.089	168.9	Templeton 2006
Coliphage T4	dsDNA	22	0.10700	W	Wat	Y	1	-	0.089	168.9	Winkler 1962
Coliphage T4	dsDNA	12	0.20000	W	Wat	N	2	40	0.089	168.9	Bohrerova 2008
Coliphage T7	dsDNA	7	0.33000	Air	Lo RH	N	1	12	0.063	39.937	Tseng 2005
Coliphage T7	dsDNA	10	0.22000	Air	Hi RH	N	1	12	0.063	39.937	Tseng 2005
Coliphage T7	dsDNA	95	0.02420	W	Wat	N	1	-	0.063	39.937	Benzer 1952
Coliphage T7	dsDNA	53	0.04320	W	Wat	N	1	180	0.063	39.937	Peak 1978 (B)
Coliphage T7	dsDNA	41	0.05600	W	Wat	Y	1	200	0.063	39.937	Bohrerova 2008 (LP)
Coliphage T7	dsDNA	38	0.06100	W	Wat	Y	1	200	0.063	39.937	Bohrerova 2008 (MP)
Coliphage T7	dsDNA	23	0.10000	W	Wat	N	1	-	0.063	39.937	Ronto 1992
Coliphage T7	dsDNA	11	0.20470	W	Wat	N	1	45	0.063	39.937	Peak 1978 (Bs-1)
Coronavirus	ssRNA	3	0.37700	Air	50	N	1	6	0.113	30.738	Walker 2007
Coronavirus	ssRNA	7	0.32100	W	Wat	-	-	-	0.113	30.738	Weiss 1986
Coronavirus (SARS)	ssRNA	226	0.01000	W	Wat	N	2	1200	0.113	29.751	Kariwa 2004
Coronavirus (SARS)	ssRNA	3046	0.00076	W	Wat	N	2	14458	0.113	29.751	Darnell 2004
Coxsackievirus	ssRNA	21	0.11100	Air	60	N	1	68	0.027	7.413	Jensen 1964

APPENDIX B: UV Rate Constants for Viruses

Microbe	Type	D₉₀ J/m²	UVGI k m²/J	Media	RH %	Sh	St	UL J/m²	Dia. μm	Base Pairs kb	Source (see Chapter 4 Refs)
Coxsackievirus	ssRNA	128	0.02000	W	Wat	N	1	348	0.027	7.413	Hill 1970
Coxsackievirus	ssRNA	86	0.02684	W	Wat	N	1	300	0.027	7.413	Havelaar 1987
Coxsackievirus B3	ssRNA	80	0.02878	W	Wat	N	1	400	0.027	7.413	Gerba 2002
Coxsackievirus B4	ssRNA	60	0.03840	W	Wat	N	1	30	0.027	7.413	Shin 2005
Coxsackievirus B5	ssRNA	95	0.02424	W	Wat	N	1	400	0.027	7.413	Gerba 2002
Coxsackievirus B5	ssRNA	72	0.03180	W	Wat	N	1	200	0.027	7.413	Battigelli 1993
CSV (fish virus)	dsRNA	501	0.00460	W	Wat	-	-	1000	-	-	Yoshimizu 2005
Echovirus (Parechovirus)	ssRNA	106	0.02190	W	Wat	N	1	348	0.024	7.354	Hill 1970
Echovirus 1	ssRNA	80	0.02878	W	Wat	N	1	400	0.024	7.354	Gerba 2002
Echovirus 2	ssRNA	70	0.03289	W	Wat	N	1	400	0.024	7.354	Gerba 2002
Encephalomyocarditis virus	ssRNA	50	0.04650	W	Wat	N	1	21	0.025	7.835	Ross 1971
Encephalomyocarditis virus	ssRNA	52	0.04460	W	Wat	N	1	900	0.025	7.835	Rauth 1965
Encephalomyocarditis virus	ssRNA	65	0.03550	W	Wat	Y	1	8	0.025	7.835	Zavadova 1968
Epstein-Barr virus (EBV)	ssDNA	162	0.01420	W	Wat	N	1	15000	-	-	Henderson 1978
Equine Herpes virus	dsDNA	25	0.09210	W	Wat	-	-	-	0.105	145.597	Weiss 1986
EVA (fish virus)	ssRNA	5	0.46050	W	Wat	-	-	-	0.06	12.7	Yoshimizu 2005
EVEX (fish virus)	ssRNA	5	0.46050	W	Wat	-	-	-	0.06	11	Yoshimizu 2005
Feline Calicivirus (FeCV)	ssRNA	434	0.00530	W	Wat	N	2	1300	0.034	7.683	Nuanualsuw an 2002
Feline Calicivirus (FeCV)	ssRNA	80	0.02880	W	Wat	N	1	200	0.034	7.683	Thurston-Enriquez 2003
Feline Calicivirus (FeCV)	ssRNA	40	0.05760	W	Wat	N	1	800	0.034	7.683	deRodaHusman 2004
Feline Calicivirus (FeCV)	ssRNA	44	0.05270	W	Wat	N	1	140	0.034	7.683	Tree 2005
Friend Murine Leukemia v.	ssRNA	320	0.00720	W	Wat	N	1	1200	0.094	8.323	Yoshikura 1971
Frog virus 3	dsDNA	25	0.09210	W	Wat	N	1	25	0.167	105.903	Martin 1982
Hepatitis A virus	dsDNA	40	0.05760	W	Wat	N	1	150	0.027	7.478	Battigelli 1993
Hepatitis A virus	dsDNA	45	0.05120	W	Wat	N	1	180	0.027	7.478	Wang 2004
Hepatitis A virus	dsDNA	50	0.04610	W	Wat	-	-		0.027	7.478	Weidenmann 1993
Hepatitis A virus	dsDNA	92	0.02500	W	Wat	N	1	368	0.027	7.478	Wang 1995
Hepatitis A virus	dsDNA	98	0.02340	W	Wat	-	-	-	0.027	7.478	Wilson 1992
Hepatitis A virus	dsDNA	307	0.00750	W	Wat	N	2	1300	0.027	7.478	Nuanualsuw an 2002
Herpes simplex virus (HRE)	dsDNA	40	0.05760	W	Wat	N	1	80	0.18	152.261	Pow ell 1959
Herpes simplex virus Type 1	dsDNA	71	0.03260	W	Wat	N	2	450	0.184	152.261	Bockstahler 1976
Herpes simplex virus Type 1	dsDNA	110	0.02090	W	Wat	N	2	200	0.184	152.261	Selsky 1978
Herpes simplex virus Type 1	dsDNA	25	0.09330	W	Wat	N	2	300	0.184	152.261	Lytle 1971
Herpes Simplex virus Type 1	dsDNA	35	0.06540	W	Wat	N	2	19	0.184	152.261	Ross 1971
Herpes Simplex virus Type 1	dsDNA	21	0.11050	W	Wat	N	2	40	0.184	152.261	Albrecht 1974
Herpes Simplex virus Type 1	dsDNA	41	0.05680	W	Wat	N	1	20	0.184	152.261	Henderson 1978
Herpes Simplex virus Type 2	dsDNA	40	0.05756	W	Wat	-	-	-	0.173	154.746	Wolff 1973
Herpes Simplex virus Type 2	dsDNA	41	0.05650	W	Wat	N	2	19	0.173	154.746	Ross 1971
Herpes Simplex virus Type 2	dsDNA	75	0.03070	W	Wat	N	2	80	0.173	154.746	Ryan 1986
Herpes Simplex virus Type 2	dsDNA	20	0.11800	W	Wat	N	2	40	0.173	154.746	Albrecht 1974
HIV-1	ssRNA	280	0.00822	W	Wat	Y	1	400	0.125	9.181	Yoshikura 1989
HIRRV (fish virus)	ssRNA	5	0.46050	W	Wat	-	-	-	0.06	11	Yoshimizu 2005
HP1c1 phage	dsDNA	40	0.05760	W	Wat	N	2	180	0.062	32.35	Setlow 1972
HTLV-1	ssRNA	20	0.11510	W	Wat	N	1	35	0.102	8.507	Shimizu 2004
Human Cytomegalovirus	dsDNA	658	0.00350	S	-	Y	1	1950	0.1	-	Hirai 1977
Human Cytomegalovirus	dsDNA	50	0.04605	S	-	N	2	-	0.1	-	Albrecht 1974
Influenza A virus	ssRNA	19	0.11900	Air	68	N	1	68	0.098	13.498	Jensen 1964
Influenza A virus	ssRNA	20	0.11700	W	Wat	N	1	9	0.098	13.498	Ross 1971
Influenza A virus	ssRNA	48	0.04800	W	Wat	Y	1	-	0.098	13.498	Hollaender 1944
Influenza A virus	ssRNA	17	0.13810	W	Wat	N	1	14	0.098	13.498	Abraham 1979
IHNV (fish virus)	ssRNA	5	0.46050	W	Wat	-	-	-	0.09	12	Yoshimizu 2005
IHNV (fish virus)	ssRNA	7	0.34500	W	Wat	-	-	-	0.09	12	Sako 1985
IPNV (fish virus)	dsRNA	397	0.00580	W	Wat	N	2	2000	0.06	6	Oye 2001
IPNV (fish virus)	dsRNA	407	0.00566	W	Wat	N	2	1220	0.06	6	Liltved 1995
IPNV (fish virus)	dsRNA	501	0.00460	W	Wat	-	-	-	0.06	6	Yoshimizu 2005
IPNV (fish virus)	dsRNA	626	0.00368	W	Wat	-	-	1500	0.06	6	Ahne 1982
IPNV (fish virus)	dsRNA	583	0.00395	W	Wat	-	-	2000	0.06	6	Sako 1985
Iridovirus (Bohle) (fish virus)	dsDNA	83	0.02760	W	Wat	-	-	-	-	-	Miocevic 1993
ISAV (fish virus)	ssRNA	11	0.20900	W	Wat	N	2	70	-	12.7	Oye 2001
ISAV (fish virus)	ssRNA	26	0.08970	W	Wat	-	-	-	-	12.7	Liltved 1995
JF-LCDV (fish virus)	dsDNA	5	0.46050	W	Wat	-	-	-	0.14	102.6	Yoshimizu 2005
Kemerovo (R-10 strain)	dsRNA	230	0.01000	W	Wat	N	1	900	0.075	-	Zavadova 1975
Kilham Rat Virus (parvovirus)	ssDNA	30	0.07650	W	Wat	-	-	-	0.022	5	Proctor 1972

APPENDIX B: UV Rate Constants for Viruses

Microbe	Type	D₉₀ J/m²	UVGI k m²/J	Media	RH %	Sh	St	UL J/m²	Dia. μm	Base Pairs kb	Source (see Chapter 4 Refs)
Lipovnik (Lip-91 strain)	dsRNA	299	0.00770	W	Wat	N	2	200	0.075	-	Zavadova 1975
LLE46 (SV/Adeno hybrid)	dsDNA	606	0.00380	W	Wat	N	1	2376	-	-	Defendi 1967
Measles virus	ssRNA	22	0.10510	W	Wat	N	1	48	0.329	15.894	DiStefano 1976
Mengovirus	dsRNA	162	0.01420	W	Wat	N	2	70	-	6.1	Miller 1974
Minute Virus of Mice (MVM)	ssDNA	28	0.08200	W	Wat	N	1	8	0.022	5.081	Vos 1981
Minute Virus of Mice (MVM)	ssDNA	17	0.13500	W	Wat	N	1	70	0.022	5.081	Rommelaere 1981
Murine Cytomegalovirus	dsDNA	46	0.05000	W	Wat	N	2	116	0.104	230.278	Shanley 1982
Moloney Murine Leukemia v.	ssRNA	115	0.02000	W	Wat	N	1	330	0.094	8.332	Nomura 1972
Moloney Murine Leukemia v.	ssRNA	370	0.00622	W	Wat	N	1	1000	0.094	8.332	Guillemain 1981
Moloney Murine Leukemia v.	ssRNA	280	0.00822	W	Wat	-			0.094	8.332	Yoshikura 1989
Murine Norovirus (MNV)	ssRNA	76	0.03040	W	Wat	N	1	250	0.032	7.382	Lee 2008
Murine sarcoma virus	ssRNA	237	0.00970	W	Wat	N	1	432	0.12	5.833	Nomura 1972
Murine sarcoma virus	ssRNA	144	0.01600	W	Wat	N	1	74	0.12	5.833	Kelloff 1970
Murine sarcoma virus	ssRNA	299	0.00770	W	Wat	N	1	300	0.12	5.833	Yoshikura 1971
Mycobacteriophage D29	dsDNA	16	0.14300	W	Wat	N	2	120	0.065	49.136	David 1973
Mycobacteriophage D29	dsDNA	324	0.00710	W	Wat	N	2	950	0.065	49.136	Sellers 1970
Mycobacteriophage D29A	dsDNA	268	0.00860	W	Wat	N	2	950	0.065	49.136	Sellers 1970
Mycobacteriophage D32	dsDNA	354	0.00650	W	Wat	N	1	950	-	-	Sellers 1970
Mycobacteriophage D4	dsDNA	245	0.00940	W	Wat	N	1	950	-	-	Sellers 1970
Mycoplasmavirus MVL2	dsDNA	154	0.01500	W	Wat	Y	1	600	-	-	Das 1977
Mycoplasmavirus MVL51	ssDNA	79	0.02900	W	Wat	Y	1	250	-	-	Das 1977
Newcastle Disease Virus	ssRNA	8	0.27600	W	Wat	-	-		0.212	15.186	vonBrodorotti 1982
Newcastle Disease Virus	ssRNA	45	0.05110	W	Wat	N	1	90	0.212	15.186	Levinson 1966
Newcastle Disease Virus	ssRNA	16	0.14400	S	-	N	1	50	0.212	15.186	Rubin 1959
OMV (fish virus)	ssRNA	5	0.46050	W	Wat	-	-		0.06	-	Yoshimizu 2005
Parvovirus H-1	ssDNA	25	0.09200	W	Wat	N	1	55	0.022	6.194	Cornellis 1982
PFRV (fish virus)	ssRNA	5	0.46050	W	Wat	-	-		0.06	11	Yoshimizu 2005
phage GA	ssRNA	200	0.01150	W	Wat	N	1	1500	-	-	Simonet 2006
phage phi 6	dsRNA	5	0.43000	Air	Lo RH	N	1	12	-	-	Tseng 2005
phage phi 6	dsRNA	7	0.31000	Air	Hi RH	N	1	12	-	-	Tseng 2005
phage B40-8 (B. fragilis)	dsDNA	67	0.03450	W	Wat	Y	1	400	-	-	Sommer 2001
phage B40-8 (B. fragilis)	dsDNA	86	0.02690	W	Wat	N	1	280	-	-	Sommer 1998
Poliovirus	dsRNA	44	0.05230	S	-	N	1	220	0.0248	7.44	Ma 1994
Poliovirus type 1	dsRNA	41	0.05620	S	-	N	1	-	0.0248	7.44	Meng 1996
Poliovirus	dsRNA	71	0.03250	W	Wat	Y	1	216	0.0248	7.44	Helentjaris 1977
Poliovirus	dsRNA	75	0.03070	W	Wat	N	1	30	0.0248	7.44	Shin 2005
Poliovirus	dsRNA	95	0.02420	W	Wat	N	1	54	0.0248	7.44	Bishop 1967
Poliovirus	dsRNA	52	0.04460	W	Wat	N	1	900	0.0248	7.44	Dulbecco 1955
Poliovirus type 1	dsRNA	67	0.03450	W	Wat	N	2	300	0.0248	7.44	Chang 1985
Poliovirus type 1	dsRNA	72	0.03200	W	Wat	-	-	-	0.0248	7.44	Wilson 1992
Poliovirus type 1	dsRNA	96	0.02400	W	Wat	N	1	480	0.0248	7.44	Wetz 1982
Poliovirus type 1	dsRNA	100	0.02300	W	Wat	-	-	-	0.0248	7.44	Thompson 2003
Poliovirus type 1	dsRNA	125	0.01840	W	Wat	-	-	-	0.0248	7.44	Oppenheimer 1997
Poliovirus type 1	dsRNA	224	0.01030	W	Wat	N	1	1165	0.0248	7.44	Nuanualsuwan 2003
Poliovirus type 1	dsRNA	240	0.00960	W	Wat	N	1	1300	0.0248	7.44	Nuanualsuwan 2002
Poliovirus type 1	dsRNA	111	0.02080	W	Wat	N	1	348	0.0248	7.44	Hill 1970
Poliovirus type 1	dsRNA	77	0.03000	W	Wat	N	1	-	0.0248	7.44	Harris 1987
Poliovirus type 1	dsRNA	80	0.02878	W	Wat	N	1	400	0.0248	7.44	Gerba 2002
Poliovirus type 1	dsRNA	83	0.02760	W	Wat	N	1	1500	0.0248	7.44	Simonet 2006
Poliovirus type 1	dsRNA	57	0.04010	W	Wat	N	1	270	0.0248	7.44	Tree 2005
Poliovirus type 2	dsRNA	121	0.01910	W	Wat	N	1	348	0.0248	7.44	Hill 1970
Poliovirus type 3	dsRNA	103	0.02240	W	Wat	N	1	348	0.0248	7.44	Hill 1970
Polyomavirus	dsDNA	480	0.00480	W	Wat	N	1	240	0.0424	5	vander Eb 1967
Polyomavirus	dsDNA	640	0.00360	W	Wat	N	1	2376	0.0424	5	Defendi 1967
Polyomavirus	dsDNA	696	0.00331	W	Wat	N	1	900	0.0424	5	Rauth 1965
Polyomavirus	dsDNA	501	0.00460	W	Wat	-	-	-	0.0424	5	Latarjet 1967
Polyomavirus (ssDNA)	ssDNA	120	0.01920	W	Wat	N	2	240	0.045	5	vander Eb 1967
Porcine Parvovirus (PPV)	ssDNA	23	0.10230	W	Wat	N	1	90	0.021	6.194	Wang 2004
Pseudorabies (PRV)	dsDNA	34	0.06760	W	Wat	N	2	15	0.194	-	Ross 1971
Rabies virus (env)	ssRNA	10	0.21930	W	Wat	-	-	-	0.07	11.932	Weiss 1986
Rauscher Murine Leukemia v.	ssRNA	157	0.01470	W	Wat	N	1	74	0.094	8.282	Kelloff 1970
Rauscher Murine Leukemia v.	ssRNA	480	0.00480	W	Wat	N	1	200	0.094	8.282	Lovinger 1975
Rauscher Murine Leukemia v.	ssRNA	959	0.00240	S	-	N	2	4800	0.094	8.282	Stull 1976

APPENDIX B: UV Rate Constants for Viruses

Microbe	Type	D_{90} J/m²	UVGI k m²/J	Media	RH %	Sh	St	UL J/m²	Dia. μm	Base Pairs kb	Source (see Chapter 4 Refs)
Reovirus	dsRNA	175	0.01316	W	Wat	N	1	348	0.075	11	Hill 1970
Reovirus	dsRNA	186	0.01240	W	Wat	N	1	740	0.075	11	Wang 2004
Reovirus	dsRNA	69	0.03358	W	Wat	-	-	-	0.075	11	vonBrodorotti 1982
Reovirus	dsRNA	245	0.00940	W	Wat	N	2	990	0.075	11	Shaw 1973
Reovirus	dsRNA	121	0.01910	W	Wat	N	1	900	0.075	11	Rauth 1965
Reovirus	dsRNA	270	0.00853	W	Wat	N	2	1080	0.075	11	McClain 1966
Reovirus	dsRNA	174	0.01320	W	Wat	N	1	348	0.075	11	Hill 1970
Reovirus type 1	dsRNA	153	0.01508	W	Wat	N	1	-	0.075	11	Harris 1987
Reovirus 3	dsRNA	334	0.00690	W	Wat	N	2	300	0.075	11	Zavadova 1975
Rotavirus	dsRNA	200	0.01150	W	Wat	N	2	200	0.07	-	Caballero 2004
Rotavirus SA11	dsRNA	89	0.02600	W	Wat	-	-	-	0.07	-	Wilson 1992
Rotavirus SA11	dsRNA	75	0.03070	W	Wat	N	1	750	0.07	-	Meng 1987
Rotavirus SA11	dsRNA	105	0.02190	W	Wat	N	2	250	0.07	-	Battigelli 1993
Rotavirus SA11	dsRNA	100	0.02300	W	Wat	N	1	350	0.07	-	Chang 1985
Rotavirus SA11	dsRNA	84	0.02740	W	Wat	N	1	380	0.07	-	Sommer 1989
Rous Sarcoma virus (RSV)	ssRNA	720	0.00320	W	Wat	N	1	36	0.127	9.392	Levinson 1966
Rous Sarcoma virus (RSV)	ssRNA	240	0.00960	W	Wat	-	-	-	0.127	9.392	Golde 1961
Rous Sarcoma virus (RSV)	ssRNA	200	0.01150	S	-	N	1	700	0.127	9.392	Rubin 1959
SBNN (fish virus)	ssRNA	698	0.00330	W	Wat	Y	2	2640	-	-	Frerichs 2000
Semliki forest virus	ssRNA	25	0.09210	W	Wat	-	-	-	0.061	11.442	Weiss 1986
Simian virus 40	dsDNA	2503	0.00092	W	Wat	N	1	2500	0.045	5.243	Bourre 1989
Simian virus 40	dsDNA	1599	0.00144	W	Wat	N	2	8000	0.045	5.243	Seemayer 1973
Simian virus 40	dsDNA	1439	0.00160	W	Wat	N	1	1500	0.045	5.243	Cornelis 1981
Simian virus 40	dsDNA	1245	0.00185	W	Wat	-	-	-	0.045	5.243	Bockstahler 1977
Simian virus 40	dsDNA	886	0.00260	W	Wat	N	1	2376	0.045	5.243	Defendi 1967
Simian virus 40	dsDNA	650	0.00354	W	Wat	N	1	1300	0.045	5.243	Sarasin 1978
Simian virus 40	dsDNA	443	0.00520	W	Wat	-	-	240	0.045	5.243	Aaronson 1970
Simian virus 40	dsDNA	23	0.10040	W	Wat	N	1	55	0.045	5.243	Cornelis 1982
Simian virus 40	dsDNA	17	0.13160	W	Wat	N	1	70	0.045	5.243	Wang 2004
Sindbis virus	ssRNA	22	0.10400	Air	62	N	1	68	0.075	11.703	Jensen 1964
Sindbis virus	ssRNA	60	0.03864	W	Wat	-	-	-	0.075	11.703	vonBrodorotti 1982
Sindbis virus	ssRNA	113	0.02030	W	Wat	N	1	400	0.075	11.703	Wang 2004
Sindbis virus	ssRNA	50	0.04610	W	Wat	N	1	200	0.075	11.703	Zavadova 1975
S. aureus phage	dsDNA	82	0.02800	S	-	N	1	30	-	-	Gates 1934
S. aureus phage	dsDNA	77	0.03000	S	-	N	1	4	-	-	Sturm 1932
S. aureus phage A994	dsDNA	65	0.03542	W	Wat	-	-	-	-	-	Sommer 1989
SVCV (fish virus)	ssRNA	10	0.46050	W	Wat	-	-	-	0.06	11.1	Yoshimizu 2005
Vaccinia virus	dsDNA	1	2.54000	Air	60	N	1	3	0.307	195.815	McDevitt 2007
Vaccinia virus	dsDNA	15	0.15300	Air	65	N	1	68	0.307	195.815	Jensen 1964
Vaccinia virus	dsDNA	7	0.34900	W	Wat	N	1	10	0.307	195.815	Galasso 1965
Vaccinia virus	dsDNA	14	0.16450	W	Wat	N	1	30	0.307	195.815	Bossart 1978
Vaccinia virus	dsDNA	14	0.16040	W	Wat	N	1	6	0.307	195.815	Ross 1971
Vaccinia virus	dsDNA	18	0.12792	W	Wat	N	1	20	0.307	195.815	Klein 1994
Vaccinia virus	dsDNA	22	0.10500	W	Wat	N	2	8	0.307	195.815	Zavadova 1971
Vaccinia virus	dsDNA	28	0.08290	W	Wat	N	1	900	0.307	195.815	Rauth 1965
Vaccinia virus	dsDNA	715	0.00322	W	Wat	N	2	70000	0.307	195.815	Davidovich 1991
Vaccinia virus	dsDNA	677	0.00340	W	Wat	N	1	4300	0.307	195.815	Collier 1955
VEE	ssRNA	55	0.04190	W	Wat	-	-	-	0.065	11.444	Smirnov 1992
Vesicular Stomatitis virus	ssRNA	13	0.18060	W	Wat	N	1	900	0.104	11.161	Rauth 1965
Vesicular Stomatitis virus	ssRNA	12	0.19000	W	Wat	-	-	-	0.104	11.161	Helentjaris 1977
Vesicular Stomatitis virus	ssRNA	100	0.02300	W	Wat	N	1	10	0.104	11.161	Bay 1979
Vesicular Stomatitis virus	ssRNA	6	0.38400	W	Wat	N	1	-	0.104	11.161	Shimizu 2004
VHSV (fish virus)	ssRNA	3	0.87400	W	Wat	-	-	20	0.07	11.158	Oye 2001
WEE	ssRNA	54	0.04300	W	Wat	N	1	83	0.07	11.484	Dubinin 1975

NOTES for Appendices A, B & C:

Type: Sp = Spore, Veg = Vegetative, VegY = Vegetative yeast

D_{90}: UV Dose for 90% inactivation (10% survival)

UVGI k: UV rate constant at the given D_{90} (and below the UL)

UL: Upper Limit within which D_{90} and rate constants are applicable

Media: A = Air, S = Surface, W = Water   RH = Relative Humidity

Sh = Shoulder in decay curve (shoulder is ignored for k and D_{90} values)

St = Number of stages in decay curve (k & D_{90} only applies to first stage)

Dia.: Logmean diameter in microns, including envelope for viruses if any

MP: Medium Pressure UV lamp, LP: Low Pressure UV lamp

APPENDIX C: UV Rate Constants for Fungi and Other Microbes

Microbe	Type	D_{90} J/m²	UVGI k m²/J	Media	RH %	Sh	St	UL J/m²	Dia. μm	Base Pairs kb	Source (see Chapter 4 Refs)
Aspergillus amstelodami	Sp	700	0.00329	W	Wat	-	-	-	3.354	35900	Jepson 1973
Aspergillus amstelodami	Sp	258	0.00892	S	-	N	1	336	3.354	35900	Luckiesh 1949
Aspergillus amstelodami	Sp	669	0.00344	Air	67	N	2	870	3.354	35900	Luckiesh 1949
Aspergillus flavus	Sp	349	0.00660	S	-	N	1	35	4.24	35900	Green 2004
Aspergillus flavus	Sp	600	0.00384	-	-	-	-	-	4.24	35900	Nagy 1964
Aspergillus flavus	Sp	853	0.00270	W	Wat	N	1	13932	4.24	35900	Begum 2009
Aspergillus fumigatus	Sp	535	0.00430	S	-	N	1	54	4.24	35900	Green 2004
Aspergillus fumigatus	Veg	560	0.00411	S	-	-	-	560	24.5	35900	Chick 1963
Aspergillus fumigatus	Sp	2240	0.00103	S	-	-	-	2240	2.64	35900	Chick 1963
Aspergillus glaucus	Sp	440	0.00523	-	-	-	-	-	3.354	35900	Nagy 1964
Aspergillus niger	Sp	1771	0.00130	S	Lo RH	Y	2	1600	3.354	35900	Zahl 1939
Aspergillus niger	Sp	1439	0.00160	S	-	-	2	3384	3.354	35900	Fulton 1929
Aspergillus niger	Veg	4480	0.00051	S	-	-	-	4480	3.354	35900	Chick 1963
Aspergillus niger	Sp	1000	0.00230	W	Wat	-	-	-	3.354	35900	Jepson 1973
Aspergillus niger	Sp	315	0.00350	S	-	Y	2	18	3.354	35900	Kowalski 2001
Aspergillus niger	Sp	1387	0.00166	S	-	N	1	1800	3.354	35900	Luckiesh 1949
Aspergillus niger	Sp	750	0.00386	S	-	-	-	3000	3.354	35900	Gritz 1990
Aspergillus niger	Sp	4480	0.00051	S	-	-	-	4480	3.354	35900	Chick 1963
Aspergillus niger	Sp	3984	0.00058	Air	55	N	2	5400	3.354	35900	Luckiesh 1949
Aspergillus niger	Sp	1320	0.00174	-	-	-	-	-	3.354	35900	Nagy 1964
Aspergillus niger	Sp	1681	0.00137	W	Wat	N	2	9288	3.354	35900	Begum 2009
Aspergillus versicolor	Sp	384	0.00600	Air	85	N	1	32	3.354	35900	vanOsdell 2002
Aspergillus versicolor	Sp	768	0.00300	Air	55	N	1	32	3.354	35900	vanOsdell 2002
Aspergillus versicolor	Sp	139	0.01660	Air	50	N	1	32	3.354	35900	vanOsdell 2002
Aspergillus versicolor	Veg	96	0.02400	Air	(Lo RH)	N	1	940	3.354	35900	Nakamura 1987
Blastomyces dermatitidis	VegY	140	0.01645	S	-	-	-	140	11.000	23000	Chick 1963
Botrytis cinerea	Sp	250	0.00920	S	-	N	1	1000	11.180	42660	Marquenie 2002
Candida albicans	VegY	230	0.01100	W	Wat	N	2	440	4.899	20000	Dolman 1989
Candida albicans	VegY	447	0.00515	W	Wat	N	1	453	4.899	20000	Abshire 1981
Candida albicans	VegY	750	0.00407	S	-	-	-	3000	4.899	20000	Gritz 1990
Candida albicans	VegY	280	0.00822	S	-	-	-	280	4.899	20000	Chick 1963
Candida parapsilosis	VegY	98	0.02360	W	Wat	N	1	390	-	-	Severin 1983
Cladosporium herbarum	Sp	500	0.04605	W	Wat	-	-	-	8.062	36000	Jepson 1973
Cladosporium herbarum	Sp	189	0.01220	S	-	N	1	246	8.062	36000	Luckiesh 1949
Cladosporium herbarum	Sp	622	0.00370	Air	53	N	2	810	8.062	36000	Luckiesh 1949
Cladosporium trichoides	Veg	560	0.00411	S	-	-	-	560	8.062	36000	Chick 1963
Cladosporium trichoides	Sp	1120	0.00206	S	-	-	-	1120	8.062	36000	Chick 1963
C. sphaerospermum	Sp	1439	0.00210	Air	50	N	1	32	8.062	36000	vanOsdell 2002
Cladosporium wernecki	Sp	4480	0.00051	S	-	-	-	4480	8.062	36000	Chick 1963
Cladosporium wernecki	Veg	560	0.00411	S	-	-	-	560	8.062	36000	Chick 1963
Cryptococcus neoformans	Sp	138	0.01670	-	-	N	1	400	4.899	23000	Wang 1994
Cryptococcus neoformans	VegY	280	0.00822	S	-	-	-	280	4.899	23000	Chick 1963
Curvularia lunata	Veg	560	0.00411	S	-	-	-	560	17.100	29700	Chick 1963
Eurotium rubrum	Sp	434	0.00531	W	Wat	N	2	4644	5.612	-	Begum 2009
Fusarium oxysporum	Sp	260	0.01420	W	Wat	Y	2	600	11.225	43000	Asthana 1992
Fusarium solani	Sp	313	0.00735	W	Wat	N	2	960	11.225	43000	Asthana 1992
Fusarium spp.	Sp	560	0.00411	S	-	-	-	560	11.225	43000	Chick 1963
Fusarium spp.	Veg	1120	0.00206	S	-	-	-	1120	34.300	43000	Chick 1963
Histoplasma capsulatum	Veg	140	0.01645	S	-	-	-	140	2.550	23000	Chick 1963
Monilinia fructigena	Sp	167	0.01380	-	-	N	1	500	10.300	-	Marquenie 2002
Mucor mucedo	Sp	600	0.00384	W	Wat	-	-	-	7.071	39000	Jepson 1973
Mucor mucedo	Sp	180	0.01280	S	-	N	1	234	7.071	39000	Luckiesh 1949
Mucor mucedo	Sp	577	0.00399	Air	63	N	2	750	7.071	39000	Luckiesh 1949
Mucor racemosus	Sp	170	0.01354	-	-	-	-	-	7.071	39000	Nagy 1964
Mucor spp.	Sp	140	0.01645	S	-	-	-	140	7.071	39000	Chick 1963
Mucor spp.	Veg	280	0.00822	S	-	-	-	280	31.600	39000	Chick 1963
Oospora lactis	Sp	28	0.08370	-	-	-	1	110	-	-	Nagy 1964
Penicillium chrysogenum	Sp	400	0.00576	W	Wat	-	-	-	3.262	34000	Jepson 1973
Penicillium chrysogenum	Sp	148	0.01560	S	-	N	1	192	3.262	34000	Luckiesh 1949
Penicillium chrysogenum	Sp	1645	0.00180	Air	50	N	1	32	3.262	34000	vanOsdell 2002
Penicillium chrysogenum	Sp	531	0.00434	Air	41	N	2	690	3.262	34000	Luckiesh 1949
Penicillium corylophilium	Sp	381	0.00604	W	Wat	N	2	4644	3.262	34000	Begum 2009
Penicillium digitatum	Sp	321	0.00718	W	Wat	Y	1	960	3.262	34000	Asthana 1992

APPENDIX C: UV Rate Constants for Fungi and Other Microbes

Microbe	Type	D_{90} J/m²	UVGI k m²/J	Media	RH %	Sh	St	UL J/m²	Dia. μm	Base Pairs kb	Source (see Chapter 4 Refs)
Penicillium digitatum	Sp	440	0.00523	-	-	-	-	-	3.262	34000	Nagy 1964
Penicillium expansum	Sp	130	0.01771	-	-	-	-	-	3.262	34000	Nagy 1964
Penicillium italicum	Sp	321	0.01140	W	Wat	Y	2	590	3.262	34000	Asthana 1992
Penicillium roquefortii	Sp	130	0.01771	-	-	-	-	-	3.262	34000	Nagy 1964
Penicillium spp.	Sp	2240	0.00103	S	-	-	-	2240	3.262	34000	Chick 1963
Penicillium spp.	Veg	280	0.00822	S	-	-	-	280	8.800	34000	Chick 1963
Rhizopus nigricans	Sp	3000	0.00077	W	Wat	-	-	-	6.928	54178	Jepson 1973
Rhizopus nigricans	Sp	267	0.00861	Air	62	N	2	348	6.928	54178	Luckiesh 1949
Rhizopus nigricans	Sp	1110	0.00207	-	-	-	-	-	6.928	54178	Nagy 1964
Rhizopus nigricans	Sp	173	0.01330	S	-	Y	2	200	6.928	54178	Kowalski 2001
Rhizopus oryzae	Sp	4480	0.00051	S	-	-	-	4480	6.928	-	Chick 1963
Rhodotorula spp.	VegY	1120	0.00206	S	-	-	-	1120	5.900	-	Chick 1963
Saccharomyces spp.	VegY	44	0.05230	-	-	-	1	176	-	-	Nagy 1964
Saccharomyces ellipsoideus	VegY	33	0.06980	-	-	-	1	132	-	-	Nagy 1964
Scopulariopsis brevicaulis	Sp	650	0.01840	W	Wat	-	-	-	5.916	-	Jepson 1973
Scopulariopsis brevicaulis	Sp	226	0.01020	S	-	N	1	294	5.916	-	Luckiesh 1949
Scopulariopsis brevicaulis	Sp	2890	0.00344	Air	79	N	2	870	5.916	-	Luckiesh 1949
Sporotrichum schenkii	VegY	280	0.00822	S	-	-	-	280	5.500	-	Chick 1963
Stachybotrys chartarum	Sp	5575	0.00041	S	-	N	1	1440	5.623	-	Green 2005
Torula bergeri	Veg	4480	0.00051	S	-	-	-	4480	40	-	Chick 1963
Torula sphaerica	VegY	23	0.09986	Air	65	N	2	30	40	-	Luckiesh 1949
Torula sphaerica	VegY	78	0.02940	S	-	N	1	102	40	-	Luckiesh 1949
Trichophyton rubrum	Veg	560	0.00411	S	-	-	-	560	4.899	-	Chick 1963
Trichophyton rubrum	Sp	560	0.00411	S	-	-	-	560	4.899	-	Chick 1963
Ustilago zeae	VegY	1120	0.00206	S	-	-	-	1120	5.916	20500	Chick 1963
Ustilago zeae	Sp	35	0.06580	-	-	-	-	-	5.916	20500	Sussman 1966
Yeast	VegY	40	0.05756	W	Wat	-	-	-	-	-	Jepson 1973
Yeast (Brewer's)	VegY	100	0.02303	W	Wat	-	-	-	-	-	Jepson 1973
Protozoa and Other Microbes											
Acanthameoba	Rhizopod	999	0.02100	W	Wat	-	-	-	-	-	Maya 2003
Acanthameoba castellani	Rhizopod	992	0.00232	S	-	-	-	3000	-	-	Gritz 1990
Algae	Algae	1000	0.00230	-	-	-	-	-	-	-	Summer 1962
Algae, blue-green	Algae	450	0.00512	-	-	-	-	-	-	-	Jepson 1973
Cryptosporidium hominis	Protoz	30	0.07800	-	-	-	-	-	-	-	Johnson 2005
Cryptosporidium parvum	Protoz	7	0.31400	W	Wat	N	1	20	3	-	Oguma 2001
Cryptosporidium parvum	Protoz	20	0.11500	W	Wat	-	-	30	3	-	Zimmer 2003
Cryptosporidium parvum	Protoz	10	0.23030	W	Wat	N	1	30	3	-	Shin 2001
Cryptosporidium parvum	Protoz	50	0.04605	W	Wat	N	2	100	3	-	Craik 2001
Cryptosporidium parvum	Protoz	10	0.23220	W	Wat	N	1	40	3	-	Bukhari 2004
Cryptosporidium parvum	Protoz	5	0.45830	W	Wat	N	1	22	3	-	Morita 2002
Encephalitozoon intestinalis	Protoz	29	0.07830	W	Wat	N	2	50	-	-	......
Encephalitozoon intestinalis	Protoz	15	0.15350	W	Wat	N	-	-	-	-	Huffman 2002
Encephalitozoon cuniculi	Protoz	43	0.05310	W	Wat	N	1	130	-	-	Marshall 2003
Encephalitozoon hellem	Protoz	80	0.02880	W	Wat	N	2	120	-	-	Marshall 2003
Giardia lamblia cysts	Protoz	50	0.04610	W	Wat	-	-	20	-	-	Campbell 2002
Giardia lamblia cysts	Protoz	3	0.92100	W	Wat	N	1	10	-	-	Shin 2005
Giardia lamblia cysts	Protoz	20	0.11500	W	Wat	N	1	40	-	-	Li 2007
Giardia muris cysts	Protoz	10	0.23020	W	Wat	N	2	10	-	-	Craik 2001
Giardia muris cysts	Protoz	7	0.34130	W	Wat	N	1	23	-	-	Hayes 2000
Protozoa	Protoz	80	0.02878	-	-	-	-	-	-	-	Jepson 1973
Protozoa	Protoz	240	0.00959	-	-	-	-	-	-	-	Summer 1962
Prions (scrapie)	Prion	24315	0.00009	-	-	-	-	-	-	-	Bellinger-Kawahara 1987
Prions (scrapie)	Prion	55618	0.00004	-	-	-	-	-	-	-	Alper 1967

Appendix D: UVGI Lamp Data and Ratings

LAMP	Mfr. Ref.	Output UV W	# of coils	Arclength cm	Dia. cm	Radius cm	Area cm²	Surface I W/m²	Rating µW/cm²	Type MP/LP
782H10	AU	2.8	1	22.2	1.58	0.79	110	254	28	LP
782H20	AU	5.5	1	47.6	1.58	0.79	236	233	52	MP
782H30	AU	5.2	1	73	1.58	0.79	362	144	46	MP
782H30	AU	8.3	1	73	1.58	0.79	362	229	73	LP
782L10	AU	2	1	22.2	1.58	0.79	110	181	20	LP
782L10	AU	2.9	1	22.2	1.58	0.79	110	263	28	MP
782L10 (cc)	ATL	2.9	1	37.46	1.58	0.79	186	156	29	LP
782L20	AU	3.9	1	47.6	1.58	0.79	236	165	35	LP
782L20 (cc)	ATL	5.8	1	62.86	1.58	0.79	312	186	55	LP
782L25_1/2	AU	7.3	1	64.8	1.58	0.79	322	227	75	MP
782L30	AU	5.2	1	73	1.58	0.79	362	144	46	LP
782L30	AU	8.3	1	73	1.58	0.79	362	229	73	LP
782L30 (cc)	ATL	8.7	1	88.26	1.58	0.79	438	199	77	LP
782VH10 (cc)	ATL	2.9	1	37.46	1.58	0.79	186	156	29	LP
782VH20 (cc)	ATL	5.8	1	62.86	1.58	0.79	312	186	55	LP
782VH29	AU	5.7	1	70.8	1.58	0.79	351	162	50	LP
782VH29	AU	9.1	1	70.8	1.58	0.79	351	259	80	LP
782VH30 (cc)	ATL	8.7	1	88.26	1.58	0.79	438	199	77	LP
83A-1	IESNA	3.1	1	27.3	1.75	0.875	150	207	35	LP
84A-1	IESNA	4.1	1	62.8	1.3	0.65	256	160	46	MP
86A-45	IESNA	1.4	1	11.4	1.75	0.875	63	223	16	LP
87A-45	IESNA	4.3	1	26.7	1.75	0.875	147	293	47	LP
88A-45	IESNA	10.4	1	62.2	1.75	0.875	342	304	113	MP
93A-1	IESNA	1.9	1	29.2	1.3	0.65	119	159	21	LP
94A-1	IESNA	7.2	1	62.8	1.75	0.875	345	209	80	LP
AC4-100LL	UVS	280	1	420	1.8	0.9	2375	1179	1023*	MP
AC4-150LL	UVS	420	1	620	1.8	0.9	3506	1198	1065*	LP
AC4-25LL	UVS	80	1	120	1.8	0.9	679	1179	666*	LP
AC4-50LL	UVS	150	1	220	1.8	0.9	1244	1206	921*	LP
AGHO287T5L	LSI	6.5	1	20.6	1.6	0.8	104	628	70	LP
C24T6L (cc)	ATL	5	1	55.56	1.58	0.79	276	181	47	LP
C24T6VH (cc)	ATL	5	1	55.56	1.58	0.79	276	181	47	LP
CC12T6L (cc)	ATL	1.4	1	25.08	1.58	0.79	124	112	14	LP
CC12T6VH (cc)	ATL	1.4	1	25.08	1.58	0.79	124	112	14	LP
CC18T6L (cc)	ATL	3.2	1	40.32	1.58	0.79	200	160	32	LP
CC18T6VH (cc)	ATL	3.2	1	40.32	1.58	0.79	200	160	32	LP
CC36T6L (cc)	ATL	8.5	1	86.04	1.58	0.79	427	199	75	MP
CC36T6VH (cc)	ATL	8.5	1	86.04	1.58	0.79	427	199	75	LP
CC48T6L (cc)	ATL	11.2	1	116.52	1.58	0.79	578	194	98	LP
CC48T6VH (cc)	ATL	11.2	1	116.52	1.58	0.79	578	194	98	LP
G10T5_1/2H	AU	5.3	1	27.6	1.58	0.79	137	387	55	LP
G10T5_1/2L	AU	5.3	1	27.6	1.58	0.79	137	387	55	LP
G10T5_1/2VH	AU	5.3	1	27.6	1.58	0.79	137	387	55	LP
G10T51/2L	ATL	5.3	1	42.86	1.58	0.79	213	249	55	LP
G10T51/2VH	ATL	5.3	1	42.86	1.58	0.79	213	249	55	LP
G10T8	T-W	2.6	1	24.5	2.55	1.275	196	132	27	LP
G10T8	GE	1.9	1	33	2.55	1.275	264	72	19*	MP
G11T5	Philips	3	1	15.5	1.5	0.75	73	411	30	MP

Appendix D: UVGI Lamp Data and Ratings

LAMP	Mfr. Ref.	Output UV W	# of coils	Arclength cm	Dia. cm	Radius cm	Area cm^2	Surface I W/m^2	Rating μW/cm^2	Type MP/LP
G11T5	GE	2.2	1	22.63	0.63	0.3125	44	495	22.9	LP
G12T5_1/2L	AU	6	1	21.6	1.58	0.79	107	560	66	LP
G12T5_1/2VH	AU	6	1	21.6	1.58	0.79	107	560	66	LP
G12T6L	ATL	3.1	1	30.48	1.58	0.79	151	205	32	LP
G12T6VH	ATL	3.1	1	30.48	1.58	0.79	151	205	32	LP
G15T8	AU	3.6	1	36.5	2.54	1.27	291	124	38	LP
G15T8	ATL	4.8	1	43.6	2.55	1.275	349	137	49	MP
G18T6L	ATL	5.8	1	40.32	2.54	1.27	322	180	59	MP
G18T6L/U	ATL	5.8	2	21	2.54	1.27	168	346	59	LP
G18T6VH	ATL	5.8	1	40.32	2.54	1.27	322	180	59	LP
G18T6VH/U	ATL	5.8	2	21	2.54	1.27	168	346	59	MP
G20T10	GE	7.5	1	58	3.25	1.625	592	127	75.8	LP
G20T10	-	4.1	1	58.85	3.25	1.625	601	68	40*	MP
G20T10	-	4.2	1	58.85	3.25	1.625	601	70	41*	LP
G24T6L	ATL	8.5	1	55.56	2.54	1.27	443	192	82	LP
G24T6L/U	ATL	8.5	2	28.6	2.54	1.27	228	372	82	LP
G24T6VH	ATL	8.5	1	55.56	2.54	1.27	443	192	82	MP
G24T6VH/U	ATL	8.5	2	28.6	2.54	1.27	228	372	82	LP
G25T8	Philips	6.6	1	35.56	2.54	1.27	284	233	65.5	LP
G25T8	ATL	6.9	1	43.6	2.55	1.275	349	198	70.4	LP
G25T8(W)	AU	5	1	36.5	2.54	1.27	291	172	54	LP
G30T6L	ATL	11.2	1	70.8	2.54	1.27	565	198	101	LP
G30T6L/U	ATL	11.2	2	36.2	2.54	1.27	289	388	101	LP
G30T6VH	ATL	11.2	1	70.8	2.54	1.27	565	198	101	LP
G30T6VH/U	ATL	11.2	2	36.2	2.54	1.27	289	388	101	LP
G30T8	T-W	11.6	1	81.3	2.54	1.27	649	179	117	LP
G30T8	ATL	11.6	1	89.3	2.55	1.275	715	162	117	MP
G30T8(W)	AU	8.3	1	81.3	2.54	1.27	649	128	85	LP
G36T5L	Philips	13.9	1	75.9	1.5	0.75	358	389	181	LP
G36T5L	GE	12	1	86.55	1.5	0.75	408	294	109.5	LP
G36T6	AU	12.7	1	76.2	1.9	0.95	455	279	110	LP
G36T6H	AU	13.8	1	76.2	1.58	0.79	378	365	120	LP
G36T6L	LSI	13.8	1	76.2	1.5	0.75	359	384	176	MP
G36T6L	ATL	13.8	1	86.04	1.58	0.79	427	323	120	LP
G36T6L(W)	AU	13.8	1	76.2	1.58	0.79	378	365	120	LP
G36T6L/U	ATL	13.8	2	43.8	2.54	1.27	350	395	120	LP
G36T6VH	AU	15.2	1	78.7	1.58	0.79	391	389	124	MP
G36T6VH	ATL	13.8	1	86.04	1.58	0.79	427	323	120	LP
G36T6VH/U	ATL	13.8	2	43.8	2.54	1.27	350	395	120	LP
G37T6L	ATL	14.3	1	88.58	1.58	0.79	440	325	124	LP
G37T6VH	IESNA	14.3	1	78.5	1.58	0.79	390	367	130	LP
G37T6VH	ATL	14.3	1	88.58	1.58	0.79	440	325	124	LP
G40T10	T-W	11.5	1	119.8	3.25	1.625	1223	94	96*	LP
G40T10	GE	19.8	1	121.3	3.25	1.625	1238	160	200	LP
G48T6L	ATL	19.3	1	116.52	1.58	0.79	578	334	164	MP
G48T6L/U	ATL	19.3	2	59.1	2.54	1.27	472	409	164	LP
G48T6VH	ATL	19.3	1	116.52	1.58	0.79	578	334	164	LP
G48T6VH/U	ATL	19.3	2	59.1	2.54	1.27	472	409	164	LP

Appendix D: UVGI Lamp Data and Ratings

LAMP	Mfr. Ref.	Output UV W	# of coils	Arclength cm	Dia. cm	Radius cm	Area cm²	Surface I W/m²	Rating μW/cm²	Type MP/LP
G4S11	AU	0.1	1	0.95	3.49	1.745	10	96	1.1	LP
G4S11	-	0.18	1	5.7	3.5	1.75	63	29	2*	LP
G4T4	AU	0.7	1	7.8	1.35	0.675	33	212	10	LP
G4T4/1	AU	1.1	1	15	1.3	0.65	61	180	12	MP
G4T5	ATL	0.8	1	8.1	1.55	0.775	39	203	8.3	MP
G4T5	GE	0.9	1	15.01	1.55	0.775	73	123	8.3	LP
G4T5	GE	0.8	1	15.01	1.5	0.75	71	113	8.3	LP
G4T5(W)	AU	0.5	1	6.3	1.58	0.79	31	160	5.4	LP
G55T8HO	GE	18	1	90.88	3.25	1.625	928	194	194	LP
G64T5L	AU	26.7	1	147.3	1.58	0.79	731	365	190	LP
G64T5L	ATL	25	1	157.16	1.58	0.79	780	320	200	MP
G64T5VH	AU	26.7	1	147.3	1.58	0.79	731	365	190	MP
G64T5VH	ATL	25	1	157.16	1.58	0.79	780	320	200	MP
G64T6	IESNA	25	1	147	1.9	0.95	877	285	200	LP
G64T6L	AU	25.5	1	147.3	1.9	0.95	879	290	208	LP
G67T5L	AU	25.6	1	152.5	1.58	0.79	757	338	205	LP
G67T5VH	AU	25.6	1	152.5	1.58	0.79	757	338	205	LP
G6T5	T-W	1.6	1	15.7	1.55	0.775	76	209	16.7	LP
G6T5	ATL	1.6	1	21.05	1.55	0.775	103	156	16.7	LP
G6T5	GE	1.7	1	22.63	0.63	0.3125	44	383	17.7	LP
G6T5(W)	AU	1	1	14	1.58	0.79	69	144	11	MP
G8T5	ATL	2.5	1	23.3	1.55	0.775	113	220	26	LP
G8T5	GE	2.3	1	30.25	0.63	0.3125	59	387	24	LP
G8T5(W)	AU	1.6	1	21.6	1.58	0.79	107	149	17	LP
GBX11/UVC	GE	3.6	2	21.5	1.5	0.75	101	355	33	LP
GBX13/UVC	GE	3.6	2	17	1.5	0.75	80	449	31	LP
GBX18/UVC/2G11	GE	5.5	2	22.5	1.5	0.75	106	519	51	LP
GBX36/UVC/2G11	GE	12	2	41.5	1.5	0.75	196	614	110	MP
GBX5/UVC	GE	1	2	8.5	1.5	0.75	40	250	9	LP
GBX55/UVC/2G11	GE	17	2	53.5	1.5	0.75	252	674	156	LP
GBX9/UVC	GE	2.4	2	14.5	1.5	0.75	68	351	22	LP
GCC369H	AU	2.4	1	22.9	1.58	0.79	114	211	23.7	LP
GCC369N	AU	1.6	1	22.9	1.58	0.79	114	141	16	LP
GHO287T5L	LSI	3.2	1	20.6	1.6	0.8	104	309	35	LP
GHO36T5/L/4PSE	ATL	28	1	75.5	1.5	0.75	356	787	260	LP
GHO36T5/VH/4PSE	ATL	28	1	75.5	1.5	0.75	356	787	260	LP
GHO64T5/L/4PSE	ATL	45	1	142.1	1.5	0.75	670	672	380	LP
GHO64T5VH/4PSE	ATL	45	1	142.1	1.5	0.75	670	672	380	LP
GPH212T5L	LSI	2.9	1	13.21	1.5	0.75	62	466	29.5	LP
GPH212T5L	ATL	2.3	1	21.2	1.5	0.75	100	230	24	LP
GPH212T5VH	ATL	2.3	1	21.2	1.5	0.75	100	230	24	LP
GPH287T5-H	IESNA	3	1	20	1.6	0.8	101	298	33	LP
GPH287T5L	LSI	3.2	1	20.57	1.5	0.75	97	330	35	LP
GPH287T5L	ATL	3.7	1	28.7	1.5	0.75	135	274	38	LP
GPH287T5LVH	ATL	3.7	1	28.7	1.5	0.75	135	274	38	LP
GPH287T5-VH	IESNA	3	1	20	1.6	0.8	101	298	33	LP
GPH330T5L/4	ATL	4.5	1	33	1.5	0.75	156	289	44	LP
GPH330T5VH/4	ATL	4.5	1	33	1.5	0.75	156	289	44	LP

Appendix D: UVGI Lamp Data and Ratings

LAMP	Mfr. Ref.	Output UV W	# of coils	Arclength cm	Dia. cm	Radius cm	Area cm^2	Surface I W/m^2	Rating μW/cm^2	Type MP/LP
GPH357T5L/4	ATL	5	1	35.7	1.5	0.75	168	297	51	LP
GPH357T5VH/4	ATL	5	1	35.7	1.5	0.75	168	297	51	LP
GPH435T5	LSI	9.6	1	35	1.5	0.75	165	582	88	LP
GPH436T5/HO/4C	LSI	14	1	29.2	1.5	0.75	138	1017	140	LP
GPH436T5/VH/HO/4PSE	ATL	13	1	36	1.5	0.75	170	766	120	LP
GPH436T5L	ATL	6.4	1	43.6	1.5	0.75	205	311	59	LP
GPH436T5L/HO/4PSE	ATL	13	1	36	1.5	0.75	170	766	120	LP
GPH436T5LHO/4P	LSI	8	1	36	1.5	0.75	170	472	75	LP
GPH436T5VH	ATL	6.4	1	43.6	1.5	0.75	205	311	59	MP
GPH450T5L/4	ATL	6.6	1	45	1.5	0.75	212	311	62	MP
GPH450T5VH/4	ATL	6.6	1	45	1.5	0.75	212	311	62	MP
GPH463T5L/4	ATL	6.9	1	46.3	1.5	0.75	218	316	63	LP
GPH463T5VH/4	ATL	6.9	1	46.3	1.5	0.75	218	316	63	LP
GPH620T5L/4	ATL	9.7	1	62	1.5	0.75	292	332	87	LP
GPH620T5VH/4	ATL	9.7	1	62	1.5	0.75	292	332	87	LP
GPH793T5L	ATL	12.8	1	79.3	1.5	0.75	374	343	112	LP
GPH793T5VH	ATL	12.8	1	79.3	1.5	0.75	374	343	112	LP
GPH810T5L/4	ATL	13.1	1	81	1.5	0.75	382	343	115	LP
GPH810T5VH/4	ATL	13.1	1	81	1.5	0.75	382	343	115	LP
GPH846T5/L/HO/4PSE	ATL	29	1	76.7	1.5	0.75	361	802	265	MP
GPH846T5/VH/HO/4PSE	ATL	29	1	76.7	1.5	0.75	361	802	265	LP
GPH846T5LHO/4P	LSI	17	1	76.71	1.5	0.75	361	470	165	LP
GPH893T5/L/HO/4PSE	ATL	30	1	81.5	1.5	0.75	384	781	270	LP
GPH893T5/VH/HO/4PSE	ATL	30	1	81.5	1.5	0.75	384	781	270	LP
GPH893T5LHO/4P	LSI	17	1	81.53	1.5	0.75	384	442	156	LP
GTL2	-	0.12	1	5.5	2	1	35	35	1.2	LP
GTL3	ATL	0.18	1	6.3	2	1	40	45	1.8	LP
GUL4	-	0.7	1	15	1.35	0.675	64	110	10	LP
LSI436T5	LSI	8	1	36	2	1	226	354	80	LP
LTC18W	SXX	5.5	2	17.2	1.25	0.625	68	814	51	LP
LTC24W	SXX	7	2	17.2	1.25	0.625	68	1036	65	LP
LTC36W	SXX	12	2	17.2	1.25	0.625	68	1777	110	LP
LTC55W	SXX	17	2	17.2	1.25	0.625	68	2517	156	LP
LTC95W	SXX	32	2	17.2	1.25	0.625	68	4738	304	LP
LTCPL11	SXX	3.6	2	12.5	1.25	0.625	49	733	33	LP
LTCPL5	SXX	1	2	12.5	1.25	0.625	49	204	9	LP
LTCPL7	SXX	1.8	2	12.5	1.25	0.625	49	367	16	LP
LTCPL9	SXX	2.4	2	12.5	1.25	0.625	49	489	22	LP
OZ4S11	AU	0.1	1	0.95	3.49	1.745	10	96	1.1	LP
OZ4T5	AU	0.6	1	15.2	1.58	0.79	75	80	6.5	LP
OZ6T5	AU	1.2	1	22.9	1.58	0.79	114	106	13	LP
OZ8T5	AU	1.8	1	21.6	1.58	0.79	107	168	19.5	MP
PL-L35W/TUV	Philips	11	1	22.7	0.9	0.45	64	1714	111*	LP
PL-L36W/TUV	Philips	12	1	41.75	0.9	0.45	118	1017	119*	MP
PL-L60W/TUV	Philips	17.5	1	41.75	0.9	0.45	118	1482	173*	MP
SA5000*	SA	16	1	43.18	1.8	0.9	244	655	158	MP
SPECIAL 1	Philips	34	1	46.7	1.8	0.9	264	1287	335	LP
TUV10W	Philips	2.2	1	33.15	2.8	1.4	292	75	22	LP

486

Appendix D: UVGI Lamp Data and Ratings

LAMP	Mfr. Ref.	Output UV W	# of coils	Arclength cm	Dia. cm	Radius cm	Area cm^2	Surface I W/m^2	Rating μW/cm^2	Type MP/LP
TUV115W	Philips	33.5	1	119.94	4.05	2.025	1526	220	300	LP
TUV115WVHO	Philips	38.8	1	119.94	4.05	2.025	1526	254	300	LP
TUV11W	Philips	2.2	2	42.42	0.9	0.45	120	183	22	LP
TUV11WPL-S	Philips	3.6	2	39.6	0.9	0.45	112	322	36	MP
TUV15W	Philips	4.7	1	43.74	2.8	1.4	385	122	47	MP
TUV16T5	Philips	5	1	22.86	1.5	0.75	108	464	46	MP
TUV16W	Philips	3.2	1	28.83	1.6	0.8	145	221	32	LP
TUV18W	Philips	5.5	2	39	0.9	0.45	110	499	90	LP
TUV18WPL-L	Philips	4.6	2	31.75	1.8	0.9	180	256	46	MP
TUV30W	Philips	11.2	1	89.46	2.8	1.4	787	142	102	MP
TUV36W	Philips	14	1	112.5	2.6	1.3	919	152	110	LP
TUV36WPL-L	Philips	12	2	73	0.9	0.45	206	581	90	LP
TUV4W	Philips	0.9	1	8.6	1.6	0.8	43	208	7	LP
TUV55W	Philips	18	2	178.92	0.9	0.45	506	356	163	LP
TUV55WHO	Philips	17	1	78.2	2.6	1.3	639	266	158	LP
TUV55WPL-L	Philips	17	2	94.65	1.8	0.9	535	318	290	LP
TUV59W	Philips	9	2	21.6	0.9	0.45	61	1474	92	LP
TUV6W	Philips	1.5	1	16.2	1.6	0.8	81	184	14	LP
TUV75W	Philips	26	1	119.94	2.8	1.4	1055	246	217	MP
TUV75WHO	Philips	25	1	112.5	2.6	1.3	919	272	213	LP
TUV8W	Philips	2.1	1	23.8	1.6	0.8	120	176	17	LP
TUV9WPL-S	Philips	2.4	2	25.8	0.9	0.45	73	329	24	LP

AU = American Ultraviolet
ATL = Atlantic Ultraviolet
UVS = UV Systec
SXX = Shangyu Xin Xin

(cc) = Cold cathode
*(computed estimate)
**(output varies)

MP = Medium Pressure
LP = Low Pressure

APPENDIX E: C++ Source Code
for Lamp UV Field Average Irradiance

// Author's note: This source code is the same as the source code provided in Kowalski (2001) and Kowalski (2003) except that the UV lamp(s) in this version may be located at any position (x1, y1, z1, x2, y2, z2) where the coordinates represent the endpoints of the lamp. In this version, the x1 coordinate need not be zero, but may be any value.

// Global variables and arrays (these values can be output to a text file)

```
double Average;                    // Average UV Irradiance
double DirectField[51][51][101];   // Direct Irradiance Field
double DistanceMtx[51][51][101];   // matrix of distances to lamp axis
double PositionMtx[51][51][101];   // matrix of position along lamp axis
```

// The following values can be set with a separate input routine or text file

```
int NUMLAMPS = 1;                  // number of lamps in system (set value to 1 or more)
double SurfInt[1] = 10000.0;       // Lamp surface irradiance for lamp #1, µW/cm2,
       surface irradiance may be computed from lamp wattage (see Chapter 7),
       (set value)
double arclength[1] = 35.0;        // Lamp arclength for lamp #1, cm (set value)
double radius[1] = 1.0;            // radius of lamp # 1, cm (set value)
```

```
brun(){        // this routine calls the computational subroutines in C++ and
               // computes the direct irradiance field for all lamps in an enclosure
               // and fills the matrix DirectField[][][] with the irradiance values
DirectIntField();       // Compute the average irradiance
AverageDirect();        // the result, Average, can be printed here or output to a file
}
```

```
void DirectIntField()
{        // Computes 50x50x100 UV Irradiance Matrix of Direct Irradiance Field
   int i, j, k, l;
   double tempsum = 0.0, x, paxis, db;
   for (i=0; i<=50; i++){
      db=0.0;
      for (j=0; j<=50; j++){
         for (k=0; k<=100; k++){
            for (l=0; l<NUMLAMPS; l=l+1) {
               x = Distance(i,j,k,l);            // Compute distance x to lamp axis
               DistanceMtx[i][j][k]=x;
               paxis = Position(i,j,k,l);        // Compute position on lamp axis
               PositionMtx[i][j][k]=paxis;
               if (paxis < arclength[l]){        // Is it within lamp arclength?
                     // Compute Intensity within Lamp arclength
                  tempsum = Intensity(SurfInt[l],arclength[l],radius[l],x,paxis);
               }
               else {    // Compute Intensity beyond Lamp end
                  db = paxis-arclength[l];
                  tempsum = IBeyondEnds(SurfInt[l],arclength[l],radius[l],x,db);
               }
                  // Add irradiances for all lamps
               DirectField[i][j][k] = DirectField[i][j][k] + tempsum;
                tempsum = 0.0;
            }
```

488

```
        }
      }
    }
}


double Distance(int i, int j, int k, int l)
{         // Compute shortest Distance to Lamp Axis
   double xi, yj, zk;
   xi = i;
   yj = j;
   zk = k;
   double x = xi*xincr;
   double y = yj*yincr;
   double z = zk*zincr;
   double dist = PointLine(x, y, z, l);
   return dist;
}

double PointLine(double x, double y, double z, int l)
{         // Compute Distance from a Point to a Line (lamp axis)
      double x1=x-lampx1[l];
      double y1=y-lampy1[l];
      double z1=z-lampz1[l];
      double x2=lampx2[l]-lampx1[l];
      double y2=lampy2[l]-lampy1[l];
      double z2=lampz2[l]-lampz1[l];
      double dist, DotProd, a;
      double p1=x1*x1+y1*y1+z1*z1;
      double p2=x2*x2+y2*y2+z2*z2;
      if (p1*p2>0){
         DotProd = (x1*x2+y1*y2+z1*z2)/sqrt(p1*p2);
         a = acos(DotProd);
         dist=fabs(sin(a))*sqrt(p1);
      }
      else {
         dist = 0;
      }
   return dist;
}

double Position(int i, int j, int k, int l)
{         // Compute Position along lamp axis
      double xi, yj, zk, p1, p2, posit, p3, p4, a, x;
      double DotProd, y, z, x1, y1, z1, x2, y2, z2;
      double pc, pa, pd, p5, posit1, posit2;
      xi = i;
      yj = j;
      zk = k;
      x = xi*xincr;
      y = yj*yincr;
      z = zk*zincr;
      posit = 1;
      x1=x-lampx1[l];
      y1=y-lampy1[l];
      z1=z-lampz1[l];
```

```
        x2 = lampx2[I]-lampx1[I];
        y2 = lampy2[I]-lampy1[I];
        z2 = lampz2[I]-lampz1[I];
        p1 = x1*x1+y1*y1+z1*z1;
        p2 = x2*x2+y2*y2+z2*z2;
        pc = x1*x2+y1*y2+z1*z2;
        pa =p1*p2;
        if (pa>0){
           DotProd = pc/sqrt(pa);
           a = acos(DotProd);
           posit1 = cos(a)*sqrt(p1);
        }
        else {
           posit1 = 0.000001;
        }
        x1=x-lampx2[I];
        y1=y-lampy2[I];
        z1=z-lampz2[I];
        x2 = lampx1[I]-lampx2[I];
        y2 = lampy1[I]-lampy2[I];
        z2 = lampz1[I]-lampz2[I];
        p3 = x1*x1+y1*y1+z1*z1;
        p4 = x2*x2+y2*y2+z2*z2;
        pd = x1*x2+y1*y2+z1*z2;
        p5 =p3*p4;
        if (p5>0){
           DotProd = pd/sqrt(p5);
           a = acos(DotProd);
           posit2 = cos(a)*sqrt(p3);
        }
        else {
           posit2 = 0.000001;
        }
        posit = max(posit1,posit2);
     return posit;
}

double Intensity(double IS, double arcl, double r, double x, double l)
{        // Compute Irardiance Field
         // IS=Surface Irradiance, arcl=arclength, r=radius,
         // x=distance from axis, l = distance along axis
     double intense;
     double VF, VF1, VF2;
         // Compute VF Lamp segment 1
     VF1 = VFCylinder(l,r,x);
         // Compute Lamp segment 2
     VF2 = VFCylinder(arcl-l,r,x);
         // Total VF for Lamp
     VF = VF1 + VF2;
         // Compute intensity at the point
     intense = IS*VF;
     return intense;
}

double IBeyondEnds(double IS, double arcl, double r, double x, double db )
{        // Compute Irradiance field beyond the ends of the lamp
```

490

```
        // IS=Surface Irradiance, arcl=arclength, r=radius
        // x=distance from axis, db=distance beyond lamp end
    double intense;
    double VF, VF1, VF2;
     VF1 = VFCylinder(arcl+db,r,x);      // Compute Lamp + Ghost Lamp segment
     VF2 = VFCylinder(db,r,x);  // Compute Ghost Lamp segment
     VF = VF1 - VF2;    // Compute Lamp VF
     intense = fabs(IS*VF);       // Compute irradiance at the point
     return intense;
}

double VFCylinder(double l, double r, double h)    // View Factor #15 per Modest (1993)
{          // l=length, r=radius, h=height above axis
    double H, L, X, Y, p1, p2, p3, VF;
    if (h<r) h=r+0.000001;  // Not inside lamp
    H = h/r;
    L = l/r;
    if (L==0) L=0.000001;
    if (H==1) H=H+0.000001;
    X = (1+H)*(1+H)+L*L;
    Y = (1-H)*(1-H)+L*L;
        // Compute Parts of View Factor
    p1 = atan( L/sqrt(H*H-1) )/L;
    p2 = (X-2*H)*atan( sqrt( (X/Y)*(H-1)/(H+1) ))/sqrt(X*Y);
    p3 =  atan( sqrt((H-1)/(H+1)) );
    VF = L*(p1+p2-p3)/(Pi*H);
    return VF;

}

double AverageDirect()
{       // compute average irradiance field
   double total = 0;
   double Avg = 0;
   for (int i=0; i<=50; i++)
     for (int j=0; j<=50; j++)
       for (int k=0; k<=100; k++)
          total = total + DirectField[i][j][k];
   Average = total/(51*51*101);
   return Avg;
}
```

NOTE: For the complete source code implementation, which includes reflectivity subroutines for diffuse surfaces, see Kowalski (2001). For specular reflectivity modeling, see Kowalski et al (2005).

References and Additional Information:

Kowalski WJ, Bahnfleth WP. 2000. Effective UVGI system design through improved modeling. ASHRAE Transactions 106(2):4-15.

Kowalski WJ. 2001. Doctoral Thesis -- Design and optimization of UVGI air disinfection systems [PhD]. State College: The Pennsylvania State University.

Kowalski WJ. 2003. Immune Building Systems Technology. New York: McGraw-Hill.

Kowalski WJ, Bahnfleth WP, Mistrick RG. 2005. A specular model for UVGI air disinfection systems. IUVA News 7(1):19-26.

Appendix F: Ultraviolet Material Reflectivities (UVC/UVB Range)

MATERIAL	ρ %	MATERIAL	ρ %	MATERIAL	ρ %
ePTFE (tm WLGore)	99+	Nickel	37-38	Tin	16
Spectralon (PTFE)	95+	Steel	37	Carborundum	14.5
Smoked magnesium oxide	93	Zinc	37	Tantalum	13
Aluminum, etched	88	Cadmium	34	Tungsten	13
Evaporated aluminum	87	S.W. white Decotint paint	33	Kalsomine white water paint	12
Alzak sheet aluminum, brightened	87	Ivory wallpaper	31	Medusa cement	11
Alzak sheet aluminum	84	Pink figured wallpaper	31	Alabastine white water paint	10
Magnesium oxide	81	Speculum	30	White baked enamel	9
Magnesium carbonate (commercial)	81	Bleached cotton, white cotton	30	White oil paints (5-10)	8
Aluminum - sputtered on glass	80	Magnesium	29	Lithopone	8
Zirconium oxide, C.P.	78	Tin plated steel	28	Wall tiles	7-8
Pressed calcium carbonate	78	Stainless steel (20-30)	28	Black paint	7
Pressed magnesium oxide	77	Tin plated steel	28	Brass	7
Calcium carbonate	75	Palladium	28	Brown wrapping paper	7
Magnesium carbonate	75	Copper	25-31	Titanox C	7
Aluminum - treated surface	74	Lead	27	Titanium dioxide	7
Aluminum foil	73	Ivory figured wallpaper	26	Titanium oxide	6
Silicon	62-73	Molybdenum	25	AZO photo paper, exp. black	6
Aluminum hydroxide	67	White paper	25	Celluloid	6
Aluminum paint	65	White blotting paper	25	Pongee silk	6
Barytes	65	Magnalium	19	Brown baked enamel	6
Basic carbonate white lead (Dutch)	62	Antimony	17-32	Casein vehicle	6
New plaster	58	AZO photo paper, unexposed	24	Zinc sulfide	6
Galvanized duct - smooth	57	Tellurium	23	Titanox B	6
Aluminum oxide	55	Silver	22-23	Lead titanate	6
China clay	54	Chromium	16-31	Flat black Egyptian lacquer	5
Galvanized duct - rough	53	Gold	23	Lithopone	5
Aluminum - untreated surface	50	White water paint	23	Zinc oxide in clear lacquer	5
Basic sulfite white lead	48	White wallpaper	22	Black lacquer paint	5
White wall plaster (40-60)	46	Bismuth	22	Zinc oxide paint	5
Stellite	46	Water paints (10-30)	20	White porcelain enamel	4.7
Diatomaceous silica (Celite 10)	45	Concrete	19 (max)	Wood	4-5
Chromium	44-45	Brownish figured wallpaper	18	Lantern slide glass	4
Cobalt	42	Tungsten	18	Glass	4
Zirconium oxide (commercial)	41	Selenium	18	China clay	4
Stellite	41	Linen	17	Zinc oxide casein paint	4
Platinum	40	Fluorescent lamp phosphors	17	35% leaded zinc oxide	4
AZO photo paper, white back	39	Antimony oxide	17	Pressed zinc oxide	2.5-3
Chrome steel	39	Carbon	16	Lead-free zinc oxide	3
Rhodium	38	Duralumin	16	Open air	~0

See Section 6.4 in Chapter 6 for References

Appendix G: UVGI Rating Values (URV)

URV	Dose J/m^2	Dose μW-s/cm^2	Mean Dose, J/m^2	Notes
1	0.01	1	0.055	
2	0.10	10	0.15	
3	0.20	20	0.25	
4	0.30	30	0.4	
5	0.50	50	0.63	
6	0.75	75	0.88	
7	1.0	100	1.25	
8	1.5	150	2	
9	2.5	250	3.75	
10	5	500	7.5	
11	10	1000	12.5	Normal UV
12	15	1500	17.5	System
13	20	2000	25	Design
14	30	3000	35	Range
15	40	4000	45	
16	50	5000	55	
17	60	6000	70	
18	80	8000	90	
19	100	10000	150	
20	200	20000	250	
21	300	30000	350	
22	400	40000	450	
23	500	50000	750	
24	1000	100000	1500	
25	2000	200000	-	

NOTE: URV 21 - 25 are newly appended URV definitions

Index

Note – The letter "*f*" following the locators refer to figures

Breinigsville, PA USA
23 April 2010
236700BV00005B/94/P